Biology of C Reactive Protein in Health and Disease

Waliza Ansar • Shyamasree Ghosh

Biology of C Reactive Protein in Health and Disease

Springer

Waliza Ansar
Department of Zoology
Behala College
Kolkata, West Bengal, India

Shyamasree Ghosh
School of Biological Sciences
National Institute of Science, Education
and Research (NISER)
Bhubaneswar, Odisha, India

ISBN 978-81-322-2678-9 ISBN 978-81-322-2680-2 (eBook)
DOI 10.1007/978-81-322-2680-2

Library of Congress Control Number: 2015960243

Springer New Delhi Heidelberg New York Dordrecht London

Printed on acid-free paper

Springer (India) Pvt. Ltd. is part of Springer Science+Business Media (www.springer.com)

Dedicated to our family, PhD guide,
and
our adorable children

Preface

Good science always demands independent replication of new thoughts and experimental results and desertion of accepted or the so-called theories in light of more and more reliable evidence. Failure to comply with this trend leads to a tremendous damaging bad science, as with the falsely claimed association/relations between facts, observations, and results. The progress of good science also often requires providence, to make discoveries by accident and prudence of things not sought. Work on the pentraxin proteins, like C-reactive protein (CRP) since the 1970s, has benefited from abundant scientific serendipity, leading to modern-day routine clinical use of CRP measurements and the ongoing diverse research projects to explore the pathophysiological role of CRP in various diseases, be it cardiovascular disease, prominently different cancers, rheumatoid arthritis, infectious diseases, and many others. Some emerging progress has been evolving to extrapolate the role of CRP in new field of disease biology. Works are published regarding the role of CRP in mumps, testicular diseases, and different damaging conditions and some other works that need some prominence also. CRP is a very renowned old molecule. It is the age-old practice to add some more dimensions to its function or importance in clinical medicine. Also targets have always been there for adding new concepts by providing models with CRP being a new therapeutic target, a new drug, a new in-house antibody, and in nanoscience and other fields as a promising molecule of pharmaceutical integrity. The research works are expanding and continuously new bubbles of work on CRP are added in the sea of CRP science. So CRP is always a hot topic.

CRP since its discovery in 1974 is known as a classical acute phase plasma protein which increases in concentration manyfold completely nonspecifically in response to most forms of tissue injury, burn, trauma, infection, and inflammation. In 1994, a new report in the *New England Journal of Medicine* revealed the prognostic significance of even a minor increase in CRP concentration/values in severe unstable angina patients. Later, aftermath works observed the prognostic significance for coronary heart disease of increased baseline CRP values in patients with angina as compared to general population. These findings shoot up an inundation of interest in CRP in cardiovascular disease. The consequence is the subsequent propositions of various thoughts in the functionality of CRP in myocardial infarction, coronary heart disease, ischemia, and others. The field is also very controversial. But for the sake of best or better scientific tradition, the accumulation or the avalanche of robust evidence coming out will eventually and most probably quite soon settle the disputed (if any) issues by establishing reproducibly valid actual relationships, the genuine pathophysiological effects of CRP, and the clinical utility of its measurement values in disease biology. Thus, the clinical context of CRP in clinical medicine is ever expanding. Meanwhile, the media has also taken it up enthusiastically. Articles are coming out where CRP was pointed as a real cause of heart attacks propounding the thought that CRP could be more dangerous than cholesterol. Thereafter, peer-reviewed papers about CRP were just hyping from prominent researchers contributing the relation or contradiction between CRP and cholesterol. Many researchers were tempted to rename CRP from a nonspecific inflammatory marker to a cardiac risk marker.

But various published works reassign or confidently reassure that CRP is not more dangerous than cholesterol. Still in experimental models, CRP can exacerbate the ischemic tissue damage of stroke and acute myocardial infarction. Through rigorous reproducible evidence, emerging and extinguish erroneous beliefs are coming out. The causal role of CRP should be validated by its epidemiological relationship between baseline CRP values and cardiovascular disease risk. At present, immunoassays and techniques were encouraged for the now universal use of CRP assays in screening for different organic diseases and in monitoring disease activity and response to therapeutic approaches and also for detection of intercurrent infection.

However, measurement of this very sensitive but completely nonspecific biomarker of inflammation and tissue damage is thoroughly useful across the field of clinical medicine, but its baseline values are not that helpful for assessment of cardiovascular disease risk. There are so many compelling evidences that CRP contributes to the pathogenesis of atherosclerosis as CRP is known to bind to low-density and very-low-density lipoproteins.

Commercially available protein preparations used for immunochemical assays are routinely shipped with sodium azide, a potently toxic bacteriostatic compound, added to CRP to prevent bacterial growth. Furthermore, CRP expressed by recombinant bacteria are inevitably contaminated with bacterial endotoxin, lipopolysaccharide, and other potentially bioactive bacterial during pharmaceutical production. These commercially produced CRP preparations containing sodium azide and endotoxin definitely produce varied stimulatory and toxic effects on cell culture, which may be wrongly ascribed to CRP during experiments, whereas authentic pure human CRP itself does not produce any of these effects.

Our goal should be now to project CRP in optimizing some drugs with clinical testing in the hope that this therapeutic approach will provide significant cardioprotection after AMI and in acute coronary syndrome, potentially reduce tissue damage after trauma, confer neuroprotection after stroke, and act as a protective molecule in a wide range of infectious and inflammatory diseases.

Our passions run high in this field when our laboratory published in 2003 that CRP is a glycosylated molecule [Das et. al. 2003]. It breaks the age-old myth that CRP was glycosylated. There was a furious response from some quarters of research. CRP was differentially glycosylated in different diseases or acute inflammatory conditions encompassing nearly sixteen diseases [Das et al 2004a; Das et. al. 2004b]. We were all bubbling with new thoughts to explore. We explore the immunomodulatory role of CRP in malaria, tuberculosis, and visceral leishmaniasis. These three diseases are not only a problem in India, but the burden of these diseases is increasing worldwide. The effect of CRP on parasitized erythrocytes, on different complement receptors of erythrocytes, helping in clearance of damaged erythrocytes through the circulation is modulated by its glycosylated moiety [Ansar et. al. 2006; Ansar et. al. 2009a; Ansar et. al. 2009b]. Within the short span of this book, we tried to just touch and grease some of the aspects of CRP and its relevance in health and diseases. In the greater hypervolume of CRP, there is so much to enlist or describe, but so little is done. This is our first endeavor to express the multitude facets of this "dynamic biomarker protein" or "constant relevant prevalence (CRP)" in diseases and other acute conditions. We hope to extend our thoughts with more research findings, updates, and experiments in the process of giving birth to our second brainchild in the womb of clinical medicine.

Kolkata, West Bengal, India — Waliza Ansar
Bhubaneswar, Odisha, India — Shyamasree Ghosh

The original version of the affiliation of Prof Ghosh on page iv was corrected. An erratum can be found at DOI 10.1007/978-81-322-2680-2_12.

Acknowledgments

As our first book, which has taken more than 3 years to be prepared and published, we take this special moment to express our sincere appreciation and gratitude for all those who have contributed in making this book take its present shape.

We thank the "Almighty" for providing us the courage and showing us this day.

We would like to express our deepest gratitude to Dr. Mrs. Chitra Mandal, Ex-Director and Outstanding Scientist, Indian Institute of Chemical Biology, CSIR, Kolkata, and our research supervisor, for giving us the opportunity for shaping our research career. She has not only been instrumental in channeling our thoughts but has also constantly supported us and encouraged us throughout our academic journey. She has taught us the intricacies of scientific research and introduced us to the exciting world of immunobiology and proteomics. She guided us to explore the different avenues of the exciting scientific world. Apart from her, there are friends who shouldered our woes and our happiness whenever needed. We are proud to have met them all in my life. It is not possible to acknowledge them individually in this short span.

We express our special gratitude to Dr. Sheikh Hasan Habib, husband of Dr. Waliza Ansar, Anesthesiologist and Senior Consultant, Critical Care specialist, Kolkata, for his constant help and support during our clinical studies. We are grateful to all other doctors and team members of the hospital administration of MB Nursing Home, Park CirCus, Kolkata, India for their kind help in our clinical study.

We are grateful to our family for their continuous support and encouragement. We treasure their help, cooperation, support, and encouragement. No word is enough to define the constant support and blessings of our parents. Our parents truly defined for us the word "hardworking."

Finally, lots of love to our children, perhaps the best in the whole world, Master *Saptarshi* and Baby *Tanaz*.

It is worth mentioning our indebtedness to all researchers in the field of C-reactive protein from across the globe. We express our gratitude to all journals of national and international repute, for accepting and publishing our work mentioned in this book. It is the scientific research world that helps us to develop confidence, clear vision, strength for struggle, and courage and stamina in search of the truth and to undertake research touching the human lives.

We are grateful to the organizations and institutions of Govt. of India including Council for Scientific and Industrial Research (CSIR), India; Indian Institute of Chemical Biology (IICB), Kolkata; National Institute of Science, Education and Research (NISER), Bhubaneswar, DAE, Govt. of India, Jatni, Orissa, and Academic Institutes under Govt. of West Bengal; and Behala College, Kolkata, for providing us the necessary support.

Our heartfelt thanks to the reviewers who reviewed the proposal submitted to Springer; we are thankful for their inputs.

We are grateful to Springer, India, and their entire team for accepting our proposal to this book. Thank you all for your constant help and cooperation, which is what has made us complete this work.

Contents

About the Authors

Waliza Ansar, PhD She is an alumnus of the erstwhile Presidency College, Kolkata. She mastered in Zoology with specialization in Fisheries and Aquaculture from Calcutta University. With interest in the field of immunology, she started her research, in 2003, in Indian Institute of Chemical Biology, Kolkata, and earned her PhD degree working on various biological roles of CRP in malaria, leishmania, and tuberculosis. After her postdoctoral research in IICB working on host–parasite interactions, she started her teaching career as a lecturer of Zoology in Presidency College, Kolkata, in 2002. In 2009 she joined Asutosh College, Kolkata, as a lecturer of Zoology for postgraduate students. Currently, she is working as an Assistant Professor of Zoology, Behala College, Kolkata, under Calcutta University. Her current research interest involves various clinical roles of CRP in diseases, pathophysiology of some acute diseases, nanotechnology and therapeutic applications, and stem cell therapy. She has presented her work in international and national symposiums. She has published a number of papers in journals of national and international repute. She also has authored a number of publications and book chapters.

Shyamasree Ghosh, PhD Dr. Shyamasree Ghosh is currently working in the capacity of Scientific Officer (E), School of Biological Sciences, National Institute of Science Education and Research (NISER), Bhubaneswar, an autonomous institute under the Department of Atomic Energy, Govt. of India, currently located in Jatni, Orissa. She graduated from the prestigious Presidency College Kolkata in 1998 in Zoology and was awarded the prestigious National Scholarship from the Govt. of India in the same year. She postgraduated in Biotechnology from Calcutta University ranking second in the university in the year 2000.

With her immense interest in the field of science, touching human lives, she continued her PhD from Indian Institute of Chemical Biology, CSIR, Kolkata, qualifying the National Eligibility Test conducted by Ministry of HRD, Govt. of India. Her PhD work in the field of childhood acute lymphoblastic leukemia (ALL) was accepted in various national and international journals of repute and is cited by the young researchers across the globe. She worked in the field of stem cells and nanotechnology in her postdoctoral research. Later she joined as a faculty in postgraduate colleges affiliated to Bangalore University. She has authored several publications and book and encyclopedia chapters in reputed journals and books.

Research that tries to decode the complexity of disease with hopes for a better living forms her research interest.

Abbreviations

αGalNAc	Alpha-N-acetyl galactosamine
αGlcNAc	Alpha-N-acetyl glucosamine
αGlcNAc	Alpha-N-acetyl glucosamine
α-L-Fuc	Alpha-L-fucose
α1-AGP	α1-Acid glycoprotein
β-ME	β-Mercaptoethanol
β2-M	β2-Macroglobulin
μg	Microgram
AAG	Alpha 1-acid glycoprotein
AAT	Alpha 1-antitrypsin
ABG	Arterial blood gas
ABGA	Arterial blood gas analysis
ABTS	2,2′-Azino-bis(3-ethylbenzthiazoline-6-sulfonic acid) diammonium salt
AcSGs	Acetylated sialoglycans
ACT	Alpha 1-antichymotrypsin
ACTH	Adrenocorticotropic hormone
ADCC	Antibody-dependent cell-mediated cytotoxicity
ADMA	Asymmetric dimethylarginine
AF	Atrial fibrillation
AFB	Acid-fast bacilli
AFP	Alpha fetoprotein
Ag	Antigen
AGP	Alpha 1-acid glycoprotein
AIF	Apoptosis inducing factor
ALI	Acute lung injury
ALL	Acute lymphoblastic leukemia
ALT	Alanine aminotransferase
AMI	Acute myocardial infarction
AML	Acute myelogenous leukemia
ANS	8-Anilino-1-napthalenesulfonic acid
AP	Acute pancreatitis
APC	Antigen-presenting cell
API	Acute-phase index
APPs	Acute-phase proteins
APR	Acute-phase reactant

APR	Acute-phase response
APRF	Acute-phase response factor
APTT	Activated partial thrombin time
ARDS	Acute respiratory distress syndrome
ARF	Acute respiratory failure
ASO	Antistreptolysin O
AST	Aspartate aminotransferase
B2M	Beta 2-microglobulin
BA	Bronchial asthma
BAL	Bronchoalveolar lavage
BCR	B-cell receptor
BH4	Tetrahydrobiopterin
BP	Blood pressure
BNP	Brain natriuretic peptide
BPB	Bromophenol blue
BPH	Benign prostatic hyperplasia
BSA	Bovine serum albumin
BSM	Bovine submaxillary mucin
BS-TFA	Bis-(trimethylsilyl)-trifluoroacetamide
C. catla	*Catla catla*
C/EBP	CCAAT/enhancer-binding proteins
C3	Complement factor-3
C5a	Complement component 5a
Ca	Calcium
CABG	Coronary artery bypass grafting
$CaCl_2$	Calcium chloride
CAD	Coronary artery disease
CB	Chronic bronchitis
CBG	Capillary blood glucose
CC-16	Club cell secretory protein
CD	Crohn's disease
CD	Cluster of differentiation
CDG	Congenital disorders of glycosylation
CDNA	Complementary DNA
CDR	Complementarity determining regions
CGD	Chronic granulomatous disease
CGT	Conglutinin
CHAPS	3-[(3-Cholamidopropyl)dimethylammonio]-1-propanesulfonic acid
CLIP	Class II-associated invariant chain peptide
CLL	Chronic lymphocytic leukemia
CM	Cerebral malaria
CMI	Cell-mediated immunity
CNS	Central nervous system
CNTF	Ciliary neurotrophic factor
COE	Coeruloplasmin
Con A	Concanavalin A
COPD	Chronic obstructive pulmonary disease

COX-2	Cyclooxygenase 2
CPL	Ceruloplasmin
CPM	Counts per minute
CPS	C-polysaccharide
CQ	Chloroquine
CR	Complement receptor
CRH	Corticotropin-releasing hormone
CRP	C-reactive protein
CRP	C-reactive proteinase
CRP_{VL}, CRP_{TB} and CRP_{SLE}	C-reactive protein purified from visceral leishmaniasis, tuberculosis, and systemic lupus erythematosus
CRP_{VL}	C-reactive protein purified from visceral leishmaniasis
CRRT	Continuous renal replacement therapy
CS	Cigarette smoke
CSF	Cerebrospinal fluid
CTLs	Cytotoxic or cytolytic T lymphocytes
Cu	Copper
CVD	Cardiovascular disease
CVP	Central venous pressure
CVS	Cardiovascular system
DAB	3, 3-Diaminobenzidine
DBA	Dolichos biflorus agglutinin
DCs	Dendritic cells
DHMEQ	Dehydroxymethylepoxy-quinomicin
DIG	Succinyl-ε-amido caproic acid hydrazide digoxygenin
DIP	Distal interphalangeal
dl	Deciliter
DLC	Differential leukocyte count
DM	Diabetes mellitus
DMB	1, 2-Diamino-4, 5-methylenedioxy-benzene
DMSO	Dimethyl sulfoxide
DNA	Deoxyriboneuclic acid
DNP	Dinitrophenol
DPH	1,6-Diphenyl-1,3,5-hexatriene
DSA	Datura stramonium agglutinin
dsRNA	Double-stranded RNA
DSS	Dextran sodium sulfate
DTNB	5′5′-Dithio(bis)-2-nitrobenzoic acid
DVT	Deep vein thrombosis
ESR	Erythrocyte sedimentation rate
E	Erythrocytes
E	Eosinophil
ECG	Electrocardiogram
EDTA	Ethylenediaminetet-raacetic acid
EF	Ejection fraction

EGC	Enteric glial cells
EGTA	Ethylene glycol tetraacetic acid
ELISA	Enzyme-linked immunosorbent assay
EMP	Endothelial microparticle
eNOS	Endothelial nitric oxide synthase
EONS	Early-onset neonatal sepsis
EPC	Endothelial progenitor cell
EPO	Erythropoietin
ER	Endoplasmic reticulum
ESR	Erythrocyte sedimentation rate
ESRF	End-stage renal failure
ET	Endotracheal
E_{TB}, E_{VL} and E_N	Erythrocytes from TB, VL, and normal (N) individuals
Fab	Antigen-binding fragments
FACS	Fluorescence-activated cell sorter
FBS	Fasting blood sugar
FcγR	Fcγ receptor
FcR	Fc receptor
FCS	Fetal calf serum
Fe	Iron
FEV	Forced expiratory volume
FFAs	Free fatty acids
FITC	Fluorescein isothiocyanate
FMD	Flow-mediated dilation
FNG	Fibrinogen
FoxP3	Forkhead box P3
Gal	Galactose
GalNAc	*N*-acetyl galactosamine
GalNAc	2-Acetamido-2-deoxi-galactopyranose
GalNAc T	*N*-Acetyl galactosamine transferase
GCS	Glasgow coma scale
GCSF	Granulocyte colony-stimulating factor
GFR	Glomerular filtration rate
GLC	Gas liquid chromatography
GlcNAc	*N*-Acetylglucosamine
GlcNAc	2-Acetamido-2-deoxy-glucopyranose
GlcNAc	N-Acetyl-D-glucosamine
GlcNAc T	*N*-Acetylglucosamine transferase
GlcNAc	2-Acetamido-2-deoxi-glucopyranose
GM-CSF	Granulocyte macrophage colony-stimulating factor
GNA	Galanthus nivalis agglutinin
GPI	Glycosylphosphatidylinositol anchor
GR	Glutathione reductase
GSH	Reduced glutathione
GSSG	Oxidized glutathione
GTN	Nitroglycerine
GTN	Glyceryl trinitrate

GVB	Gelatin veronal buffer
H_2O_2	Hydrogen peroxide
HA	Hemagglutination assay
Hb	Hemoglobin
HDL	High-density lipoprotein
HDL-Chol	HDL cholesterol
HDLs	High-density lipoproteins
HEPES	N-2-Hydroxyethyl piperazine-N′-2-ethanesulfonic acid
HGB	Haptoglobin
HGF	Hepatocyte growth factor
HLA	Human leukocyte antigen
HLA-DM	Human leukocyte antigen DM
Hp	Haptoglobin
HPA	Helix pomatia agglutinin
HPF	High-power field
HPT	Haptoglobin
HRP	Horseradish peroxidase
HSA	Human serum albumin
HSAP	Female protein of Syrian hamster
hsCRP	High-sensitivity C-reactive protein
HU	Hemagglutination unit
i.m.	Intramuscular
i.v.	Intravenous
I	Iodine
IBD	Inflammatory bowel disease
IC	Infected control
ICAM	Intercellular adhesion molecule
ICAM-1	Intercellular adhesion molecule 1
ICU	Intensive care unit
IEF	Isoelectric focusing
IFA	Indirect fluorescent antibody
IFN	Interferon
IFN-gamma	Interferon gamma
Ig	Immunoglobulin
IgE	Immunoglobulin E
IgG	Immunoglobulin G
IgM	Immunoglobulin M
IGT	Impaired glucose tolerance
IL	Interleukin
IL-1ra	IL-1 receptor antagonist
INF-γ	Interferon-γ
iNOS	Inducible nitric oxide synthase
IQR	Interquartile range
IRS1	Insulin receptor substrate 1
ITU	Intensive care unit
JAK/STAT	Janus kinase (JAK) and signal transducer and activator of transcription (STAT)
JVP	Jugular venous pressure

*K*a	Affinity constant
KA	Kala azar
kDa	Kilodalton
L. donovani	*Leishmania donovani*
L. rohita	*Labaeo rohita*
L	Lymphocyte
LBP	Lipopolysaccharide binding protein
LD body	*Leishmania donovani* body
LDH	Lactate dehydrogenase
LDL	Low-density lipoprotein
LIF	Leukemia inhibitory factor
LPA	Limulus polyphemus agglutinin
LPG	Lipophosphoglycan
LPS	Lipopolysaccharide
LTβR	Lymphotoxin-β receptor
LVH	Left ventricular hypertrophy
Ly-6G	Lymphocyte antigen 6 complex, locus G
M	Monocyte
MAA	*Maackia amurensis* agglutinin
mAb	Monoclonal antibody
MAC	Membrane attack complex
Mal	Malaria
MALDI	Matrix-assisted laser desorption ionization
MALDI-TOF	Matrix-assisted laser desorption ionization time of flight
MAP kinase	Mitogen-activated protein kinase
MASP	MBL-associated serine protease
MBL/MBP	Mannose-binding lectin/protein
MBL	Mannose-binding lectin
MBP	Mannose-binding proteins
MBP	Myelin basic protein
MCP	Metacarpophalangeal
MCP-1	Monocyte chemoattractant protein-1
MCV	Mean corpuscular volume
MDA	Malondialdehyde
mDC	Myeloid dendritic cell
MDR	Multidrug resistant
mEq	Milliequivalents
MET	Metformin
MFI	Mean fluorescence intensity
mg	Milligram
MHC	Major histocompatibility complex
MI	Myocardial infarction
MIP1a	Macrophage inflammatory protein-1 alpha
MIP1b	Macrophage inflammatory protein-1 beta
miR	Micro-ribonucleic acid
MMP	Matrix metalloproteinase
Mol. Wt	Molecular weight

MP	Microparticle
mPCa	Metastatic prostate cancers
MPs	Microparticles
MRI	Magnetic resonance imaging
mRNA	Messenger ribonucleic acid
MS	Mass spectroscopy
MSC	Mucosal stem cell
MSCs	Mesenchymal stem cells
MSP	Modified seminal plasma
MW	Molecular weight
N	Neutrophil
NA	Not applicable
NaCl	Sodium chloride
NADPH	Nicotinamide adenine dinucleotide phosphate hydrogen
NAPI	Nutritional and acute-phase indicator
NCQ	Nanochloroquine
ND	Not determined
Neu5,9Ac$_2$	9-O-Acetylneuraminic acid
Neu5Ac	Neuraminic acid
NF-κβ	Nuclear factor kappa-light-chain-enhancer of activated B cells
NF	Nuclear factor
NF-L	Neurofilament light chain
NHS	Normal human serum
NK cells	Natural killer cells
NM	Non-malarial
NO	Nitrogen oxide
NO	Nitric oxide
NOS	NO synthase
OD	Optical density
O-AcSA	O-acetyl sialic acid
O-AcSG	O-acetyl sialoglycoconjugate
°F	Degree Fahrenheit
O-GalNAc	O-N-Acetylgalactosamine
O-GlcNAc	O-N-Acetylglucosamine
OGT	GlcNAc transferase
OPD	O-diamisidine
OS	Osteogenic sarcoma
OSA	Obstructive sleep apnea
OSM	Oncostatin M
oxLDL	Oxidized low-density lipoprotein
p65	Transcription factor p65 also known as nuclear factor NF-kappa-B p65 subunit
PAF	Platelet-activating factor
PAGE	Polyacrylamide gel electrophoresis
PAI	Plasminogen activator inhibitor
PAI-1	Plasminogen activator inhibitor type 1

PAMPs	Pathogen-associated molecular patterns
PAP	Pulmonary artery pressure
PB	Phosphate buffer
PBL	Peripheral blood lymphocytes
PBMC	Peripheral blood mononuclear cells
PBMs	Peripheral blood-derived macrophages
PBS	Phosphate-buffered saline
PC	Phosphorylcholine
PCa	Prostate cancers
PCT	Procalcitonin
PTCA	Percutaneous transvenous coronary angiography
pDC	Plasmacytoid dendritic cell
PDGF	Platelet-derived growth factor
PE	Phosphoethanolamine
PEF	Peak expiratory flow
PET	Positron emission tomography
Pf	*Plasmodium falciparum*
pg	Pictogram
PGN	Peptidoglycan
PI	Propidium iodide
pI	Isoelectric point
PINI	Prognostic inflammatory and nutritional index
PKDL	Post-kala-azar dermal leishmaniasis
PLA_2	Phospholipase enzyme
PMA	Phorbol myristate acetate
PMN elastase	Polymorphonuclear elastase
PNA	Peanut agglutinin
PON1	Paraoxonase
PRE	Prealbumin
pro-BNP	Pro-brain natriuretic peptide
PRRs	Pattern recognition receptors
PSA	Prostate-specific antigens
PSD	Post-source decay
PTT	Prothrombin time test
PTX3	Pentraxin 3
PUFA	Polyunsaturated fatty acids
pz complex	Properdin–zymosan complex
RA	Rheumatoid arthritis
RAGEs	Receptor for advanced glycation endproducts
RBBB	Right bundle branch block
RBC	Red blood cells
RBC-CRP complex	RBC were complexed with CRPMal
RBC_{Mal}	Erythrocytes from malaria
RBC_N	Erythrocytes from normal individual
RBP	Retinol binding protein
RBS	Random blood sugar
RCA	*Ricinus communis* agglutinin

Rf	Rate of front mobility
RNA	Ribonucleic acid
RNI	Nitrogen intermediates
ROS	Reactive oxygen species
rRAP-1	*Plasmodium falciparum* rhoptry-associated protein-1
RT-PCR	Reverse transcription polymerase chain reaction
RWMA	Resting wall motion abnormality
S.D	Standard deviation
S_1	First heart sound
S_2	Second heart sound
SA	Sialic acid
SAA	Serum amyloid A
SAG	Sodium antimony gluconate
SAP	Serum amyloid P component
sCr	Serum creatinine
SD	Standard deviation
SDS	Sodium dodecyl sulfate
SDS-PAGE	Sodium dodecyl sulfate polyacrylamide
SEM	Standard error of mean
SEM	Scanning electron microscopy
SGOT	Serum glutamate-oxalate transaminase
SGPT	Serum glutamate pyruvate transaminase
SialT	Sialyltransferases
SLE	Systemic lupus erythematosus
SLED	Sustained low-efficiency dialysis
SMA	Severe malarial anemia
SNA	*Sambucus nigra* agglutinin
SNPs	Single-nucleotide polymorphisms
snRNPs	Small nuclear ribonucleoprotein particles
SOS	As and when required
SP-D	Surfactant protein-D
$sPLA_2$	Secretory phospholipase A_2
$sPLA_2$-IIA	Secreted group IIA phospholipase A_2
SpO_2	Peripheral capillary oxygen saturation
SPR	Surface plasmon resonance
SSM	Sheep submaxillary mucin
ST	The ST segment represents the period when the ventricles are depolarized in ECG
STAT	Signal transducers and activators of transcription
STAT3	Signal transducer and activator of transcription 3
sTNFR75	Soluble receptor for TNF alpha
T2D	Type 2 diabetes mellitus
TAP	Transporter associated with antigen processing 1

TB	Tuberculosis
TBA	Thiobutiric acid
TBARS	Thiobarbituric acid reactive substance
TBS	Tris-buffered saline
TCA	Trichloroacetic acid
TCR	T-cell receptor
TEC	Tubular epithelial cells
TEMED	N,N,N′,N′-Tetrameth-ylethylenediamine
TFA	Trifluoroacetic acid
TfR	Soluble transferrin receptors
TfR	Transferrin receptor
TGF-β	Transforming growth factor beta
T_H cell	T helper cell
TIBC	Total iron-binding capacity
TLC	Thin-layer chromatography
TLC	Total leukocyte count
TLR	Toll-like receptors
TMCS	Trimethylchlorosilane
TMS	Trimethylsilyl
TNF	Tumor necrosis factor
TOF	Time of flight
TRF	Transferrin
TSST-1	Toxic shock syndrome toxin-1
TTR	Transthyretin
UC	Ulcerative colitis
uCRP	Urinary CRP
UEA	*Ulex europaeus* agglutinin
USG	Ultrasonography
VAP	Ventilator-associated pneumonia
VCAM	Vascular cell adhesion molecule
VCAM-1	Vascular adhesion molecule-1
VEGF	Vascular endothelial growth factor
VFD	Ventilator-free days
VIP	Vasoactive intestinal peptide
VL	Visceral leishmaniasis
VL	Visceral leishmaniasis
VLDL	Very low-density lipoprotein
VPF	Vascular permeability factor
vWF	Von Willebrand factor
WBC	White blood corpuscles
WGA	Triticum vulgaris agglutinin
WHO	World Health Organization
WNL	Within normal limit
WT	Wild type
Zn	Zinc
ZPP	Zinc protoporphyrin

Immune System: Freedom from the Burden of Diseases

1

Abstract

In the daily life of a healthy individual, microorganisms are encountered at random, causing diseases only occasionally. The defense mechanisms mostly detect these microbes and destroyed them within minutes or hours through mechanisms of *innate immunity. Adaptive immunity* acts through antigen-specific lymphocytes to culminate pathogens. Both the innate and adaptive immune systems, to recognize pathogens, distinguish between self and nonself particles. The innate immunity discriminates very successfully between host cells and pathogens, providing earlier defenses and also induces adaptive immune responses. Defects in the components of innate immunity, in very rare cases, can lead to amplified susceptibility to infection, even in the presence of an intact adaptive immunity.

Keywords

Immune system • Innate • Adaptive • Lymphocytes • Macrophages • T cell • B cell • Antigen • Antibody • Phagocytosis

Chapter Highlights

1. The immune system has evolved through the involvement of as complex mesh of molecules, cells, secretory products and organs to defend against various pathogenic microbes.
2. The immune system also defends against noninfectious foreign substances.
3. The immune system helps the host in protection, maintains tissue homeostasis, regulates tissue repair and also screens cell surface expression of specific molecules.
4. Cells of the immune system also locate and remove injured, dead or malignant cells.
5. Immune cells are derived from different hematopoietic stem cells, circulate in the fluidic blood and lymph medium, form complex aggregation in particular lymphoid organs and infiltrate virtually all tissue.
6. Host defense and host protection is achieved by two interdependent and closely interlinked types of immunity, innate and adaptive.
7. The older system is the *innate immune system*, the first line of defense, is present in all verte-

W. Ansar, S. Ghosh, *Biology of C Reactive Protein in Health and Disease*,
DOI 10.1007/978-81-322-2680-2_1

brates, mediates nonspecific protection and is widely conserved in species.

8. A diverse set of cells, like monocytes, macrophages, dendritic cells, natural killer (NK) cells, eosinophils, basophils, neutrophils and mast cells along with a variety of chemical mediators, including complement system, acute phase reactants and cytokines, amplifies inflammatory responses to prevent tissue invading pathogens.
9. The highly specific adaptive mechanisms are the immediacy of the responses which is highly effective against microbial invasion through cell mediated and humoral responses by T and B lymphocytes.

1.1 Introduction

Diseases caused by infections agents like viruses, fungi, bacteria, protozoa, worms, and toxins—is kept at bay by the physiological role of the immune system. A broader definition of the immune system would be its ability to identify *self* and thus recognize *nonself*. The self–nonself-recognition is relatively primordial: ctenophores also have the capacity to recognize and react to nonself. The immune system in humans is often challenged by nonself, including pathogens, organs transplanted from unrelated donors. Protection from these is carried by complex and destructive processes, the understanding of which forms the basis of immunology.

From our own everyday experience, we can easily understand concepts of immunity and infection. For example, exposed tissues are subjected to infection and loss of physical barriers lowers immunity. We also know that some infections are easily combated by the immune system, while others are not.

We consider the newborn, old-age people, or pregnant women to be at overall greater risk of infection, i.e., they have less immunity as compared to adult individuals. Childhood infections like chicken pox leave us immune against further infection. On the other hand, having had chicken pox would not stop a child catching mumps. Thus, conclusively, we are born with some immunity and the rest may be acquired or adapted during our lifetime. Immune responses can be highly specific for a pathogen: it may be learned and retained in an "immunological memory" or may be generalized at first hand.

1.2 Immune System and Its Branches

The human immune system has evolved as a multifaceted network of cells, molecules, and organs to protect against pathogenic microbes and non-infectious foreign antigens. In addition to its function in host protection, it controls tissue repairing and homeostasis by selecting cell surfaces for the expression of specific molecules. Cells of the immune system recognize and eliminate malignant cells or some injured, dead, apoptotic, or necrotic cells. An immune cell derived in the bone marrow from hematopoietic stem cells matures and is anatomically organized in specialized lymphoid organs, circulates in both blood and lymph, and infiltrates the tissues. Host defense is pronounced by two types of immunity, basically the innate and adaptive system. These two arms of the immune system are not only dependent on each other but are also closely interlinked.

1. *Innate immunity*

Immunity which is inborn, i.e., present at birth, is termed *innate*. The innate immune system is the early first-stroke resistance against invading organisms. Innate immunity is without any specificity and lacks memory, and characteristically it is present throughout the life. But still, there are some protective antibodies that babies acquire from their mothers (Box 1.1).

2. *Acquired immunity*

In contrast, there is an immune response which is not present at birth but is gained or acquired by as part of adaptations during our growth and development. The acquired or specific immune response is the antithesis of innate immunity, which increases with age and has specificity and memory; hence, it may also be

Box 1.1: Functions of the Innate Immune System

The major functions of the vertebrate innate immune system include:

- Recruiting immune cells to sites of infection
- Through the production of chemical factors, including specialized chemical mediators, called cytokines
- Activation of the complement cascade to identify bacteria
- Activation of cells and promotion of clearance of dead cells or antibody complexes
- Identification and removal of foreign substances present in organs, tissues, blood, and lymph, by specialized white blood cells
- Activation of the adaptive immune system through a process known as antigen presentation

termed *adaptive*. To distinguish pathogens, both the early innate and late adaptive immune systems can discriminate between self- and nonself-entities, but they always differ in their modus operandi (Table 1.1).

Innate immunity functions via a restricted number of receptors and soluble factors that are scripted in the germline and that identify some generalized molecules to many pathogens. In contrast, adaptive immunity uses an enormous range of antigen receptors that are proficient in discriminations between closely associated molecules generated due to somatic cell gene rearrangement. The innate immune system provides initial protection and also contributes to the introduction of adaptive immune responses. Paralysis of one component of either of these two forms of immunity can have a profound effect on the host defense against infection. The importance of innate immunity remains in the fact that fallacies in its components, which are quite rare, can lead to aggravated vulnerability to infection, even in

Table 1.1 Basic differences between an innate and adaptive immunity (Janeway et al. 2001)

Properties	Innate	Adaptive
Definition	First line of defense	Specific reaction to each infectious agent
Memory	Lacks memory	Keeps memory of the infectious agent and can prevent disease later, e.g., measles, kala-azar, diphtheria
Cells	NK cells, macrophages, dendritic cells, granulocytes	T lymphocytes B lymphocytes
Soluble mediators	Cytokines, chemokines, Pentraxins, complement, etc.	Cytokines, chemokines Antibodies
Protects from	Bacteria, fungi, worms	Bacteria, intracellular infection, viruses, and protozoa
Recognition	Conserved pattern molecules on pathogen (LPS, mannans, phosphorylcholine)	Detailed molecular structure (proteins, peptides, carbohydrates)
Receptors	Germline-encoded receptors	Receptors are not encoded in the germline
Specificity of receptors	Specificity of each receptor is genetically predetermined	Not predestined to recognize any particular antigen; extremely diverse repertoire of receptors is generated randomly
Receptor rearrangement	Rearrangement is not necessary	Rearrangement necessary
Self–nonself-discrimination	Selected over evolutionary time	Selected in individual somatic cells
Action time	Immediate, fast	Delayed, slow
Distribution	Non-clonal (all cells of a class identical)	Clonal (all cells of a class distinct)

the attendance of an intact adaptive competent immune system.

Innate immunity has three components, namely, the physicochemical, the humoral, and the cellular barriers.

Physical barriers are the *skin* and *mucosae*; some *secretions*, which continually wash and cleanse mucosal surfaces; and *cilia*, which help the removal of debris and foreign matter from the body. Immunologically active factors/molecules present in blood, in mucosal secretions, and in the cerebrospinal fluid (the *humors*) are known as humoral. The most important of these is the *complement* and the *mannan-binding lectin*, as well as other *opsonins* (an opsonin aids digestion of bacteria by neutrophils), such as *C-reactive protein*, and proteolytic *enzymes* (like *lysozyme*). There are some generalized components of innate immunity (Box 1.2). Cellular components are the *neutrophil*, the *eosinophil* and the *mast cell*, as well as the *NK cell* (Table 1.2, Box 1.3). The functions of NK cells are tabulated in Box 1.4.

The innate immune system is represented by the handling of a restricted number of germline-encoded receptors that is familiar with diverse antigens which are attacking the host. The immune system in invertebrates is a bit different from that of the vertebrate system (Box 1.5). The immune system has actually first (physical and biochemical), second (inflammation), and third (adaptive) lines of defense system (Box 1.6). This type of immune system pinpoint a set of molecular structures that are deprived from host cells which nevertheless are inimitable to microbes and communal by diverse pathogens. The immune system is capable of initiating autoimmune responses by identifying the *pathogen-specific* patterns. C-reactive protein is also a pattern recognition molecule. Its function as a family member of pentraxin is tabulated in Box 1.7.

The different receptor classes exploited during phagocytosis, apart from the pattern recognition receptors, are tabulated in Box 1.8. The chief functions of phagocytosis include pattern recognition of microbes through surface pattern recognition receptors, secretion and synthesis of cytokines or the chemokine network, initiation of the process of phagocytosis by binding and engulfment of particles, intralysosomal digestion and killing of digested particle, and antigen presentation to lymphocytes and chemotaxis. Phagocytosis or cellular ingestion of particles is of two types—opsonic and non-opsonic phagocytosis. In non-opsonic phagocytosis there is direct engulfment of particles through pattern recognition molecules, and the process is quite slow, limited, and inefficient. In opsonic phago-

Box 1.2: Components of Innate Immunity

Components of Innate Immunity

- **Physical/Chemical barriers**
  - Intact skin, epithelial layer, cough, fever
- **Nonspecific chemical factors**
  - Antimicrobial peptides and fatty acids, gastric pH, lysozyme
- **Inflammation**
  - phagocytes (engulf & digest microbes)
  - proinflammatory factors (cytokines, complement proteins)
- **Natural killer cells (nonspecific cytotoxic cells)**
- **Interferon (produced by virus-infected cells and induces anti-viral state in neighboring uninfected cells)**

Table 1.2 Cells of Innate Immunity

Cells	Function	Product	Location	Reported markers	Activation
Mast cells	Against pathogens, heal the wound, allergy, and anaphylaxis Recruit neutrophils and macrophages Dilate blood vessels	Histamine, heparin, chemokines, or chemotactic cytokines	Connective tissue and mucous membrane	CD9, CD33, CD43, CD15, CD54, CD117	IgE: allergen attachment
Macrophages	Chemotaxis, phagocytosis, obliterate bacteria through respiratory burst, inflammation, apoptosis, antigen presentation	ROS, cathepsins, lysozymes, PLA_2, iNOS, chemokines/cytokines (IL-8, MIP1a, MIP1b, TGFβ, PDGF), etc.	Migrate to the areas between cells in quest of invading pathogens	FcγRs, CD11c, CD14, CD68, CD82, CD163 HLA-DM, MHC class	LPS, LBP, IFNγ, MSP, IL-4, IL-13 .
Neutrophils	Phagocytic cells, chemotaxis, phagocytosis, demolish bacteria via respiratory burst	Oxidizing agents (H_2O_2), free oxygen radicals, serine proteases, neutrophil elastase, cathepsin G	First cells to disembark at the site of an infection	CD11b, CD15, CD66b, Ly-6G (applicable for mouse)	IL-8, IFNγ, C5a
Dendritic cells (mDC, pDC)	Phagocytic cells, antigen presentation, act as a bridge between innate and adaptive immune systems	Cytokines	Skin, mucosa	CD1a, CD11c, CD40, TLR2 and TLR4 (mDC), TLR7 and TLR9 (pDC)	TLRs, LTβRs
Basophils	In opposition to parasites, allergic reactions like asthma	Histamine	Infection site	Ly-6G	
Eosinophils	Destroy bacteria and parasites, lead to allergic reactions	Toxic proteins and free radicals	Infection site	CD15, Ly-6G, CD67	

Box 1.3: Leukocytes

- All white blood cells (WBC) are known as leukocytes.
- Leukocytes are different from other cells of the body.
- Not tightly associated with a particular organ or tissue.
- Function similar to independent, single-celled organisms.
- Leukocytes are able to move freely and interact with and capture cellular debris, foreign particles, or invading microorganisms.
- Unlike many other cells in the body, most innate immune leukocytes cannot divide or reproduce on their own but are the products of pluripotent hematopoietic stem cells present in the bone marrow.
- The innate leukocytes include natural killer cells, mast cells, eosinophils, basophils, and phagocytic cells including macrophages, neutrophils and dendritic cells and function within the immune system by identifying and eliminating pathogens that might cause infection.

Box 1.4: NK Cells

- *Natural killer cells, or NK cells, are a component of the innate immune system.*
- NK cells attack host cells that have been infected by microbes but do not directly attack invading microbes.
- NK cells attack and destroy tumor cells and virus-infected cells, through a process known as "missing self."
- "Missing self"—cells with low levels of a cell surface marker called MHC I (major histocompatibility complex)—a situation that can arise in viral infections of host cells.
- They were named "natural killer" because of the initial notion that they do not require activation in order to kill cells that are "missing self."

Box 1.5: Host Defense in Invertebrates

- Invertebrates do not possess lymphocytes or an antibody-based humoral immune system, and it is likely that a multicomponent, adaptive immune system arose with the first vertebrates.
- Nevertheless, invertebrates possess mechanisms that appear to be precursors of these aspects of vertebrate immunity.
- *Pattern recognition receptors* are proteins used by nearly all organisms to identify molecules associated with microbial pathogens. *Toll-like receptors* are a major class of a pattern recognition receptor that exists in all coelomates (animals with a body cavity), including humans.
- The complement system is a biochemical cascade of the immune system that helps clear pathogens from an organism and exists in most forms of life. Some invertebrates, including various insects, crabs, and worms, utilize a modified form of the complement response known as the proPO system.
- *Antimicrobial peptides* are an evolutionarily conserved component of the innate immune response found among all classes of life and represent the main form of invertebrate systemic immunity. Several species of insect produce antimicrobial peptides known as *defensins* and *cecropins*.

cytosis engulfment of complement-coated or antibody-coated particles takes place through complement receptors (CR) or antibody receptors (FcR), and the process is rapid and very efficient. Neutrophils, one of the eminent professional phagocytes, initiate the process of phagocytosis through respiratory burst, degranulation, and formation of neutrophil extracellular traps. Phagocytosis may be through an oxygen-dependent (production of reactive oxygen and reactive nitrogen intermediates) mechanism or oxygen-independent (production of lysosomal hydrolases, lysozyme, lactoferrin, defensins, acid

Box 1.6: Three Lines of Defense of Immunity

- First line of defense
  - Innate resistance
  - Physical (skin/epithelial layer, GI and resp. tract) and mechanical (cough, sneeze, vomit, cilia action in the trachea)
  - Biochemical barriers (antimicrobial peptides, lung secretions, mucus, saliva, tears, earwax)
- Second line of defense
  - Inflammation—vascular response—dilation, histamines increasing vessel leakage, WBC action, cytokines, leukokines, and fever. Usually redness and heat with swelling
- Third line of defense
  - Adaptive (acquired) immunity—antibody production

Box 1.7: Pentraxin as a Pattern Recognition Molecule

Pentraxin

- Members include:
  - Serum amyloid protein (SAP)
  - C-reactive protein (CRP)
- Recognizes phosphorylcholines on microbes
- Functions as an opsonin
- Binds to C1q and activates the classical complement pathway

pH) mechanisms. Macrophages are another group of potent phagocytic cells whose functions are tabulated in Box 1.9.

Some molecules like lipopolysaccharide (LPS), many bacterial lipoprotein, the peptidoglycan (PGN), lipoteichoic acid (LTA), and also double-stranded RNA (dsRNA) are primarily synthesized by viruses or bacteria, but not by the host cells. Furthermore, carbohydrate entities like *N*-acetylglucosamine and mannose on the external surface of pathogens are also symbolized targets for recognition.

Box 1.8: Different Receptors Exploited by Macrophages for Identification of Microbes

Phagocytic cells have a variety of receptors on their cell membranes through which infectious agents bind to the cells. These include:

1. Fc receptors—bacteria with IgG antibody on their surface have the Fc region exposed, and this part of the Ig molecule can bind to the receptor on phagocytes. Binding to the Fc receptor requires prior interaction of the antibody with an antigen. Binding of IgG-coated bacteria to Fc receptors results in enhanced phagocytosis and activation of the metabolic activity of phagocytes (respiratory burst).
2. Complement receptors—phagocytic cells have a receptor for the third component of complement, C3b. Binding of C3b-coated bacteria to this receptor also results in enhanced phagocytosis and stimulation of the respiratory burst.
3. Scavenger receptors—scavenger receptors bind a wide variety of polyanions on bacterial surfaces resulting in phagocytosis of bacteria.
4. Toll-like receptors—phagocytes have a variety of toll-like receptors (pattern recognition receptors or PRRs) which recognize broad molecular patterns called PAMPs (pathogen-associated molecular patterns) on infectious agents. Binding of infectious agents via toll-like receptors results in phagocytosis and the release of inflammatory cytokines (IL-1, TNF-alpha, and IL-6) by the phagocytes.

1.2.1 Recognition of Microbes by the Innate Immune System

The nonspecificity of the innate immune system is diverse and mediated by cells and soluble mediators and different from the adaptive immunity where the specificity of lymphocytes

Box 1.9: Macrophages: Removal of Dead Cellular Debris

- Removal of necrotic cellular debris in the lungs in chronic inflammation, as the early stages of inflammation are dominated by neutrophil granulocytes.
- Handled by *fixed macrophages* at the lungs, liver, neural tissue, bone, spleen, and connective tissue.
- Ingesting foreign materials such as pathogens, recruiting additional macrophages if needed.
- The pathogen becomes trapped in a, which then fuses with a lysosome.
- Within the phagolysosome, enzymes and toxic peroxides digest the pathogen.
- Some bacteria, such as *Mycobacterium tuberculosis*, have become resistant.
- As secretory cells, macrophages are vital to the regulation of immune responses and the development of inflammation.
- They produce an amazing array of powerful chemical substances (monokines) including enzymes, complement proteins, and regulatory factors such as interleukin-1.
- At the same time, they carry receptors for lymphokines that allow them to be "activated" into single-minded pursuit of microbes and tumor cells.

were found. The constituents of innate immunity distinguish structures/molecules that are shared by diverse classes of pathogens and are not simultaneously present on host cells. Each component of innate immunity may perhaps be acquainted with many bacteria, or viruses, or fungi. For example, phagocytes convey receptors for bacterial external lipopolysaccharide (LPS), which is also known as endotoxin, which is present as the components of cell wall of many bacteria but not in the mammalian cells. Other receptors of phagocytes recognize terminal mannose residues which are specific for bacterial glycoproteins but not for the mammalian glycoproteins. Phagocytes distinguish and counter to double-stranded RNA (present in viruses, not in mammalian cells) and to unmethylated CpG oligonucleotides (present in microbial DNA, not in mammalian DNA).

The microbial molecules which are the targets of this immune system are known as *pathogen-associated molecular patterns*. The receptors of innate immune pathways that identify these structures are known as *pattern recognition receptors*.

Some components of the innate immune system are able to bind with the host cells but are prohibited from activation by these cells. For example, if plasma proteins are deposited on host cells, the activation of these complement proteins is uncreative by dictatorial molecules on host cells.

The constituents of the innate immunity network have evolved to identify structures of pathogens that are often indispensable for the infectivity and survival of these microorganisms. The properties of the innate immune system make it an exceedingly effective defense mechanism because microbes can't able to elude innate immunity by mutating themselves. The receptors of the innate immunity are encoded in the germline factors and are not essentially produced by the somatic recombination of genes. The germline-scripted pattern recognition receptors have evolved as a defending adaptation against potentially detrimental microbes. The innate immune system also does not react against the host.

The incapability of this innate immune system to retort against an individual's "self-"cells and molecules is due to the inherent specificity for pathogen structures and various regulatory molecules, articulated by mammalian cells that avert innate immune reactions. The innate immune mechanism responds in the same way with simultaneous repeated attacks by the microbe, whereas the adaptive immune network responds more efficiently to each successive encounter with a microbe.

1.2.2 The Complement Cascade

The complement system is a series of some 20 proteins, which generates one of the triggered enzyme systems, found in plasma in addition to

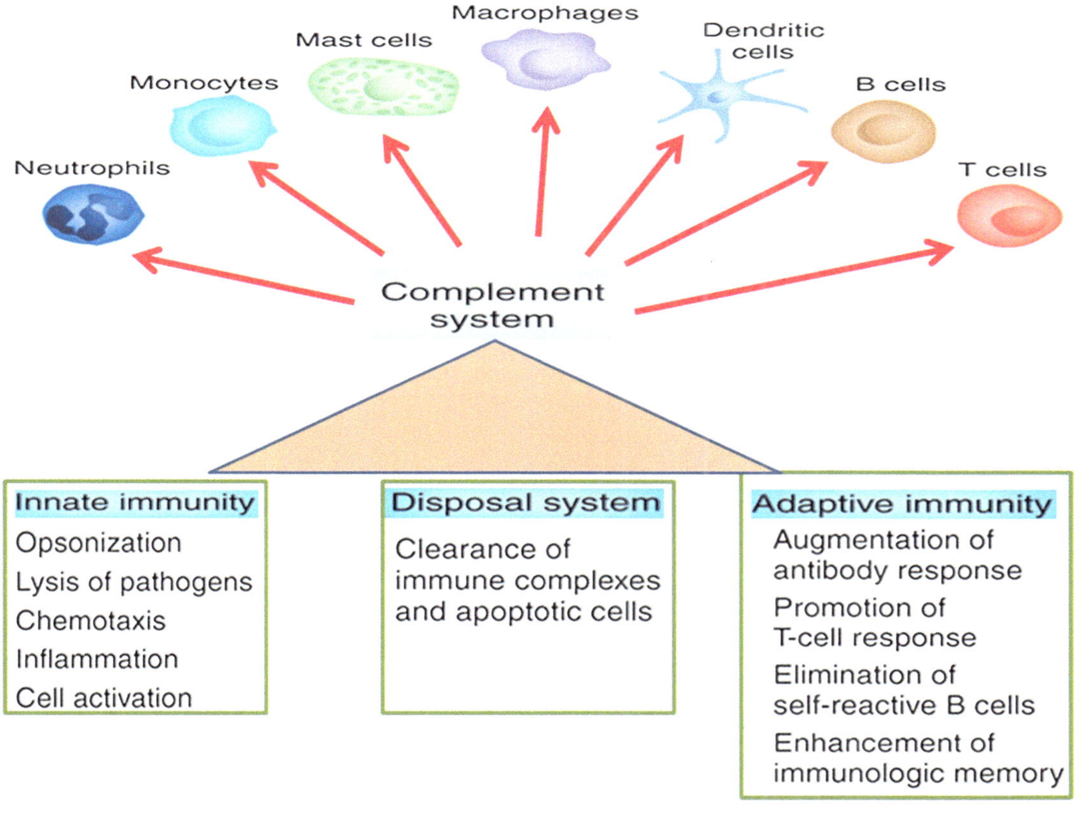

Fig. 1.1 Functions of the complement system

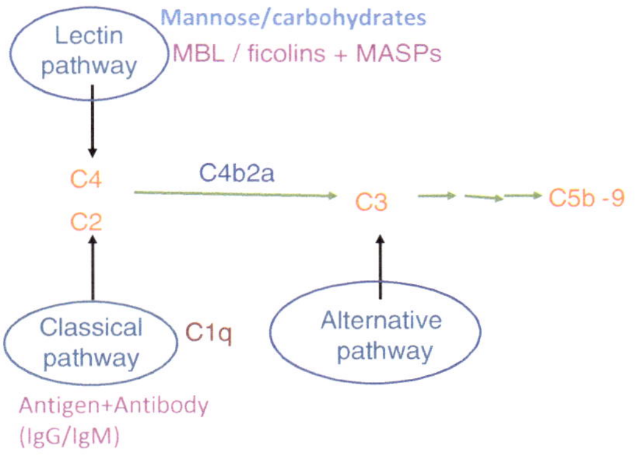

Fig. 1.2 Activation of the complement system

blood clotting fibrinolysis and kinin formation, and these pathways produce a rapid, amplified response to a stimuli. They are secreted by the hepatocyte cells of the liver. There are *three* such types, namely, the classical pathway, the alternative pathway, and the lectin pathway. The function of the complement system includes enhancement of phagocytosis, helps in lysis of microorganism directly, and induces peptide fragment that regulates inflammation and immune responses (Fig. 1.1).

Among the three equivalent preliminary complement pathways, each stimulates the final common pathway (Fig. 1.2). Due to the evolutionary changes, the pathways are evolved with new features, such as the "alternative pathway" which is moderately primitive and becomes a part of the innate immune system. The "classical pathway" is rather recent and binds with an antibody to commence activation, and therefore, it is an appendage to the acquired immune system. The "mannose-binding lectin pathway" interacts with the pathogen directly and becomes a part of innate immune system, but structurally it possesses the primitive components of the classical pathway.

The classical pathway is started with an interaction between the antigen and antibody by forming an "immune complex." Antibodies are capable of binding or fixing complements only after reacting with the antigen. This complex leads a conformational change in the antibody molecule which closes the binding site of the first complement component C1. C1 is a multimeric compound which is composed of six molecules known as C1q, two each of C1s and C1r. C1q, an elongated protein, is a rodlike stem that is made up of a triple helical structure and a globular head. This globular head helps to bind an antibody. Six C1q molecules assemble themselves in a "bunch" and the four *C1r* and *C1s* molecules affix in a calcium dependent interaction. C1r is cleaved to give an active molecule after the binding of an antibody with two or more heads of C1q. This active molecule cleaves C1s which continues the activation process by cleaving the next complement components C4 to C4b (extending the reaction process) and C4a (other biological properties). Stable C4b binds to C2 in a magnesium-dependent reaction. The C2 is then cleaved by C1s to form C4b2b, which is known as the C3 convertase. Next C3 is cleaved to form smaller C3a (having powerful biological properties) and larger C3b which show the labile binding site. The last enzyme of classical pathway, C4b2b3b (C5 convertase), generates due to the proximity of C3b to C4b2b and then cleaves a component of the membrane attack pathway, C5.

The mannose-binding lectin or MBL is activated by binding to microbes and closely resembles C1q in structure. It also forms a "bunch of tulips" structure and allows two serine proteases for binding to the stalks (MBL-associated serine proteases 1 and 2, MASP-1 and MASP-2). This activates the MASPs, which undergo to *stimulate C4*, and the leftovers of the classical pathway extend as described above.

The alternative pathway is started with the activation of C3 by PZ complex (properdin–zymosan complex). PZ functions as proteinase and cleaves C3 to smaller C3a and larger C3b. C3b then binds to carbohydrate moiety on the cell surface of microbes or autologous cell. Then the factor B binds to C3b to form C3bB, which is stabilized by factor P. Factor D cleaves the C3bB complex and releases C3 cleavage enzyme C3bBb (C3 convertase). This activated C3bBb splits C3 into C3a and C3b. The C3b is either deposited on the surface of microbes or interacts with factor "B" and "D." C3b may activate C5 and enters into the classical pathway.

The membrane attack complex forming pathway is the final and common complement pathway which forms a biologically active component, C5, as well as a "killer molecule" of the system that is known as the "membrane attack complex" or "MAC complex." It facilitates membrane damage. The cleavage of C5 forms the smaller C5a and larger C5b by the classical and alternative pathway convertases. This C5b then binds to C6 and stirs it up to express a labile reactive site for C7. As this C5b67 complex is highly lipophilic, it binds to membranes where it remains as a high-affinity receptor for C8. C8 is composed of three chains; among them, one inserts into the membrane for anchoring the C5b678 complex. This complex binds and polymerizes C9 for the formation of MAC, the ultimate component of complement cascade. Twelve to fifteen C9 molecules may huddle around one C5b678 complex.

1.2.3 Mechanism of the Innate Immune System

The innate immune recognition is accomplished by three different mechanisms. They are the lectin–complement pathway, the toll-like receptor pathway, and the cytoplasmic pattern recognition receptors. The summary of the different recognition mechanisms of innate immunity is tabulated in Box 1.10.

The lectin–complement pathway is primarily mediated by different lectins and circulatory complement proteins which are secreted/released into the extracellular space. When a lectin binds to a microbe, a chain of reaction triggers, on the facade of that microbes which causes the demolition/cell lysis or "opsonization" of that microbe. The *toll*-like receptor (TLR) network (Box 1.11) is intervened by transmembrane receptors on the

Box 1.10: Different Recognition System of Innate Immune System

- *Toll-like receptors:*
  Bacterial cell wall components and viral nucleic acids
- *Collectins and mannose receptor:*
  Distinctive cell surface polysaccharides
- *Alternative pathway of complement:*
  Cell surfaces lacking protective anti-complement proteins
- *Antimicrobial peptides:*
  Acidic phospholipids outside of the membrane
- *Interferon induction:*
  Double-stranded RNA (replication of viral genome)
- *Virus replication-induced cell stress:*
  Induction of apoptosis and expression of stress-induced molecules that alert NK cells

innate immune phagocytic cells like macrophages and dendritic cells (DCs). When TLRs reveal microbes, the cells become activated to fabricate chemokines and pro-inflammatory cytokines to stir up inflammation and to conscript immune cells to the infection site.

This pathway can able to provoke maturation of DCs to inculcate and satisfactorily stimulate acquired immunity. The cytoplasmic pattern recognition receptors are intervened by cytoplasmic pattern recognition receptors that perceive viruses and bacteria that have lucratively attacked cells.

The lectin–complement pathway is an imperative mechanism in a mammalian innate immune system for identifying a wide variety of pathogens present at the outer surface of the host cells. This pathway is intervened by a number of proteins which are concealed into the host serum. This pathway possesses carbohydrate recognition with the help of mannose-binding protein (MBL) and ficolins. These MBL and ficolins explicitly distinguish N-acetyl-D glucosamine (GlcNAc),

Box 1.11: Toll-Like Receptors and Recognition of Pathogens (Takeda et al. 2003)

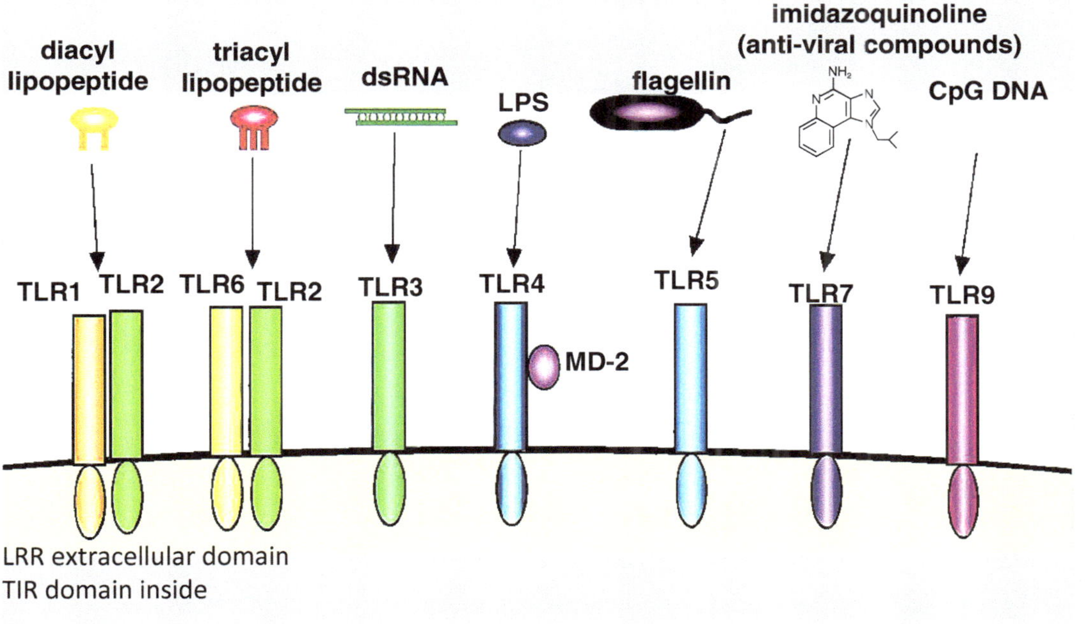

mannose, and fucose but don't interrelate with D-galactose or sialic acid present on mammalian cells. As soon as the MBL and ficolins bind with microbial carbohydrates, they activate the MBL-associated serine protease (MASP) family on the bacterial cell surface. By forming a complex with MBL, MASPs (enzymes) smite the complement components for activation of complement pathway. This cleavage proceeds to the direct destruction of the invading bacteria or leads to their opsonization and facilitates phagocytosis by macrophages.

In humans the deficiency of MBL leads to augment susceptibility to a wide range of infectious diseases. Among all MASPs, MASP-2 helps in activation of C2 and C4, which are accountable for the formation of C4bC2a enzyme complex. This complex furthermore acts as a C3 convertase which cleaves the C3 for generating the ultimate products that can destroy bacteria. In disparity to MASP-2, MASP-1 is capable of cleaving C3 directly. The lectin–complement pathway plays a pivotal role in distinguishing microbes in the host bloodstream as well as in the extracellular space.

1.3 Adaptive Immunity

How can we remain healthy, for the whole of our life, when we have to survive amid an environment clouded with infectious and opportunistic microbes, potential carcinogens, and allergens? The secret lies within our immune system that is boosted to protect us, against all pathogens (Fig. 1.3). It is therefore worthwhile to understand the organization of the immune machinery, which is specifically characterized by its two arms: innate and adaptive.

The *innate immunity* provides the early first line of defense against invading infection, a set of disease-fighting mechanisms that are not very specific to a particular intruding pathogen but identify a pattern of molecules expressed extracellularly by frequently encountered pathogens.

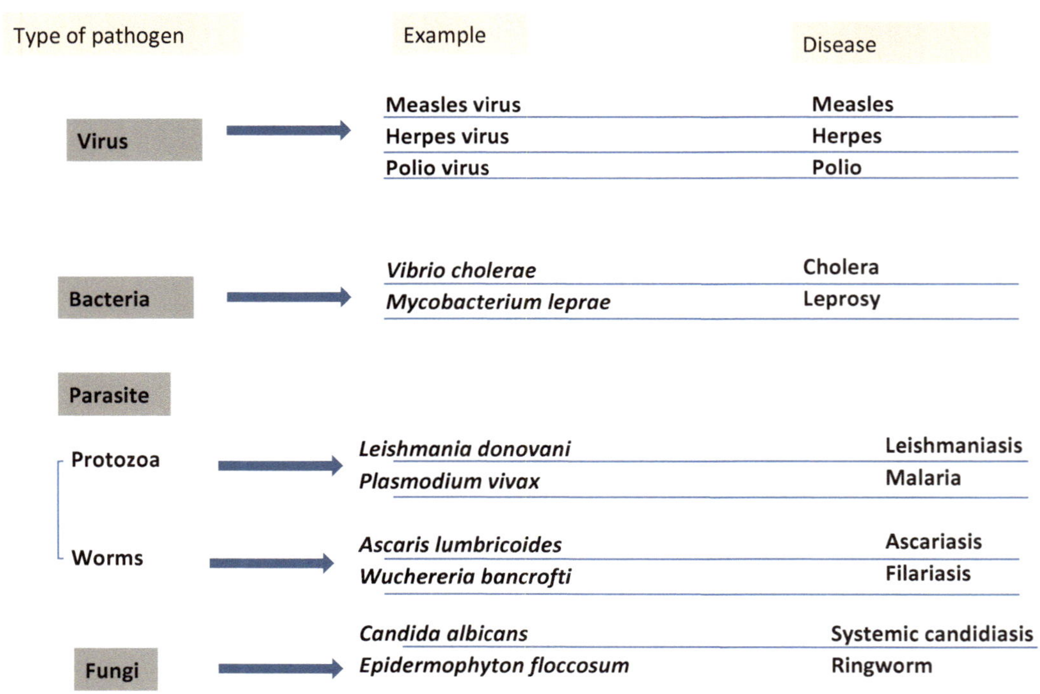

Fig. 1.3 Diversified pathogens and their role in the immune system

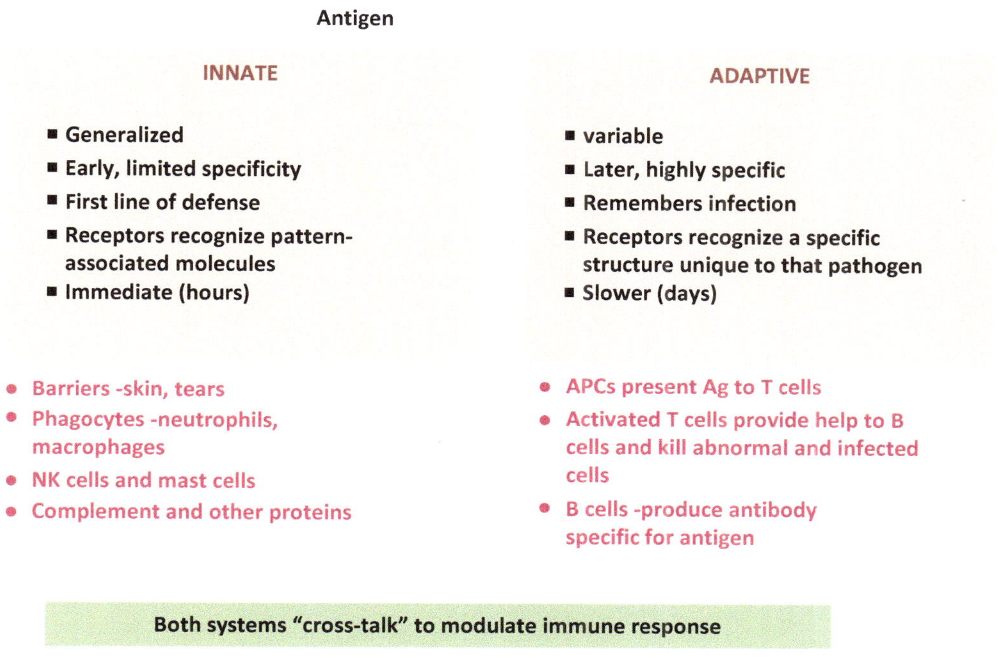

Fig. 1.4 Cross talk between an innate and adaptive immunity

The specific component, *adaptive immunity* (also called *specific or acquired immunity*), comes into the scenario when the organism faces an antigenic challenge. The nonspecificity of the innate immune system is counterbalanced by the high degree of specificity of the adaptive immune response which works on the remarkable property of "memory" (Fig. 1.4).

The adaptive immune system is extraordinarily complex, involving a meshwork of cells, cytokines, and other compounds at its different levels. In fact, immunologists are still working out on many of its unrevealed secrets. In this chapter, we will first cover the general strategies the adaptive immune response uses to eliminate invading microbes encompassing various cells and molecules. This will be followed by development of the immune system, concentrating on how the cells involved in adaptive immunity gain the specificity required to respond to enormous, diverse, and ever-changing cocktails of microbes. We will also describe some of the mechanisms used by the adaptive immune system to build *tolerance* against few relevant diseases.

1.4 Cell Network and the Immune System

The cells of the immune system consist of lymphocytes including B, T, and NK cells, antigen-presenting cells (APCs), and effector cells including macrophage, granulocyte, etc. (Box 1.12).

Lymphocytes are the only cells that produce specific receptors for antigens and are thus the unique mediators of adaptive immunity. Lymphocytes though extremely heterogeneous in lineage, phenotype, and function are morphologically similar and are capable of varied complex biologic responses. Lymphocytes are distinguishable by surface proteins or CD (cluster of differentiation) (Box 1.13) that may be identified by anti-CD monoclonal antibodies.

1. *B lymphocytes* are the mediator of humoral immunity through the production of antibodies. Soluble antigens or microbial surface antigens may bind to B-lymphocyte antigen receptors (B-cell receptor or BCR) and educe humoral immune responses.

Box 1.12: Cell Network of the Adaptive Immune System

1. Antigen-presenting cells (APCs)
   - Dendritic cells; follicular dendritic cells, macrophages, and B cells
   - Capture, process, and present antigens to lymphocytes for recognition
2. Lymphocytes
   - T and B lymphocytes and NK cells
   - Cells with specific receptors for antigens
   - The key component of adaptive immune responses
3. Effector cells
   - Lymphocytes, neutrophils, eosinophils, and macrophages
   - Help in eliminating pathogen

Box 1.13: Cluster of Differentiation

By Definition

- Leukocyte surface molecule, expressed on cells of a particular lineage (or "differentiation")
- Recognized by a group (or "cluster") of monoclonal antibodies which is called a member of a cluster of differentiation (CD)
- CD markers (or CD antigens) are:
  - Identified by numbers (e.g., CD4, CD8, etc.)
  - Used to classify functionally distinct subpopulations of leukocytes (e.g., helper T_H cells are CD4+CD8-)

Anti-CD antibodies are used to:

- Identify and isolate leukocyte subpopulations and particular cell populations
- Study different functions of leukocytes

2. *T lymphocytes* are the cells of the cell-mediated immune system. The antigen receptors of most T lymphocytes (or T-cell receptors) (Fig. 1.5) only identify different peptide fragments of various protein antigens that are bound to the *major histocompatibility complex (MHC)*; the specialized peptide displays molecules (Fig. 1.5, Box 1.14) on the surface of specialized cells called antigen-presenting cells (APCs).

Among *T lymphocytes*, *helper T cells* (T_H) or CD4+ T cells help B lymphocytes to produce antibodies and help different phagocytic cells to destroy ingested microbes. *Regulatory T lymphocytes* are a specialized CD4+ T subset that limits or prevents immune responses. *Cytotoxic, or cytolytic, T lymphocytes (CTLs) are* CD8+ T lymphocytes because they lyse or kill cells docking intracellular microbes (Figs. 1.6 and 1.7).

The differentiation of T cells from either CD8+ or CD4+ lineage depends on definite selection process B and T lymphocytes mature in the generative lymphoid organs, namely, the bone marrow and thymus, respectively (Figs. 1.8 and 1.6).

Mature and developed lymphocytes leave the generative lymphoid organs and enter the body circulation and the allied peripheral lymphoid organs, where they may come across the antigen for which they express definite and specific receptors (Fig. 1.9).

3. *Natural killer (NK) cells* are termed after their ability to kill infected host cells, but they do not express or represent any kinds of clonally distributed antigen receptors like what T cells and B cells do. All lymphocytes develop from stem cells in the body's bone marrow by the process of *hematopoiesis* (Fig. 1.10).

1.4.1 Antigen-Presenting Cells

Antigen-presenting cells (APCs) are located in the epithelium of the common entry portals for microbes (the skin, respiratory and gastrointestinal tract), capture the antigens, transport them to peripheral lymphoid tissues, and display them to lymphocytes (Box 1.15).

1. *Dendritic cells*, a specialized APC, capture the antigen and present it to T lymphocytes by its long processes for antigen recognition.

- Structurally similar to Immunoglobulins (similar to a single F_{ab} fragment).
- One T cell expresses only one specific type of TCR.
- Composed of two glycoprotein chains (α/β or γ/δ). Most mature T cells have TCRs composed of tan α chain and a β chain (α/β T cells).
- Similar to an antibody light chain, each chain has a constant region and a variable region.
- A TCR recognizes small (8-13 aa) peptide epitopes presented on MHC.
- A TCR complex is composed of one heterodimeric TCR (usually α/β), and a CD3 complex which helps in activating the cell signalling for T cell activation.
- Binding of antigen/MHC to the TCR stimulates CD3. CD3 then sends an activation signal to the inside of the T cell.

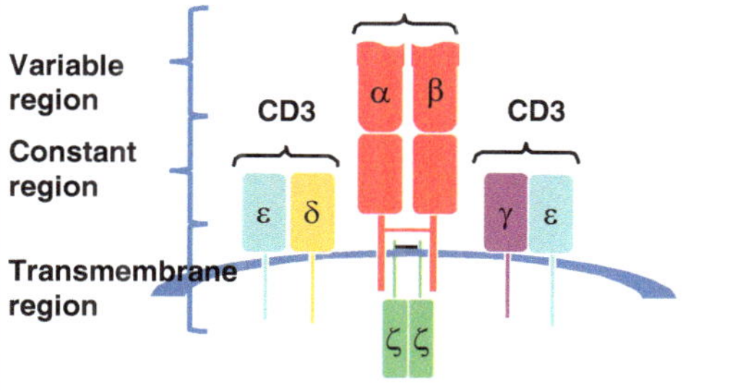

Fig. 1.5 T-cell receptor (TCR)

Box 1.14: Location, Structure, and Functions of MHC

Human MHC=Human Leukocyte Antigen (HLA)

a. MHC Class I , II and III genetic loci (short arm of chromosome 6)

Major Class I genes:	*HLA-A, B, C*
Minor Class I-like genes:	*HLA-E, F, G, H, J, X*
Major Class II genes:	*HLA-D* region
DP (A1, A2, B1, B2), *DQ* (A1, A2, B1, B2, B3), *DR* (A, B1, B2, B3)	
Major Class III genes:	*Diverse (non-antigen presenting functions)*

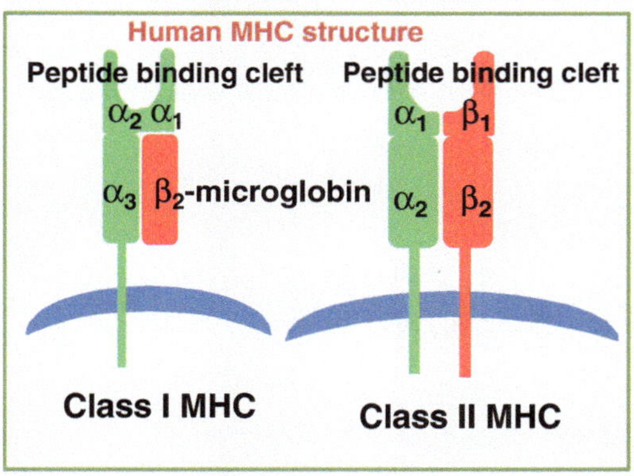

MHC classes I and II have an almost identical 3-D structure.

Locations of MHC:

Class I MHC found on *all nucleated cells* (all cells need to be prepared to be killed in case of a viral take-over or tumorigenic transformation).
Class II MHC found only on antigen presenting cells (cells that present antigen to CD4+ T cells --> Macrophages, activated B-cells, dendritic cells.

Functions of MHC:

class I MHC:

Displays peptides derived from antigen originating inside the cell (endogenous antigen).
Important in cytotoxic responses (eg, $CD8^+$-killing of virus-infected cells).

Class II MHC:

Displays antigen derived from ingested antigens (exogenous antigen).
Important in humoral (antibody) responses as well in fighting as some intracellular parasites (eg. *Mycobacterium tuberculosis and M. leprae*)

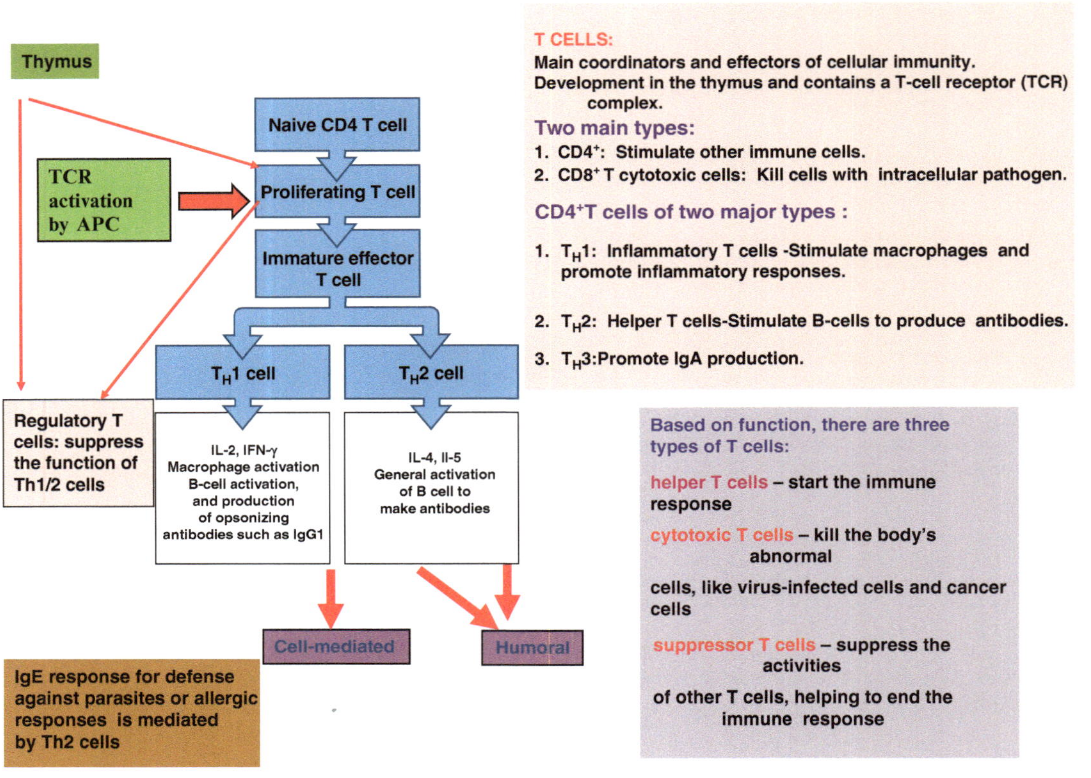

Fig. 1.6 T-cell differentiation to Th1 and Th2 subset (Parham P. The immune system, Garland Science, 3rd edition, 2009)

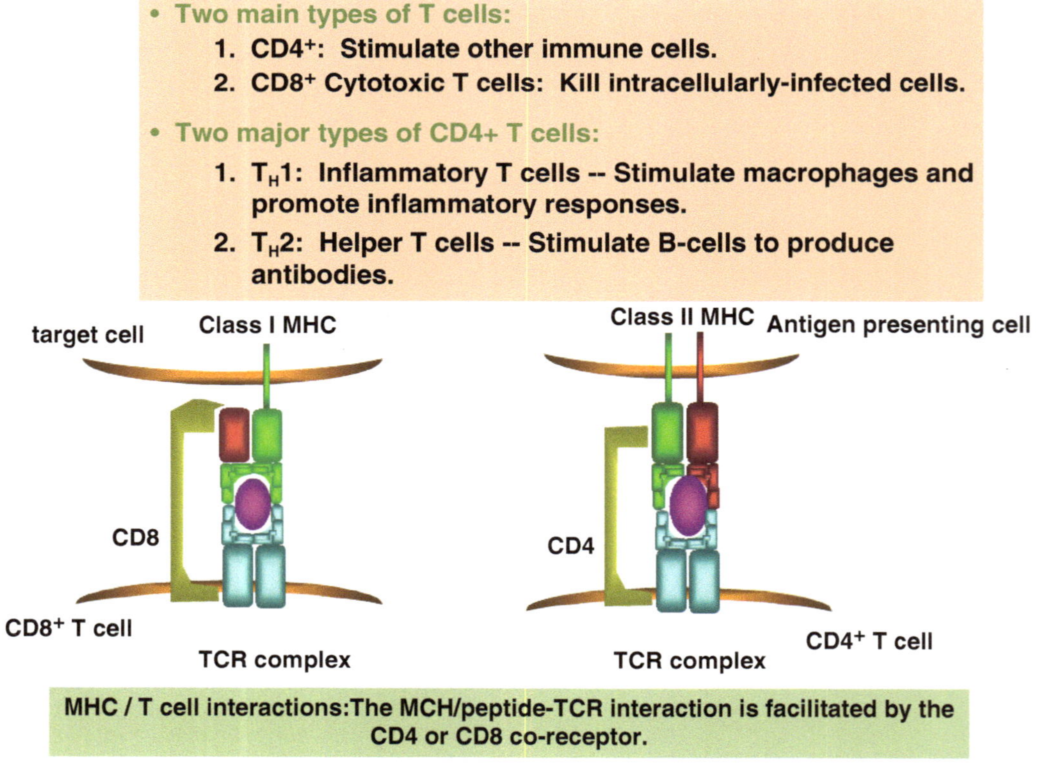

Fig. 1.7 T-cell and MHC interaction

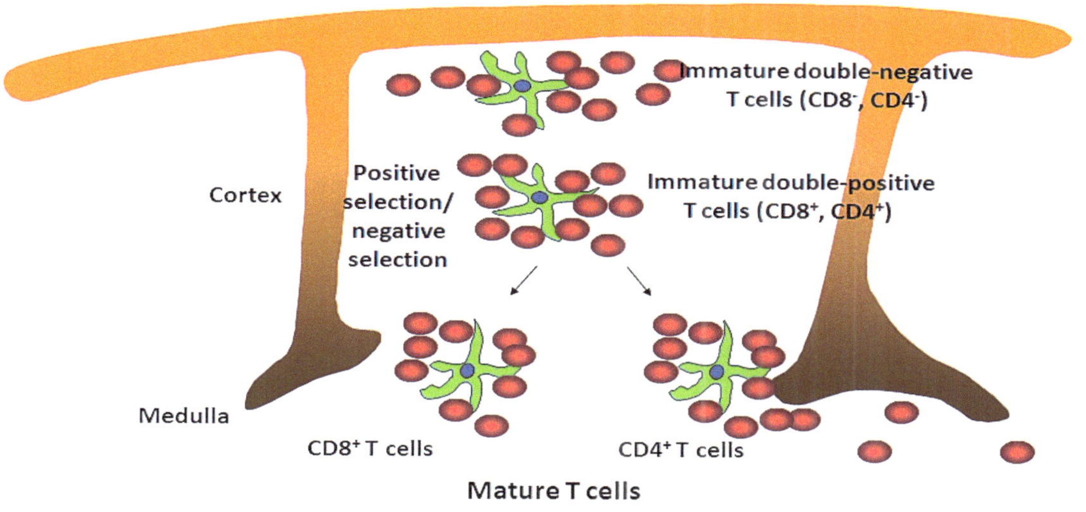

Fig. 1.8 Thymic selection and T-cell development in the thymus

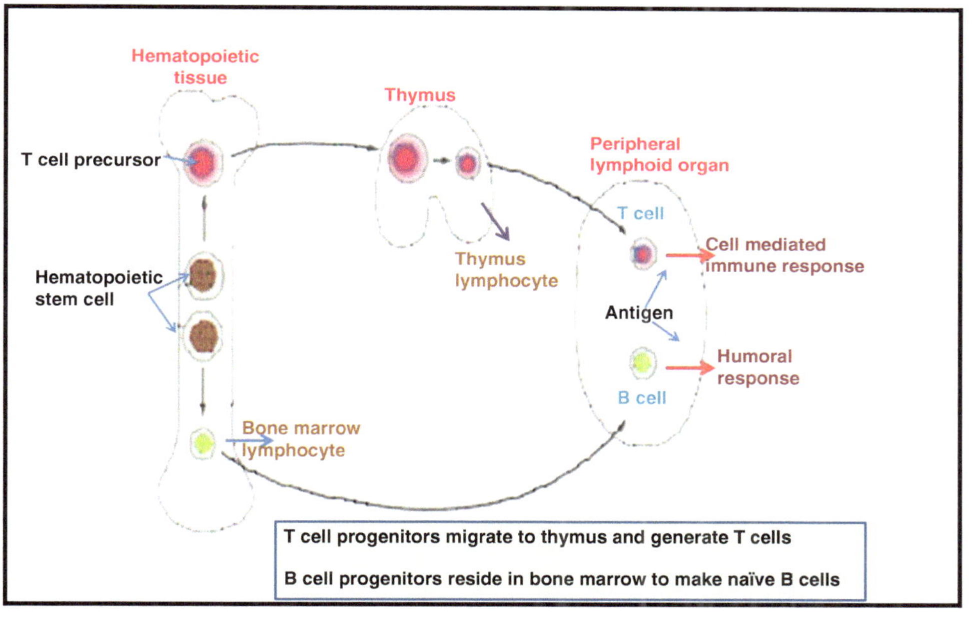

Fig. 1.9 Origin of lymphocytes

2. Invading epithelial microbes are phagocytosed by another APC, the macrophages, which are also capable of presenting protein antigens to T cells. These APCs apart from displaying and presenting antigens to T lymphocytes are able to trigger T-cell responses. The prototypical professional APCs are dendritic cells, but macrophages and a few other cell types may serve the same function.
3. *Effector cells*: lymphocytes and other leukocytes are called effector because these cells can eliminate microbes differentially (Fig. 1.11). The elimination of microbes by effector cells often requires the participation

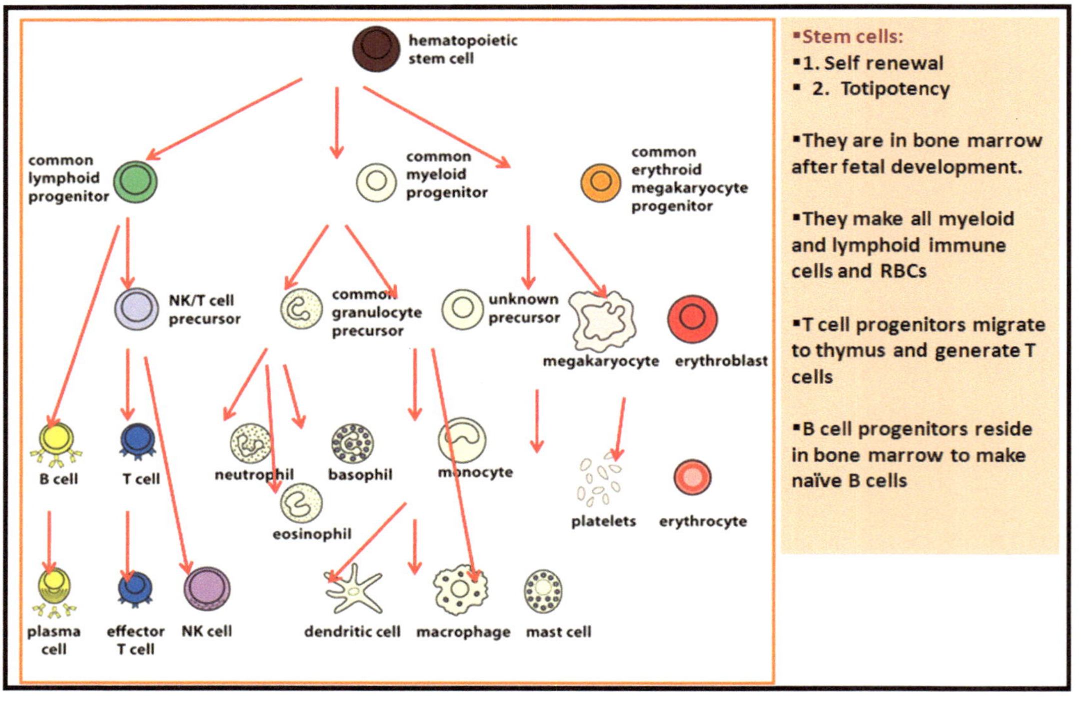

Fig. 1.10 Hematopoiesis generates immune cells

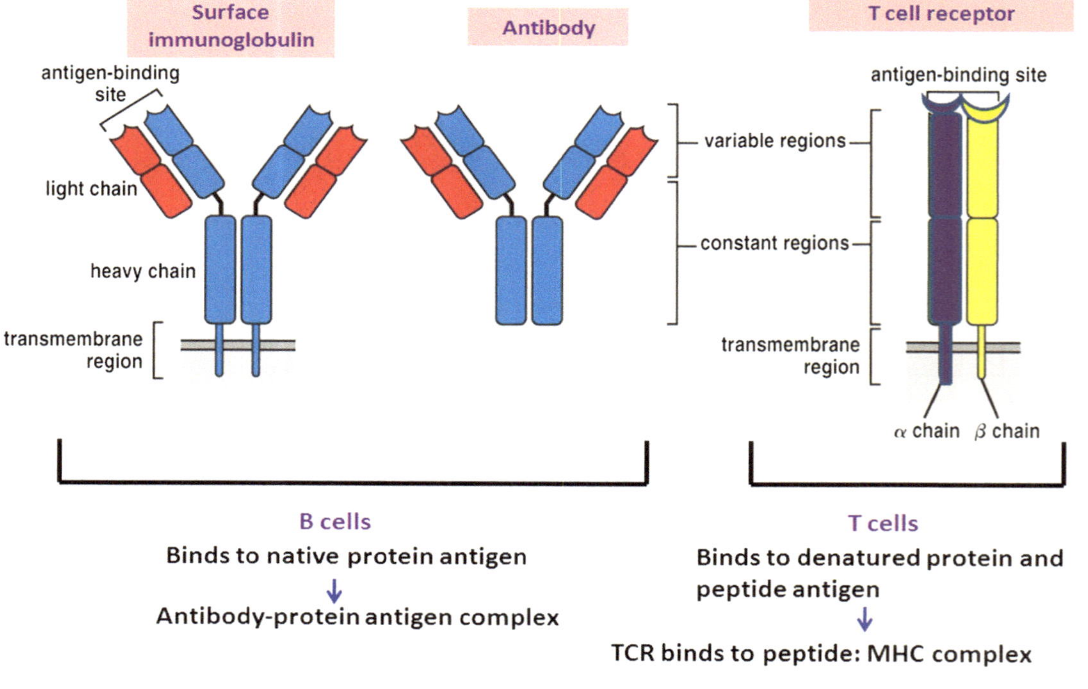

Fig. 1.11 Recognition modules of adaptive immunity

Box 1.15: Antigen-Presenting Cells

- Three types of APCs are found in the lymph nodes:
  - Dendritic cells—constitutively express MHC I and MHC II (can stimulate both CD4$^+$ and CD8$^+$ T cells) as well as B7 (the co-stimulatory signal). Antigen presentation appears to be the sole purpose of dendritic cells, and these cells can be infected by a wide variety of viruses. Dendritic cells are *not phagocytic.* They can present some viral peptides on their MHC II and contribute to the induction of an antibody against viruses. They are *very* efficient at stimulation of cytotoxic responses.
  - Macrophages—resting macrophages express *little MHC II or B7* but have receptors for bacterial cell wall components which, upon binding, activate the macrophage to express high levels of B7 and MHC II. Once activated, macrophages are efficient at stimulating CD4$^+$ T cells, both for inflammatory responses and helper (antibody) responses.
  - B cells—B cells express high levels of MHC II, *but not B7*. Microbial cell wall components can induce B7 expression by B cells (like macrophages). Once induced to express B7, B cells can activate helper T cells. B cells can take up a *soluble* antigen through their Ig receptors (unlike dendritic cells or macrophages).

of other nonlymphoid leukocytes, such as granulocytes and macrophages. These leukocytes are effector cells of both the immune system. In an innate immune system, some granulocytes and macrophages directly recognize pathogens and eliminate by phagocytosis, whereas in an adaptive immune system, the end products of lymphocytes in concoction with other leukocytes are activated to kill microbes.

1.5 Adaptive Immunity at a Glance

The third line of human defense system is the *adaptive (acquired) immunity* or adaptive *immune response*. Once the body's external barriers have been confronted and inflammation of innate immunity has been stimulated, the adaptive immunity comes into scenario. It propagates more slowly and specifically than nonspecific inflammatory response and has memory. Thus, it serves two major purposes: destroying infectious pathogens that are resistant to inflammatory cascades and providing highly sensitive long-term protection against future microbial exposure to the similar microorganism.

The function of the adaptive immune system is to destroy attacking pathogens and any released toxic molecules produced by it. The fundamental feature of the adaptive immune system is to distinguish the *foreign* from the *self*-molecules. When the system fails to make this difference, it reacts viciously against the host's own molecules ushering to a fatal *autoimmune disease*.

Adaptive immune responses function through *white blood cells* (or *lymphocytes*). Adaptive immune responses work bilingually through either *antibody responses* or *cell-mediated immune responses*. These are carried out by different sets of lymphocytes, known as B cells and T cells, respectively. In *antibody responses*, B cells are stimulated to secrete some functional proteins which are called *immunoglobulins* or antibodies.

The antibodies circulate in the blood and penetrate the body fluids, where they bind with multiple specificities (*polyclonal antibodies*) to the foreign exposed antigen that stimulated their production. Binding of antibody inactivates toxins

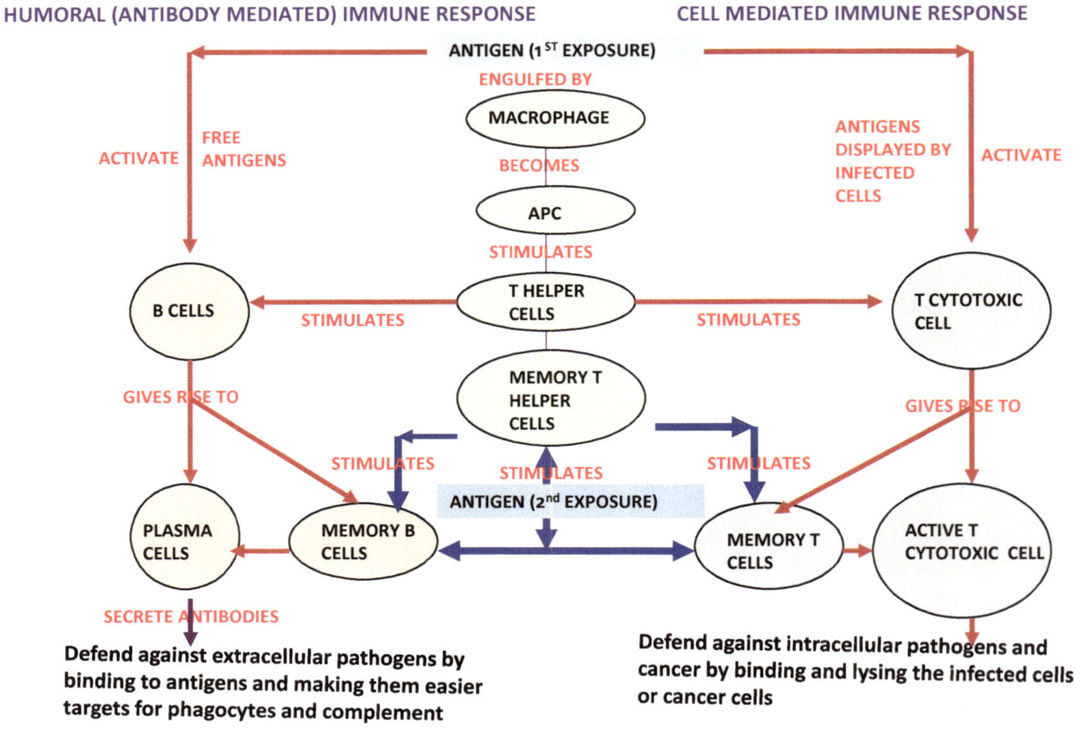

Fig. 1.12 Pathways of successive immune responses through the humoral and cellular pathway

(such as tetanus toxin) and microbes by inhibiting their ability to recognize and bind to receptors on host cells. This is the clinical basis of many present-day diagnostic kits which have been implemented to detect antigen-specific serum antibodies Antibody binding also tags invading pathogens to different phagocytic cells of the innate immune network to ingest them. Thus, components of innate resistance and adaptive immune response complement or supplement each for complete protection against infectious disease.

Cell–cell contact mediated by *T lymphocytes* is the prerequisite for *cellular immunity*. T cells express the *T-cell receptor (TCR)* which recognizes short peptides (derived from self- or microbial proteins, allergens, or tumor-specific proteins) presented by *major histocompatibility complex (MHC)* expressed on the cell surface of *antigen-presenting cells (APCs)* (Fig. 1.11). In addition, *natural killer (NK) cells* (a separate group of lymphocytes) are the key molecule for immunity against tumors and also in "antibody-dependent cell-mediated cytotoxicity" (ADCC).

Finally, the achievement of an adaptive immune response depends on both the functioning of cellular and humoral responses (Fig. 1.12) and their proper interactions, and along with it the specialized functions of *memory cells*, which are capable of *remembering* the specific antigen, respond more efficiently and rapidly against probable infections.

1.6 Different Modes of Adaptive Immunity: Active and Passive Immunity

Adaptive immunity can be of two kinds, active or passive, which may be either long-lived or temporary. *Active acquired immunity or active immunity* is produced by any subject after natural exposure to antigens via environment or through immunization, whereas in *passive acquired immunity (passive immunity)*, preformed antibodies or T cells produced by an individual (donor) in response to antigens are transferred directly from the donor to a recipient who can act against an antigen without stimulating its own immune response. Passive immunity can occur naturally during the pregnancy period when maternal antibodies cross the placenta to the fetus to evade against a specific disease. The donor's antibodies or T cells are eventually destroyed; thus, it is always a temporary response.

1.7 Characteristics of Adaptive Immunity

For proper effectiveness of adaptive immune responses, several properties are crucial for combating infections:

1. *Diversity and its association with specificity*

The adaptive immune system is capable of analyzing millions of different antigens from portions of antigens. Specificity for many different antigens toward the total lymphocyte repertoire is extremely diverse. The lymphocyte diversity means that very few cells (~one in 100,000 lymphocytes) are specific for any one antigen. To carry effective defensive mechanism against microbes, these few cells have to proliferate profusely to generate a large number of cells capable of combating the microbes. Thus, remarkable expansion of the lymphocyte pool specific for an antigen after encountering the antigen, few positive feedback loops that amplify the immune responses, and selection mechanisms (memory cells) that preserve the most useful lymphocytes form the total effective package for adaptive immunity.

The basis of these diversities lies in the underlying mechanism of clonal selection.

2. *Clonal selection*

When lymphocytes are activated by diverse antigens, they undergo a proliferation process, generating many thousands of mature lymphocytes (B cells) which bear extracellular receptors for many antigens even before encountering these antigens. A *clone* refers to a population of lymphocytes with identical antigen receptors and specificities, and all these cells are presumably derived from one precursor cell. Each antigen selects a preexisting clone of specific lymphocytes and thus stimulates the proliferation and differentiation of that unique clone. B lymphocytes give rise to the clone of antibody-secreting effector cells, and also T lymphocytes follow the same principle (Fig. 1.13).

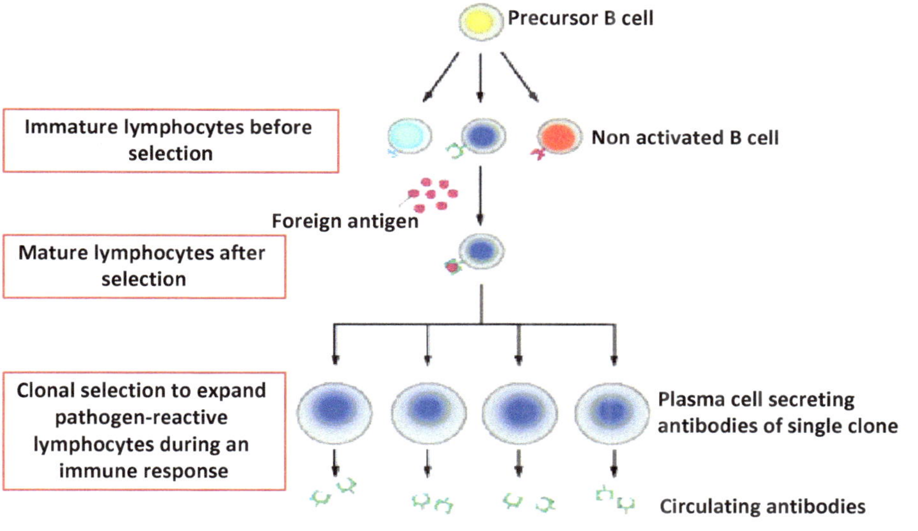

Fig. 1.13 Clonal selection mechanism

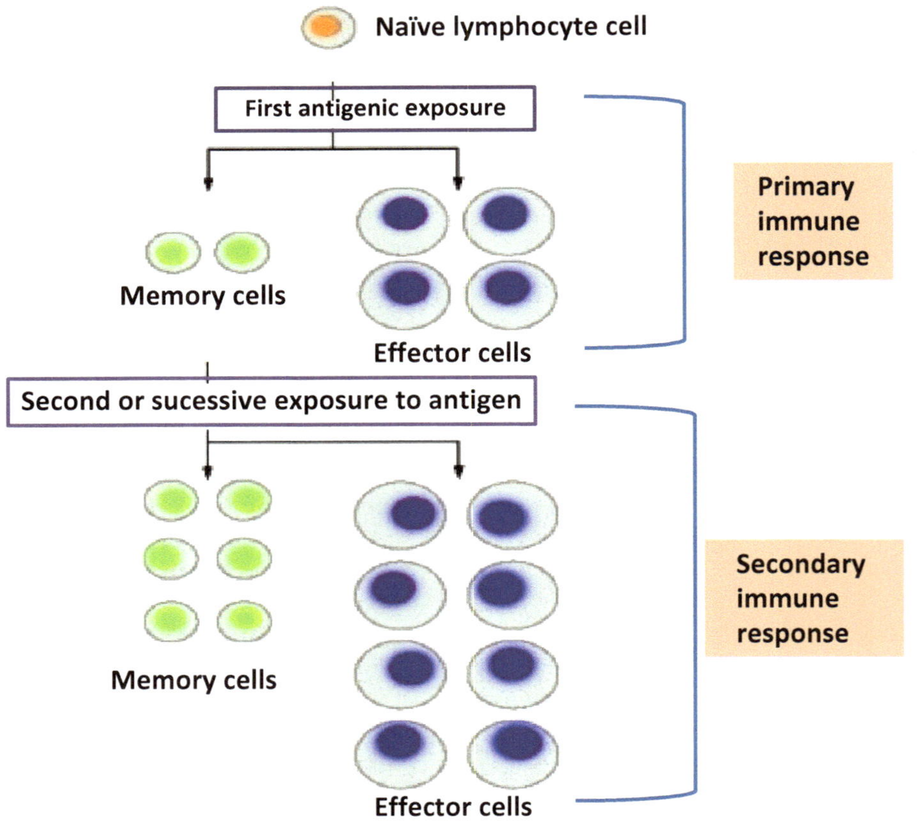

Fig. 1.14 Primary and secondary immune responses

3. *Memory*

The effectiveness of the immune system increases more and more with repeated exposures to the same antigen. The response to the first antigen exposure is called the *primary immune response*, which is mediated by *naive* lymphocytes, as they are seeing an antigen for the first time. Subsequent encounters with the same antigen lead to responses; usually more rapid, larger, and better than the primary responses are called *secondary immune responses* (Fig. 1.14). Secondary responses are the result of the activation of long-lived *memory lymphocytes*, which were induced during the primary immune response. Immunologic memory optimizes the ability of the immune system to combat persistent and recurrent infections, because after each microbial encounter, more memory cells are activated than the previously generated memory cells. A vaccine has long-lasting protection against infections due to activation of memory cells.

1.8 Primary and Secondary Immune Responses

There are two phases of immune responses to antigens—the primary and secondary responses—that are demonstrated by measuring most easily the values of circulating antibody concentrations over time (Fig. 1.14).

1. After the first initial exposure to most antigens, there is a latent period (or lag phase), during which the antigen is processed and B-cell differentiation and proliferation occur.

Then the circulating IgM antibody is detected after approximately 5–7 days. This lag phase is necessary for the process of clonal selection. This is the *primary immune response*, typically characterized by initial production of IgM followed by equal to or less amount (titer) of IgG antibody against the same antigen. If no further exposure to the antigen occurs, the circulating antibody is catabolized (broken down) and the titer (measurable amount) falls.

2. A second exposure by the same antigen results in the *secondary immune response*, which is characterized by a more rapid production of larger quantities of antibodies than the primary response. The rapid response of the secondary immune response is due to the action of memory cells that do not require further differentiation. These memory cells are already activated during the clonal selection of the primary immune response. In the secondary immune response, there may be production of IgM, but IgG production is increased considerably. During vaccination or natural infection, in response to these antigenic challenges, the level of protective IgG may remain elevated for decades.

1.9 Antigens or Immunogen

Any foreign substances that react with molecules of the immune system (antibodies, receptors on T and B lymphocytes) are called *antigens* (Fig. 1.3).

Antigens may be infectious microbes (e.g., bacteria, viruses, fungi, or parasites), on a noninfectious environmental component (e.g., pollens, foods, or bee venoms), or toxins, drugs, vaccines, and transfused and transplanted elements.

1. *Immunogens* are antigen molecules that will *induce* an immune response. So theoretically, all immunogens are antigens but not all antigens are immunogens. Generally, large molecular (above 10,000 Da) antigens, such as carbohydrates, proteins, and nucleic acids, are mostly immunogenic.
2. Antigens that induce an allergic response are also called *allergens*. Thus, the immunogenicity of an antigen depends on its larger size, its proficient quantity, and its foreignness to the host. These qualities will pave for the development of vaccines, highly immunogenic and protective immune responses against infectious pathogens.
3. Many low-molecular-weight *hapten* molecules are too small to be immunogenic by themselves but become immunogenic after combining with larger molecules that act as carriers for hapten. A very high or low amount of antigens, in some cases, may induce a state of tolerance rather than immunity. Other factors also may pose for immunogenicity of an antigen. For example, the route of *antigen entry or administration* (intravenous, intraperitoneal, intranasal, subcutaneous, and oral) has clinical implications. Each route preferentially stimulates either cell-mediated or humoral immune responses. A vaccine turns out to be effective if administered through a definite route but may prove ineffective in another route.
4. *Adjuvants* are substances that stimulate the immune response when administered along with an antigen and convert the antigen to an immunogen.
5. *Haptens* are a low-molecular-weight molecule that is not immunogenic by itself in nature but when coupled to a carrier can elicit anti-hapten antibodies. For example, dinitrophenol (DNP) is a common hapten. Finally, an individual's immune response to an antigen can also be affected by its genetic constitution, age, nutritional content, or reproductive capability or a background of exposure to trauma, recurring diseases, and immunosuppressive status.
6. A *self-antigen* normally does not elicit an immune response because it is not foreign to the host. Thus, most individuals are *tolerant* to their own antigens. Some pathogens successfully mimic self-antigens and avoid induction of an immune response. In contrast, in autoimmune diseases, a breakdown of tolerance occurs that leads to an individual's immune system attacking its own antigens.

Box 1.16: Antibody Diversity and the Number Dilemma

- Millions of different types of Ag react with only about a trillion different antibodies.
- Only about 30,000–60,000 genes code for all our proteins of our entire body, most of which are not Ab.
- So there cannot be one coding gene for one antibody!

Antibody Variability

- Different combinations or recombinations of heavy and light chains which are encoded by different genes.

Antibody Genes

Genes for antibodies aren't like most other genes—they come in pieces ("gene-lets"):

- Variable segments (V)—many different versions.
- Diversity segments (D)—several different versions.
- Joining segments (J)—a few different versions.
- Constant segments (C)—a few different versions that are nearly identical; each B cell combines these gene segments to make an Ab chain like shuffling a deck of cards.
- V, D, and J are joined to C for the heavy chain; V and J are joined to C for the light chain.

1.10 Antibodies (or Immunoglobulin)

An antibody (immunoglobulin) (Ig) is a serum glycoprotein produced by *plasma cells* of B lymphocytes (or B cells), in response to an antigen. An immunoglobulin is a generalized term used for all antibodies, whereas the term *antibody* refers to a particular set of immunoglobulins with known antigenic specificity. There are five classes of immunoglobulins (IgG, IgM, IgA, IgD, and IgE), which are characterized by differences in structure and function (Table 1.3). The huge variety of antigens encountered by few thousands of antibodies leads to antibody diversity (Box 1.16).

1.11 Antibody: Molecular Structure

G. M. Edelman and R. R. Porter were awarded the Nobel Prize (1972) for discovering the structure of the antibody molecule. *Generally*, all immunoglobulins have a similar *Y*-shaped overall structure consisting of two polypeptides: *heavy chain* (higher molecular weight) and *light chain* (lower molecular weight). Typically, an antibody molecule is composed of two heavy and two light chains that are held together by both covalent (disulfide bonds) and non-covalent forces. There are two types of light chains, κ (*kappa*) and λ (*lambda*), and five types of heavy chains, α (*alpha*), δ (*delta*), € (*epsilon*), γ (*gamma*), and μ (*mu*) (Fig. 1.15).

In a single Ig molecule, one type of heavy chain is associated with either light chain but never both. The five classes of human immunoglobulins are grouped depending on the type of heavy chain they bear (Table 1.3). IgG and IgA are further subdivided into subclasses depending on minute differences in the amino acid. The serum concentration of each antibody varies both in normal and diseased conditions. Each heavy and light chain can be further divided into constant and variable regions. The antigen-binding pocket of different immunoglobulins is flexible and represented by the first 110 amino acids which differ in their sequence in each Ig type. The sequence variability from different immunoglobulin shows three hypervariable regions or *complementarity determining regions* (CDR), within the variable region. In contrast, there is not much variability in the "constant region" of antibody molecules. To form the heavy and light chain, there is profuse gene rearrangement (Box 1.17, and 1.18).

Table 1.3 Comparative differentiation of the five classes of immunoglobulin molecules (Data collected from Kuby)

Abundance					
Attributes	IgG	IgM	IgA	IgD	IgE
Total serum immunoglobulin (%)	80	5–10	10–15	0.2	0.002
Mean serum concentration (mg/ml)	13.5	1.5 mg/ml	3.5	0.03	0.0003
Physical properties					
Molecular weight ($\times 10^3$)	150	900	160 and 320	150	190
Heavy chain	γ	μ	α	δ	ε
Light chain	κ, λ	κ, λ	κ, λ	κ, λ	κ, λ
Subclasses	γ 1, γ 2, γ3, γ 4	None	α1, α2	None	None
Molecular formula	$\gamma_2\ \kappa_2$, $\gamma_2\ \lambda_2$	$(\mu_2\ \kappa_2)_n$, $(\mu_2\ \lambda_2)_n$, n = 1or 5	$(\lambda_2\ \kappa_2)_n$, $(\lambda_2\ \lambda_2)_n$, n = 1,2,3 Or 4	$\delta_2\ \kappa_2$, $\delta_2\ \lambda_2$	$\varepsilon_2\ \kappa_2$, $\varepsilon_2\ \lambda_2$
Half-life (days)	23	5	6–8	3	2
Subunits	Monomer	Pentamer	Monomer, dimer, trimer, or tetramer	Monomer	Monomer
Carbohydrate (%)	3	12	8	13	12
Sedimentation coefficient	7S	19S	7S, 9S, 11S	7S	8S
Valency for Ag binding	2	5 (10)	2, 4	2	2
Biological functional properties					
Distribution	Internal body fluid	Serum	Seromucosal secretion	Lymphocyte surface of newborn	External body surface
Major function	Combats extravascular microbes and their toxins	Effective first line of defense	Protects entry of microbes via intestinal mucosa, tears, saliva, respiratory mucosa, gastrointestinal mucosa	Antigen receptor on lymphocyte	Protects against allergy, parasitic infection by recruiting antimicrobial peptides
Crosses placenta	++	–	–	–	–
Presence on mature B-cell membrane	–	+	–	+	–
Classical complement activation	++	+++	–	–	–
Alternate complement activation	-	+	–	–	–
Binds to Fc receptors of phagocytes	+/++	?	–	–	–
Mucosal transport	–	+	++	–	–
Induces mast cell degranulation	–	–	–	–	+

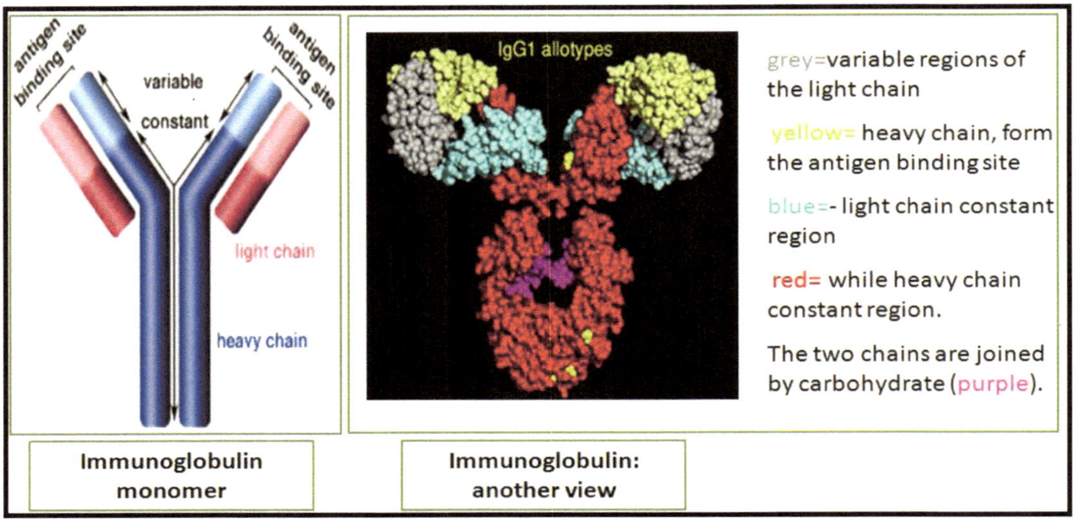

Fig. 1.15 Immunoblogulin [Mike Clark, www.path.cam.ac.uk/~mrc7/]

Box 1.17: Immunoglobulin Gene Rearrangement

- Ig rearrangement occurs before antigen exposure.
- Allelic Exclusion: only one of two parental alleles of Ig is expressed on a B cell.
- Light chain isotype exclusion: either κ or λ light chain is expressed by a B cell.
- Heavy chain rearranges before light chain.
- First D–J joining followed by V–DJ joining.
- Production of μ heavy chain: rearrangement of one allele inhibits rearrangement on other allele.
- If rearrangement on the first allele is nonproductive, then rearrangement on the second allele is stimulated.
- Thus, in any antibody-producing B cell, one allele is productively rearranged, and the other is either not rearranged or is aberrantly rearranged.

Depending on the action of papain enzyme to digest IgG, an antibody molecule was fragmented into two parts—the *antigen-binding fragments (Fab)* named on its ability to bind an antigen and the other crystallized fragment termed the *crystalline fragment (Fc)*. The Fab portions contain the sites (receptors) to bind for specific antigen. The Fc portion is responsible for most of the biologic functions of antibodies. The junction of Fab and Fc is represented by the *hinge region* whose flexibility is maintained by disulfide linkages between the heavy chains. The precise area of the molecule (a few amino acids or sugar residues) that is recognized by an antibody is called its *epitope* (or *antigenic determinant*), and the matching portion on the antibody is sometimes referred to as the *paratope* (or *antigen-binding site*). The antigen–antibody lock and key fit reaction depend on the chemical nature of the particular amino acids in the CDRs and the shape of the binding site. The number of functional binding sites on a molecule is called its *valence* which differs among different antibody classes.

1.11.1 Generalized Function of Antibodies

The functions of antibodies in *antibody-mediated immune response or AMI* including the host defenses against pathogenic microbes, are summarized below.

Box 1.18: Steps in Immunoglobulin Gene Rearrangement at Which the Cell Can Be Lost (Elsevier Science, 1999)

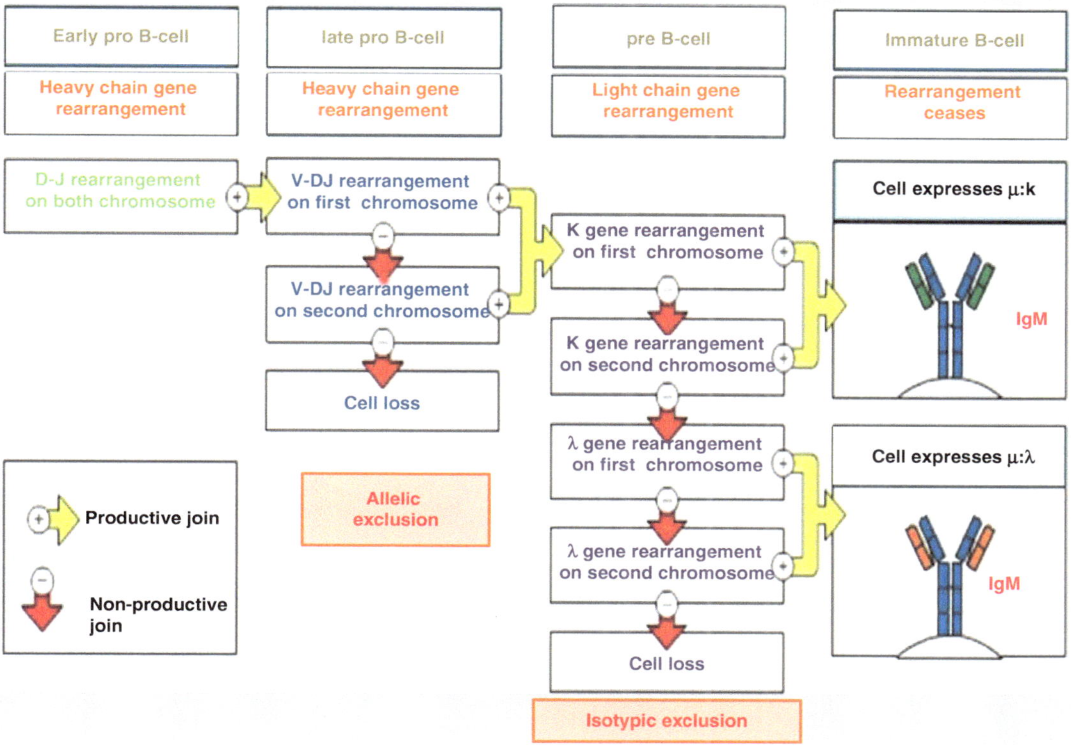

1. *Opsonization*: antibodies (IgG and IgM) enhance phagocytic engulfment of microbial antigens by binding with the antigen and develop a cytophilic affinity toward phagocytes. Bacteria and viral particles are ingested with increased efficiency.
2. *Toxin neutralization*: toxin-neutralizing antibodies (antitoxins) react with a soluble bacterial toxin and block the interaction of the toxin on a specific target cell or a substrate in the host.
3. *Agglutination and precipitation*: antibodies agglutinate or precipitate microorganisms or soluble antigens or aggregates of neutralized toxin by combining with its surfaces. Thus, a number of infectious antigens are more readily phagocytosed due to the clumping of particles being larger in size.
4. *Steric hindrance*: antibodies block or prevent the attachment of microorganisms to their susceptible cells or mucosal surfaces by combining with their surface components. Thus, viral infectivity may be reduced by blocking the viral component of the virus to attach to susceptible host cells. Secretory IgA can block the attachment of pathogens to mucosal surfaces.
5. *Antibody-dependent cell cytotoxicity (ADCC)*: IgG can bind with antigen and opsonize it and thus help certain cells like natural killer (NK) cells, monocytes, and neutrophils to recognize and kill the target antigen.
6. *Activation of complement*: antibodies combined with the surface antigens (opsonization) of microbes and activate the complement cascade during the inflammatory response (Fig. 1.16).

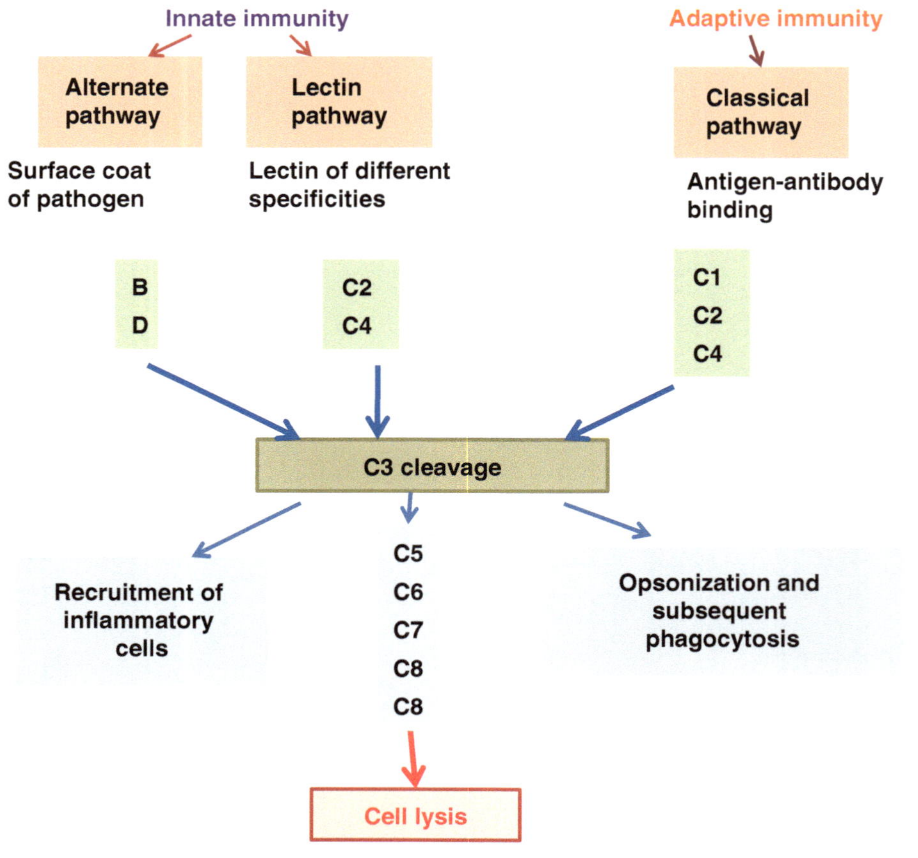

Fig. 1.16 Activation of complement cascades

1.12 Antigen–Antibody Interaction

The interaction of the antigen with the antibody by nature is basically of three types:

1. *Lock and key concept*: like the lock and key model of an enzyme–substrate reaction, the epitopes of the antigen bind with the hypervariable region of the heavy and light chain of the Fab segment of a specific antibody. This interaction was confirmed by *X-ray crystallographic analysis* where the antigen acts as a key that fits into the lock, i.e., the antibody.
2. *Non-covalent bonding*: for an effective and specific antigen–antibody interaction, the closeness of the antigen and antibody is a prerequisite. Due to this closeness, an antigen can interact with an antibody through non-covalent bonding like hydrogen bonds, ionic bonds, van der Waals bonds, electrostatic bonds, and hydrophobic interactions.
3. *Reversible*: the antigen–antibody interaction is mostly reversible in nature.

1.12.1 The Antigen–Antibody Interaction Is of the Following Types:

1. *Antibody affinity*: the *affinity* of an antibody for an antigenic determinant (epitope) describes the strength of binding of a single copy of the antigenic determinant to a single antigen-binding site (*idiotope*), and it is independent of the number of antigen-binding sites. The reversible binding reaction between

an antigen (Ag) and an antibody (Ab) can be expressed as

$$Ag + Ab \leftrightarrow Ab - Ag$$

The equilibrium point depends both on the concentrations of Ab and Ag and on the strength of their non-covalent interaction. A larger fraction of Ab will become associated with Ag as the concentration of Ag increases. The strength of the interaction is generally expressed as the *affinity constant* (Ka), where

$Ka = [AgAb]/[Ag]\ [Ab]$ (the square brackets indicate the concentration of each component at equilibrium).

One can determine the affinity constant (or association constant), by measuring the concentration of free Ag required to fill half of the antigen-binding sites on the antibody. When half the sites are filled, $[AgAb] = [Ab]$ and $Ka = 1/[Ag]$.

Thus, the reciprocal of the antigen concentration that produces half the maximum binding is equal to the affinity constant of the antibody for the antigen.

2. *Antibody avidity*: Most antigenic macromolecules have many different antigenic determinants and are said to be *multivalent;* if two or more of the determinants are identical (as in a polymer with a repeating structure), the antigen is said to be *polyvalent.* Even a bivalent IgG molecule can bind at least 100 times more strongly to a polyvalent antigen if both antigen-binding sites are engaged than if only one site is engaged. The total binding strength of a bivalent or polyvalent antibody with a polyvalent antigen is referred to as the *avidity* of the interaction. If the affinity of the antigen-binding sites in an IgG and an IgM molecule is the same, the IgM molecule with ten binding sites will have a much greater avidity for a polyvalent antigen than an IgG molecule which has two binding sites. This difference in avidity, often 10^4-fold or more, is important because antibodies produced early in an immune response usually have much lower affinities than those produced later. Because of its high total avidity, IgM—the major Ig class produced early in primary immune responses—can function effectively even when each of its binding sites has only a low affinity.
3. *Cross-reactivity*: the antigen–antibody reaction is very specific, but sometimes a few antigens show cross-reactivity with some unrelated antibodies. Cross-reactivity occurs when two different antigens share an identical epitope or a specific antibody can bind with other antigens with more or less similar epitopic structure.
4. *Precipitation reaction*: the interaction between a soluble antigen with an antibody in an aqueous solution forms a lattice which ultimately develops a visible precipitate. The reaction is time dependent. Antibodies which can precipitate an antigen are called *precipitin*. The valency of the antigen and antibody helps in the formation of Ag–Ab lattice.

1.13 Antibody-Mediated Immune Response (Humoral Response)

B cells are developed in the bone marrow of humans and mice and in the bursa of Fabricius of chickens (B cells derive its name from here) (Fig. 1.9).

B cells generate a humoral response, i.e., produce antibodies against a specific antigen. Each B cell expresses a unique antigen-binding receptor (or *BCR*) on its cell surface. Naïve B cells are circulating B cells that have not been previously exposed to antigens but in due course get activated upon binding with an antigen with its BCR. Activated B cells/APCs can internalize antigens, digest them, and present peptides derived from these antigens together with MHC molecules to T cells. Activated T cells induce the expression of cell surface proteins (e.g., CD40L) and secrete cytokines which in turn stimulate the activated B cells to divide rapidly and differentiate into plasma cells and memory B cells. The whole process occurs in germinal centers of lymph nodes or the spleen. Increase in the numbers of germinal centers in lymph nodes and the spleen has a positive correlation with an active humoral response. Plasma cells or effector B cells secrete large numbers of immunoglobulin molecules (~ more than 2000 Ig/s). IgM was the first immunoglobulin to be secreted in this response which subsequently instruct B cells via a stimulation by different cytokines to secrete different classes of Ig (IgG, IgA, and IgE), with the same antigenic specificity. Memory cells have a longer life span and they respond quickly upon a second exposure to the same antigen.

1.13.1 Cell-Mediated Immunity (CMI)

Cell-mediated immunity is mainly comprised of antigen-specific T cells and a network of other immune cells. It is of two types—T-cell-dependent immune responses and T-cell-independent immune responses. Both these two types are mediated through an activation phase and an effector phase (Box 1.19).

Box 1.19: Cell-Mediated Immune Response (CMI): A Quick Look

CMI are responsible for resistance to:

- Intracellular pathogens
- Fungal and protozoal infections
- Tumors

Mediated by:

- T-cytotoxic cells
- Natural killer cells
- Activated macrophages

Functions by:

1. Activating antigen-specific cytotoxic T-lymphocytes that are able to induce apoptosis in body cells displaying epitopes of foreign antigen on their surface (e.g., virus-infected cells, cancer cells displaying tumor antigens)
2. Activating macrophages and natural killer cells, enabling them to destroy intracellular pathogens
3. Stimulating cells to secrete a variety of cytokines that influence the function of other cells involved in adaptive immune responses and innate immune responses

Roles in some harmful conditions:

- Hypersensitivity reactions type IV (contact dermatitis)
- Graft rejection
- Autoimmune diseases

T-cell-dependent immune responses:

A. *Processing and presentation of endogenous antigen via the MHC class I pathway (endogenous pathway):*

I. *Activation Phase*

1. An intracellular antigen (viral protein or an abnormal cellular protein) is ubiquitinated, degraded, hydrolyzed, and processed to peptides in a proteasome (a barrel-shaped protein complex composed of 28 subunits) and is transported to the endoplasmic reticulum via TAP (*transporter associated with antigen processing*).
2. MHC I proteins are synthesized and assembled in ER and associated with TAP with the help from calnexin chaperone.
3. MHC I proteins bind peptides, vesicles fuse with plasma membrane, and MHC I/peptide complexes are expressed on the cell surface and presented to CD8 cells (also called cytotoxic T cells, Tc, or CTL).
4. MHC I cannot leave ER without loaded peptides.
5. In the absence of infection, self-peptides are presented on MHC but do not activate T cells.
6. MHC class I molecules present antigens to $CD8^+$ Tc cells.
7. Co-stimulatory signals from infected cells cause Tc cells to proliferate and clones of Tc cells are produced.

II. Effector Phase

1. $CD8^+$ T cells need the CD28 signal from APCs at the priming stage to become effector $CD8^+$ cells
2. But effector $CD8^+$ cells don't need the CD28 signal (at the effector stage) to kill target cells by apoptosis.
3. Tc-cell receptors again recognize MHC I-presented peptides and started releasing a perforin or Fas receptor on the target cell.
4. The target cells are lysed by degranulation by a perforin or granzyme and/or Fas-induced apoptosis.
5. The cytolytic activity of CTL is promoted by IL-2, IL-12, and IFNγ.
6. This pathway is also responsible for graft rejection as Tc cells can identify both self- and nonself-endogenous antigens.

B. *Processing and presentation of an exogenous antigen via the MHC class II pathway (exogenous pathway):*

I. *Activation Phase:*

1. Professional antigen-processing cells internalize antigens during phagocytosis and are degraded in phagolysosome.
2. Antigens are internalized into endosomes and degraded into peptide fragments.
3. MHC class II proteins are synthesized in ER, and the peptide binding site is protected by Ii (invariant chaperone; CLIP region of Ii protects the site).
4. Exocytic vesicles (from Golgi) containing MHC II proteins fuse with endosome containing peptides.
5. Ii is degraded, and CLIP is removed by HLA-DM protein.
6. MHC II proteins bind peptides, vesicles fuse with the plasma membrane, and MHC II/peptides are expressed on the cell surface and presented to CD4 cells.
7. Certain pathogens (e.g., mycobacteria), when engulfed, prevent the fusion of phagosomes and lysosomes and persist in phagosomes.
8. MHC class II molecules present antigens to CD4+ T_H cells.
9. A T-cell receptor recognizes processed and MHC II-presented antigen peptides.
10. Cytokines released by T_H and IL-1 cytokines released by macrophages stimulate T_H cells to produce clone of B cells.

II. *Effector Phase:*

1. These B cells started differentiating into plasma cells.
2. Plasma cells start secreting specific antibodies against the antigen.
3. B cells are also antigen-presenting cells. Binding of an antigen to an IgM receptor triggers receptor-mediated endocytosis, degradation, and presentation of the processed antigen.

III. *T-cell-independent immune responses:*

1. Phagocytosis usually occurs before cells kill microorganisms.
2. By the process of chemotaxis, phagocytes move toward the released microbial component (endotoxin, formyl peptides) at the site of infection.
3. The phagocyte then recognizes the pathogen/microbes through specific receptors.
4. Phagocytic cells have a variety of receptors on their cell membranes through which infectious agents bind to the cells. These receptors include Fc receptors, complement receptors, scavenger receptors, and toll-like receptors.
5. Phagocytosis (cellular ingestion of microbes) occurs in two ways:
6. Opsonic engulfment of complement-coated or antibody-coated microbes via complement receptors or FcR receptors in a faster and efficient way
7. Non-opsonic direct engulfment via an innate pattern recognition receptor in a slower and inefficient way.

1.14 Discussions

The host organism generally develops an immunity appropriately adapted toward the invading pathogen. This inevitably involves the risk to mount diverse host responses directed against different self-antigens. Actually, the adaptive immune system differentiates between self- and nonself-molecules. Thus, to prevent injury to the host, the organism has gradually actively acquired nonreactivity to self and tried to maintain this collective *self-tolerance* by several mechanisms. For each host, discrimination between self- and nonself-molecules is host specific and requires the assortment of an individual set of nonself-reactive receptors. Subsequently the outcome of self-/nonself-discrimination in a host organism is not only transferred from one generation to other generations but at the same time is devoid of evolutionary pressure. Moreover, innate immunity mostly relies on genetically programmed recognition patterns/structures that restrictedly respond to various foreign antigens. The identification and recognition of pathogens, but not any self-molecules, through an evolutionary conserved set of receptors are done to potentially avoid autoimmunity. The likelihood of developing an autoimmune disease is generally believed to result from an indefinitely higher failure rate in the adaptive immune system mechanisms.

The adaptive immune system together with its capability of generating tremendous specificity and diversity has the ability to regain homeostasis and an inbuilt ability to self-limit responses. This mechanism provides space for budding lymphocytes that require a highly specific immune response and thus crucially prevent excessive immunity.

Although the adaptive immune system is phylogenetically younger, the innate immune system is the more ancient form. The most recent evolution added was the formation of different lymphoid organs. In the greatly specialized microenvironment of lymphoid tissues, both branches of the immune system can cooperate intimately and collaborate proficiently to boost and optimize different immune reactions. The gut-associated lymphoid tissues are the earliest detected organized lymphoid structures during evolution. The spleen, thymus, and lymph nodes, the more sophisticated secondary lymphoid organs, are found only in higher vertebrates.

Suggested Readings

Asma Ahmed, Banishree Saha, Anand Patwardhan, Shwetha Shivprasad and Dipankar Nandi, The Major Players in Adaptive Immunity1.Humoral Immunity resonance, may 2009,451–471

Basic Immunology: Functions and Disorders of the Immune System chapter 1, pp 1–22

Bonilla FA, Oettgen HC (2012) Adaptive immunity. J Allergy Clin Immunol 125(2):S33–S40

Fundamental Immunology (5th Edn) Lippincott, Williams & Wilkins Chapter 17 (pp 497–517)

Immunobiology (6th Edn) Janeway, Travers, Walpert & Capra Chapter 2 (pp 37–100), and Chapter 6 (pp 209–212)

Janeway Jr CA, Paul T, Mark W, Shlomchik MJ (2001) Immunobiology. In: The immune system in health and disease, 5th edn. Garland Science, New York. ISBN 10: 0-8153-3642-X

Neal s Rote. Adaptive immunity. Chapter 6, pp 143–165

Owen J, Punt J, Stranford S (2012) Kuby immunology, Seventh edn. Macmillan Higher Education

CRP: Historical Perspective, Structure, Evolution, Synthesis, Clinical and Biological Functions

2

Abstract

The C-reactive protein (CRP) is a plasma protein belonging to pentraxin family and has hepatic origin. It is the chief component of any inflammatory reaction. A key mediator of the innate immune system, an inflammatory biomarker whose concentration rapidly increases to nearly or more than 1000-folds above the normal values during tissue injury or infection. CRP functions through interaction of components of both humoral and cellular effector mechanisms of inflammation. Although considered as an acute-phase protein in tissue injury, infection and inflammation has now attained a distinct status of inflammatory disease marker not only in cardiovascular diseases but has well-known clinical and pathological relevance. The present study encompassing a number of diseases and associated conditions has highly illuminated the research field with the therapeutic role of CRP in health and disease. The future prospect of this chapter lies in the monitoring and understanding the applicability of CRP in the biology of various diseases.

Keywords

C-reactive protein • Disease • Acute phase protein • Evolution • History • Biology

Chapter Highlights

1. CRP is a highly conserved and phylogenetically ancient major acute-phase plasma protein in human. It belongs to the family of "pentraxin" and its discovery dates back in 1930.
2. Its serum concentration increases to up to 3000-fold in response to infection or tissue injury or other inflammatory conditions.
3. CRP is synthesized in the liver and secreted in the plasma, mainly regulated by cytokine IL-1 and IL-6 and functions in the mediation of innate immune system.
4. CRP has the initial ability to recognize pathogens by binding to the surface components of microbes, the PC thereby activating the complement cascade and causing opsonization of pathogen leading to phagocytosis.

W. Ansar, S. Ghosh, *Biology of C Reactive Protein in Health and Disease*,
DOI 10.1007/978-81-322-2680-2_2

5. CRP has been studied in research of clinical sciences as a screening device for inflammation, an inflammatory biomarker, a marker for disease activity and also as a diagnostic adjunct.
6. The biological significance of CRP in different diseases was elucidated.

2.1 Introduction

The human C-reactive protein (CRP) is a highly conserved and phylogenetically ancient major acute-phase plasma protein. Discovered in 1930, it belongs to the family "pentraxin." Its concentration in the serum rises from few folds to nearly 3000-fold in response to various infections or tissue injuries. It shows a pronounced and rapid rise in its concentration. CRP is an important component of innate host immunity. It is capable of recognizing pathogens by initially binding to their surface components, phosphorylcholine (PC). CRP recognizes many foreign pathogens as also their phospholipid component of various damaged cells. It can activate the complement cascade when bound to one of its several ligands and can also bind to different phagocytic cells. Thus, it can initiate the elimination of complement triggered or opsonized targeted cells as it can interact with varied cells and molecules of both the humoral and cellular pathways of inflammation (Volanakis 2001). Since its discovery, CRP has been considered as a screening device for various inflammatory conditions, also as a marker for a number of disease activities, and as a diagnostic accessory (Clyne and Olshaker 1999). The immunomodulatory role of CRP is mediated by activating the complement system or triggering opsonization, which leads to phagocytosis. Several such functional properties make CRP an important component of our first line of nonspecific innate host defense (Li et al. 1994).

2.2 CRP: From the Historical Perspective

CRP was first isolated from serum of patients infected with pneumococcus bacteria late back in 1930 by William Tillett and Thomas Francis working in the Rockefeller University. Initially, it was termed as "fraction C" as it was the third component that was isolated. Nearly a decade later, CRP was described as an "acute-phase reactant" by Oswald Avery and Maclyn McCarty as it showed increased concentration in serum of patients suffering from a range of diseases, including inflammation connected with rheumatic fever and myocarditis. Abernethy and Avery coined the term CRP to this serum protein responsible for precipitin with CPS (Abernethy and Avery 1941). The crystallization of CRP was done first by M. McCarty (2004). Studies between the period of the 1940s and the 1950s confirmed the importance of CRP as a biomarker of inflammation as observed in patients suffering from cardiac disorders like atherothrombosis, acute myocardial infarction, coronary ischemia, and myocardial necrosis. J. Hurlimann in 1966 first described that the liver is the site of CRP formation (Hurlimann et al. 1966). However, extrahepatic expression of CRP has also been documented (Dong and Wright 1996). Around the middle of the 1980s, a research done by John Volanakis, Mark Pepys, Irving Kushner, and many others revealed CRP as a circulating pentraxin molecule composed of 5 identical subunits arranged in pentameric symmetry. CRP is a hepatically originated, non-glycosylated acute-phase protein. Das et al. in 2003 reported differential glycosylation of CRP in 16 different pathological conditions (Pepys 1981; Das et al. 2003, 2004a, b). CRP has a characteristic calcium-dependent binding affinity to specific ligands like LDL cholesterol.

2.3 Synthesis

J. Hurlimann in 1966 first described that the liver is the site of CRP formation (Hurlimann et al. 1966). However, extrahepatic synthesis of CRP has also been reported (Dong and Wright 1996; Ridker 2009; McCarty 2006). Human CRP can also be produced by smooth muscle-like cells and macrophages, present in the human atherosclerotic plaques (Yasojima et al. 2001). While such synthesis may be physiologically significant at local sites, it is unlikely that it contributes substantially to increased plasma concentration. Probably due to some local immune response from a systemic inflammatory pathway, an acute-phase reaction takes place. Consequently renal cortical tubular epithelial cells (TEC) of inflamed kidneys and human respiratory tract epithelial cells express CRP mRNA genes and thereof CRP expression (Gould and Weiser 2001 and Jabs et al. 2003). The plasma half-life is about 18 h, and complexed CRP is catabolized by hepatocytes in vivo and rapidly cleared from the circulation (Hutchinson et al. 1994 and De Beer et al. 1982a, b).

In normal healthy subjects CRP is a trace plasma protein, ranging between 0.1 and 0.5 μg/ml (Das et al. 2003). The serum level of CRP increased rapidly about 1000-folds by 48 h in response to various bacterial infections, traumatic conditions, necroses of tissue, parasitic invasions, and malignant neoplasias and in most types of inflammation during an acute reaction. In some chronic infections and various inflammatory disorders, plasma CRP levels may remain persistently high. Gradually, with the resolution or remission of the pathologic condition or either spontaneously or in response to therapeutic treatments, the CRP level falls rapidly to normal concentration. Generally bacterial infections are a much less potent stimulus for CRP synthesis than localized viral infections. CRP levels are usually normal or only modestly increased in some severely active tissue-damaging pathological conditions or in some important disorders. These include scleroderma, dermatomyositis, ulcerative colitis, Sjögren's syndrome, and leukemia. Active systemic lupus erythematosus (SLE)-suffering patients do not have high plasma levels of CRP but their concentration increases during the bacterial infection. Nowadays, obesity has also been regarded as a low-grade systemic inflammatory disease. Elevated serum levels of CRP have been seen in both overweight and obese children and adults (Das 2001), and also in caloric-restriction-induced weight loss individuals, decreased plasma concentration of CRP was observed (Techernof et al. 2002). Mean concentration of CRP is substantially higher in smokers than in nonsmokers (Haverkate et al. 1997).

Specifically, the rate of synthesis is greatly increases during the acute phase (Macintyre et al. 1985). Under physiologic conditions, CRP is synthesized at quite a low rate. The pentameric CRP is mostly retained by two resident carboxyl esterases during its constitution in the endoplasmic reticulum (ER) (Macintyre et al. 1994). During the acute-phase reaction, the half-life needed for the exit of pulse-tagged CRP from the ER is reduced to nearly 18 h to 75 min. The marked acceleration of secretion is actually due to reduced affinity for CRP of one of the esterases, which has been accredited to a conformational change (Yue et al. 1996).

2.3.1 Structure

"Pentraxin," the structure assigned for some acute-phase proteins like CRP, SAP, and the female protein of Syrian hamster (HSAP), is a rare configuration due to their conserved pentameric arrangement of their protomers.

Most of these proteins exhibit the common characteristics but few of them showed some variations in their structure and functions. The phylogenetically oldest member of the pentraxin family is the CRP of an invertebrate arachnid, the horseshoe crab *(Limulus polyphemus)* which differs from other vertebrate counterparts as it is hexameric. The three major acute-phase proteins (CRP, SAP, and HSAP) differ significantly in their calcium-binding and calcium-dependent ligand-binding properties (Srinivasan et al. 1994).

Electron microscopic studies have demonstrated that the five identical polypeptide subunits held together by non-covalent bonds arranged in a disklike configuration with cyclic symmetry (Osmand et al. 1977). The molecular weight of a pentamer ranges between 110 and 144 kDa (Oliveira et al. 1979), and the subunit molecular weight estimated by various ways varies from 20 to 30 kDa (Volanakis et al. 1978 and Woo et al. 1985 and Lei et al. 1985). Each subunit contains a single, intramolecular disulfide bond. The stability of purified CRP is dependent upon temperature, pH, and protein concentration (Fig. 2.1). Solutions of pH 7–8 containing CRP at concentrations from 50 μg/ml to 1 mg/ml are stable at

PHYSICOCHEMICAL CHARACTERISTICS:

1.Molecular Weight (kDa)(Polymeric)

a. Molecular weight (range)-110-144

b. By sedimentation equilibrium-118

c. From sequence-120

2.**Molecular Weight (kDa)(Subunit)**

(a) Sedimentation equilibrium-24.3

(b) Gel filtration-20

(c) Acrylamide gel electrophoresis-24/23.3

(d) Derived from sequence-24

3.Symmetry- Cyclic pentameric

4.Subunit- 5

5.Location- Chromosome 1

6.Number of nucleotides- 2263

7.Sedimentation coefficient($s^{\circ}_{w.20}$)(S)- 6.5

8. Partial specific volume(v) (ml/g)- 0.735

9. Isoelectric point (p*I*)- 5.4-7.9

10. Binding constant (K) for phosphorylcholine, (M)- 1.9×10^{-5}

11. Extinction coefficient ($E^{1\,cm}_{1\%}$)- 19.5

12.Signal peptide- 18

13.Protein- 206 amino acids

14.mRNA (Northern Blotting)- 2.3kb

15.Number of introns- Single

16.Intron (sequence)- 278bp

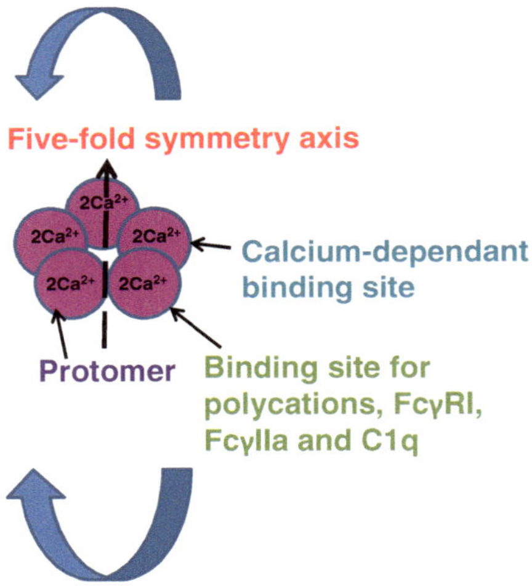

Fig. 2.1 *Physicochemical characteristics of human CRP*: the binding pockets of CRP in each protomer are noted along with its physicochemical properties (Osmand et al. 1977)

4 °C for weeks and at −20 °C for months and few years. Under conditions of lower pH or higher protein concentration, CRP molecules may aggregate. Dissociation of CRP into subunits or intermediate forms (e.g., trimers) may be appreciable in more alkaline solutions or with repeated freeze thawing (Macintyre 1988).

Alternations in glycosylation, often associated with the acute-phase reactant (APR), generate microheterogeneous changes in acute-phase proteins (APP). This is normally occurred by differing the arrangement or composition of their glycan side chains due to differences in activity of various glycosylating enzymes (Raynes 1982). The microheterogeneity of APPs frequently differs in acute and chronic types of inflammation.

There are hardly a few reports to show the microheterogeneity of CRP. Three types of CRP have been identified in the plasma samples of the Japanese horseshoe crab (*Tachypleus tridentatus*, CRP designated as tCRP), based on various affinities against different ligands, named as tCRP-1, tCRP-2, and tCRP-3, and exhibited varied calcium-dependent hemolytic and sialic acid binding and also hemagglutinating activities. The molecular heterogeneity of tCRPs exits both in their amino acid sequences and in N-linked glycosylation and indicates the possibility of isolectins (Iwaki et al. 1999). Similar findings have been demonstrated in freshwater fishes (e.g., *Labaeo rohita* and *Catla catla*) that increased CRP levels under different toxic aquatic conditions giving rise to different molecular variants of CRP in an agent-specific manner showing differences in amino acid and carbohydrate compositions (Sinha and Mandal 1996 and Sinha et al. 2001 and Paul et al. 1998, 2001).

There are only a few extensive studies on human CRP in various aspects and in various diseases. The microheterogeneity of human CRP has not been reported yet and this could theoretically be the result of altered synthesis, synthesis outside the liver, structural alterations, cleavage, or complication of the circulating CRP which may play diverse physiological roles by altering its binding properties (Lasson and Göransson 1999].

2.4 The Immune System and CRP

The human immune system has developed refined defense mechanisms which are highly specific for the diverse invading pathogens. Most of the cells of the innate immune processes are highly conserved and bear pattern recognition receptors (PRRs) that recognize pathogens through pathogen-associated molecular patterns (PAMPs) and trigger a plethora of mechanisms of pathogen removal. Major PRR proteins include plasma pentraxin families of proteins. Pentraxins are a prototypic apparatus of the innate immune system. The C-reactive protein was known as the first identified PRR which forms an important clinical inflammatory biomarker.

CRP is a positive acute-phase protein whose plasma concentration increases during inflammatory disorders (Morley and Kushner 1982). It is important to emphasize that there is great variability in acute-phase behavior of CRP from one species to another (Kushner 1988). Of at least 30 acute-phase proteins, CRP and serum amyloid A (SAA) are the two proteins whose concentration arise more than 1000-fold and even up to 3000-fold within 48 h following injury (Morley and Kushner 1982).

A number of wide-ranging biological functions have been assigned to acute-phase reactants, CRP. Besides CRP, the other major APP, SAA becomes one of the major apolipoprotein of high-density lipoprotein (HDL) during inflammation, and its rapid binding to HDL may influence cholesterol metabolism in inflammatory states. It is also reported to have a function in adhesion and chemotaxis of phagocytic cells and lymphocytes, in inflammation, inside atherosclerotic coronary arteries, and in oxidation of low-density lipoprotein (LDL) (Gabay and Kushner 1999).

A few APP-like fibrinogen like plasma proteins, plasminogen activator–inhibitor type-1 (PAI-1), and most possibly CRP are known to possess procoagulant function (Munford 2001). This action is required for the formation of abscess, for wading off invading pathogens, and for various delayed hypersensitivity reactions. CRP, fibrinogen, and haptoglobin play some roles in wound healing (Cid et al. 1993).

It also found that CRP shares many properties with the immunoglobulins, such as the ability to promote agglutination, complement fixation, bacterial capsular swelling, phagocytosis, and precipitation of polycationic and polyanionic compounds (Marnell et al. 1995). However, it is structurally distinguishable from the immunoglobulins on the basis of its antigenicity and its possession of five apparently identical subunits. Other distinctive characteristics of CRP are its binding specificities and its site of synthesis. Thus, CRP has been assigned to a new superfamily of proteins.

2.5 CRP: Its Clinical and Biological Role

2.5.1 Biological Role

The first line of defense toward diverse environmental agents or foreign particles is the preexisting natural antibodies or immunoglobulins and the non-antibody proteins present in the body. CRPs belong to this group of non-antibody proteins and are well involved to build the body's first immunity. CRP shares many properties with the immunoglobulins, such as the ability to promote agglutination, complement fixation, bacterial capsular swelling, phagocytosis, and precipitation of polycationic and polyanionic compounds (Marnell et al. 1995). However, it is structurally distinguishable from the immunoglobulins on the basis of its antigenicity and its possession of five apparently identical subunits. Other distinctive characteristics of CRP are its binding specificities and its site of synthesis. Thus, CRP has been assigned to a new superfamily of proteins.

Human CRP binds to a wide array of ligands including damaged tissues, diverse bacteria, and different fungi (yeast) (Kindmark 1971; Richardson et al. 1991). The known ligands for CRP can be classified into three major groups—(1) compounds that contain PC or related structures (Volanakis and Kaplan 1971), (2) polycationic compounds (Dicamelli et al. 1980), and (3) carbohydrates that contain D-galactose-related structures (Lee et al. 2002) (Fig. 2.1).

PC is the fully defined calcium-dependent ligand for CRP. The other ligands are the sphingomyelin and the polar head groups of the prevalent phospholipids phosphatidylcholine found in the outer surface of cell membranes which becomes exposed after cell damage (Volanakis and Kaplan 1971; Culley et al. 2000). Interaction of CRP with lipids (Kaplan and Volanakis 1974), liposomes (Mold et al. 1981), and phosphatidylcholine liposomes in the presence of lysophosphatidylcholine perturb the bilayer (Volanakis and Wirtz 1979), lysophosphatidylcholine or phospholipase A_2-treated rabbit and human erythrocytes (Volanakis and Narkates 1981) LDL (Fu and Borensztajn 2002) where unesterified cholesterol in LDL provides the ligand was shown.

CRP has also been reported to exhibit significant calcium-dependent and PC-inhibitable binding to eukaryotic constituents such as chromatin (Robey et al. 1984), histones (H1, H2A, and H2B) (Du Clos et al. 1988) under nonphysiological conditions, small nuclear ribonucleoprotein particles (snRNPs) (Du Clos 1989 and Pepys et al. 1994) under relatively physiological conditions, and galactose-containing polysaccharides (Ulenbruck et al. 1981). Ligand-complexed CRP is recognized and bound by molecules like C1q, FcγRI, and FcγRIIa receptors (Agrawal et al. 2001) (Fig. 2.2). CRP binds to matrix proteins laminin and fibronectin (Salonen et al. 1984 ; Tseng and Mortensen 1988). Surface-

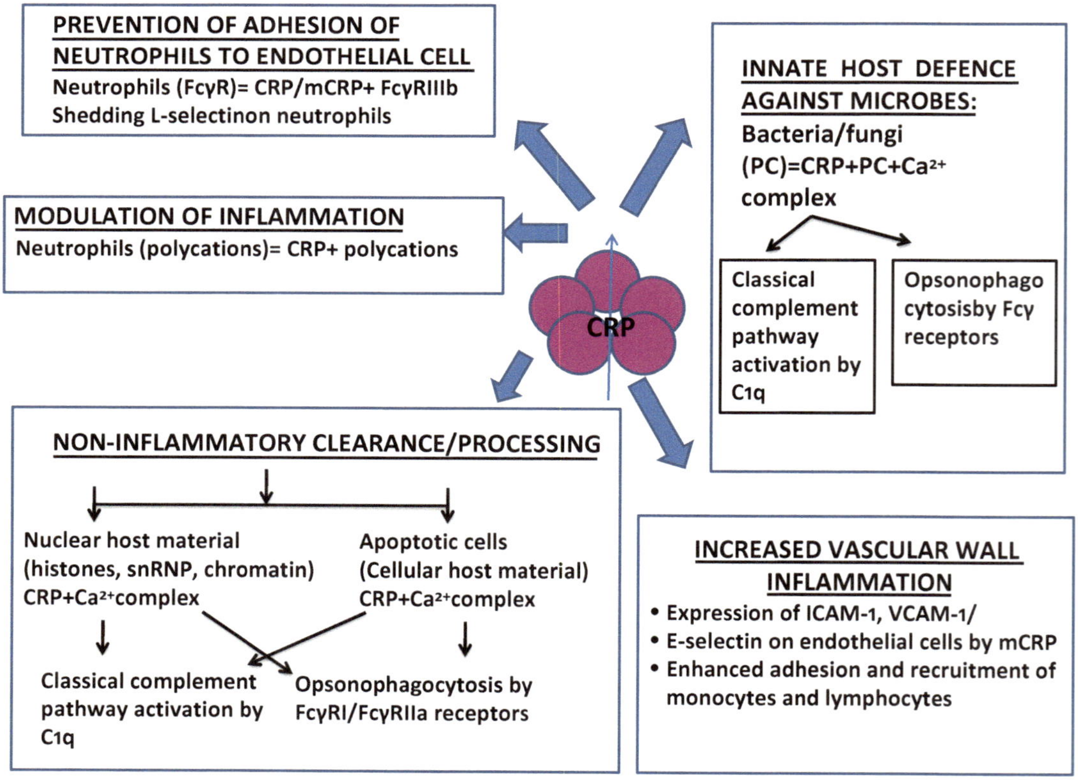

Fig. 2.2 *Biological functions of CRP*: CRP binds with a variety of ligands and activates the complement cascade or mediates opsonophagocytosis to help in the clearance of microbes and nuclear materials (Ablij and Meinders, EJIM, 2002)

immobilized CRP (neo-CRP) expresses a novel set of antigenic determinant molecules (Ying et al. 1989). Surface-adsorbed CRP also binds to lactosylated human serum albumin (HSA) protein but does not bind with derivatives containing other terminal sugars, like glucose, mannose, sialic acid, and N-acetylglucosamine (Kottgen et al 1992). A recent finding revealed that this immobilized CRP binds to different carbohydrate ligands in a much better way in absence of calcium. Thus, the specificity of this neo-CRP in accordance to the sugar structure is quite broad, and the presence of a different negatively charged group, such as a carboxylic acid or a phosphate group on a sugar residue, greatly stimulates the affinity of the present sugar (Lee et al. 2002). Human CRP immobilized onto surface or onto latex beads binds distinct plasma glycoproteins including IgG, IgA, IgM, fetuin, asialofetuin, etc., and likewise, synthetic glycoproteins, as a lectin, exhibit binding specificity for terminal galactosyl residues of the glycoprotein glycans.

Evidence from researches was presented that CRP binds to lipophosphoglycan (LPG), the major cell surface glycoconjugate on *Leishmania* to coat the parasites (Culley et al. 1996). A different observation that recognition of a pathogen group by CRP takes place via a phosphorylated carbohydrate contributes toward the carbohydrate specificity of CRP, which was demonstrated using synthetic LPG oligosaccharides and a range of monosaccharides, like phosphorylated, amino, and sulfated carbohydrates (Culley et al. 2000). CRP also binds to leishmanial excreted factors in presence of calcium, almost similar to that occurring between CRP and various galactans and quite different from the reaction occurring between pneumococcal excreted factors and CRP (Pritchard and Volanakis 1985). A recent study revealed that CRP binds to the apoptotic cells. This binding is specific, calcium dependent, and restricted for the surface membrane of intact apoptotic cells. Kinetic studies and confocal microscopy with annexin V staining suggested that CRP binds to a ligand other than snRNPs or chromatin on the apoptotic cell surface, most likely, to lysophospholipids on apoptotic cells (Gershov et al. 2000). CRP can also bind to human neutrophils and small fraction of peripheral blood lymphocytes (PBL) in the presence of CPS (Fig. 2.2).

2.5.2 Clinical Importance

Despite the lack of unique diagnostic specificity, CRP is highly useful in clinical practice because changes in their serum or plasma level in a broad range reflect the presence and strength of diverse inflammatory processes. The estimation may help to differentiate inflammatory from noninflammatory conditions, reflect response to and need for treatment, assess prognosis, and predict future risk. Moreover, CRP level is not directly affected by many anti-inflammatory or immunosuppressive drugs unless these drugs affect activity of underlying disease. Therefore, recently, CRP has been established as a more important marker of acute-phase reactions than others, like ESR, and widely used in clinical practice.

Plasma/serum CRP concentration showed marked elevation in various conditions like bacterial infection, abscess, Crohn's disease, connective tissue disorders (except SLE), neoplasia (except leukemia), trauma, and Necrosis, whereas it showed slight elevation in viral infection, steroid and estrogen therapy, ulcerative colitis, and SLE. CRP is also known to be elevated in many other disease and pathological conditions and the spectrum of its clinical application is increasing day by day. It is useful in the diagnosis of neonatal septicemia and meningitis, postoperative infection, postoperative thromboembolic complication, intercurrent infection of SLE, intercurrent infection of leukemia, acute pancreatitis, acute appendicitis, acute osteomyelitis, acute rheumatic fever, insulin resistance, etc. and in the prognosis of rheumatoid arthritis, juvenile rheumatoid arthritis, ankylosing spondylitis, Reiter's syndrome, psoriatic arthropathy, vasculitic syndrome, Crohn's disease, rheumatic fever, familial Mediterranean fever, myocardial infarction, stable and unstable angina, and visceral leishmaniasis. It is also used as a risk assessment of different diseases like myocardial infarction, angina, cardiovascular mortality, morbidity and mortality in the elderly, metabolic syndrome, and PKDL.

Most normal subjects have the average plasma CRP concentrations of 2 μg/ml or less, but some individuals have plasma concentrations as high as 10 μg/ml as most of us are under some regular stress and inflammatory responses. This may also be due to the limited activation by minimally apparent or some unapparent low-grade process. Thus, the 10 μg/ml level of CRP is regarded as clinically unimportant. However, according to recent findings, slightly elevated concentration within the normal range may predict future coronary events. The basis for the differences observed in CRP responses in some apparently similar inflammatory disease is not known. Differential behavior of CRP may be contributed toward differential glycosylated molecules present in CRP (Ansar et al. 2009a; b). The differential glycosylation is induced in various acute-phase conditions (Das et al. 2003). The differential behavior/role may attribute the nature and/or degree of tissue damage or may be due to some genetic modulations in the potential to produce mediators, to respond to some mediators, and/or to produce CRP protein itself.

Increased CRP production is a very sensitive early indicator of most forms of microbial infection. Accordingly, CRP is useful in the diagnosis of neonatal septicemia and meningitis. Similarly, CRP is elevated in most cases of acute osteomyelitis including those early cases where constitutional symptoms, leukocytosis, and plain radiographic abnormalities are absent. It also facilitates the recognition of intercurrent infection in SLE and leukemia. But this application has been somewhat limited by the finding that active lupus serositis and chronic synovitis also induce elevation of CRP. Any surgery is known to be associated with elevation of CRP, but the absence or the fall in CRP level after 7–10 days or secondary rise provides early warning of postoperative infection or thromboembolic complications. Recently, CRP is regarded as the only useful marker that shows great promise in diagnosing acute pancreatitis.

CRP has some definite value in evaluating clinical course and response to treatment in connective tissue disorders. It is also related to the prognosis of acute myocardial infarction in relation to its long-term outcome.

Higher level of CRP in treated visceral leishmaniasis (VL) patients predicts higher future occurrence of post kala-azar dermal leishmaniasis (PKDL) (Gasim et al. 2000). CRP measurement for clinical study in the high normal range values predicts an increased/prolonged long-term risk of angina, myocardial infarction, and death. Indeed, a high normal baseline plasma CRP level has been correlated with an increased risk of some coronary events even in normolipidemic persons. The relationship between elevated CRP and adiposity suggests its predictive value in development of metabolic syndrome.

Recently, researches are going on regarding measurements of plasma cytokines, cytokine receptors, and their clinical significance. But, their measurements have no apparent advantage over CRP measurements. The definitive role of CRP estimation in some diseases has been well established and in some it is not. The clinical significance of CRP has come a long way from its quantitative estimation in different diseases in the light of inflammatory processes to that in the light of pathogenesis other than inflammatory in nature.

2.6 CRP and Evolution

The immune system of humans has evolved from both invertebrate and vertebrate organisms for over a million of years to develop into a full-proof defense mechanism which is highly specific for invading microbial pathogens. The human innate immune system has inherited from the invertebrates, using a system of germline-encoded proteins. This ancient defense system helps to recognize pathogens. The cells of innate immune system are mostly conserved highly. This system includes pattern recognition receptors (PRRs) that recognize microbial pathogens and trigger a plethora of mechanisms of pathogen abolition. The host proteins like PRRs recognize pathogen-associated molecular patterns (PAMPs) and some human molecules whose ancestors are also evolutionary ancient and constant. Thus, recognition of surface pathogen molecules leads to activation/production of the effector molecules like complement pathways, diverse cytokines, and antimicrobial peptides. Major PRR families of proteins include plasma pentraxins and others.

The nonspecific innate immune responses begin much more earlier than the specific adaptive system. Indeed, the innate immune system therefore forms the first line of defense after the host has been attacked by a pathogen. In cooperation with the nonspecific mechanism, the body mounts an adaptive immune response. Innate immunity can be seen to be mainly composed of anatomic, physiologic, phagocytic, and inflammatory types of four defensive barriers.

The first line of defense toward diverse environmental agents or foreign entities is preexisting natural antibodies or immunoglobulins and the non-antibody proteins present in host body. CRPs belong to this group of non-antibody proteins and are well involved to build the bodies' first immunity. It also found that CRP shares many properties with the immunoglobulins, such as the ability to promote agglutination, complement fixation, bacterial capsular swelling, phagocytosis, and precipitation of polycationic and polyanionic compounds (Marnell et al. 1995). However, it is structurally distinguishable from the immunoglobulins on the basis of its antigenicity and its possession of five apparently identical subunits. Other distinctive characteristics of CRP are its binding specificities and its site of synthesis. Thus, CRP has been assigned to a new superfamily of proteins.

CRPs are highly conserved evolutionarily members of the pentraxin family, indicating their important biological role across species. They exhibit high sequence homologies with other members of the family, found in all mammals, birds, amphibians and marine teleosts, sweet water fishes (Pepys et al. 1978 and Hokama and Nakamura 1987 and Sinha and Mandal 1996 and Paul et al. 1998). The most primitive creature in which CRP is known is the arachnid *Limulus polyphemus* (Robey and Liu 1981). This evolutionary conservation of sequence and binding specificity over a diverse range of species suggests that CRPs have an important biological role.

CRP is a normal component of the invertebrate hemolymph, whereas in humans and other mammals, CRP level increased significantly from its basal level under inflammatory conditions. As, for example, in *Limulus* (Robey and Liu 1981), rat (Nagpurkar and Mookerjea 1981 and De Beer et al. 1982a, b), and fish (Sinha and Mandal 1996 and Paul et al. 1998), CRP is constitutively expressed and a major normal plasma component present at a concentration of 1–5 mg/ml. In contrast, CRP is a trace plasma protein in humans. It is expressed dramatically and rapidly as part of the diverse acute-phase response to injury or infection. It has been suggested that CRP might be serving as a primitive form of immunoglobulin in those groups, which have high concentration of CRP as a constitutive plasma protein.

2.7 CRP: A Different State-of-the-Art Research

The state of the art of research in CRP encompasses different areas, including:

1. *Epidemiological and clinical research study*
   - In the CVD risk prediction beyond the known traditional risk factors, the individual and pooled data from various existing and new epidemiological studies were analyzed. This is done to examine whether diagnostic CRP measurement can be improved and can help target individuals who should be treated.
   - To examine the correlation between CVD and CRP level and to check whether this association is truly independent of other known CVD risk factors.
   - In clinical practice, is it worthwhile to add CRP to the basic risk assessment category in sufficient population of patients from one treatment regime to another (reclassification) group.
   - Is it worthwhile to replace a traditional cardiovascular risk factor(s) with CRP (or any other newly discovered biomarker as risk predictors) to improve the CVD risk pre-

diction assessment for use in clinical science.

- Possible heterogeneity observed by population subgroups (like age, ethnicity, etc.) and inclusive prospective research studies in minority groups to assess the correlation of CVD prevalence with CRP and other inflammatory biomarkers and to ascertain whether CRP cutoff values should vary by sex or ethnicity parameter.
- Whether CRP is superior as a CVD risk predictor than other inflammatory biomarkers, other novel biochemical procedures or various subclinical disease markers, or traditional CVD risk predictors, either singly or in combination mode, in cost-effectively predicting CVD risk.

2. Clinical studies to further demarcate whether CRP is a causal risk predictor for CVD and the efficacy of measuring CRP values in clinical practice.
    - Observational research studies correlating CVD with CRP genetic polymorphisms that persuade CRP level.
    - Development and trial of CRP-inhibiting drugs for utilization in clinical trials that could resolve the usefulness of such drugs in curbing CVD events.
    - Clinical testing to trial usefulness of CRP lowering in diminishing mortality rate and infarct size in the situation of various acute coronary syndromes.
    - To significantly lower CVD events, is it necessary to measure CRP levels in clinical trials for better targeting of therapy? To establish by all these trials, whether a high or low CRP level should influence different preventive therapies, in monitoring response to therapy, or to rouse modifications of therapy.
3. *Primary prevention clinical studies*
    - Clinical trials of various principal prevention approaches in which increased CRP is used to recognize individuals at high risk, for example, trials of hypertension patients or diabetes prevention in subjects with high CRP. Analyses could be potentially done using stored samples already completed trials in a post hoc basis.
    - To identify and characterize changes in pathway biology in the general individuals, in the machinery of innate immunity cascades, various cellular and molecular epidemiological studies need to be addressed.
    - To prevent the process of elevated CVD risk in the first position, effective studies need to be developed on individual or population-wide strategies, inclusive of elevated CRP levels and its chief behavioral determinants, fatness and exercise.
    - Valid clinical trials to resolve whether calculating and measuring CRP levels can motivate subjects or clinicians to pursue CVD prevention strategy.
    - Observational or various experimental studies pinpointing the extent to which dietary symphony or physical activity manipulates CRP levels irrespective of obesity.
4. *Relationship of CRP to other conditions*
    - Diverse studies of the interconnections of obstructive sleep apnea or sleep deprivation with CRP, obesity, and CVD events
    - Prospective studies of different inflammatory conditions (like psoriasis, rheumatoid arthritis, etc.) and the risk of consequent CVD and whether the CRP elevation may explain the probable associations
5. *Other clinical trial studies*
    - Clinical studies to determine various genetic polymorphisms in the known CRP gene that affect plasma levels and acute-phase variation of CRP. Verification of the CRP level in young people would be of special mention.
6. *Studies to correlate the interrelationship between the plasma CRP levels and the degree of the atherosclerosis*
    - Research to assess the importance of measuring the CRP level in acute coronary syndromes
    - Analyses of various clinical trial data to examine whether any anti-inflammatory effects of curing lipids or blood pressure

may contribute any difference in the treatment outcome

7. *Development of various methods*
    1. Development of various improved biostatistical protocols for clinical decision-making treatment when the aim is to predict the risk which can be used for screening and treatment targeting. The application to CRP and diverse inflammatory markers is needed.
    2. Research to develop improvised biomarkers that attribute pathology-associated alterations in inflammation. Use of biomarker in various imaging measures like PET, MRI, ultrasound, and many others is potentially validated. This might be done by an inflammation-centered clinical trial web that could evaluate various surrogate markers and other new candidate targets and therapeutics.
    3. Development of database study to establish various permissible laboratory errors for CRP and its measurement standards. It is imperative to develop a CRP concentration measurement standardization protocols.
8. *Basic research areas*
    1. To develop appropriate animal models to study the effect of CRP and atherosclerosis. To find out appropriate genes and develop some study knockout model animals.
    2. The mechanisms by which CRP may influence ischemic CVD and atherosclerosis should be studied by using animal models. To examine and determine the pathological and physiological functions of human CRP in atherosclerosis, atherogenesis, and thrombosis. Moreover, the impacts of CRP inhibitory mechanisms need to be studied.
    3. Various animal models were used to test budding therapeutic approaches before conducting differential CRP-targeted human clinical trials.
    4. Basic and clinical researches to identify various specific markers of atherosclerosis and to develop immunological assays. These are needed to help characterizing the role of different inflammatory process and its effector molecules in the development of atherosclerosis and consequent CVD events.
    5. Researches on the sources of CRP elevation and also sources of various stimuli for CRP production.

2.7.1 Discussions: CRP, Research and Challenges Ahead

Research to understand the epidemiology and assess the quality and consistency of different epidemiological data and the association of CRP with incident cardiovascular disorders is on track. Understanding the basic biological mechanisms relating CRP to CVD is being carried on. Questions on the role of CRP as a causal player for CVD or a biomarker of CVD risk or whether it is a nonspecific marker of inflammation in CVD are still unanswered. In-depth studies to understand the proper function of CRP measurement in the CVD avoidance and management remain the daring act ahead. Studies showed strong association between the CRP measurement during CVD etiology, and sorting in order to notify the screening of subjects for risk of future CVD and in pinpointing preventive therapeutic approaches needs to be highlighted. More research studies and investigation are needed to determine the question of clinical efficacy of CRP.

Acute-Phase Proteins and Responses and Their Application in Clinical Chemistry

3

Abstract

The host innate responses to immunological stress initiate with a nonspecific response followed by specific responses. Host homeostasis is affected due to injury, infection, surgery, trauma, neoplastic growth, or immunological disorders and is combated by a prominent systemic reaction of the organism known as acute-phase response (APR). At the tissue injury sites and sites of invasion of microbes, a series of tissue responses are itself initiated. The activation of vascular system and inflammatory cells and release of pro-inflammatory cytokines are the key steps of these responses. These responses are then amplified by the overproduction of cytokines and different inflammatory mediators released in blood circulation. The acute-phase response is also characterized by hepatic secretion of some proteins termed as acute-phase proteins (APP). APR characterized by alteration in blood plasma composition is beneficial to the host in preventing microbial propagation and thus helps in restoring homeostasis. Multifunctional APPs can opsonize the invading microbes and activate the complement system and remove dead cells and free radicals. In this chapter, the changes induced by acute-phase response disruption of host body homeostasis are discussed. In addition, the diagnostic, prognostic, and therapeutic use of APPs in assessing health in animals as well as human patients is also described.

Keywords

Acute phase protein • Acute phase response • Inflammation • Inflammatory mediators • Acute inflammation • Chronic inflammation • Disease biomarker • Cytokine

W. Ansar, S. Ghosh, *Biology of C Reactive Protein in Health and Disease*,
DOI 10.1007/978-81-322-2680-2_3

Chapter Highlights

- The acute-phase response of the host disrupts the normal homeostatic balance contributing to defensive or adaptive capabilities.
- These changes are of variable and diverse nature that either play a role to (1) initiate, (2) sustain, (3) modulate the inflammatory responses, or have (4) adaptive roles.
- These changes are enabled by signaling events mainly triggered by cytokines associated with an inflammation process.
- Several cytokines, like IL-6, stimulate the production of acute-phase proteins as a response to stimulation.
- Cytokines produced in APR differ with types of inflammatory conditions.
- Acute-phase responses are indicative of inflammation and its intensity, and they find application in clinical diagnosis and management of APR.
- Serum concentration of CRP offers advantages over the traditional method of measuring the erythrocyte sedimentation rate as a measure of acute-phase responses.

3.1 A Brief History

Human CRP is one of the best-studied major acute-phase proteins (APP). It was first discovered by Tillet and Francis in 1930 (Tillett and Francis 1930). They were studying serological reactions with different pneumococci extracts. They found that a non-type-specific somatic polysaccharide fraction, termed by them as fraction C, showed precipitation by the serum of acutely ill patients. The capacity of the patient's sera to precipitate C-substance, which was termed as C-polysaccharide (CPS) later, rapidly disappeared, after the crisis is over. In 1941, Macleod and Avery first isolated the C-reactive material and characterized it as a protein. This protein requires the presence of calcium ions in order to react with CPS. Macleod and Avery in 1941 coined the term "acute phase." In the same year, Abernethy and Avery coined the term CRP to this serum protein which is responsible for precipitin with CPS (Abernethy and Avery 1941). In 1966, J. Hurlimann first discovered the hepatic site for CRP formation (Hurlimann et al. 1966). However, the extrahepatic expression of CRP has also been documented (Dong and Wright 1996).

3.2 Acute-Phase Response (APR)

The complex series of reactions initiated in response to (1) infection, (2) injury, (3) wound, (4) physical trauma, or (5) malignancy is termed as the acute-phase response (APR). These reactions function to bring about (1) preventing ongoing tissue damage, (2) isolating and destroying pathogenic infectious organism, and (3) activating the repair mechanism required to restore the host's/organism's normal function (Baumann and Gauldie 1994). Thus, APR helps in restoring the homeostasis of the host organism. It is characterized by (1) leukocytosis, i.e., increase numbers of circulating neutrophils and their precursors,(2) fever, (3) anorexia, (4) somnolence, (5) lethargy, (6) alterations in the cellular metabolism, and (7) alteration of the plasma APP concentration (Hack et al. 1997b, Gabay and Kushner 1999). APR is broadly divided into induction of APPs and acute-phase phenomena. Bacterial products including (1) peptidoglycans, (2) lipoteichoic acid, (3) exotoxins, (4) lipoproteins, and (5) glycolipids can lead to the initiation of local inflammatory processes (Paul et al. 2008). Post-bacterial invasion into cells and many cell types on the mucosa or skin may produce infection controlling molecules (Baumann and Gauldie 1994) (Table 3.1).

3.3 Synthesis of APPs

APPs are synthesized by hepatocyte cells in the liver. The synthesis is stimulated by the cytokines released including IL-1, IL-6, and tumor necrosis factor (TNF)-α from the macrophage and monocyte cells at inflammation sites.

Table 3.1 Different APPs in human (Morley and Kushner 1982)

Acute-phase proteins	Group	Function
CRP, SAA, serum amyloid P	Major APR	Positive APRs
Complement components—C2, C3, C4, C5, C9, factor B, C1 inhibitor, C4 binding protein	Complement proteins	
Fibrinogen, von Willebrand factor	Coagulation protein	
α1-Antitrypsin, α1-Antichymotrypsin, α2-Antiplasmin, heparin cofactor II, plasminogen-activator inhibitor I	Proteinase inhibitors	
Haptoglobin, hemopexin, ceruloplasmin, enzyme: manganese superoxide dismutase	Metal-binding proteins	
α1-Acid glycoprotein, lipoprotein (a), heme oxygenase, lipopolysaccharide-binding protein mannose-binding protein, leukocyte protein I	Other proteins	
Albumin, prealbumin, apo AI, apo AII, HS glycoprotein, histidine-rich glycoprotein, transferrin, inter-α-trypsin inhibitor		Negative APRs

These cytokines activate their specific receptors on different target cells that lead to downstream signaling events leading to a series of systemic reactions. The reactions include (1) the hypothalamic–pituitary–adrenal gland activation, (2) the reduced secretion of growth hormone (Gruys et al. 1999), and (3) a number of physical changes.

The physical changes show clinical symptoms of fever and anorexia (Dinarello 1983; 1989; Ingenbleek and Carpentier 1985; Ingenbleek and Young 1994; Kraft et al. 1992; Kushner et al. 1981; Langhans 1996; van Miert 1995). Other measurable changes are (1) decreased concentration of high- and low-density lipoprotein-bound cholesterol in blood plasma; (2) decreased leukocyte numbers in blood; (3) complement cascade and blood clotting system activation; (4) increased concentration of adrenocorticotrophic hormone (ACTH) and glucocorticoids; (5) reduced serum concentration of ions like zinc, iron, and calcium; (6) decreased α-tocopherol and vitamin A level; and (7) most importantly alteration in concentration of several plasma proteins, the APPs (Dinarello 1983, 1989; Gruys et al. 1994). The concentrations of APPs are mostly due to a changed hepatic metabolism. Thus, within hours of infection, the protein synthesis pattern of the hepatocytes is altered drastically that results in an increase of positive APPs in the blood proteins (Blackburn 1994; Dinarello 1983, 1989; Gruys et al. 1994; Ingenbleek and Young 1994; Kushner et al. 1981). CRP, SAA, haptoglobin, and many others represent the positive APPs synthesized by the liver after cytokine stimulation (Heinrich et al. 1998, 1990). The APPs whose synthesis decreases in blood plasma are the negative APPs. The examples of negative APPs include albumin, retinol binding protein (RBP), transthyretin (TTR), transferrin, and cortisol-binding globulin (Ritchie et al. 1999).

3.4 Cytokines in the Synthesis of APPs

Macrophages and other leukocytes when activated release inflammatory cytokines (TNF-α, IL-1, and IL-6). The pattern recognition receptors (PRRs) on leucocytes and macrophages bind pathogen-associated molecular patterns (PAMPs). Exposure to released bacterial products stimulates mast cells to release pro-inflammatory cytokines that recruits neutrophils to the inflammation sites (Abbas et al. 2012). Major cytokines in APR include interleukin-1, interleukin-6, interleukin-8, and interleukin-1β, TNF-α, interferon-γ (INF-γ), and transforming growth factor β (TGF-β) which have a profound effect on the behavior, neuroendocrine system, and metabolic effect (Kushner 1993; Wigmore et al. 1997; Gabay and Kushner 1999) on acute and many systemic acute-phase effects. There are reports of the presence of the cytokines and their specific receptors in the neuroendocrine system and brain. In experimental animals, IL-1, IL-6, and TNF-α have been shown to modulate carbohydrate, fat, and protein metabolism and regulate hypothalamic–pituitary secretions (Johnson 1997; Johnson et al. 1993a). These cytokines circulate through the blood and on reaching the liver stimulate liver hepatocytes to synthesize and secrete APPs.

The liver plays an important role in the immune system. Both innate and adaptive immunity functions of the liver are carried by different groups of cells with highly specialized functions. Liver secretes many acute-phase proteins, like short pentraxins, pathogen recognition receptors (PRRs), constituents of the complement cascade, and different regulators that control iron metabolism. Thus, liver hepatocytes contribute largely to the control of responses of systemic inflammation. APP synthesis from the hepatocytes is regulated by a number of cytokines released during the inflammation. Cytokines like IL-1 and IL-6 type act as additive, inhibitory, or synergistic regulators of APP expression. These prime cytokines operate both as a cascade and as a network. But, a cytokine like IL-1β substantially modifies IL-6-induced APP production. Sometimes IL-1β almost completely reduces the production of APPs like α(1)-antichymotrypsin, γ-fibrinogen, and α(2)-macroglobulin or it may upregulate the production of CRP, SAA, and hepcidin. The entire switch-like regulation of IL-1β on IL-6 induces production of acute-phase protein through a complex intracellular signaling process. In response to the activation of IL-6 and/or IL-1β, the NF-κB and STAT3-mediated signal transduction, their cross talk and the complex formation between STAT3 and the p65 subunit of NF-κB play an important role in controlling gene transcription of acute-phase genes (Bode et al. 2012).

Cytokines initiate a signaling cascade culminating in synthesis of APP within 12–48 h of its stimulation (Gabay and Kushner 1999; Li et al. 1994). The cytokine-regulated and hormone-dependent hepatic production of APP results from the induction of a common transcription factor. Within hepatocytes a broad spectrum of genes code for secretary proteins during APR (Baumann and Gauldie 1994). The expression of genes coding for APP is regulated at the transcriptional level. Posttranslational events participate in alteration of APP. Posttranslational modifications of plasma proteins like glycosylation during inflammation include (1) altered pattern of oligosaccharide branching, (2) overall increase in the sialylation in orosomucoid (α_1-acid glycoprotein), and (3) decreased galactosylation in IgG (Gabay and Kushner 1999).

3.5 Hormonal Regulation of APR and Synthesis of APP

Synthesis of APP is mostly mediated by the pituitary–adrenal endocrinal activation. The pituitary–adrenal axis is stimulated by cytokines and also involves the production of the glucocorticoid group of hormones. Glucocorticoid hormones like adrenocorticotrophic hormone (ACTH) and hydrocortisone were produced at the early stage and cortisol at the later stage. IL-6 receptor-mediated expression on liver cells was enhanced by cortisol thereby stimulating IL-6-mediated synthesis of APPs. During APR, insulin-like growth factor I synthesis in downregulated. Cortisol also negatively regulates the synthesis and secretion of IL-6, IL-1, and TNF-α cytokines by inhibitory activity. Inflammation-associated cytokines stimulate directly the adrenal gland-related catecholamine and the production of corticotropin-releasing hormone and also simultaneously stimulated the production of corticotrophin. In some inflammatory disorders, the occurrence of hyponatremia can be explained by the IL-6-stimulated production of arginine vasopressin (Gabay and Kushner 1999).

3.6 Symptoms of APR

APR involves a series of complex reactions with both localized and systemic effects. A large variety of biochemical, physiologic, behavioral, and nutritional changes occurred during APR. As mentioned in the earlier paragraph, neurological changes are the production of different classes of hormones. Inhibitory or stimulated productions of hormones are coupled with marked induction of fever, solmonescence, anorexia, and lethargy. Leukocytosis, thrombocytosis, and anemia in some chronic disorders are the chief hematological changes observed during APR. The major metabolic changes are cachexia, osteoporosis, increased adipose tissue lipolysis and hepatic lipogenesis, and reduced gluconeogenesis and show reduction in lipoprotein lipase activity in muscle and adipose tissue, negative nitrogen balance, and loss of muscles. The hepatic changes of APR are hypoferremia, hypercupremia, and hypozincemia, increased metallothionein, heme oxygenase, plasma retinol, tissue inhibitor of metalloproteinase-1, manganese superoxide dismutase, inducible nitric oxide synthase and glutathione concentrations, decreased phosphoenol pyruvate carboxykinase activity, and also changes in non-protein plasma constituents. Apart from all these, increased synthesis of a large number of APP by the liver leads to increased plasma level of APP (Dinarello 1983, 1984, 1996, 1997; Gabay and Kushner 1999).

3.7 Local Inflammation and APR

Inflammation involves a biochemical reaction against infection and or tissue injury, initiated by the influx and activation of leukocytes and plasma proteins at the sites of inflammation. The pathogenic infectious organism initiates local inflammation in the body during infection (Gabay and Kushner 1999). Followed by a systemic response of the body collectively termed as APR (Gabay and Kushner 1999). In infection pathophysiology such reactions lead to elimination of infectious agents (Gabay and Kushner 1999).

APR response are manifested by (1) the induction of fever, anorexia, lethargy, solmonescence; (2) increased expression of ACTH and hydrocortisone hormones (Gabay and Kushner 1999) ; (3) leukocytosis; and (4) altered synthesis of proteins from the liver (Zheng et al. 1995). These proteins with altered concentration during inflammation are termed as APPs (Gabay and Kushner 1999). Bacterial components like peptidoglycans, lipoteichoic acid, exotoxins, lipoproteins, and glycolipids initiate the local inflammatory processes (Paul et al. 2008). After bacterial invasion, mucosal and skin cells produce molecules that control infections (Gabay and Kushner 1999) Mast cells store histamine, serotonin (Moshage 1997), preformed TNF, and various pro-inflammatory cytokines (Arnett and Viney 2010). These cells when exposed to bacterial products release these pro-inflammatory cytokines. This then leads to neutrophil influx in the inflammation sites (Abbas et al. 2012). IL-1, IL-6, and TNF-α are important cytokines that exert a behavioral, neuroendocrine, and metabolic effect (Gabay and Kushner 1999). Cancer and infection cause local infection and show development of APR and cytokine (Cavaillon and Duff 1999). APR may continue for a few days. In chronic or recurring inflammation, continuous APR may lead to damage of underlying tissues that leads to disease progression and further complications, like cardiovascular or protein deposition diseases such as reactive amyloidosis. The concentration gradient of released tissue products activates the vascular system and the cells that cause inflammation. These responses in turn lead to the production of pro-inflammatory cytokines and mediators that diffuse into circulation (Gabay and Kushner 1999).

3.8 Biomolecules and Pathways that Alter during Acute-Phase Response

Other than the cytokine-mediated rise of clinical symptoms, alteration in biomolecular concentration of macromolecules and a series of biochemical pathways occur during APR like (1) alteration of the APP concentration; (2) complement cascade activation; (3) increased ACTH and glucocorticoids; (4) decreased calcium, zinc, iron, vitamin A, and α-tocopherol (Moldawer and Copeland 1997; Gruys et al. 2005); and (5) altered biosynthetic pattern of the liver through its altered metabolism and gene regulation. Under normal circumstances, the liver synthesizes plasma proteins which are altered during the acute-phase response after an inflammatory stimulus, synthesizing APPs. APPs are also synthesized by monocytes, endothelial cells, fibroblasts, and adipocytes and are largely controlled by inflammatory mediators, cytokines (TNF-α, IL-1, IL-6, IL-11, IFN-γ, LIF, OSM, CNTF, TGF-β) and hormones (glucocorticoids) secreted by polymorphonuclear leukocytes, fibroblasts, endothelial cells, monocytes, and lymphocytes by regulating their transcription (Fig. 3.1a). Insulin and okadaic acid are recently reported to act as inhibitors of the cytokine regulation of APP synthesis.

An important feature of the APR remains in the fact that IL-1 and TNF-α stimulate the adrenal glands to synthesize glucocorticoids through the central nervous system which in turn enhances IL-1 and TNF-α-mediated induction of liver APP synthesis. Glucocorticoids also cause downregulation of IL-1 synthesis by macrophages, forming a negative-feedback loop between the nervous and immune system in order to control cytokine synthesis. Some APPs are induced also by CNTF (ciliary neurotrophic factor). Most of the increase or decrease in APR biosynthesis is through altered gene transcription.

3.9 Acute-Phase Proteins

During tissue injury, burns, acute infections, or chronic inflammation, the plasma concentrations of the APPs increase or decrease. After the onset of a systematic inflammatory reaction, the levels of APPs can either increase collectively termed as positive APPs or decrease collectively termed as negative APPs to several folds. CRP, fibrinogen, mannose-binding proteins (MBP), SAA, complement components, alpha 1-acid glycoprotein (AGP), etc. are the different APPs. Following an inflammatory stimulus, the biosynthetic profile of the liver is radically altered and affects APP synthesis by hepatocytes. Up to 50 % above-normal level induction of positive APPs over normal levels are observed, while decreases of up to 25 % are observed in the negative APP s during inflammatory disorders (Morley and Kushner 1982). It is important to emphasize that there is great variability in an acute-phase behavior from one species to another (Table 3.2). In general, increased synthesis of fibrinogen, AGP, and haptoglobin and decreased albumin synthesis are observed in all mammalian species. In contrast, there are sharp differences between the species in the acute-phase behavior of CRP, SAA, serum amyloid P component (SAP), and α_2-macroglobulin (Kushner 1988).

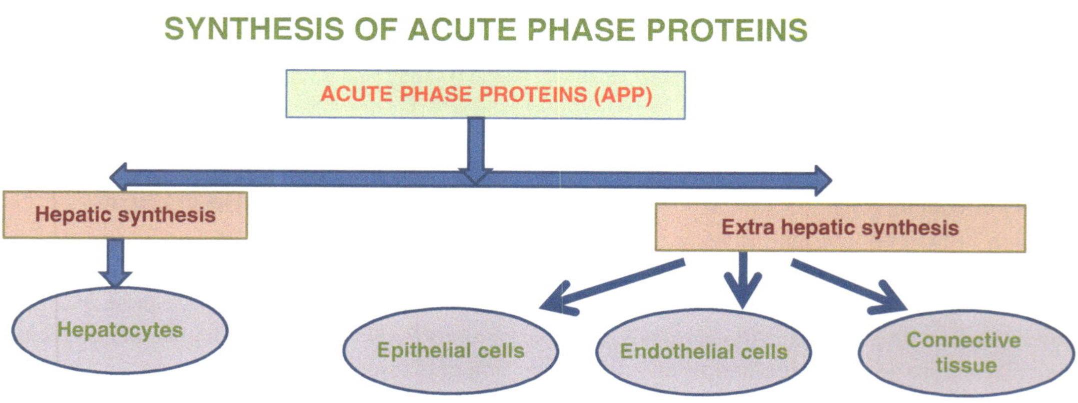

Fig. 3.1 Synthesis of acute-phase proteins [www.jpbsonline.org]

Table 3.2 Some species differences in acute-phase proteins (Kushner 1988)

Protein	Human	Rabbit	Mouse	Rat
α-1 Acidic glycoprotein	↑↑	Unknown	↑↑↑	↑↑↑
α-1 Proteinase inhibitor	↑↑	Unknown	Unknown	↑↑
α-1 Antichymotrypsin	↑↑	Unknown	↑	**Absent**
α-2-Macroglobulin	NO	↑↑	Unknown	↑↑
CRP	↑↑↑	↑↑↑	↑↑	↑
Ceruloplasmin	↑	Unknown	Unknown	↑
Fibrinogen	↑↑	↑↑	↑↑	↑↑
Haptoglobin	↑↑	↑↑	↑↑	↑↑
Hemopexin	↑	Unknown	↑	↑↑
Serum amyloid A	↑↑↑	↑↑↑	↑↑↑	**Absent**
Serum amyloid P	0	Unknown	↑↑	**0**
Transferrin	**Absent**	↑	**Absent**	Unknown

↑↑↑: elevation>=100-fold; ↑↑: elevation two- to fivefold; ↑: elevation<twofold; **NO**: no significant change

Most of these proteins exhibit the common characteristics, but a few of them showed some variations in their structure and functions. The oldest member by evolution in this group is CRP of an arachnid, *Limulus polyphemus* (the horseshoe crab), which differs from its vertebrate counterparts as being hexameric, and those three major acute-phase proteins (CRP, SAP, and HSAP) differ significantly in their calcium-binding and calcium-dependent ligand-binding properties (Srinivasan et al. 1994).

3.10 Pleiotropism of Acute-Phase Proteins

Most of the APPs showed multifunctionality. Different research data confirmed that APPs act on varied types of cells and play role in both the early and late stages of inflammation. CRP, an APP, acts on red blood cells, white blood cells, parasites like *Leishmania*, and other cell types (Ansar et al. 2006, 2009a, b; Das et al. 2003, 2004a, b). Interestingly, the concentrations of APPs and their functional effects are time dependent and affected by their molecular conformation (Das et al. 2003) (Fig. 3.1). Many APPs amplify the inflammatory responses when the infecting pathogen is residing within the host body and also downregulate the response after eradication of the pathogen. Thus, these APPs have a dual function (Figs. 3.2 and 3.3).

3.11 Different APPs

In many papers the role of different APPs was mentioned. Their concentration increases in different diseased conditions. Interestingly, their concentration also varies among species. A brief account of some of the APPs has been discussed here.

1. *C-reactive protein*

CRP is one of the positive APPs which are overexpressed in connective tissue disorders, inflammatory disorders, e.g., inflammatory arthritis, vasculitis, Crohn's disease, tissue injury, or necrosis, e.g., burns, necrosis, myocardial infarction, pulmonary embolus, neoplastic disease, bacterial infections, but reveal less elevation in viral infections. CRP is a better diagnostic marker than the erythrocyte sedimentation rate (ESR) for monitoring inflammation-induced rapid changes in the host physiology. CRP does not depend on different immunoglobulin or fibrinogen levels for its functionality. CRP unlike ESR is also not affected by the number or shape of RBC (erythrocytes). CRP is named after its

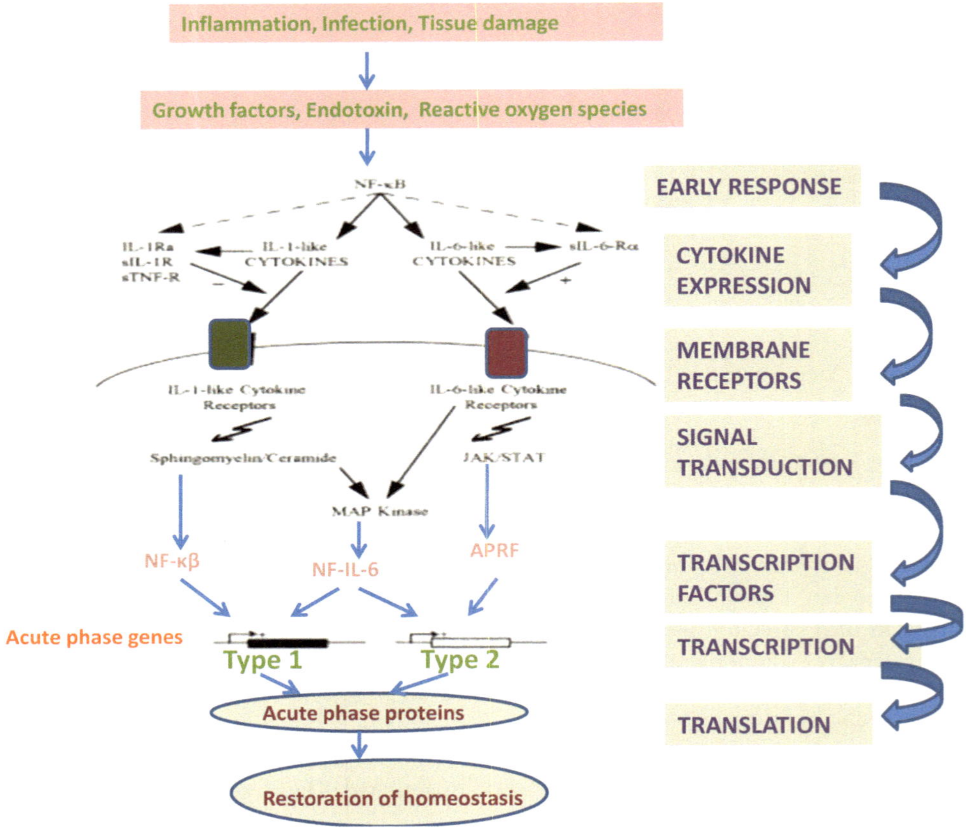

Fig. 3.2 Events that lead to acute-phase response [www.intechopen.com]. ((1) Initial activation of NF-kB, a transcription factor. ((2) Activation of cytokine expression. ((3) Cytokines on binding to their specific receptors initiate signaling events, activating downstream transcription factors. ((4) They then bind to their response elements in the acute-phase promoter region genes. ((5) Acute-phase genes are then transcribed leading to acute-phase protein secretion

ability to react with the C-polysaccharide of *Streptococcus pneumonia* (Ansar et al. 2006, 2009a, b; Das et al. 2003, 2004a, b). CRP has the property to bind to nuclear chromatin-basic histone–protein complexes and also to other cell types. Once bound, CRP is able to engulf the pathogen by phagocytosis or may lyse the pathogen by activating the classical complement pathway (Bandyopadhyay et al. 2004). If no complications develop, within a week of appropriate bacterial meningitis treatment, CRP concentrations characteristically return to its normal level. Serum and cerebrospinal fluid (CSF) concentrations of CRP and their monitoring may be important from a clinical and diagnostic point of view. CRP is a nonspecific inflammatory biomarker, and its clinical usefulness in diseases was therefore limited, especially in differential diagnosis. However, in the last few years, CRP has been used in monitoring different disease activities in inflammatory conditions like rheumatoid arthritis, cardiac disorders, infections, or malignancy. CRP is used as a prognostic marker for conditions like acute pancreatitis. It is known that CRP levels rise significantly during acute inflammation and therefore can be used for indicating the presence of significant infectious diseases or inflammatory conditions, especially in children. Low specificity of CRP may indicate its drawback as a biomarker of adult sepsis but finds importance in screening the early onset in neonatal sepsis. Interestingly, CRP being an acute-phase protein, its concentration rises negligibly to a minute detectable rise in autoimmune disor-

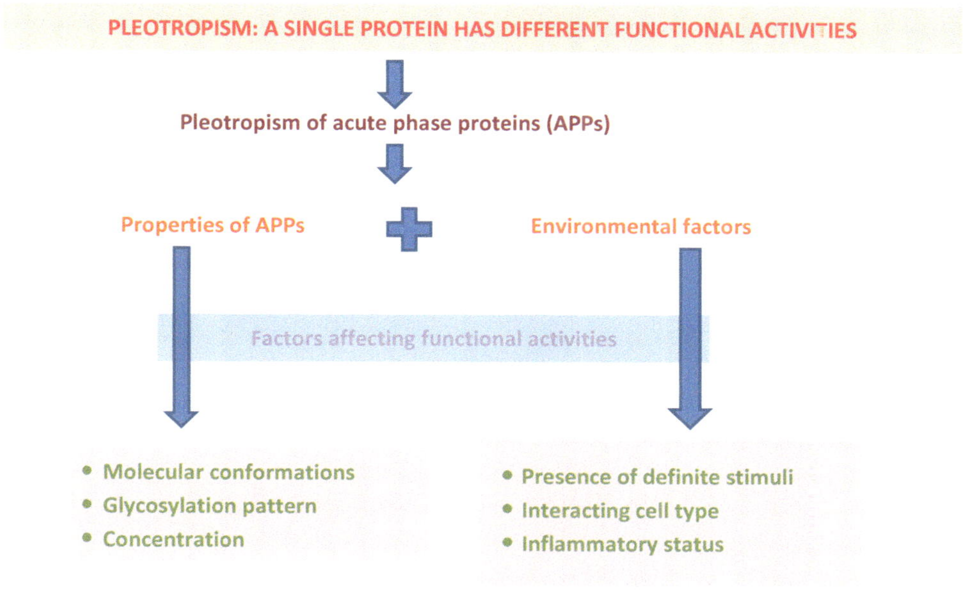

Fig. 3.3 Pleiotropism of acute phase proteins [www.intechopen.com]

ders systemic lupus erythematosus (SLE), osteoarthritis, anemia, leukemia, virus infection, polycythemia, pregnancy, ulcerative colitis, estrogen or steroid treatments. But researches showed that increased CRP concentration is an important risk factor for atherosclerotic cardiovascular disorders. In an acute coronary syndrome, high CRP levels correlate with a poor prognosis for patients. However, the role played by CRP in atherosclerosis is quite controversial. CRP with a predictive value of the development of type 2 diabetes, even after patient's body weight adjustment, is important in diabetes management. CRP also increases each symptom of the metabolic syndrome (Ansar and Ghosh 2013; Das et al. 2003) (Fig. 3.4).

2. *Ferritin*

Ferritin is an iron–protein complex found in most tissues but shows predominance in the bone marrow and reticuloendothelial system. Under normal physiological conditions, it acts as a primary iron-storage protein and its level indicates the patient's iron status. It is an APP and shows increased levels of expression during inflammation, malignancy, and liver disease conditions. Nearly 10 % of elevated plasma ferritin concentration in blood is due to iron overload. To test the iron stores of the body, it is not always appropriate to measure the increased concentrations of ferritin. The normal concentration values of ferritin ranges from 27 to 329 ng/ml in men and 9 to 125 ng/ml in women. However, during inflammation, it increases due to APR. The concentration of ferritin increases in cases of liver damage and malignancies. Nearly 90 % of the reported elevated serum ferritin in individuals are due to chronic alcohol consumption, metabolic syndrome, obesity, diabetes, malignancy, infection, and different inflammatory conditions (Doğanavşargil and Gümüşdiş 2003; Kilicarslan et al. 2013).

3. *Haptoglobin*

Haptoglobin (HGB) is an α-2 globulin, which has a normal physiological function of removing free plasma hemoglobin. The levels of haptoglobins are downregulated during any cause of hemolysis in blood. Haptoglobin is also an APP, the levels of which increase in malignancy especially in cases of bone secondary infection, inflammation, trauma, surgery, steroid or andro-

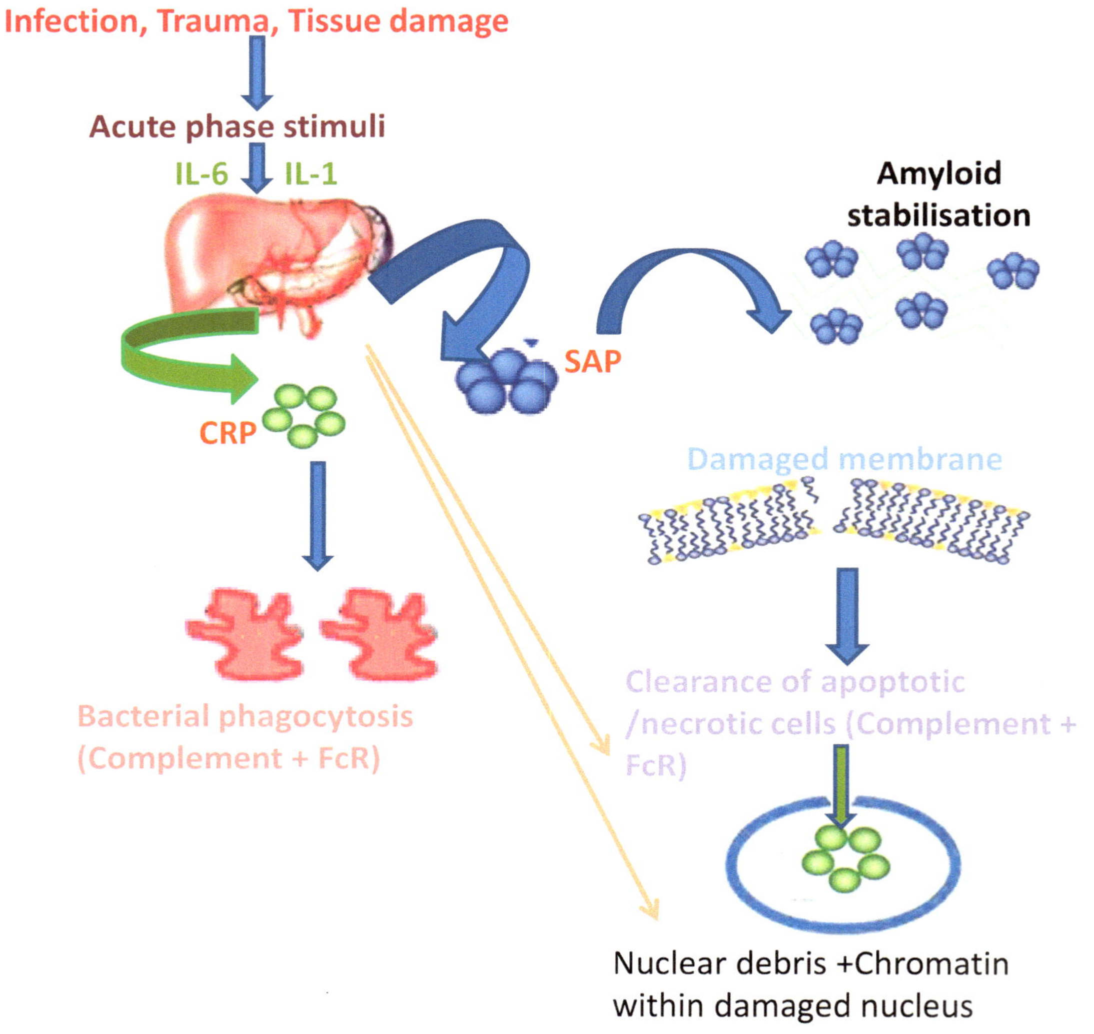

Fig. 3.4 CRP: Biological role [www.medscape.com]

gen therapy, and diabetes. Haptoglobin inhibits bacterial propagation in the host body by binding iron molecules, depriving bacteria of this element.

4. *Fibrinogen (FNG)*

Fibrinogen is an important APP having implicated in coagulation and wound healing. Higher concentration of fibrinogen revealed erythrocyte agglomeration and fast deposition and also increases the sedimentation velocity of erythrocytes. It takes part in the APR after tissue damage and inflammation. In circulatory system disorders or in heart diseases, high fibrinogen levels are also seen. High fibrinogen levels may be observed in the stomach, in the renal system and breast malignancies, in the exogenous use of estrogen and oral contraceptives, and in inflammatory diseases like rheumatoid arthritis. In a familial Mediterranean fever, the fibrinogen finds application as a disease marker. Low fibrinogen levels are observed in liver disorders, obstetric complications, and traumas, in response to blood transfusion, bone lesions, malnutrition, some bleeding disorders, and prostate and lung cancers. Additionally, lack or low levels of fibrinogen are the characteristic features of afibrinogenemia, hypofibrinogenemia, and dysfibrinogenemia like congenital diseases. Use of steroids, phenobarbital, androgens, urokinase, streptokinase, and valproic acid can lower the levels of fibrinogen in blood. The normal plasma fibrinogen level is nearly 200–400 mg/dl. The APR not initially reflected by the fibrinogen test. The disadvantages of the fibrinogen test are its unstable plasma when frozen and preserved plasma, delayed increase, long half-life, and continuous high levels even after inflammation (Doğanavşargil and Gümüşdiş 2003; Kilicarslan et al. 2013; Guzel et al. 2012; *Balci* et al. 2002; *Livneh* 2006; Cathcart et al. 1967).

5. *Conglutinin (CGT)*

CGT is a plasma protein. CGT binds with the C3b complement component in multiple combination sites and thereby activates the complement clearance of pathogens. CGT also contributes in agglomeration or co-agglutination of C3B opsonized pathogens or pathogen-derived debris, followed by their removal from the host circulation by phagocytosis (Tirziu 2009).

6. *Serum amyloid A (SAA)*

SAA has varied functions in the host immune system. Sometimes its functional behavior resembles that of a CRP molecule. SAA decreases IL-1 and TNF-α-induced fever, inhibits aggregation of thrombocytes, and also inhibits oxidative reaction in the neutrophils. Higher plasma concentration of SAA determined the amyloidosis condition. Amyloidosis is a disease characterized by deposition of this APP in various tissues of the body and in the fibrils; thus, it interferes with the normal functioning of the organs. Amyloidosis interferes normal functions like myocardial contraction and glomerular filtration (Tirziu 2009).

7. *α1-Acid glycoprotein (α1-AGP)*

α1-Acid glycoprotein suppresses antibody synthesis and lymphocyte blastogenesis response.

8. *β2-Macroglobulin (β2-M)*

β2-Macroglobulin neutralizes macrophages and neutrophil lysosomal hydrolases and thus is a proteases inhibitor. It behaves similarly to other antiproteases.

9. *Ceruloplasmin (CPL)*

Ceruloplasmin helps in eliminating neutrophil superoxides (Tirziu 2009).

3.12 Positive Acute-Phase Proteins

In innate immunity, positive acute-phase proteins serve different physiological functions. Some of these proteins inhibit or destroy microbial growth, while others contribute to a negative-feedback mechanism on the inflammatory response. It is known that the levels of elevated APP expression can differ widely with species. APP in one species may not be the same in other/related species.

Positive acute-phase proteins can increase to a maximum 1000-fold over normal levels. Three major groups of positive acute-phase proteins: (1) with two- to threefold, elevated haptoglobin, fibrinogen, α-globulins with antiprotease activity, and lipopolysaccharide-binding protein; (2) with an increase of about 50 %, ceruloplasmin and complement factor-3 (C3); and (3) with a rapid increase of up to fivefold to 1000-fold of CRP and SAA. These proteins differ in their physiological functions: (1) opsonization and trapping of microorganisms and their products, (2) negative feedback to inflammatory response, (3) activating complement, (4) binding cellular remnants like nuclear fractions, (5) neutralizing enzymes, (6) scavenging free hemoglobin and radicals, and (7) modulating the host's immune response. Examples of positive APP include CRP, SAA, mannose-binding protein (MBP), haptoglobin, complement components, ferritin, ceruloplasmin, α 2-macroglobulin, etc.

3.13 Negative APP

During inflammatory disorders, negative acute-phase proteins are downregulated by 25 % (Morley and Kushner 1982) to increase the capacity of the liver to synthesize the induced APRs, e.g., albumin, transferrin, retinol, binding protein, etc. Their reduction indicates a transient increase in free hormones bound to these proteins. The negative APPs are therefore also termed as "acute booster reactants" (Ingenbleek and Young 1994).

3.14 Species Variation in Acute-Phase Proteins

Some positive APPs show species specificity; serum amyloid P component (SAP) is an APP in mice, but not in men. CRP on the other hand reacts as APP in several monogastric species but is not very well in small ruminants (Gruys et al. 1994). Transferrin, in majority of the mammalian species, shows features as a negative APP while behaving as a positive APP in chickens (Hallquist and Klasing 1994; Tohjo et al. 1995, 1996).

The major positive acute-phase reactants in mammals are TNF-α, IL-1 and IL-6, cortisol, SAA, CRP, haptoglobin, α1-acid glycoprotein (AGP), fibrinogen, ceruloplasmin, and copper. But the major positive acute-phase reactants in birds are TNF-α, IL-1, IL-6, CRP, SAA, cortisol, hemopexin, AGP, fibrinogen, ceruloplasmin, transferrin, copper, and calcium (Gruys et al. 2005).

The major negative acute-phase reactants in mammals are transthyretin (TTR), retinol binding protein (RBP), albumin, transferrin, iron, zinc, and calcium. The major negative acute-phase reactants in birds are haptoglobin, albumin, zinc, and unbound serum iron (Gruys et al. 2005).

3.15 Responses by Acute-Phase Proteins in Various Species

The natural defense mechanisms in animals in relation to APR are also studied in great relevance. The APR and the production of APPs are partially similar among related species and also some differences between species APP molecules and their functions are also seen. In a pathological condition, APPs are intervened like innate immunity factors as described by some authors. In general increased synthesis of fibrinogen, α_1-acid glycoprotein (AGP), and haptoglobin and decreased albumin synthesis are observed in all mammalian species. In contrast, there are sharp differences between the species in the acute-phase behavior of CRP, SAA, serum amyloid P component (SAP), and α_2-macroglobulin (Kushner 1988).

1. *Cattle*—CGT is one of the main APPs in a ruminant. In cattle, APP synthesis is regulated and initiated by IL-6, IL-1, and TNF. Haptoglobin, fibrinogen, SAA, ceruloplasmin, α1-antitrypsin, α1-antichemotrypsin, β2-macroglobulin, and α1-acid glycoprotein are the main plasma proteins whose concentrations increase during APR in cattle (Alsemgeest and Taverne 1993; Funke et al. 1997; Godson et al. 1995; Hirvonen et al. 1999; McNair et al. 1997; Tirziu 2009).

Haptoglobin is a paraclinical biomarker for different inflammatory process severity like acute inflammations, acute infections, and traumatic reticulitis in cattle. Increased concentration of haptoglobin is seen in defective nutrition, mastitis, and hepatic lipidosis; in experimental inflammation; and in dexamethasone treatment. Haptoglobin forms polymers in association with albumin and binds free hemoglobin (Alsemgeest and Taverne 1993; Funke et al. 1997; Godson et al. 1995; Hirvonen et al. 1999; McNair et al. 1997; Tirziu 2009).

In cattle, fibrinogen forms the fibrin precursor in coagulation processes. SAA in cattle is associated with serum high-density lipoproteins, and it is the fibrillar amyloid proteins precursor. Ceruloplasmin participates in oxygen radical elimination and can bind copper ions in the blood. The protease inhibitor functions in cattle are played by APPs like α1-antitrypsin, α1-antichemotrypsin, and β2-macroglobulin. The protein transport function is mainly carried by α1-acid glycoprotein. In cattle APR is characterized by fever, hypoalbuminemia, and decreased concentrations of zinc and iron molecules. During pneumonia, increased concentration of SAA in cows fed with endotoxin was also seen (Alsemgeest and Taverne 1993; Funke et al. 1997; Godson et al. 1995; Hirvonen et al. 1999; McNair et al. 1997; Tirziu 2009).

2. *Sheep*—the concentration of APPs like haptoglobin, fibrinogen, and ceruloplasmin was increased during endotoxin shock and various pulmonary diseases. In sheep dystocia prognosis, haptoglobin functions as a significant paraclinical biomarker. In infections, for auxiliary diagnosis, measuring of HGB is useful. In inflammatory diseases, the concentration of haptoglobin, fibrinogen, and ceruloplasmin decreases (Tirziu 2009; Haig et al. 1998; Pfeffer and Rogers 1993; Skinner and Roberts 1994).
3. *Goat*—in goat, during dystocia, unlike those with nonpregnant conditions and in the normal births, there is a significant increase of serum haptoglobin (Black et al. 2004; Tirziu 2009).
4. *Swine*–in sterile inflammations there is an increase in APP concentrations, like CRP, *α1-acid glycoprotein*, and haptoglobin. CRP and haptoglobin are the most useful biomarkers in swine for the diagnosis of inflammatory damages (Asai et al. 1999; Tirziu 2009; Burger et al. 1998).
5. *Horses*—during some pathological condition, there are significant changes in synthesis of many APPs. In diagnosis of horses, during inflammatory diseases, haptoglobin, CRP, and SAA concentration was measured (Cote et al. 1998; Tirziu 2009; Hulten et al. 1999).
6. *Dogs*—Dogs' CRP was described in comparison with human homologue in recent publications. The homology of dog and human CRP was established by different isolation and purification studies. In some diseases of dogs, CRP concentration was highly modified.
7. *Cat*—in feline infectious peritonitis, the synthesis of some APPs was highly increased. It was also observed that the activity of serum and ascitic fluid IL-6 is high in peritonitis (Goitsuka and Ohashi 1990; Tirziu 2009).
8. *Rat*—in inflammations of rats, the concentration of serum transferrin decreases. In rats, typical APR is induced by local administration of turpentine or complete Freud's adjuvant and also by systemic administration of LPS. APR can also be induced by glucocorticoids especially in associated with cytokine inoculation (Tirziu 2009; Loeffler et al. 1999; Wu et al. 2002).
9. *Birds*—SAA and transferrin are important APPs in birds. These APPs are important variables in diagnosis. The concentration of SAA and transferrin were measured in broilers inoculated with turpentines or infected by *Staphylococcus aureus* in comparison with un-induced control groups. During infection or inoculation, the concentration of transferrin doubled and the concentration of SAA also increases (Chamanza et al. 2000; Tirziu 2009).
10. *Fish*—CRP is the main APP in fishes and its role is established in many species. The role of CRP as APP in *Labeo rohita* was differential as exposed to metallic pollutants (Sinha and Mandal 1996; Sinha et al. 2001). The role played by CRP in fishes is macrophage activation, complement activation, and bacterial agglutination. CRP inhibits the development of some pathogen bacteria and helps in bacterial phagocytosis by using peritoneal exudate cells. CRP showed nonspecific cyto-

toxic action mostly targeted on myeloma cells (Tirziu 2009; Funke et al. 1997; Kodama and Arimitsu 1999).

3.16 Quantitation of APP Is the Resultant of Synthesis and Catabolism. Their Level Can Be Measured

1. Within 4–5 h after a single inflammatory stimulus, e.g., SAA and CRP.
2. From about 8 h onward, e.g., lipopolysaccharide-binding protein.
3. With a single stimulus the elevated level of these protein continues for a minimum 24 h and decreases after about 48 h. In cows, the endotoxin exposure keeps plasma SAA quantities to a plateau (Werling et al. 1996).
4. Permanent stimulation occurs during chronic infection, leading to elevated positive APP levels as compared to normal ones and finds application in diagnostic purposes.

The overexpressed serum levels of certain APPs are of diagnostic and prognostic relevance and enable distinction of inflammatory processes from functional disturbances with similar or identical clinical features. Under normal circumstances an APR is not observed with functional disturbances that are not the result of an inflammatory process, thereby allowing the differentiation between failure of function and organic disease. Some APRs are observed in chronic disorders like rheumatoid arthritis, but APRs are associated with malignant diseases and therefore their determination cannot be used for differential diagnoses. In many diseases, the APP synthesis rises with the degree and progression of the inflammatory processes.

3.17 Human APP

CRP, SAA, mannose-binding proteins, fibrinogen, complement components, alpha 1-acid glycoprotein (AGP), etc. APPs are known. CRP among them is the best known APP. In blood plasma, the CRP level can increase 1000-fold after infection, inflammation, surgery, trauma, and burns; whereas CRP concentration increases moderately with exercise and stress. Actually, the presence and intensity of inflammation is reflected by the marked rises in CRP concentrations. Another important APP is SAA used to detect and monitor inflammatory and infectious diseases. Ferritin, primarily an iron-storage protein, is also an APP and is often used to detect the iron status of an individual. Human APP may be divided into three groups based on the usual magnitude of the plasma changes observed (Table 3.3). Of at least 30 acute-phase proteins (Table 3.4), CRP and SAA are the two proteins whose concentrations arise more than 1000-fold and even up to 3000-fold within 48 h following injury (Morley and Kushner 1982).

Table 3.3 Classification of human acute-phase proteins[a] (Morley and Kushner 1982)

Group I (about 50 % increase)	Group II (about 2–4x increase)	Group III (up to 1000x increase)
(1) Ceruloplasmin (2) Complement components C3, C4	(1) α_1-AGP (2) α_1-Antitrypsin (3) Haptoglobin (4) Fibrinogen	(1) CRP (2) SAA

[a]Classification on the basis of usual change in plasma concentration

3.18 Factors Contributing to Alteration in the Levels of APP

Conditions leading to substantial changes in the plasma concentrations of APPs include burns, wounds, injury, infection, trauma, surgery, tissue infarction, various immunologically mediated inflammatory conditions, and advanced cancer. Moderate changes are observed due to strenuous exercise, heatstroke, and childbirth. Minor changes occur after psychological stress and several psychiatric disorders (Gabay and Kushner 1999). Advanced age is associated with increased levels of plasma APPs like CRP, SAA, and AGP in humans (Ballou et al. 1996) (Table 3.5).

Table 3.4 Acute-phase proteins in humans[a] (Gabay and Kushner 1999)

Positive acute-phase proteins		Negative acute-phase proteins
Complement system	C3, C4, C9, factor B, C1 inhibitor, C4b-binding protein, mannose-binding lectin	Albumin Transferrin Transthyretin α_1-HS glycoprotein Alpha-fetoprotein Thyroxine-binding globulin Insulin-like growth factor I Factor XII
Coagulation and fibrinolytic system	Fibrinogen, plasminogen, tissue plasminogen-activator, urokinase, protein S, vitronectin, plasminogen-activator inhibitor 1	
Antiproteases	α_1-Protease inhibitor, α_1-antichymotrypsin, pancreatic secretory trypsin inhibitor, inter-α-trypsin inhibitors	
Transport proteins	Ceruloplasmin, haptoglobin, hemopexin	
Participants in inflammatory responses	Secreted phospholipase A_2, lipopolysaccharide-binding protein, interleukin-1-receptor antagonist, granulocyte colony-stimulating factor	
Others	CRP, SAA, α_1-AGP, fibronectin, ferritin, angiotensinogen	

[a]Information acquired/obtained from Gabay and Kushner 1999

Table 3.5 Expression levels of acute-phase proteins in various diseases (Pepys and Hirschfield 2003)

Elevation in APP		
High	Negligible	**Unaltered**
Bacterial infections	Myalgia	**Leukemia**
Fractures	Arthralgia	**Mixed connective tissue disease**
Juvenile chronic arthritis	Irritable colon	**Colitis ulcerosa**
Tumors	Back pain	**Scleroderma**
Surgery		**Osteoarthritis**
Polymyalgia rheumatica		**Systemic lupus erythematosus**
Crohn's disease		
Systemic vasculitis		
Burns		
Rheumatoid arthritis		

3.19 Function of Acute-Phase Responses

A number of wide-ranging biological functions have been assigned to acute-phase reactants. Besides CRP, the other major APP, SAA becomes one of the major apolipoproteins of HDL during inflammation, and its rapid binding to HDL may influence cholesterol metabolism during inflammation. It is also reported to have function in adhesion and lymphocytes and phagocytic cell chemotaxis, in inflammation, inside atherosclerotic coronary arteries and in oxidation of low-density lipoprotein (LDL) (Gabay and Kushner 1999).

APPs (1) play a role in host defense, (2) directly neutralize inflammatory agents, (3) minimize local tissue damage, and (4) participate in tissue repair and regeneration. The APR and the production of APPs are the most important effector mechanisms of the innate immune defense. Though the reaction or response is nonspecific, simultaneously it is an effective host mechanism. APRs are mobilized within a few minutes and provide essential tools for control of aggressiveness from the pathologic organisms. It controls the pathogen-induced damage till lymphocytes and other defensive cellular mechanisms start their action within 4–6 days. APPs not only support the adaptive immunity but they also are an important component of the innate immunity. APR is thus a very complex mechanism. To restore the homeostasis and to reduce the pathologic damage, APPs carry out a wide range of defensive and repairing functions to fulfill the same goal. The agonist, antagonist, and feedback loops always control this response. APR is optimized to counteract the pathogenic or inflammatory menace, and thus it enables to control and avoid detrimental effects.

Plasma concentration of many complement cascade components occur ultimately resulting in the local accumulation of macrophages, neutrophils, and plasma proteins. They lyse infectious agents, clear the foreign and host cellular debris, and repair tissue damage. Fibrinogen promotes healing of wound. Proteinase inhibitors cause neutralization of lysosomal proteases released after infiltration of activated neutrophils and macrophages, therefore controlling pro-inflammatory enzymatic activity. The increased plasma levels of some metal-binding proteins help prevent iron loss during infection and injury and also minimize the heme iron level required for bacterial uptake and potential scavenging action by damaging oxygen free radicals.

3.20 Downstream Effects of APP

APPs synthesized in the liver by cytokines that reach the inflammation site through blood and cause pathogens by opsonization and activating complement cascade. The alterations in the APP concentrations, both for positive APPs and negative APPs, are due to the changes in their synthesis by the liver. In addition they scavenge cellular remnants and free radicals and mediate neutralization of proteolytic enzymes (Gabay and Kushner 1999). Infection-induced inflammation takes place by the cooperative cascade of cytokines and leukocytes. Tumor necrosis factor, IL-1, andIL-6 play dominant roles as pro-inflammatory cytokines thereby mediating local inflammation reaction and activate other inflammatory cells like monocytes, macrophages, and neutrophils. Over fifteen cytokines of varying but low molecular weight are released by activated leukocytes and cause triggering of APR leading to symptoms of fever, leukocytosis, increased secretion of adrenocorticotropic hormones, and production of APPs.

1. *Pro-inflammatory and anti-inflammatory action*

APPs of the classical complement cascade (Table 3.4) have overall central pro-inflammatory roles in immunity. Mannose-binding lectin/protein (MBL/MBP) plays a role in complement initiated chemotaxis, plasma protein exudation, and opsonization. Granulocyte colony-stimulating factor (GCSF) leads to enhancement of the number of granulocytes to enhance the inflammatory response (Gabay and Kushner 1999). In contrast, α_1-protease inhibitor and α_1-antichymotrypsin antagonize the proteolytic activity of enzyme and inhibit the generation of superoxide anion thus exerting an anti-inflammatory action. Haptoglobin and hemopexin have shown antioxidant properties (Kilpatrick et al. 1992).

2. *Anti-infective role*

Many APPs are thought to be anti-infective. Human secretory phospholipase A_2 ($sPLA_2$) exhibits an anti-infective role through its antistaphylococcal and antistreptococcal enzymatic action (Munford 2001). MBP has antileishmanial potential through the antibody-independent activation of complement (Green et al. 1994).

3. *Procoagulant action*

A few APPs like fibrinogen, plasminogen-activator–inhibitor type-1 (PAI-1), and possibly CRP are known to possess procoagulant functions (Munford 2001). This action is required for the formation of abscess, for walling off invading microbes, and for delayed hypersensitivity reactions.

4. *Wound healing*

CRP, fibrinogen, and haptoglobin play some roles in the healing of wound. Fibrinogen causes endothelial cell adhesion, spreading, and proliferation, and haptoglobin stimulates angiogenesis (Cid et al. 1993).

The advantages, from decreased plasma concentration of negative APPs, are not clearly known. The availability of amino acid for positive APP production or loss of inhibitory effect of negative APP (like transthyretin) on proinflammatory substances (like IL-1) may explain this (Gabay and Kushner 1999).

5. *Role of other acute-phase phenomena*

The advantages of other acute-phase phenomena are biologically significant. Somnolence reduced demand for energy; fever enhances immunity and stabilizes cell membrane. Glucocorticoids maintain hemodynamic stability, modulate the immuno-inflammatory response (Gabay and Kushner 1999). Increased hepatic lipogenesis and increased lipolysis in adipocytes provide nutrients to cells and substrate for regeneration of membrane. Circulating lipoprotein can bind and decrease the toxicity of lipopolysaccharide (Hardardottir et al. 1994).

In conclusion, it can be said, the highly conserved adaptations that make up the APR serve important survival functions. Like all phenomena associated with inflammation, APR is not always beneficial. It can be fatal in severe response and can lead to various complications in persistent response.

3.21 APP and Diseases

Dysregulation in the APP synthesis and function is observed in diseases related to APPs. A key role in APP-related disorders are due to defects in the CRP-mediated complement activation especially in disorders like cardiac infarction. Elevated SAA levels are seen in chronic arthritis and tuberculosis. Other APPs show more moderate rise, usually less than fivefold.

Functions of different acute-phase proteins

Proteins	Important functions
Heme oxygenase	**Heme degradation**
Heparin cofactor-2	**Proteinase inhibitor**
Hepcidin	**Iron hemostasis**
Kallikreins	**Vascular permeability and dilatation**
LPS-binding protein	**Macrophage cell activation**
Manganese superoxide dismutase	**Copper–zinc binding protein Formation of reactive oxygen species**
Prothrombin	**Clotting formation of fibrin matrix for repair**
IL-1ra (IL-1 receptor antagonist)	**Receptor antagonist for IL-1**
Alpha-1 acid glycoprotein	**Interaction with collagen Promotion of growth of fibroblasts Binding of certain steroids**
Alpha-2 antiplasmin	**Modulation of coagulation cascade**
Antithrombin-3	**Modulation of coagulation cascade**
Apolipoprotein H (β-2-glycoprotein 1)	**Complement component**
B C1 inhibitor	**Negative control of complement cascade**
LPS-binding protein	**LPS binding to phagocytes via CD14**
Transcortin (**negative acute-phase protein)**	**Decreased binding of cortisol, upregulation of inflammation**

3.22 Acute-Phase Proteins and Their Clinical Significance

(I) *In humans*

The function of most APPs still remains unknown. Like all inflammation-associated events, the APR does not provide uniform benefit. Estimation of APP changes, even lacking diagnostic specificity, is of clinical importance as such changes are indicative of the occurrence and intensity of an inflammation reaction in process. APP determination can help in monitoring the health of individuals especially with several combined acute-phase variables in an index. An accu-

Functions of important positive APP

Acute-phase proteins		Immune system function
Proteinase inhibitors	Alpha 1-antitrypsin	Serpin, downregulates inflammation, neutralizes trypsin that has leaked from the digestive system, resolution of emphysema
	Alpha 1-antichymotrypsin	Serpin, downregulates inflammation
Coagulation factor	Fibrinogen, prothrombin, factor VIII, von Willebrand factor, plasminogen	Role in coagulation
Complement factors	C2, C4, C5, C9, C4 binding protein	Complement system
Metal-binding protein	Ceruloplasmin	Oxidizing iron, facilitating for ferritin, inhibiting microbe iron uptake, copper transport protein
	Haptoglobin	Binds hemoglobin, inhibits microbe iron uptake, hemoglobin scavenger, bacteriostatic effect (Delange et al. 1998), stimulation of angiogenesis (Cid et al. 1993), role in lipid metabolism, causes development of fatty liver in cattle (Katoh and Nakagawa 1999), immunomodulatory effect (Murata and Miyamoto 1993), inhibition of neutrophil respiratory burst activity
Serum protein	CRP	Acts as opsonin on microbes (Lippincott's Illustrated Reviews 2007), binds to membrane phosphorylcholine, activates complement, interacts with T and B lymphocytes, modulates monocytes and macrophages, produces cytokines, enables chromatin binding, prevents tissue migration by neutrophils
	D-dimer protein	Product of fibrin degradation
Serum lectin	Mannose-binding protein	Complements MBL pathway
Inhibitor of serum proteases	Alpha 2-macroglobulin	inhibitor of coagulation by inhibiting thrombin (Boer et al. 1993), inhibitor of fibrinolysis by inhibiting plasmin
Iron transport protein	Ferritin	Iron binding, inhibits microbe iron uptake
Serum protein	Serum amyloid A	Immune cell recruitment as inflammatory sites; enzymatic induction of degradation of extracellular matrix; cholesterol and HDL scavenger; cholesterol transport from dying cells to hepatocytes; inhibitory effect of (1) fever; (2) oxidative burst of neutrophilic granulocytes (Linke et al. 1991); (3) in vitro immune response, chemotaxis effect on monocytes, T lymphocytes, and polymorphonuclear leucocytes; induction of calcium mobilization by monocytes (Badolato et al. 1995); inhibition of platelet activation
Fibrinolytic system	Plasminogen	Proteolytic role, activates complement, clotting, causes fibrinolysis
	Plasminogen-activator inhibitor-1	protease inhibitor activity
Transport protein	Hemopexin	Heme binding and transport protein

rate combination of different variables which may differ with species results in a nutritional and acute-phase indicator (NAPI). The APR offers an effective mechanism to include future systems for assessing health in animals and human patients. The knowledge of every APP and their very complex expression regulatory pathway should be studied essentially for the development of safe and effective anti-inflammatory therapies in different disorders. In conditions like amyloidosis or cachexia and in subsequent failure of the defensive regulatory mechanisms, these anti-inflammatory therapies can be applied:

1. During wounds, injury, surgery, trauma, burns, tissue infarction, and various immunologically mediated inflammatory conditions and in advanced stages of cancer, substantial changes in the plasma APP concentration occur. Moderate changes occur even after strenuous exercise, heatstroke, and childbirth. Small changes are observed after psychological stress and in several psychiatric disorders (Maes et al. 1997).
2. During starvation or protein malnutrition, there is a general depression of positive hepatic protein synthesis, while starvation process itself reduces the negative acute-phase reactants.
3. Both inflammation-induced anemia and hypoalbuminemia are common among patients under hospitalization conditions.
4. Invasive infections such as dysentery, pneumonia, and some chronic infections induce positive APR.
5. During the APR, an increase in plasma viscosity due to total alteration in blood protein concentration is observed. Among such reactions, an increase in plasma fibrinogen is recorded that in turn influences the erythrocyte sedimentation rate (ESR) (Majno and Joris 1996). Fibrinogen acts as a slow-reacting positive acute-phase protein. Due to a possible delay few days after infection, the ESR level elevates and then reflects the activity of the APR.
6. In bacterial infections usually there is a strong systemic acute-phase response (Alsemgeest 1994; Alsemgeest et al. 1994).
7. In viral infections, the genera response is milder (Alsemgeest 1994; Höfner et al. 1994; Kimura et al. 1995; Nakayama et al. 1993), whereas in severe cellular destruction, a complete response is observed (van Reeth et al. 1998).
8. Haptoglobin, a positive APP, has AN anti-inflammatory effect. It strongly binds to hemoglobin, with decrease in quantity during massive erythrolysis (Smith and Roberts 1994).
9. The relationship between SAA and deposition of reactive amyloid (Gruys and Snel 1994) in patients suffering from tuberculosis, chronic arthritis, or familial Mediterranean fever has been well studied.
10. During anorexia and changed metabolism, positive APPs are formed during the APR.
11. The APR has negative effects like anemia and impaired growth (Gabay and Kushner 1999). Even in extreme conditions, cytokine-induced changes associated with the APR can be fatal, like in septic shock.
12. In persisting APR in diseases like advanced stages of cancer and the AIDS due to long-term stimulation, metabolic disturbances arise that ultimately result in cachexia.
13. Quantitative estimation of plasma or serum CRP can aid to differentiate inflammatory from noninflammatory conditions and finds importance in disease management as the concentration often reflects the response to and need for therapeutic intervention. In diseases, like rheumatoid arthritis, serial measurements of CRP are of prognostic importance. Patients with active SLE do not have high plasma CRP or SAA concentration but show marked increase after bacterial infection, in patients with active lupus serositis or chronic synovitis. Also, recent research has shown that slightly elevated CRP concentration might be an important marker in the increased risk of human cardiac diseases (Ledue and Rifai 2003). CRP has also been observed in human patients

during acute infections caused by acute lobar pneumonia, active rheumatic fever, and bacteremia caused by *colon bacillus* (Abernethy and Avery 1941).

14. Stunting in a population from Nepal was significantly associated with higher plasma levels of the acute-phase protein: α1-antichymotrypsin which in turn shows an inverse relation to the albumin level in plasma (Panter-Brick et al. 2001).
15. Phospholipase enzyme (PLA_2) was found at high concentrations in the synovial fluid of patients with rheumatoid arthritis and has been found to be associated with a variety of inflammatory diseases. Serum levels of PLA_2 can rise almost 1000-fold during severe sepsis, and it has been recognized as an APP under the transcriptional control of pro-inflammatory cytokine signaling. The antibacterial properties of this highly cationic extracellular protein is a component of innate immune response and can be readily explained by its ability to both bind and hydrolyze the anionic cell membranes of bacteria (Gabay and Kushner 1999).

Thus, the APR appears to constitute an array of changes enabling the survival of organisms and to regain normal function following both major and minor threats to their integrity. It also forms a biologic phenomenon of utmost importance and forms an interesting domain of research. At present, little is known about the acute-phase response elements, their components, recruitment, and synthesis, and their biological functions. Further researches on these lines will show more light.

3.23 APR in Different Animals

CRP, haptoglobin, SAA, and many other APPs are found to be useful for assessing health in different patients (Blackburn 1994; Counotte et al. 2002; Ferard et al. 2002; Ingenbleek and Carpentier 1985; Sipe 1995; Gruys et al. 2005) and in various domestic animals (Gruys et al. 1994, 2005; Petersen et al. 2004; Pyorala 2000; Toussaint 2000; Toussaint et al. 1997, 2000a, b) and also in wildlife animal species (Duffy et al. 1996; Funke et al. 1997). These abovementioned APPs are more sensitive than the ESR, which is used in most of the developed countries for treating human individuals. In veterinary clinical chemistry, the importance of APPs finds application as nonspecific variables for monitoring inflammatory activity. In ruminants, horse, pig, and several other animal species, the role of APPs and cytokines have been reported (Gruys et al. 1994, 2005). APPs are used in differential conditions. The role of APPs varies in different species. The role of APPs in mastitis of cows was mentioned (Eckersall et al. 2001). Mastitis is caused by different types of gram-positive and gram-negative bacteria in the mammary glands of lactating dairy cattle. APP is a potent biomarker in mastitis (Gabay and Kushner 1999). During mastitis, the APP concentration of haptoglobin, SAA, and α1 acid glycoprotein has been shown previously to induce significantly in the plasma (Eckersall et al. 2001). The role of various APPs in diseases like tropical theileriasis in cattle (Glass et al. 2003) and influenza in horses (Hulten et al. 1999) was mentioned. APPs find application to assess the status of an unhealthy animal versus healthy ones. But in most of the cases, the single reactants are often not that sensitive enough to detect a special animal subject in a population of livestock animals. Thus, experimentally, the values of positive APPs combined with those of rapid and slow negative APPs form an index known as acute-phase index (API). API can be used in animals to enhance the acute-phase signal/starvation situation obtained for an individual (Toussaint et al. 1995). Along with API, the nutritional acute-phase indicator (NAPI) (Gruys 2002) can also be used. NAPI is a prognostic inflammatory and nutritional index (PINI) used for humans (Bonnefoy et al. 1998; Ingenbleek and Bernstein 1999a, b), whereas

API is mostly used for cattle (Toussaint et al. 1995). NAPI and API together remarkably enhance the sensitivity and specificity in comparison to single usage of APPs as biomarkers in the detection of unhealthy subjects among a population of normal livestock (Toussaint et al. 1995; 2000a, b). These were experimentally proved in experimental pigs infected with *Streptococcus suis* (Toussaint et al. 2000b). In bone fracture patients the values of CRP/TTR was a favored simple quotient, and its usefulness in monitoring disease status has already been proved (Ferard et al. 2002; Gruys et al. 2005).

4 Inflammation and Inflammatory Diseases, Markers, and Mediators: Role of CRP in Some Inflammatory Diseases

Abstract

The knowledge of inflammation records dates back in first century AD. Initially discovered with features of rubor, tumor, calor, and dolor, scientific investigations have revealed chemical components, cells, and pathways involved in the process of inflammation. The body's initial defense in response to infection, trauma, or inflammation is through the acute-phase response (APR). APR is a multifaceted set of systemic reactions seen shortly after the experience of a triggering event. One of the many aspects of an APR is the increased hepatic synthesis of positive acute-phase proteins (APPs) leading to increased serum concentration of these proteins. The serum level of these APPs returns to base concentration when the stimulating factor is not anymore present. Today a plethora of inflammatory diseases are causing concern to global health. All the key players and mediators of inflammation change its role with the change in setup of disease and patients. The biomarkers of inflammation and inflammatory mediators are also used as therapeutic targets in under-trial animal models. Even in clinical diagnosis of an inflammatory patient, some broad-spectrum markers were analyzed without individual dissection of each mediator or biomarker. This chapter also provides a review of the acute-phase protein C-reactive protein and its possible use as inflammatory biomarker in diseases. We have highlighted case studies of some patients from Kolkata, India, revealing inflammation from disease together with their clinical history. The question which we probe in here is that whether there is a correlation with the clinical history, C-reactive protein, and inflammation and whether CRP can act as a unique diagnostic and prognostic biomarker in some diseases. The future course of this chapter lies in identifying clinical markers for inflammation, the sequential flow of inflammatory responses for a wide spectrum of diseases and its diagnostic, and therapeutic application to screen out pro-inflammatory diseases vs. anti-inflammatory conditions.

W. Ansar, S. Ghosh, *Biology of C Reactive Protein in Health and Disease*,
DOI 10.1007/978-81-322-2680-2_4

Keywords
Acute-phase response • Acute-phase proteins • Inflammation • Cytokine inflammatory mediators • Biomarkers, disease • Inflammation • Immunomodulation • Clinical diagnosis • Therapeutic treatment

Chapter Highlights

1. Inflammation initiated by infection, trauma, or injury takes place by the coacting cascade of various leukocytes and cytokines. To mediate local inflammation, pro-inflammatory cytokines like interleukin-1, tumor necrosis factor, and interleukin-6 play important roles in activating inflammatory cells like neutrophils, macrophages, and monocytes. Activated leukocytes secrete at least 15 different low-molecular-weight cytokines and also trigger acute-phase response through the manifestation of fever and leukocytosis, increased synthesis of adrenocorticotropic hormones, and production of various acute-phase proteins.
2. Acute-phase proteins are synthesized by the liver under the influence of some cytokines, which flow through the bloodstream, reach the site of inflammation, and remove the pathogens through opsonization and activating the complement pathways.
3. The changes in the serum concentrations of negative and positive acute-phase proteins are due to the altered synthesis from the liver. Three of the best renowned acute-phase proteins are C-reactive protein, serum amyloid A, and haptoglobin. Some disease statuses are actually correlated to the concentration of acute-phase proteins.
4. C-reactive protein activates complement pathways and has a major role in some forms of tissue alteration like in cardiac infarction.
5. Diagnostic routine tests are sometimes invasive. For the care of patients, the employment of noninvasive biomarkers is needed. The new biological markers are thereafter the serologic markers and acute-phase proteins.
6. Molecular biological tools have revolutionized the field of the biomarker. For the development of biomarkers, genomics and proteomics are used. To understand the pathophysiology of a disease, the most available technique is correlating serologic markers with clinical phenotypes and genotypes.

4.1 Introduction

Inflammation is a complex dynamic protective response to cell injury, infection *via* microbes, trauma, or toxins in the vascularized tissues. The causative agent is diluted, destroyed, or isolated and a sequential cascade of molecular events is set that leads to repairing, healing, and reconstituting the damaged tissue (Li et al. 2007; Iwalewa et al. 2007; Libby 2007; Delves et al. 2006; Goldsby et al. 2007; Guyton and Hall 2001). It is thoroughly characterized by the reaction in tissues and its microcirculation as clinically reflected by redness (erythema), heat (hyperemia), swelling (exudation), pain (through nerves and chemical mediators), and loss of function (pain). Pathologically, it takes place by vasoconstriction followed by vasodilation, hyperemia, stasis, accumulation of leukocytes, exudation of fluid, and finally deposition of fibrin. The combined vascular and cellular inflammatory responses are triggered by inflammatory stimulus and mediated through chemical factors derived from some cells or blood plasma. Even the injured or dead tissues release mediators (Li et al. 2007; Iwalewa et al. 2007; Libby 2007; Delves et al. 2006; Goldsby et al. 2007; Guyton and Hall 2001).

The etiologies for inflammation are varied ranging from microbial infections (infections by bacteria, virus, fungi, etc.), physical agents (like burns, stress, trauma from cuts, radiation), and chemicals (drugs, toxins, alcohol) to immunologic reactions (e.g., rheumatoid arthritis). Inflammation involves the influx of various cells of the host immune system and release of numerous mediators to the site of assault. Assembling and regulating inflammatory responses would be impossible without the concoction of controlled leukocyte population migration, various inflammatory mediators, inflammatory biomarkers (acute or systemic inflammatory marker), and subsequent physiologic changes that carry inflammatory responses (Figs. 4.1 and 4.2). In laboratories, animal models for inflammatory diseases are being developed to understand the process of inflammation, identify inflammatory mediators, and find out their probable role in therapeutics. Currently, there are no specific markers for inflammation; rather some broad-spectrum inflammatory markers were routinely investigated in hospitals. The question we asked is, since the biochemistry of inflammation is known, is it possible to identify some potential biomarkers of inflammation for safe pharmacological treatment which may in the future be helpful for routine hospital diagnosis?

4.2 History of Inflammation

The record of inflammation dates back to nearly first century AD by Celsus. Initially it was reported as a mechanism of tissue reaction or response to injury that gave rise to *rubor* (redness, due to hyperemia), *tumor* (swelling, due to greater permeability of the microvasculature and leakage of several proteins into the interstitial space), *calor* (heat, associated with the greater blood flow and the metabolic activity of the cellular inflammatory mediators), and *dolor* (pain, in part due to alterations in the perivasculature and related nerve endings). The fifth characteristic of inflammation, known as *functio laesa* (dysfunction of the organs involved), was revealed from the writings of Rudolf Virchow in the 1850s. Then, in the late nineteenth century, Elie Metchnikoff introduced the new concept of phagocytosis, a fundamental attribute of the innate immunity, after studying the engulfment of particulate matter by protozoa and also researching the ingestion of foreign bodies by blood leukocytes. Subsequently, Metchnikoff was awarded the Nobel Prize for Physiology or Medicine in 1908, along with Paul Ehrlich, for his work and discovery of humoral immunity, a component of adaptive immunity (Li et al. 2007; Iwalewa et al. 2007; Libby 2007).

4.3 Acute and Chronic Inflammation

Inflammation is mainly divided into two types—acute and chronic inflammation. If the inflammation winds up in less than 48 h, then it is acute inflammation (e.g., abscess) (Fig. 4.3), and if it rests for greater than 48 h (i.e., weeks, months, or years), then it is chronic inflammation.

Acute inflammation is initiated by mostly resident dendritic cells (DCs), macrophages, Kupffer cells, histiocytes, and mastocytes bearing pattern recognition receptors (PRRs) on their surface, which recognize/identify pathogen-associated molecular patterns (PAMPs) expressed exclusively on the outer surface of the pathogens. These immune cells after activation release inflammatory mediators, and subsequently the cardinal signs of *rubor*, *calor*, *tumor*, and *dolor* are visible. Vasodilation and increased permeability result in an exudation/seepage (edema) of fluid and plasma proteins into the tissue space and migration and extravasations of mainly neutrophils along a chemotactic incline created by the locally inhabited cells to reach the location of injury in the tissue. Inflammatory mediators increase the sensitivity of the tissues to pain (hyperalgesia), and a resultant neurological reflex

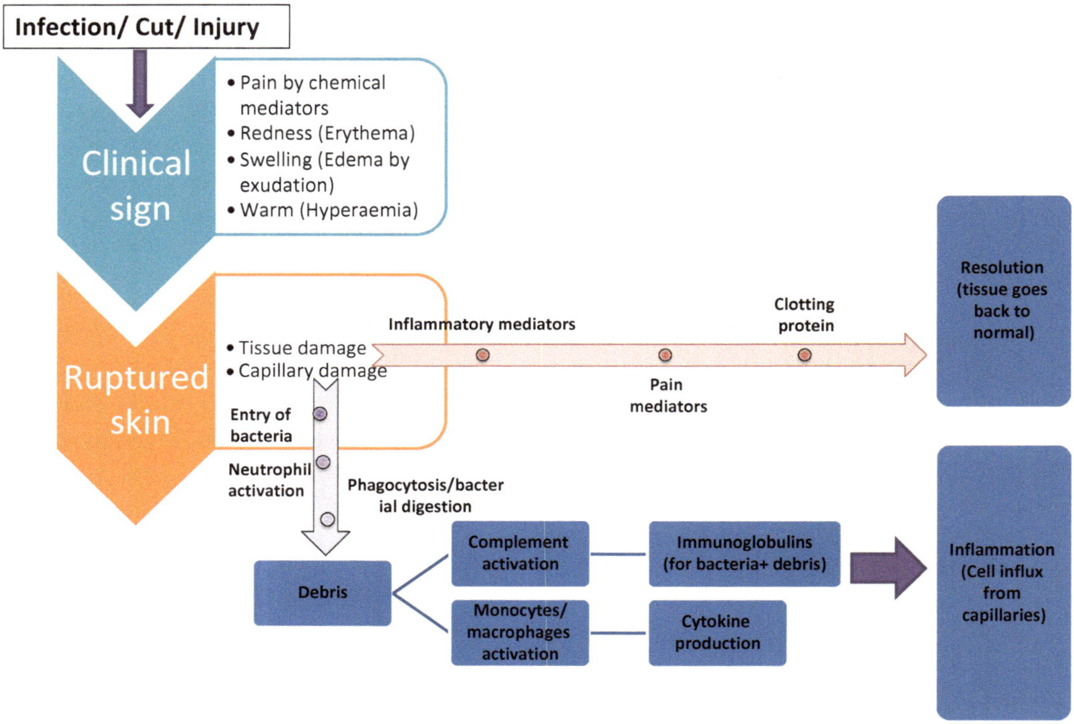

Fig. 4.1 *Mechanism of inflammation*. The process of inflammation started with physical assault of anatomical barrier with the common symptoms of pain, swelling, redness, and warmth. As the bacteria enters the host body, the activation of phagocytes, complement, and antibodies takes place in a cyclical way for both the pathogen and the debris also. At the same time, the tissue resolution goes on and further entry of pathogen is restricted

in response to pain leads to loss of function (*functio laesa*) (Fig. 4.1). A number of biochemical cascade systems, namely, the complement system (mostly activated by bacteria) and coagulation and fibrinolysis systems (mostly activated by necrosis), work in conjunction with inflammatory mediators to continue the inflammatory response (Fig. 4.2). However, the short half-life of inflammatory mediators is coterminous with the inflammation-activating signal.

Acute inflammatory outcomes like resolution, abscess, and ulcer (fistula, sinus) may lead to a chronic inflammation. An unresolved acute-phase inflammatory response leads to the development of chronic inflammation by persistent injury or infection (e.g., ulcer, tuberculosis (TB)) and prolonged exposure to toxic agents like silica and autoimmune diseases like rheumatoid arthritis (RA) and systemic lupus erythematosus (SLE). Chronic inflammation leads to tissue destruction, fibrosis, and necrosis. Neutrophils are the major cell types in acute inflammation while mononuclear cells (mostly lymphocytes, macrophages, and plasma cells) participate in chronic inflammation. The outcomes of acute inflammation are resolution, abscess, and ulcer (fistula, sinus) and may expand to a chronic inflammation. In chronic inflammation, the outcomes are tissue destruction, fibrosis, and necrosis (Fig. 4.4).

Acute inflammation is perpetuated by any immune cells previously present in the tissue. Acute inflammation is initiated mostly by the cells like DCs, macrophages, Kupffer cells, mastocytes, and histiocytes. These cells bear on their outer surfaces PRRs, which recognize the PAMPs expressed exclusively on the surface of the pathogens. These immune cells after activation with the pathogen releases inflammatory mediators and subsequently showed the cardinal signs

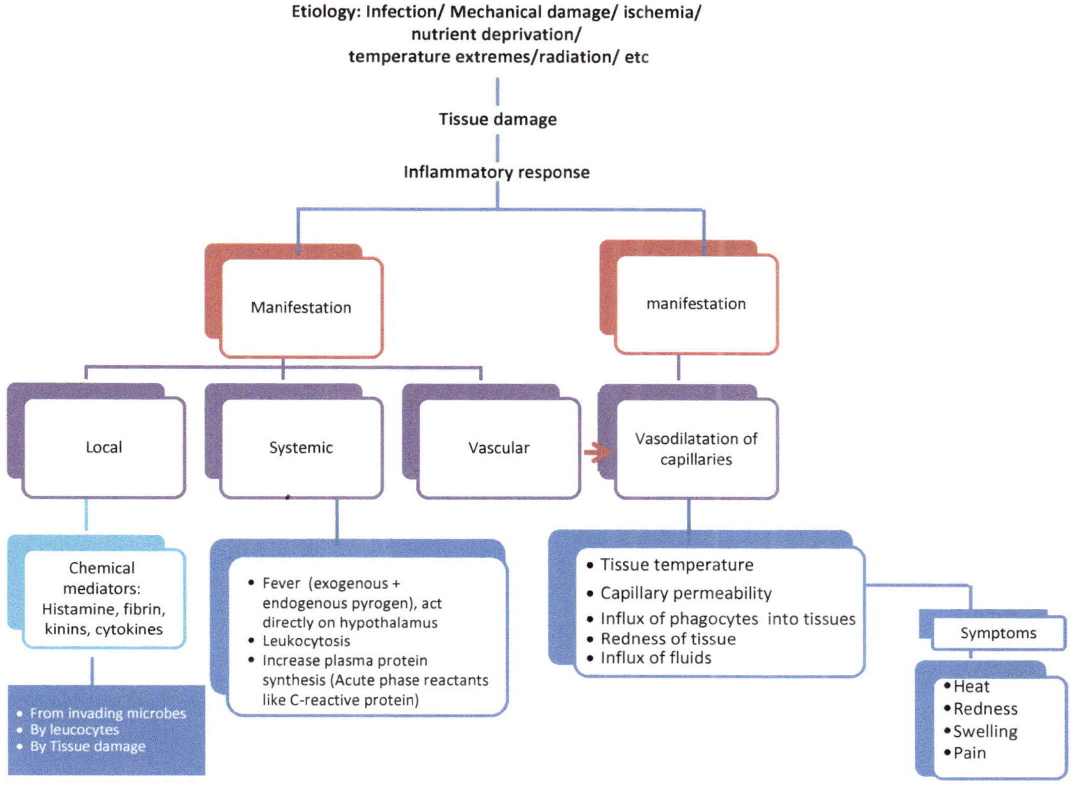

Fig. 4.2 *The mechanism of inflammatory response*. With the onset of the inflammation, the inflammatory response is manifested in local, systemic, and vascular mode. As a result, mediators are released, increasing vascularity and synthesis of plasma proteins, and finally the four symptoms of inflammation are seen. Cytokines provide the cord of connection between all these processes

(*rubor*, *calor*, *tumor*, and *dolor*) of inflammation. Vasodilation and increased permeability result in an exudation/leakage of fluid and subsequent edema. Release of plasma proteins into the tissue and migration and extravasations of leukocytes (mainly neutrophils) are the next steps of reactions. These reactions take place along a chemotactic flow generated by the locally inhabited cells to reach the locale of injury/assault in the tissue. Inflammatory mediators enhance the sensitivity of the affected tissues to pain (hyperalgesia), and a resultant neurological reflex in response to pain leads to loss of function (*functio laesa*) of the tissue (Figs. 4.1, 4.2, 4.3, 4.4, 4.5, and 4.6).

Additionally, to continue the inflammatory response with the help of the inflammatory mediators, a number of biochemical cascade systems are activated. These systems are the complement system (chiefly activated by bacteria), kinin system, and coagulation and fibrinolysis systems (chiefly activated by necrosis). Once the stimulus of acute inflammation is removed, the inflammatory mediators, having short half-lives, are disgraced in the inflamed tissue. Thus an acute inflammatory response requires the constant stimulation.

4.4 Mechanism of Inflammation

Inflammatory reactions involve a series of biochemical and cellular changes, the extent of which is associated with the spread of the initial trauma. The mechanism of inflammation takes place in four phases—vasodilation, exudation

Types of inflammation

Acute inflammation

Derived from plasma proteins

1. **Enzyme cascades/Plasma proteases**
   - Complement system
   - Coagulation system
   - Fibrinolytic system
   - Kinin system
2. **Lipids**
   - Eicosanoids (SM)
   - Platelet Activating Factor (SM)
3. **Others**
   - Histamine (SM)
   - Serotonin (SM)
   - Nitric oxide (SM)
   - Endotoxin

SM= Specific mediators
MSM= More Specific mediators

Chronic inflammation

1. **Cytokine Protein:** IL-1, IL-6, TNF-α(MSM)
2. **Complement protein**
   1. C5a:chemotaxis, phagocyte degranulation, stimulation of O_2^-
   2. C5a, C3a: mast cell and platelet degranulation.
   3. C5a, C5b-9: Enhancement of cytokine release, induction of eicosanoid synthesis.
   4. C3b: Potentiationof Abresponse, opsonisation of cells and lysis.
   5. C5b-9: Cell lysis.
3. **Coagulation system:** Coagulation Factor XII (Hageman Factor)
4. **Fibrinolytic system:** Plasmin
5. **Kinin system:** Bradykinin
6. **Eicosanoids:** Phospholipids
7. **Platelets Activating Factors**
8. **Histamine**
9. **Serotonin**
10. **Endotoxins**
11. **Nitric Oxide**
12. **Cytokines:** IL-1, TNF-α

Fig. 4.3 Types of inflammation

(edema), emigration of cells, and chemotaxis (Figs. 4.1, 4.2, 4.3, 4.4, 4.5, and 4.6). Inappropriate stimulation of inflammatory responses is the primary cause of many known diseases, and inflammatory reactions are, as a result, also an imperative target for drug development (Li et al. 2007; Iwalewa et al. 2007; Libby 2007; Delves et al. 2006; Goldsby et al. 2007; Guyton and Hall 2001).

The most prolific systemic expression of inflammation is a hike of body temperature and a number of biochemical changes known as the acute-phase reaction which steers to the production of acute-phase proteins from the liver. The local inflammatory reaction is portrayed by an early increase in blood flow to the locale of injury, increased vascular permeability, and the sequential and directional influx and careful accumulation of various effector cells from the peripheral blood flow at the place of injury. Influx of nonantigen-specific but prominent destructive cells (neutrophils) is one of the initial stages of the inflammatory response/reactions (Fig. 4.1). These cells mount a phagocytic response which is rapid but nonspecific. An exudation/leakage of plasma into the lesion in the initial stage is also seen (Li et al. 2007; Iwalewa et al. 2007; Libby 2007; Delves et al. 2006; Goldsby et al. 2007; Guyton and Hall 2001). At a subsequent stage, macrophages, monocytes, and different cells of varied lineages of lymphocytes (T and B cells of specific subsets) appear at the locale of injury. These cell types are related with more tightly regulated, antigen-specific immune responses/reactions, and once they are stimulated, they also produce several protective inflammatory molecules.

Inflammatory cells articulate greater numbers of cell adhesion molecules like glycoproteins and cell surface proteins. Endothelial cells are also stimulated during the early phase of the inflammatory response/reactions and thereafter express, among many other things, several adhesion molecule counter-receptors. The controlled expression of these molecules allows for the prominent trafficking of blood-circulating leukocytes to a locale of inflammation. Cellular attachment of some immune cells to the wall of the endothelial

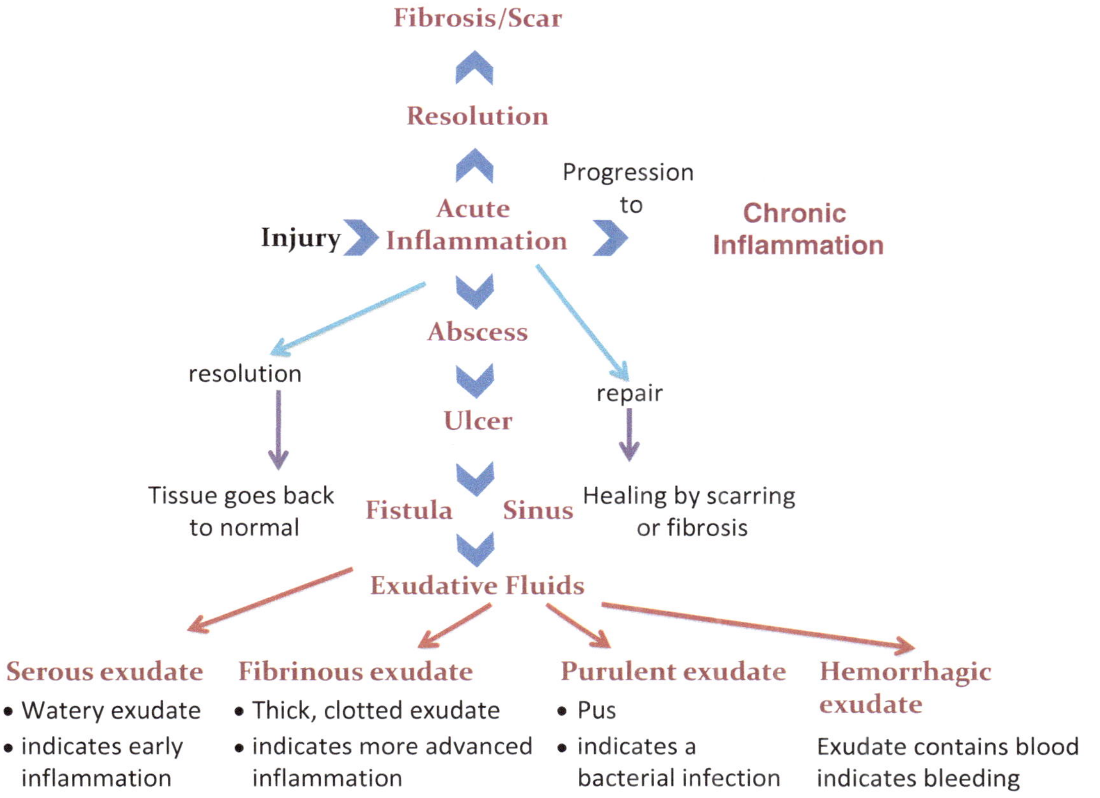

Fig. 4.4 *The outcomes of acute and chronic inflammation*. With the progression of the process of inflammation, the resolution and repair go hand in hand, and if these processes are disrupted, then it leads to exudation, pus, abscess, or its continuation as chronic inflammation

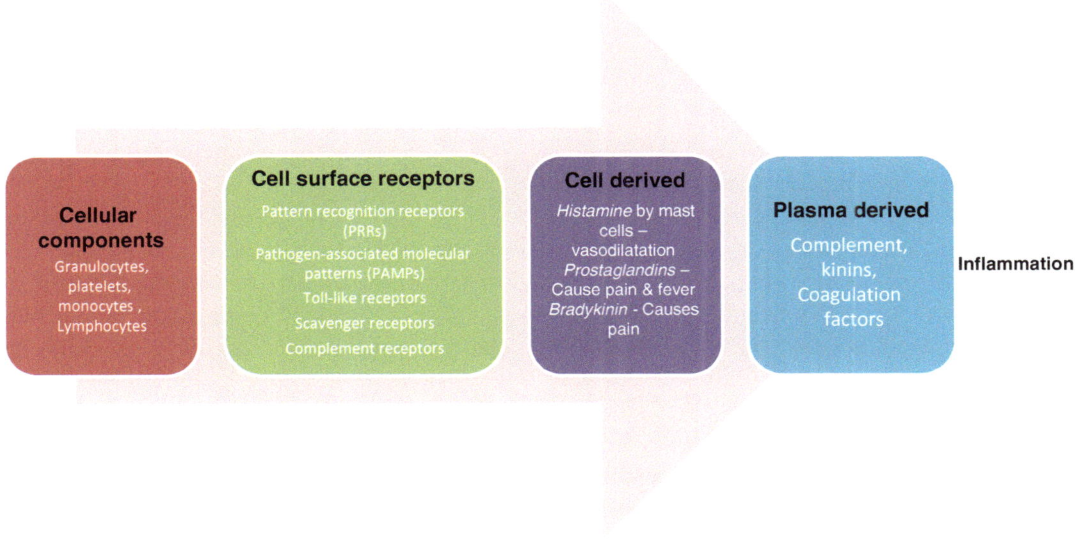

Fig. 4.5 *Components of inflammation*. The components of inflammation are different cells, cell surface receptors, and cell and plasma-derived components

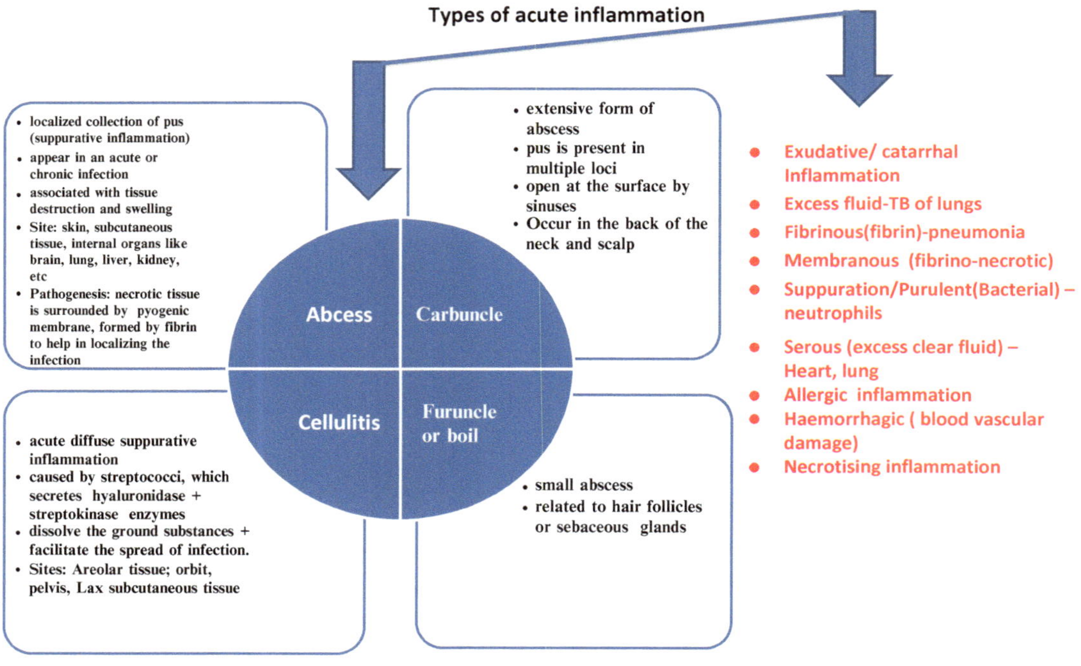

Fig. 4.6 *Types of acute inflammation*. There are different kinds of acute inflammation ranging from abscess, carbuncle, and boil to many other types

cells lining the blood vessels around the inflammatory site stops them from being swept away from the site of tissue damage or infection. This is a crucial stage required for the consequent emigration of these immune cells into the surrounding medium of inflammatory tissues (extravasation) (Li et al. 2007; Iwalewa et al. 2007; Libby 2007; Delves et al. 2006; Goldsby et al. 2007; Guyton and Hall 2001).

Once leukocytes have appeared at a locale of inflammation or infection, they release inflammatory mediators, which regulate the later accumulation and stimulation of other cells. However, in inflammatory responses initiated by the innate immune system, the final control is expressed by the antigen itself, in the identical way as it regulates the immune response itself. For this valid reason, the cellular accumulation at the locale of autoimmune reactions (antigen itself ultimately cannot be eradicated) or in chronic infection is quite distinct from that where the antigenic stimulus is rapidly cleared from the inflammatory sites (Li et al. 2007; Iwalewa et al. 2007; Libby 2007; Delves et al. 2006; Goldsby et al. 2007; Guyton and Hall 2001). In homeostasis and control of inflammation, four major plasma enzyme systems are there, which have an important role. This enzyme system includes the complement cascade, the clotting system, the plasmin (fibrinolytic) system, and the kinin system (Fig. 4.7, and 4.8).

4.4.1 Exudation of Leukocyte

The infiltration and accumulation of leukocytes past the blood vessels move through four pathways. These pathways are margination, rolling, and adhesion of leukocytes to the endothelium, diapedesis or transmigration of leukocytes across the endothelium, their migration toward a chemotactic stimulus released from the injured tissue, and finally phagocytosis.

The leukocyte starts expressing on its surface sialyl Lewis X-modified glycoproteins and integrin molecules in low-affinity state on its surface. Recruitment of leukocytes on the cell surface is through different receptors. Histamine, an inflam-

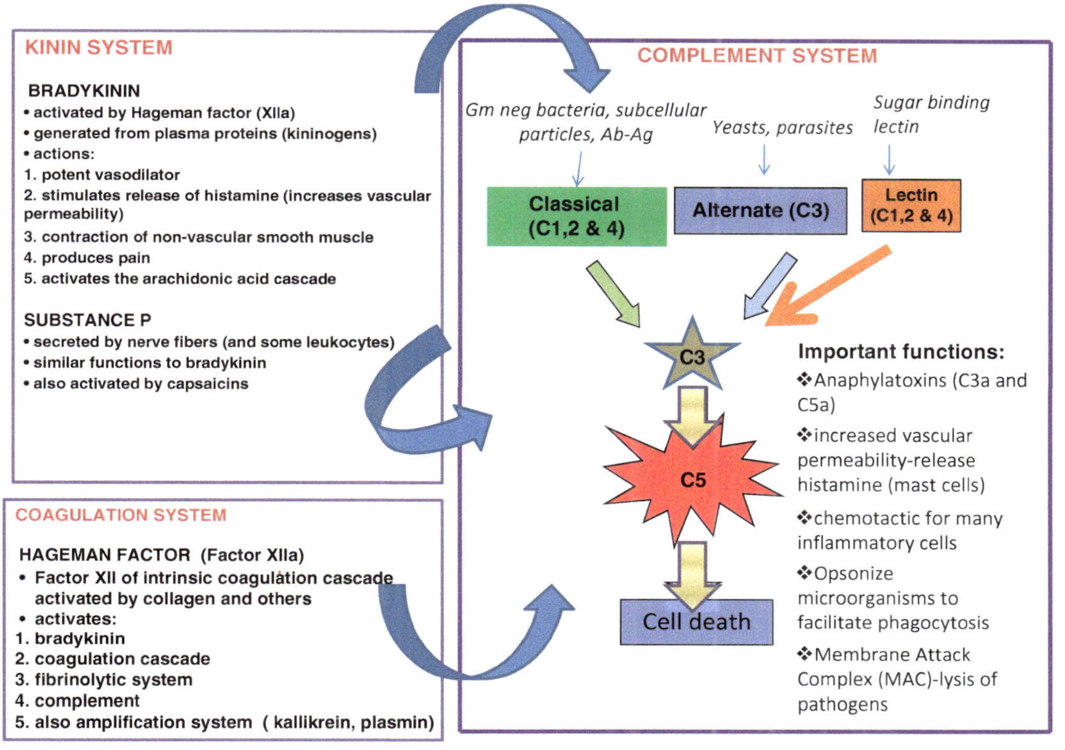

Fig. 4.7 *Relationship between kinin, clotting, fibrinolytic, and complement systems*. In inflammation, the complement, clotting, fibrinolytic, and kinin systems are activated in response to one another in an interlocked cascade

matory mediator, and cytokines liberated from injured cells promote the immediate expression of P-selectin and E-selectin of similar functions on endothelial outer cell surfaces. These receptors bind feebly to glycoprotein ligands expressed on leukocyte surfaces, and the leukocyte starts "rolling" along the endothelial surface. Other molecular expressions induced by cytokines on the outer surface of the cells are adhesion molecules, namely, vascular adhesion molecule-1 (VCAM-1) and ICAM-1 (immunoglobulin ligands), which further slow down the rolling leukocytes (Delves et al. 2006).

Leukocytes are weakly bound to these cell surface molecules and are free to detach if not properly stimulated by chemokines produced from the injured tissues. Chemokine activation starts expressing greatly the bound integrin receptors in a high-affinity state for various immunoglobulin ligands on the surface of an endothelial cell. This tightly binds the leukocytes to the endothelial surface. The chemokine gradients stimulate the adhered leukocytes to "migrate" (or transmigrate) over the endothelium and pass through the basement membrane into the tissue space through the process of diapedesis. The leukocytes start moving through the tissue through "chemotaxis." Leukocyte cells reach the tissue interstitial space and bind to extracellularly expressed matrix proteins (CD44 and integrins) to prevent their fall from the site. The flow of leukocytes toward the source of inflammation along a chemotactic gradient is mediated by these chemoattractants (Delves et al. 2006).

After reaching the locale of inflammation or infection, leukocytes released inflammatory mediators, which control the subsequent activation and later accumulation of several immune cells. Depending on the antigenic stimulus at the site of chronic infection, autoimmune reactions, or a shorter inflammatory reaction, the immune system is evoked and controlled by the antigen

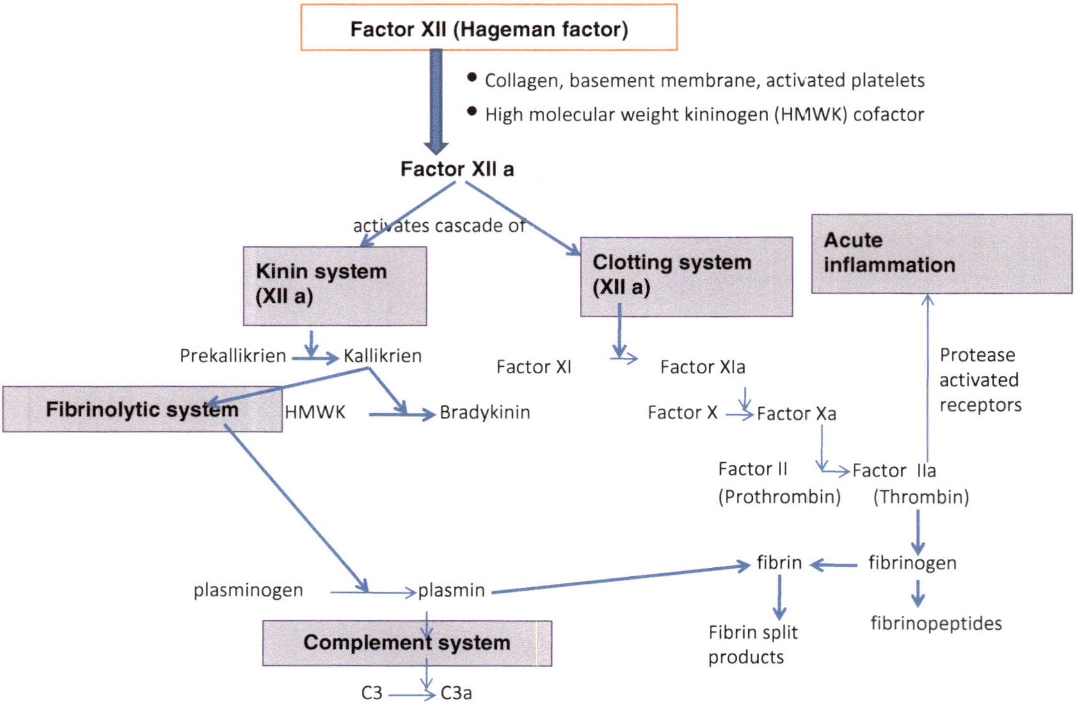

Fig. 4.8 *Three interrelated systems of inflammatory mediators*. The complement, coagulation, and kinin systems are the three interrelated and interconnected inflammatory pathways that act as mediators in both acute and chronic inflammation

itself. There are four major systems of plasma enzymes, which have a pivotal function in homeostasis and control of inflammation. These are the enzyme system of complement cascade, clotting proteins, the fibrinolytic (plasmin) system, and the kinin system (Delves et al. 2006).

Defect in leukocyte function leads to margination and adhesion of cells observed in leukocyte adhesion deficiency, whereas emigration toward a chemotactic stimulus like drugs and chemotaxis inhibitors and defective phagocytosis are observed in diseases like chronic granulomatous disease (CGD) or Chediak–Higashi syndrome.

4.4.2 Role of Lymphatics in Inflammation

Lymphatics are mainly responsible for draining the edema during inflammation. Edema is exemplified by a surplus of fluid accumulation in the interstitial tissue or serous cavities. Edema may be a transudate or exudate. A transudate is nothing but an ultrafiltrate of blood plasma and contains less plasma proteins (mostly albumin) where the permeability of endothelium is usually normal. In contrast, exudates are a filtrate of blood plasma with high plasma protein content, mixed with inflammatory cells and cellular debris; also the permeability of endothelium is usually altered. Pus is a purulent inflammatory exudate containing parenchymal cell debris and high content of leukocytes (mostly neutrophils) (Libby 2007; Delves et al. 2006).

4.5 Regulation of Inflammation

Immunoregulatory pro-/anti-inflammatory cytokines are the ones that accelerate/decelerate the process of inflammation and thus control inflammatory reactions either indirectly or directly and stimulate the production of cell adhesion molecules or several other cytokines in diverse cell types. A collection of cytokines are known as pro-/anti-inflammatory cytokines because they accelerate/decelerate the inflammatory responses either directly or indirectly by their ability to activate the synthesis of some cellular adhesion molecules or several cytokines in some definite cell types (Libby 2007; Delves et al. 2006).

Pro-inflammatory cytokines is a common term denoted for those immunoregulatory cytokines that favor inflammation. IL-1α, IL-1β, IL-6, and TNF-α are the major pro-inflammatory cytokines responsible for early responses. LIF, IFN-γ, OSM, CNTF, TGF-β, GM-CSF, IL-11, IL-12, IL-17, IL-18, IL-8, and a group of other chemokines that chemoattract several inflammatory cells are the other collections of pro-inflammatory mediators. These cytokines either behave as endogenous pyrogens (IL-6, IL-1, TNF-α), upregulate the production of pro-inflammatory cytokines and secondary mediators by both mesenchymal cells (including fibroblasts and epithelial and endothelial cells) and macrophages, activate the production of some acute-phase proteins, or attract diverse inflammatory cells. Anti-inflammatory cytokines actually counteract the synthesis of pro-inflammatory cytokines and also exert influences on several inflammatory responses in vivo (Libby 2007; Delves et al. 2006).

Anti-inflammatory cytokines is a broad-spectrum term for those immunoregulatory cytokines that actually neutralize various aspects of inflammation, namely, the cell stimulation or the synthesis of pro-inflammatory cytokines, and thus in addition also control the magnitude of the inflammatory responses/reactions in vivo. IL-4, IL-10, and IL-13 are the prominent anti-inflammatory cytokines. Soluble receptors for TNF or IL-6 as well as IL-16, IFN-α, TGF-β, IL-1ra, and G-CSF are some anti-inflammatory mediators. These mediators act chiefly by the reticence of the synthesis of pro-inflammatory cytokines or by neutralizing/balancing many biological mechanisms of pro-inflammatory mediators in several ways (Libby 2007; Delves et al. 2006).

The final effect of an inflammatory response is decided by the stability between anti-inflammatory (e.g., IL-4, IL-10, IL-13, IL-16, IFN-α, TGF-β, IL-1ra, G-CSF, soluble receptors for TNF or IL-6) and pro-inflammatory cytokines. It should be recorded that the general and clear-cut grouping of cytokines as either a pro-inflammatory or anti-inflammatory one may be quite misleading. The duration, type, pattern, and also the expansion of cellular activities activated by one definite cytokine can be regulated considerably by the property of the target cells, the surrounding microenvironment of a cell, the type of neighboring cells, the activation state and growth of the cells, different cytokine concentrations, the presence of other accumulating cytokines, and even the sequential flow of several cytokines procuring on the same cell (Li et al. 2007; Iwalewa et al. 2007; Libby 2007; Delves et al. 2006; Goldsby et al. 2007; Guyton and Hall 2001).

4.6 Mediators of Inflammation

At the inflammatory sites, different anaphylatoxins of the complement cascade, kinin proteins of the coagulation system, some prostaglandins, leukotrienes, and several other lipid mediators act in a confined manner at the site of infection and tissue damage and at more distant locale to disperse and support the inflammation process. Inflammatory mediators serve as muscle-active molecules, edema-promoting substances, chemotaxins, and cellular activators and activators of different kinds of effector cells engaged in the inflammatory response (Rang and Dale 1999; Libby 2007; Delves et al. 2006). The advanced and efficient process of cellular inflow to inflammatory sites is carried over by a plethora of mediator molecules dispersing and supporting inflammation. These inflammatory mediators are found in the tissue fluids or serum, are discharged by degranulating cells, and are also secreted by some inflammatory cells upon stimulation or triggered endothelial cells in blood vessels at the locale of inflammatory responses. They actually serve as edema-promoting substances, muscle-active molecules, chemotaxins, and cellular activators and inducers of all kinds of effector cells involved in the inflammatory response (Figs. 4.7, and 4.8) (Rang and Dale 1999; Li et al. 2007; Iwalewa et al. 2007; Libby 2007; Delves et al. 2006; Goldsby et al. 2007; Guyton and Hall 2001). Inflammatory mediators embrace some well-researched compounds like anaphylatoxins of the complement cascade, kinin proteins of the coagulation system, prostaglandins, leukotrienes, and many other related lipid mediators (Figs. 4.7, 4.8, and 4.9).

Another group of inflammatory mediators is neuropeptides like VIP (vasoactive intestinal peptide), tachykinins, and VPF (vascular permeability factor). These substances have vasodilatory and bronchoconstrictory activity with enhanced capillary permeability and also trigger increased secretion of mucus. At the early onset of inflammatory responses, at the locale of an injury, several cells contain mediators as preformed substances within their storage granules (like histamine) or may quickly switch to produce the inflammatory mediators as required to synthesized metabolites of arachidonic acid (Delves et al. 2006).

Inflammatory mediators are varied soluble and diffusible molecules that act nearby at the locale of infection and tissue damage and at more remote sites. Mediators may be an endogenous (like lipopolysaccharide of gram-negative bacteria) and exogenous (toxins and bacterial products) type. The host immune system of higher organisms can be activated by endotoxins, which simultaneously activate the Hageman factor, coagulation proteins, the complement pathway, and kinin and fibrinolytic cascades, eliciting specialized T-cell proliferation in response to superantigens (Fig. 4.1) (Libby 2007; Delves et al. 2006; Guyton and Hall 2001).

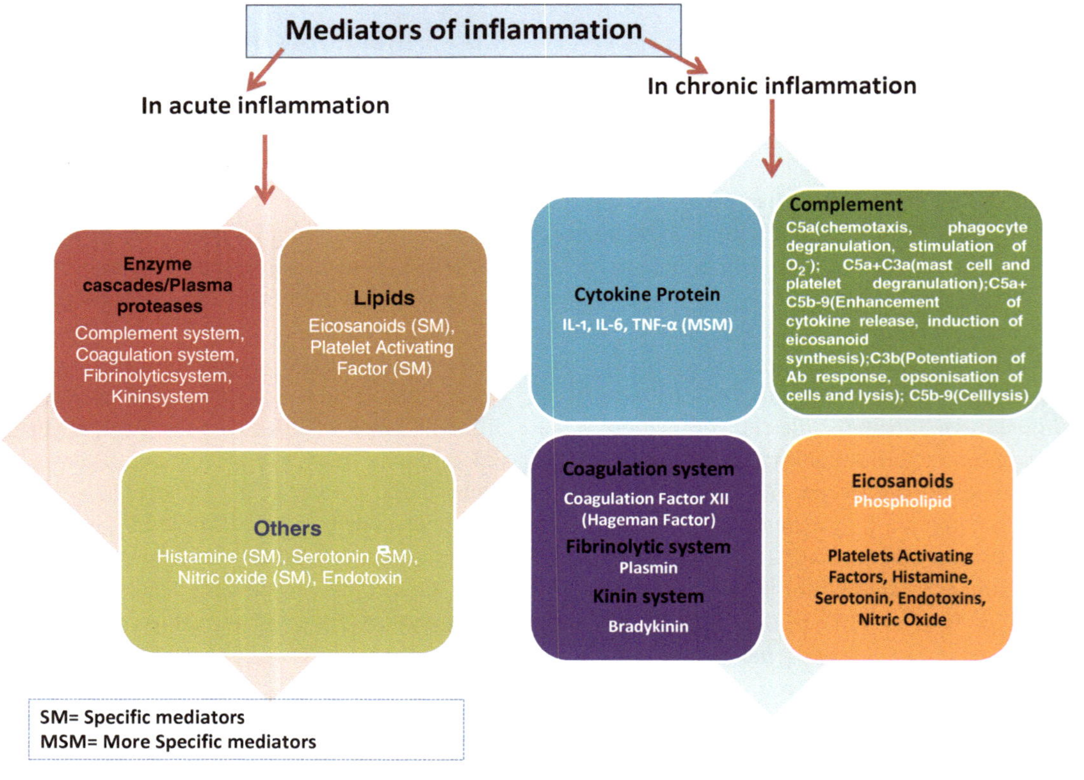

Fig. 4.9 *Types of mediators in acute and chronic inflammation*. Different subsets of mediators are released in acute or chronic inflammation. Some common mediators are present in both types of inflammation but they function differentially. *SM* specific mediators, *MSM* more specific mediators

Monocytes, macrophages, and neutrophils and a class of inflammatory mediators, controlled by cytokines, regulate different plasma enzyme systems to actively phagocytosed microbes. These cells possessing specialized granules have receptors for different complement components and also for Fc domains of immunoglobulins. NK cells and cytotoxic T lymphocytes also possess granules responsible for their cytotoxic function, involved in the adaptive and early inflammatory responses in concoction with innate immunity. Inflammatory mediators during inflammation are also discharged at the site of injury by a variety of cell types that either contain them as preformed molecules within their storage granules, like histamine, or which can quickly switch on the mechanism required to produce the mediators as and when required, for example, to produce different metabolites of arachidonic acid (Fig. 4.10) (Libby 2007; Delves et al. 2006; Guyton and Hall 2001).

Bacterial toxin products can act as exogenous mediators of inflammation. Notable among these is endotoxin or LPS of gram-negative bacteria. The immune system of higher organisms has probably evolved in a veritable sea of endotoxin, so it is perhaps not surprising that this substance evokes powerful responses. Endotoxin also activates the Hageman factor, leading to activation of the coagulation and fibrinolytic pathways as well as the kinin system. In addition, endotoxin elicits T-cell proliferation and has been described as a superantigen for T cells) (Libby 2007; Delves et al. 2006; Guyton and Hall 2001).

Endogenous mediators of inflammation are produced in both innate and adaptive immune systems. These mediators can be originated from molecules that are generally present in the plasma in an inactive form, like peptide fragments of some components of complement proteins (Figs. 4.7, and 4.8), coagulation factors, and

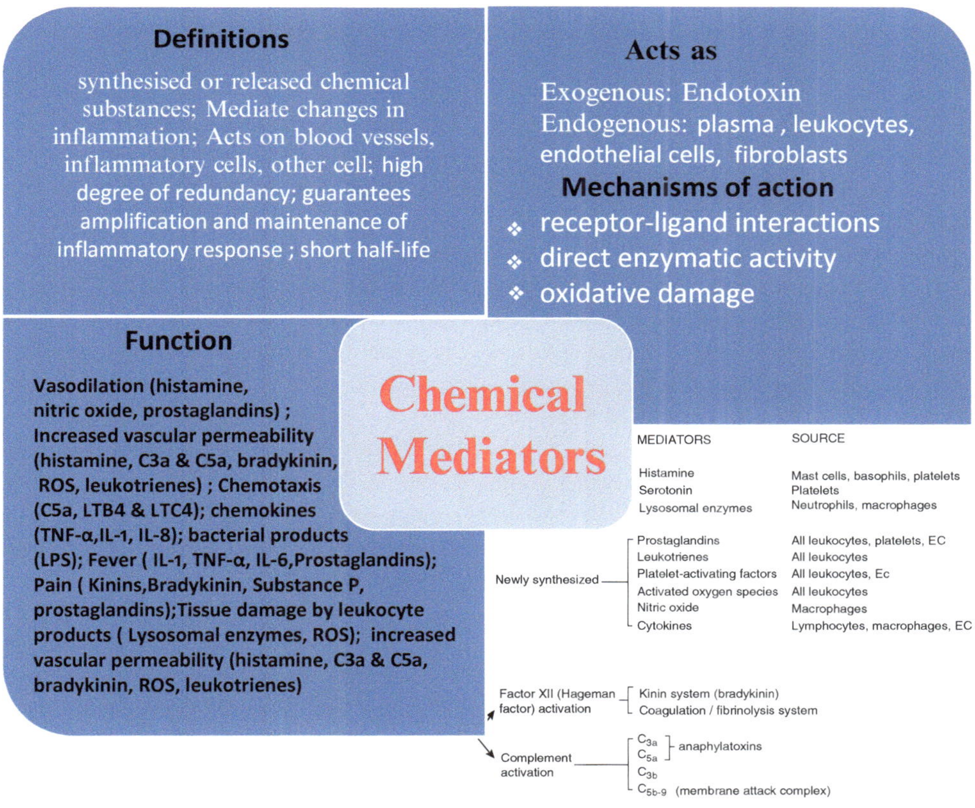

Fig. 4.10 *Chemical mediators*. The definition, types, function, and mode of action of chemical mediators are revealed

kinin systems (Bone et al. 1992; Libby 2007; Delves et al. 2006; Guyton and Hall 2001).

Mononuclear phagocytes like monocytes and macrophages play a central role in inflammation, as they produce many products which contribute in or regulate the different systems of plasma enzyme and hence control the mediators of different inflammatory responses. They are also highly phagocytic and are entailed in microbial killing, as are neutrophils. The role of these cell types is at least biased under the regulation of cytokines. All these inflammatory cells have receptors for complement proteins and for Fc domains of immunoglobulin molecules

They also possess specialized granules containing a diverse variety of products that are discharged perhaps by some general mechanisms. NK cells and cytotoxic T lymphocytes, in a wide-ranging way, also possess granules, which are significant for their cytotoxic role. In general, lymphocytes are engaged in the adaptive immune response to inflammation, and the initial events of inflammation are mediated in fraction by different molecules produced by cells of the innate immune system.

Early-phase mediators are produced generally by platelets and mast cells. They are especially significant in acute inflammation and consist mainly of serotonin, histamine, and other vasoactive substances; cytokines like IL-1, IL-6, and TNF-α; and chemoattractants like C5a. Platelets may contribute to inflammatory responses as a consequence of tissue injury, through a series of mechanisms involving: the release of permeability factors and other vasoactive amines; the release of coagulation factors; the release of lysosomal enzymes, which lead to generalized and localized fibrin deposition/accumulation; and the configuration of platelet aggregates or thrombi which causes the blocking of capillaries and vessels.

Late-phase mediators are accountable for the control of vascular events occurring later—from nearly 6–12 h after the triggering of inflammation. The later vascular events are mediated, at least in part, by various products of arachidonic acid (Bone et al. 1992; Libby 2007; Delves et al. 2006; Guyton and Hall 2001).

Mediator aggregation at local inflammatory series in skin blisters is somewhat dissimilar from the systemic effects after intravenous endotoxin. Under normal conditions, these flows of inflammatory reactions triggered by the mediators are sternly regulated. Failure to do so can direct to multiple organ failure which is known as systemic inflammatory response syndrome. Suitable inhibitors and inflammatory mediators are, therefore, of chief interest for ameliorating and modulating the effects of inflammatory responses and their sequelae (Bone et al. 1992; Libby 2007; Delves et al. 2006; Guyton and Hall 2001).

4.7 Systemic Inflammatory Response Syndrome

This syndrome is commonly seen as a systemic expression of diverse inflammatory mediators (oxygen free radicals and coagulation factors work in a juxtacrine fashion together with cytokines and related cytokine signals that generally function in an paracrine or autocrine way). Not only anti-inflammatory cytokines but also pro-inflammatory cytokines are increased in the blood and the abnormal condition has been termed as a "cytokine storm." Sepsis or septic shock syndrome is a systemic response to inflammation during an infection. It is a lethal, severe, and frequently hemodynamic breakdown caused by bacterial endotoxins during gram-negative septicemia. This toxic shock syndrome is mainly seen in younger women due to tampons contaminated with *Staphylococcus aureus* bacteria. The bacterial exotoxin, TSST-1 (toxic shock syndrome toxin-1; 23.1 kDa), triggers the synthesis of TNF and IL-1. In transgenic mice deficient in expression of CD28, the essential co-stimulatory signals of CD28 has been established in TSST-1-induced toxic shock syndrome. Chiefly due to tissue acidosis, hypoxia, and severe local modifications of metabolism, the characteristic symptoms are hypotension, insufficient tissue perfusion, and uncontrolled bleeding. The end stage of severe systemic inflammatory response syndrome is represented by multiple organ dysfunctions. The massive deterioration of homeostasis during sepsis, also known as disseminated intravascular coagulation, involves mostly the blood vessels along with fibrinolytic, complement, and blood coagulation processes, platelets, the absence or presence of inhibitors, and the kallikrein–kininogen system. Sepsis management according to symptoms is still one of the hard challenging problems faced by clinicians in intensive care patients (Bone et al. 1992; Libby 2007; Delves et al. 2006; Guyton and Hall 2001).

4.8 Inflammatory Mediators: "The Holy Grail"

In this chapter, we have tried to focus on the multiple mediators of inflammation which form a network of immune reactions. Mediators are chemical substances liberated by the endogenous trigger released from activated or injured cells during inflammatory response. They may be blocked indirectly or directly by inhibitors of inflammation. Mediators are concentrated in tissues where the observed symptoms or effects are visible. The action of mediators is the same in all species where the phenomenon occurs. It can be destroyed systemically or locally to avoid their undue concentration.

Nonimmune and immune chemical mediators are one of the many factors that control inflammation. There is long list with never-ending additions. There are mediators which suppress the inflammation whereas some others can stimulate the same. A plethora of mediators or a single mediator is responsible for most of the inflammatory signs and symptoms. By antagonizing the action or inhibiting the formation of these mediator(s), an end of treatment of most forms of inflammatory disease would be possible.

Humans and animals are continuously exposed to microbial pathogens, trauma, stress, and injury that can pose a considerable relevance in our daily livelihood. After the initiation of the etiological agent, the organism elicits a set of highly organized immunological, physiological, metabolic, and behavioral responses to represent its strategy to fight the infection. Inflammatory mediators generated at the site of assault in the peripheral tissues communicate with the brain to modify its immune response and upgrade as necessary which aid in its ability to fight and eliminate the source. The groups of mediators, pathways, and key responders/effectors during inflammation are varied and change their modus operandi in a different disease in a different organism.

As inflammation and its network of reactors form a juggling condition, thus to dissect all strings of its web, different animal models are developed in different diseases to mimic the original scenario. Inflammation is induced through varied stimulator molecules in these animals even for a single disease in a different setup to decode the mechanism of inflammation and bring out probable diagnostic and therapeutic agents.

These studies have also provided evidence of systemically generated inflammatory mediators versus local molecules both in chronic and acute inflammation. The phenomenon of inflammation is initiated by mononuclear phagocytes, which in turn synthesize diverse inflammatory mediators, the cascade of complement pathways, antibody production, and subsequent tissue repairing and resolution. Thus it traverses through both innate and adaptive immune systems. While these responses are part of our normal homeostatic mechanisms, it is quite clear that systemic inflammation has a detrimental effect in humans and also in animals.

Animal models have great inputs in our comprehending of the factors that regulate an inflammatory disease development, and they may also present valuable tools for budding novel therapeutic strategies. In inflammatory disease-induced models, chemical and cellular mediators are administered to induce acute or chronic diseases. It is useful for studying the involvement of multiple immune processes. Indeed, the diverse circulated published papers have researched on animal models revealing acute phase with mostly innate immune responses, but some are known as chronic pathologies. Thus a significant feature of these models is that the supervision of repetitive cycles of mediators leads to intestinal chronic inflammation, which allows significant revelation about the adaptive immune responses. This permits studies of inflammatory mediators and cellular inflow associated in the severity process of inflammatory disease.

The mediators of inflammation responsible for the pathobiology of an inflammatory disease are of an assorted nature, and diverse clinical studies have reported contrasting results, especially for their cell inflow and protein expression mechanisms. Such discrepancy in observations may at least in some part be explained by the definite methodologies used, varied time points at which analyses were done, and mice strain susceptibility and/or concentration (or dose) of the administered inducer. Moreover, to the superlative of our collective research knowledge, quite a few studies aimed to elucidate the kinetics of inflammatory cascades regulating acute and chronic phases of inflammatory disease.

In this framework, the ongoing chapter aimed to judge cellular inflow and inflammatory markers during the phases of acute and chronic inflammatory disease of humans induced in mice. This chapter showed very significant differences among the stimulation phase and the consequent recovery phase which may be useful for future modeling of the experimental inflammatory diseases. This may particularly provide sturdy evidence for the recognition of inflammatory biomarkers essential for the prediction of disease pathogenesis and the designing of possible therapeutic strategies.

It is quite evident from the clinical history from the hospital that no specific mediators were investigated during the routine investigations in the diagnosis process. Rather some broad-spectrum inflammatory markers (like CRP, ESR, procalcitonin, etc.) were diagnosed which will point toward the process of inflammation and not to the pathways of activation of mediators, whereas in the laboratory people are developing inflammatory animal models to understand all possible molecules involved in inflammation. Very soon in hospitals, definite inflammatory markers for each disease will be investigated for routine diagnosis.

Over the past years, much study is ongoing to highlight the significance of understanding the pathology of inflammatory disease for the procurement of safe and efficient pharmacological treatments. In this milieu, we highlighted a few of the immunological mechanisms that occur during acute and chronic periods of inflammation, with its remission/recovery period, in some of the widely used experimental model of a spectrum of diseases. These diseases are quite common ailments and are posing quite a nuisance not only to global health but also in proving obstructions in its socioeconomic developments.

During the initial recovery phase, the undertaking of the body resolves the inflammation upshot in a complete reduction of the some inflammatory mediators, while some inflammatory mediators did not even reach normal levels. Also after the repetitive phases of administration with inducer molecules, there is a greater elevation in intensity, and some different partners of inflammatory mediators were added to the sequential immune mechanisms. Although during the succeeding recovery phase, a new homeostatic mechanism of the body to reimburse and resolve inflammation was observed, but interestingly, the clinical factors were still persisting, revealing the chronic phase.

This present chapter revealed different inflammatory mediators, the cascade of inflammatory sequences, and cell inflow in the acute and chronic period of broad spectrum of diseases. The pro-inflammatory reactions were predominant in the stimulation phase; pro-resolution mediators were revealed observed during the recovery processes. Strikingly, some important mediators were still persisting even after the initial recovery phase and thus seemed to add to the magnified inflammatory reactions elicited in the subsequent induction phase. Also, our observations revealed a balance between anti- and pro-inflammatory mediators and a contrasting disparity between anti- and pro-inflammatory responses in the consequent recovery phases which dampen the course of resolution of inflammation with the repairing of tissue. Thus, this chapter contributes to improve comprehension of the inflammatory models and emphasizes the importance of understanding the fundamental mechanisms associated with the pathophysiology of experimental animals for development of appropriate therapeutic interferences in human inflammatory diseases.

What needs to be addressed in research in inflammation biology is that whether it is a cause of disease or the result of it or just a side effect of the disease. No clear idea exists as to whether inflammation actually causes chronic diseases or merely accompanies them. Further research needs to be conducted in the area of genes and molecules involved in causing inflammatory disorders thereby improving the diagnosis and treatment of such diseases. The question we pose here is that whether it is possible to screen the mediators of inflammation and their levels of expression in diseases associated with inflammation and correlate them with the clinical history. This would offer an advantage both from the point of view of diagnosis, therapy, and prognosis of inflammatory disorders. The future scope of this chapter remains in identifying such markers for inflammation and its correlation with disease condition.

4.9 Biomarkers of Inflammation

A biomarker is a parameter (chemical, physical, or biological) that can be applied to detect and compute the progress of disease or the results of treatment in preclinical research and clinical diagnosis. More significantly, a biomarker points out a change in state or expression of proteins, peptides, genes, and other factors that associate with the progression or risk of a disease, initial diagnosis, drug response, susceptibility of the patient to a given treatment, drug target identification, or disease intervention. The whole subset of biomarkers might be acknowledged using proteomics technologies, genomics, different imaging technologies, and invasive or noninvasive laboratory investigations using different physical parameters (Fig. 4.11) (Mayeux 2004; Kumar and Sarin 2009).

In laboratories, animal models for inflammatory diseases are being developed to understand

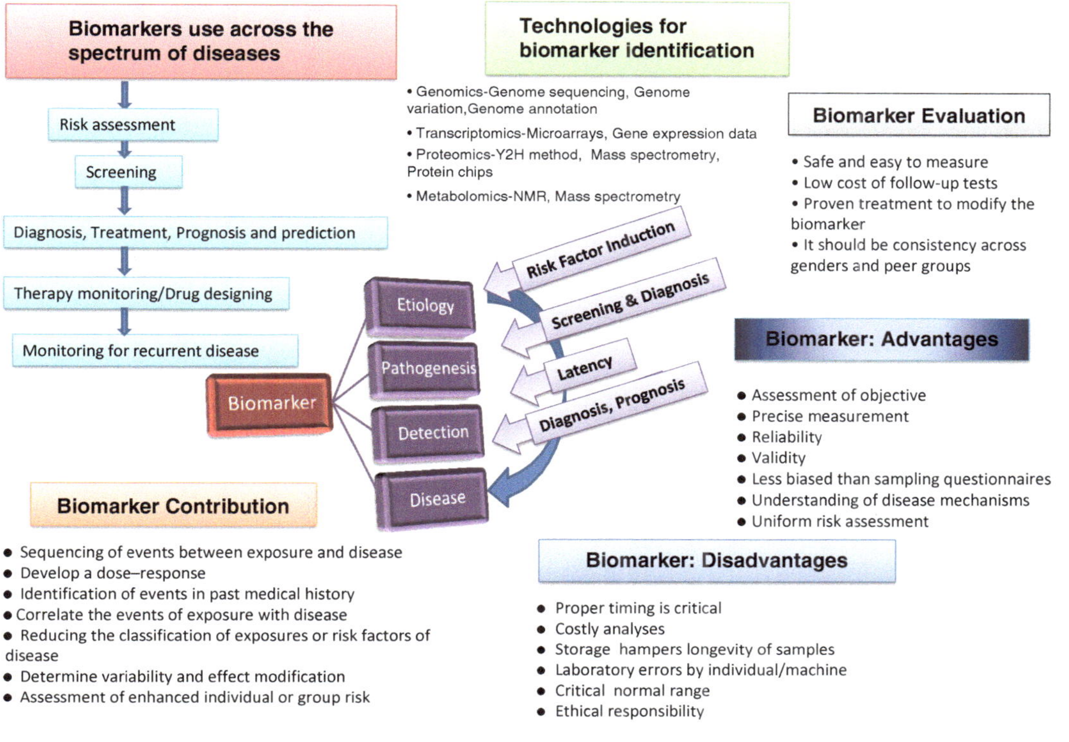

Fig. 4.11 *Disease pathogenesis and potential applications of biomarkers*. Different subsets of biomarkers are released from tissues and tumor or present in blood, urine, or other body fluids. The screening, diagnostic, prognostic, and therapeutic applications of biomarkers are presented with respect to disease pathology (Mayeux 2004; Kumar and Sarin 2009)

the process of inflammation, identify inflammatory biomarkers, and find out their probable role in therapeutics. Currently, there are no specific markers for inflammation; rather some broad-spectrum inflammatory markers were routinely investigated in hospitals. The question we asked is, since the biochemistry of inflammation is known, is it possible to identify some potential biomarkers of inflammation for safe assessment in diagnosis, treatment, and prognosis which may in the future be helpful for routine hospital diagnosis (Fig. 4.11)?

However, inflammatory biomarkers are often examined one by one; their interrelation and clear-cut aspects of their associated pathobiological mechanisms remain unclear. Explanation of these relationships could aid the suitable implementation of prognostic biomarkers in different clinical practices. Biomarkers in inflammation are gaining increasing interest given their clinical benefits. The most commonly used biomarkers are presented first, followed by a description of variable acute conditions with their relevant biomarkers. In addition to the conventional use of these biomarkers, other biomarkers are outlined in variable critical conditions that may be related to acute inflammation.

Biomarkers can be defined as any alterations in the constituents of body or tissue fluids. These markers provide a medium for uniform classification of a disease with its risk factors and can be extended in understanding the basic underlying pathophysiology of disease. Biomarkers provide a powerful and dynamic tool to grasp the spectrum of inflammatory diseases with usage in observational and analytic epidemiology, clinical trials in populations, and screening with diagnosis and prognosis. Biomarkers can also reflect the entire steps of a disease from the earliest symptoms/screening to the terminal stages. Analytical assessment of the validity of biomarkers is required to correlate with respect to the stage of disease. Variability in the measurement of biomarkers ranges from individual error in laboratory technicians, machine dysfunction, improper storage of body fluid, and other bias and confounding issues.

There has been a major sea change, in the past decade, in the way disease is diagnosed and investigated with the usage of high-throughput technologies, like proteomics, genomics, lipomics, metabolomics, microarrays, lab on a chip, etc. These advances have paved the way to the discovery of novel inflammatory disease biomarkers relating to cancers, autoimmune disorders, intestinal diseases, endocrine diseases, genetic disorders, neural damage, etc. In many cases, these developments have gone together with the discovery of biomarkers expressed via traditional or conventional methods, such as immune histopathology or clinical biochemistry. These inflammatory disease biomarkers together with microprocessor-based statistical data analysis, bioinformatics, and newer invasive/non-invasive analytical methods have been used to identify individuals with active disease versus refractory pathology versus those having distinguishing pathologies. Sometimes diagnosis relies on a cohort of biomarkers rather than a single disease biomarker. The techniques and methods of elucidating a biomarker are complex, and unfortunately an inflammatory disease biomarker cannot be readily transferable to other disease states. Thus there is a demand for a focused and comprehensive study in search of universal clinical biomarkers which is the urgent need in wide areas of cancer, nutrition, cardiology, endocrinology, immunology, addictions, birth defects, genetics, etc., to correlate with their therapeutic applications.

4.10 Inflammatory Mediators and Different Biomarkers in Different Subsets of Diseases and Related Immunomodulatory Therapies and/or Strategies

4.10.1 Intestinal Inflammation in Inflammatory Bowel Disease (IBD)

Inflammatory bowel disease (IBD) is a heterogeneous collection of inflammatory situations of the small intestine and colon which affects millions of people worldwide. These two diseases express familiar symptoms or some common extraintestinal complication of recurrent diarrhea, abdominal pain, rectal bleeding, anemia, weight loss, vomiting, and severe internal cramps or muscle spasms in the pelvis. Generally diagnosis mainly involves assessment of inflammatory markers in stool and colonoscopy and biopsy of pathological lesions. The two main forms of IBD are ulcerative colitis (UC) and Crohn's disease (CD) characterized by massive cell influx and release of varied pro-inflammatory mediators to the intestinal tissues. By contrastingly different research groups, different experimental animal models were developed where the mediators are administered to induce different forms of IBD in the animal.

In UC, the laboratory tests most used to measure the APPs in clinical practice are the serum concentration of CRP and the erythrocyte sedimentation rate. Other biomarkers of acute-phase response include platelet and leukocyte count and serum albumin and orosomucoid concentrations. There is a significant difference in the CRP response between UC and CD as reported in literatures. A clear increase in CRP is described in CD patients, whereas in UC the response is slight or absent. For these differences, there is no satisfactory explanation. Serum IL-6 levels were significantly elevated in patients with CD compared to UC patients and normal individuals. The most sensitive serologic markers of inflammation in adult population for detecting IBD are CRP as compared to other markers. The sensitivity of CRP is quite high which ranges from 70–100 % in the differential diagnosis between CD versus irritable bowel syndrome. To differentiate between CD and UC, the higher levels of CRP in active CD than in UC might be used. For differentiation between both types of IBD, the measurements of circulating levels of CRP and ESR and platelet count are not useful at all (Cioffi et al. 2015).

Inflammatory marker levels of CRP are associated with poor sleep quality, and clinical disease activity in IBD suggested a relation between circulating inflammatory markers and sleep. The key drivers of poor sleep quality are due to the common symptoms of IBD like diarrhea and abdominal pain (Wilson et al. 2015).

Dextran sodium sulfate (DSS) can induce both acute and chronic colitis in animal models. This induction with DSS was differentiated by severe disease activity, immense colonic polymorphonuclear penetration, and elevated levels of TNF-α, IL-17, VCAM-1, and keratinocyte-derived chemokine (CXCL1/KC). In the recovery phase of intestinal inflammation, marked elevation of IL-10, IL-4, TGF-β, and cyclooxygenase 2 (COX-2) as anti-inflammatory mediators was seen. In chronic experimental colitis, nuclear factor κβ (NFκβ) and regulatory T-cell marker forkhead box P3 (FoxP3) concentrations were elevated gradually representing immune disbalance in intestinal mucosal inflammation (Bento et al. 2012).

The worldwide food-borne bacterial enterocolitis infection is caused by a zoonotic pathogen *Campylobacter jejuni*. It forms an integral part of the commensal flora in many domestic and wild animals including broiler chickens which primarily transmit this pathogen to humans. Gnotobiotic IL-$10^{-/-}$ mice were generated as a suitable vertebrate model in the laboratory by quintuple antibiotic intervention to mice just after weaning to protect these animals from chronic colitis. Then oral infection of *C. jejuni* leads to colonization of the bacteria in the gastrointestinal tract and resultant acute enterocolitis within 7 days as observed by bloody diarrhea and significant histopathological changes of the colonic mucosa. The inflamed colon was immunopathologically characterized by greater numbers of B and T lymphocytes, regulatory T cells, and apoptotic cells as well as increased concentrations of IFN-γ, TNF-α, and MCP-1 mimicking severe phases of human campylobacteriosis as a perfect animal model in vivo. In control, animal models infected with another commensal bacterium, *E. coli* indicate no symptoms of the disease. The lipoproteins and lipooligosaccharides of *C. jejuni* are the inflammatory mediators that use the toll-like receptor (TLR-2 or TLR-4) signaling pathways as mice lacking these TLRs are abated from intestinal immunopathology (Haag et al. 2012).

Enteric glial cells (EGCs) have protective functions against pathogens and play an important role in the continuation of gut homeostasis to support intestinal inflammation. EGCs may be activated and proliferated in reaction to inflammation and injury undergoing reactive gliosis (enterogliosis) due to alterations in the homeostasis of the enteric nervous system. Enterogliosis is exemplified by the massive overexpression and secretion of distinct S100B protein like astroglia-derived signaling molecules. S100B is a highly diffusible; Ca^{++}/Zn^{++} and p53-binding protein coordinate different pro-inflammatory signals thus playing a vital role during intestinal inflammation. Pentamidine directly blocks S100B activity, as an antiprotozoal drug. Acute UC was induced in mice model by inflammatory mediator DSS (4 % DSS for 4 days) as compared to control mice group and colitis groups with pentamidine treatment (0.8 mg/kg and 4 mg/kg). The anti-inflammatory role of pentamidine was evaluated in colonic tissue by evaluating the disease activity index (diarrhea, blood in the feces, animal weight loss); histopathological severity; postmortem evaluation of cyclooxygenase-2, S100B, glial fibrillary acidic protein, and iNOS; p50 and p65 protein expression; phosphorylated-p38, MAP kinase, and myeloperoxidase activity; malondialdehyde synthesis; and macrophage influx in colonic tissues. Also plasma concentrations of NO and prostaglandin E2, IL-1β, TNF-α, and S100B levels were also identified in samples. Additional in vitro quantification was done on longitudinal muscle myenteric plexus (LMMP) preparations and dissected mucosa in LPS + DSS or exogenous S100B protein induction in the absence or presence of pentamidine. The effects of pentamidine intervention on UC induced by inflammatory mediators (DSS, LPS) were implicated in histological/biochemical evaluation and macroscopic observations in colonic tissues (Mishra et al. 2012; Esposito et al. 2012).

In DSS-induced acute colitis in experimental mice deficient in IL-17C, the intestinal inflammation is exaggerated in the presence of inflammatory mediator DSS. But these mice were protected against DSS-induced colon pathology. The colons of diseased IL-17C-deficient mice showed high numbers of $\gamma\delta^{+}$ and $CD4^{+}$ T cells. Mechanistically, IL-17C directly controls the expression of the occluding or tight junction molecule in colonic epithelial cells and thus helps in perpetuation of mucosal barrier integrity (Reynolds et al. 2012).

UC is less prevalent in current smokers as compared to ex-smokers and nonsmokers. Though there were reports of smokers having lowered rates of colectomy, hospitalization, and requirement for oral corticosteroids and immunosuppressant to control this disease, other potential active mediators in smoking may pose some clinical effects. Nicotine's application in therapeutic regime in ulcerative colitis is variable as compared to conventional medicines and placebo as it modifies inflammation and risk in ulcerative colitis. The adverse events of nicotine limit its clinical relevance (Lunney and Leong 2012).

In DSS-induced mice, UC was developed resulting in acute intestinal inflammation. In colon tissues, histological alterations (eosin and hematoxylin staining), neutrophil inflow (myeloperoxidase assay), levels of iNOS (immunohistochemical staining), and mRNA expression of pro-inflammatory mediators like TNF-α, IL-1β, IL-6, and IFN-γ by RT-PCR were evaluated. Aqueous extract of chaga mushroom (*Inonotus obliquus*) (IOAE) at treatment doses to UC mice indicated suppressed mucosal damage, edema, and the loss of crypts in histological examinations in colon tissues. IOAE acts as a suggestive anti-inflammatory agent at colorectal sites due to slowing down of the expression of inflammatory mediators and thus may be useful addendum in the setting of IBD (Mishra et al. 2012).

4.10.2 Inflammatory Airway Disorder

Asthma, or bronchial asthma (BA), affects millions of people worldwide. It is a chronic inflammatory disease of the airways. It causes periodic outburst of coughing, shortness of breath, wheezing, increased contractibility of surrounding smooth muscles, bouts of narrowing of the airway, and chest tightness.

BA is associated with the interplay of various inflammatory mediators, cell infiltration (T lymphocytes, eosinophils, and mastocytes), and the liberation of many membrane bound or co-stimulatory soluble molecules (CD40, CD40L, CD30, TNF receptor, and B7-H3). The concentrations of soluble OX40L (sOX40L), the vital co-stimulatory signal molecule, increases in asthmatic children and adults and also in many diseases. Pulmonary functions by spirometer and serum concentrations of sOX40L by ELISA were detected in acute asthmatic adult patients as compared to the control group. The patients were graded according to their disease severity into stable, severe, moderate, and mild asthmatic group. In adult asthmatic patients, the serum levels of sOX40L (6.80 ± 4.95 ng/L) were distinctly elevated than that of the control population, and they were negatively associated with pulmonary function indexes. Also, serum concentrations of sOX40L showed prominent differences among severe, moderate, and mild control groups, and its concentrations reduce to the same extent as the control population after therapeutic intervention of the adult asthmatic patients. Thus sOX40L level may act as a possible inflammatory mediator in the pathology of asthma (Lei et al. 2012).

In the diagnosis of asthma, the most important cells are eosinophils and sputum eosinophilia. Apart from local inflammation, systemic inflammation in asthma can be revealed by increased concentrations of CRP. By ultrasonic nebulizer, sputum was induced, and then peripheral venous blood samples were collected to calculate the concentration of CRP (by ELISA) and to count peripheral cells. Serum levels of high-sensitivity CRP (hs-CRP) were significantly elevated in the inhaled steroid nonuser group as well as user group as compared to healthy controls. Sputum and peripheral blood eosinophilia were seen in steroid nonusers as compared to the healthy individuals. In the steroid nonuser group, serum hs-CRP levels were positively associated with sputum eosinophilia, which was not statistically significant. The levels of hs-CRP did not get affected by age, sex, and atopy status in both asthmatic groups. Thus, serum hs-CRP being a biomarker can indirectly reflect the extent of airway inflammation (Halvani et al. 2012).

The transcription factor, NK-κβ, performs a crucial role in the pathogenesis of asthma. Dehydroxymethylepoxyquinomicin (DHMEQ), a compound that inhibits NF-κB activation, was reported to abate various inflammatory diseases in animal models. In ovalbumin-induced inflamed BALB/c mice, DHMEQ was administered before induction. DHMEQ significantly reduces concentrations of TH2 cytokines in bronchoalveolar lavage fluid and lowered eosinophilic airway inflammation, mucus production, peribronchial fibrosis, eotaxin-1, and the expression of α-smooth muscle actin. Thus DHMEQ inhibits allergic responses in airway inflammation; in murine models, airway remodeling of asthma may be used in therapeutic intervention (Shimizu et al. 2012).

Chronic obstructive pulmonary disease (COPD) is a very common lung disease with inflamed lung condition associated with mild to moderate breathing problem. The two main forms of COPD are chronic bronchitis (CB) or progressive pulmonary inflammation with a mucus-laden long-term cough and destruction of the lungs for a long time by emphysema. The symptoms of COPD are cough and breathlessness to an extreme of ischemic heart disease, stroke or death in severe cardiovascular complications. Actually in COPD and in other related chronic lung inflammation, atherosclerotic plaque formation and rupture of plaque lead to cardiovascular events (Man et al. 2012).

Chronic airway inflammation is a combinational indicator of several diseases like asthma, COPD, and cystic fibrosis. The symptoms occur after the failure of immune system to combat an acute inflammation spontaneously resulting in structural and functional alterations in the parenchyma and walls of the airways, uninterrupted influx of different inflammatory cells toward the locale of inflammation, and production of protein (like chemokines, cytokines, enzymes, etc.) and eicosanoids (pro-inflammatory mediators). Eicosanoid, an n-6 polyunsaturated fatty acid (PUFA), is mainly synthesized by the metabolism of arachidonic acid in the membrane phospholipids. Contrastingly, anti-inflammatory n-3 PUFA decreases the synthesis of pro-inflammatory cytokines and functions of immune cell, releases some anti-inflammatory pro-resolving mediators (protectins and resolvins), and may be used in airway disorders with an inflammatory constituent (Giudetti and Cagnazzo 2012).

hs-CRP indicates low grade of systemic inflammation in COPD. COPD associates positively with hs-CRP and smoking status and negatively related with body mass index. Apart from forced expiratory volume (FEV), the potential of CRP as biomarker to complement the present system of staging with the help of FEV can be expanded (Nillawar et al. 2012).

Acute lung injury (ALI) caused by any local or systemic inflammatory stimulus results in a dispersed heterogeneous lung injury symptomized by hypoxemia, low lung compliance, and noncardiogenic pulmonary edema with widespread capillary leakage. For treatment of ALI and pulmonary fibrosis, transplantation of bone marrow mesenchymal stem cells (MSCs) is one of the possible resorts. Experimentally, rats were exposed to cigarette smoke (CS) for nearly 11 weeks followed by administration of rMSCs into the lungs. Infusion of rMSCs mediates a reduced pulmonary cell apoptosis; a downregulation of TNF-α, IL-1β, MCP-1, and IL-6 (pro-inflammatory mediators); and an upregulation of vascular endothelial growth factor (VEGF), proteases (MMP9 and MMP12) in lung, VEGF receptor 2, and TGFβ-1 and improves destructive pulmonary function and emphysema triggered by CS exposure (Guan et al. 2013).

High plasma CRP levels within 48 h of ALI in children were tested along with its association in 28-day mortality and ventilator-free days (VFD). The CRP level in nonsurvivors was 126 mg/L which was quite high than in survivors (CRP = 56 mg/L). As cardiovascular organ failure at onset of ALI was the strongest predictor for mortality, so for every 10-mg/L rise in CRP level, mortality increased by 4.7 %. Increased CRP levels were thus associated with a decrease in VFD. Therefore, increased plasma CRP levels have no favorable outcome in ALI in children, but it is in contrast with findings in adults with ALI (Bruijn et al. 2013). CRP is used as an inflammatory biomarker to distinguish ALI or acute respiratory distress syndrome (ARDS) from cardiogenic pulmonary edema. CRP behaves as a diagnostic marker in these diseases. CRP along with brain natriuretic peptide (BNP) behaves as a stronger prognostic marker of these diseases (Komiya et al. 2011).

The ALI along with its severe form, ARDS, is exemplified by greater vascular and epithelial permeability, hypofibrinolysis, hypercoagulation, inflammation, and immunomodulation. A distinct population of an intact lipid bilayer containing small cytosolic vesicles within the microparticles (MPs) in both the alveolar and vascular compartments in respective patient groups or in ALI/ARDS animal models may serve as diagnostic and prognostic biomarkers. MPs are released in vascular, parenchymal, or blood cells containing membrane and cytosolic proteins, different organelles, lipids, and RNA derived from their relevant parental cells. MPs act as modulators, intrinsic stimulators, or even attenuators in some diseases as it can effectively interrelate with various cell types. MPs in ALI/ARDS are derived from diverse cell types of heterogeneous function and may act as a promising therapeutic agent to modulate inflammatory mechanisms by either removing/inhibiting or administrating/stimulating MPs (Mcvey et al. 2012).

Multiple-organ dysfunction syndrome and ALI are initiated by different inflammatory mediators after trauma and hemorrhagic shock (T/HS) to enter into the systemic circulation through mesenteric lymph ducts. Post-HS mesenteric lymph (PHSML) activates polymorphonuclear leukocytes (PMNs), activates red blood cell and vascular endothelial cell dysfunction, and contains biologically active lipids as pro-inflammatory mediators. In the PHSML, phosphatidylethanolamine, lysophosphatidylethanolamine (LPE), phosphatidylcholine, lysophosphatidylcholine (LPC), and sphingomyelin were detected; arachidonoyl, linoleoyl, and docosahexaenoyl LPCs and LPEs were significantly elevated in the PHSML of the T/HS group.

Elastase was also released after induction by linoleoyl and arachidonoyl LPCs. These biologically active lipids in PHSML are involved in the pathology of ALI or multiple organ dysfunction syndrome (Morishita et al. 2012).

4.10.3 Pancreatitis

Pancreatitis is an inflammatory disease of the pancreas which requires medical attention and immediate hospitalization during an attack. It occurs mainly when trypsin and other pancreatic enzymes start digesting food when activated in the pancreas instead of the small intestine. Acute pancreatitis (AP) begins suddenly and lasts for a few days, whereas in chronic pancreatitis (CP) it lasts for many years. The most common (~80 %) etiology of AP and CP is gallstones and alcohol, respectively. Pancreatitis has multiple causes including some viruses (cytomegalovirus, hepatitis B, mumps), bacteria (*Mycoplasma*, *Salmonella*), fungi (*Legionella*, *Aspergillus*), and parasites (*Ascaris*, *Cryptosporidium*) and a number of other infectious agents have been recognized.

AP if present in its severe form includes systemic organ dysfunction, local pancreatic complications (pseudocysts, necrosis, or abscess), or both. Severity of AP depends on a number of causal factors which includes both systemic organ failure and local peripancreatic necrosis. In AP, inflammatory mediators initiate the intracellular stimulation of pancreatic proenzymes and/or NF-κB. Thus, stimulated leukocytes penetrate deep into and around the pancreas and decide a central role in the severity of AP. The inflammatory reaction is initially local, which may multiply in cascade to produce excess inflammatory mediators leading to systemic and/or early organ failure (Kylänpää et al. 2012).

These immune responses are combated by the release of specific cytokine inhibitors and anti-inflammatory cytokines thus preventing the hazard for systemic infection. At present, there is no specific treatment for AP, but an enhanced understanding of the pathology of systemic inflammation and the advance of organ dysfunction may prepare for future treatment methodologies (Kylänpää et al. 2012).

In AP patients, the concentration of CRP equal or more than to 150 mg/L even at 48 h after hospital admission was relevantly associated with higher chance of receiving prophylactic antibiotics after prescription. Thus CRP was one of the most prominent biomarkers in prescribing prophylactic antibiotics in AP (Cardoso et al. 2014). In CP patients, systematic inflammation was identified. Patients with osteoporosis and higher levels of inflammatory markers had the highest systemic inflammation. Thus, in CP, the potential alteration of risk factors like CRP may provide an avenue to avert fractures and reduce bone loss in this group (Duggan et al. 2015). CRP along with serum interleukin-6 (IL-6) is used as a diagnostic marker in the differentiation between pancreatic cancer and CP (Mroczko et al. 2010).

In nonalcoholic fatty pancreas disease (NAFPD), obesity is a risk factor. During AP, IL-10 acts as an effective anti-inflammatory cytokine in downregulating the release of pro-inflammatory mediators. Obesity reduces the synthesis of pro-inflammatory cytokines in the spleen, so spleen-originated IL-10 may regulate the NAFPD pathology, caused by high-fat (HF)-diet-induced obesity. In splenectomy (SPX)-treated mice, the increased fat deposition and inflammatory responses of the pancreas in obese mice (with HF diet-induction) were reduced by systemic administration of IL-10. In IL-10 knockout mice, SPX had little effect on inflammatory responses and fat deposition in the pancreas. Thus, obesity reduces IL-10 release by the spleen which may protect against the development of NAFPD (Gotoh et al. 2012).

A biologically active portion of the plant *Nardostachys jatamansi* (NJ; NJ4) was administered intraperitoneally in mice followed by injection with the cerulein (analogue of stable cholecystokinin) for nearly 6 h. After the last cerulein injection, the morphological examination of the lung and pancreas and blood and neutrophil infiltration were done along with cytokine expression. NJ4 administration abates the AP severity and lung injury related with AP, reduces neutrophil inflow and cytokine production, and results to in vivo elevated levels of heme oxygenase-1 (HO-1). Also, NJ4 along with NJ4-2 (another active fraction) induces HO-1 in isolated pancreatic acinar cells and thus prevents the cerulein-triggered death of acinar cells. NJ4 may be a probable therapeutic agent offering defense in AP and might also lower the severity of the disease by stimulating HO-1 expression (Bae et al. 2012).

4.10.4 Kidney Disease

The kidneys are two organs that cleanse our blood by removing excess fluid and waste, maintain the salt and mineral balance of the blood, and help to regulate blood pressure. When the kidneys are damaged, fluid and waste products can accumulate in the body, causing swelling of the ankles, with weakness, poor sleep, vomiting, and shortness of breath. Loss of kidney function is a serious and potentially fatal state and if left untreated, eventually the diseased kidneys may stop functioning totally.

The most abundant leukocytes in the kidney are DCs and macrophages which are the key players in innate immune responses as they coordinate ischemia–reperfusion injury in the kidney followed by inflammation. They directly induce sterile inflammation after reperfusion by producing pro-inflammatory cytokines, soluble inflammatory mediators, and chemokines and indirectly through activation of NK cells and effector T lymphocytes. DCs possess tolerogenic functions in normal and diseased conditions and macrophages participate in tissue repair. Governing the microenvironment of the kidney and understanding the function and phenotype of DCs and macrophages will throw light on the pathogenesis of this disease and offer novel drug targets (Okusa and Li 2012).

Chronic kidney disease (CKD) or chronic renal disease is the gradual failure in kidney function for a period of few months to some years. The worsening signs of kidney function are highly nonspecific and might normally include an uneasy feeling, anemia, some cardiovascular disease, lowered appetite, and also pericarditis. Often, people with increased blood pressure or diabetes and their blood relatives are at the risk of CKD.

The progression of CKD is aggravated by inflammation and oxidative stress. Oxidative stress leads to depletion of the most prominent endogenous intracellular antioxidant, tissue glutathione (GSH), but dilapidation of oral GSH by digestive enzymes limits its therapeutic interventions. In chronic renal failure (CRF), GSH repression reduces the oral GSH precursor, F1 (contains cystine as a cysteine carrier), which attenuates oxidative stress and inflammation and restores tissue GSH and thus reduces the severity of interstitial nephropathy. Experimental male Sprague Dawley rats were broken up into CRF (rat chow with 0.7 % adenine) group, F1-treated CRF group (rat chow containing adenine with F1) group, and control group (with regular rat chow). Finally, the animals were given regular chow and then euthanized. Consumption of adenine-containing diet causes rigorous kidney swelling; azotemia; high glomerular and tubular injury; lowered urinary concentrating capacity; massive tubulointerstitial nephropathy; severe anemia; elevated levels of markers of oxidative stress, inflammatory mediators (p-IκBα, cytoplasmic NF-κB, NF-κB, cyclooxygenase-2, p65), and plasma oxidized glutathione disulfide (GSSG) 3-nitrotyrosine; and lowered GSH/GSSG ratio and manganese superoxide dismutase. F1 cotreatment caused significantly reduced tubulointerstitial edema and inflammation; higher urinary concentrating capacity, anemia, azotemia; and normalized expression of markers of tissue oxidative and nitrosative stress. Thus, F1 acts as a unique oxidative stress modulator, prominently attenuating inflammation, renal damage, oxidative stress indicators, and renal dysfunction (Nicholas et al. 2012).

Elevated pre-procedural serum hs-CRP levels were correlated with the extended clinical outcomes of subjects with stable chronic kidney disease (CKD) who were implanted with first-generation drug-eluting stent (DES) (Ortega et al. 2014; Ogita et al. 2015). Lower sodium levels in serum had been associated in patients with CKD through their increased mortality. CRP acts as the independent predictors of lower plasma sodium in CKD patients. Thus higher CRP levels are correlated with lower sodium levels. Thus inflammation could be one of the underlying confounding factors explaining the increased mortality in these CKD patients (Ortega et al. 2014).

4.10.5 Injury

Inflammation without any sterile inflammation or infection leads to either acute injury or chronic disease. In cerebral ischemia, the primary injury is caused by reduced blood supply and contributed by IL-1β. IL-1β is controlled by the protease caspase-1 and its related inflammasome NLRP3, the activating complex. In macrophages, the in vitro NLRP3 inflammasome-dependent responses require an early priming stimulus by a damage-associated molecular pattern (DAMP) or a pathogen (PAMP). In mouse, in the cultured brain-originated mixed glial cells (DAMP ATP), calcium pyrophosphate dehydrated crystals and monosodium urate had no effect on the expression of IL-1β or IL-1α and were released when stimulated by PAMP. Alternately, without priming, these DAMPs may trigger inflammation through the synthesis of CXCL1 and IL-6 and cathepsin B (lysosomal protease). Apart from PAMP, the acute-phase protein like serum amyloid A (SAA) may also behave as a priming stimulus. After cerebral ischemia, in vivo, synthesis of IL-1 increased the overproduction of CXCL1 and IL-6 in the ischemic hemispheres of IL-1α/β double KO mice though in these mice injury-induced cytokine responses were not abated. Thus DAMPs enhance brain inflammation by directly stimulating production of glia-originated inflammatory mediators and IL-1-dependent responses (Savage et al. 2012).

Through in vitro experiments, using necrotic and apoptotic cells revealed the binding of CRP to necrotic and apoptotic cells and also facilitated the removal of such cells. But in vivo experiments performed using animal models having ischemia/reperfusion (I/R) injury revealed that the binding of CRP to such damaged cells is detrimental to the tissue. Thus, the binding role of CRP with phosphocholine is quite unfavorable if it occurs on injured host cells as it causes more damage to the tissue by stimulating the complement pathway. So, in acute myocardial infarction and ischemia–reperfusion injury, the scenario is worsened by CRP. The consequence of the binding of CRP to damaged cells thoroughly depends on the type of tissue. In tissues like skin and subcutaneous tissue, CRP does not harm to bind the complement and hasten the death of the dead tissue. It is different and harmful to remove dead tissue having no regeneration property. In myocardial infarction, CRP will remove the necrotic part of the myocardium. But, in the ischemic part of the tissue, where the damage can be reversed, CRP may remove the tissue as described previously. Thus, the phosphocholine-binding function of CRP is quite defensive for the host because it leads to removal of necrotic tissue and also protection against pneumococcal infection. On the contrary, the phosphocholine-binding function of CRP is detrimental for the host when CRP binds reversibly to the damaged myocardial cells, as it causes more damage to the tissue by activating the complement pathways (Agrawal et al. 2014).

A heat shock protein family, αβ-crystallin, exerts cell protection under various stress-related conditions developed in the mouse model of multiple sclerosis, spinal cord injury brain ischemia, and Alexander disease. After contusion lesions, the levels of αβ-crystallin are lowered in spinal cord tissue. After contusion injury (for the first week), administration of recombinant human αβ-crystallin leads to increased granulocyte infiltration and continuous improvement in locomotor skills, modulates inflammatory responses in the injured spinal cord, reduces secondary tissue damage, and lowered recruitment of inflammatory macrophages. Thus, release of recombinant human αβ-crystallin promotes increased locomotor recovery after spinal cord injury suggesting its use as a better therapeutic agent for treating acute spinal cord injury because at present there is currently no effective treatment (Klopstein et al. 2012).

A sustained trauma (car accident, falls, sports injuries, gunshot wounds, stab wounds, a penetrating foreign object such as a knife, etc.) to the liver leads to liver injury, the most common abdominal injury, and constitutes 5 % of all traumas. Majority of the people who sustain this injury are also accompanied by some other injury. Nonoperative management with observation is necessary for a full recovery. Liver injury was triggered in rats by administration of acetaminophen (800 mg/kg, i.p.) followed by administration of fucoidan (extracted from various brown seaweeds, a pharmacological sulfated polysaccharide) 2 h before and after acetaminophen administration. Liver damage and cell death, overexpression of CYP2E1 (metabolizing enzymes of acetaminophen), and hepatic apoptosis (shown by the protein expression of Bax, Bcl-2, and cleaved caspase-3) induced by acetaminophen were attenuated by co-treatment of fucoidan. Fucoidan acts as an antioxidative agent and increases the production and expression of glutathione, glutathione peroxidase, and superoxide dismutase. All of them were reduced by acetaminophen. Further, fucoidan lowered the expression of inflammatory mediators (IL-1β, TNF-α, iNOS). Thus, against acetaminophen-induced liver injury, fucoidan has hepatoprotective effects through the anti-apoptotic, antioxidant, and anti-inflammatory pathways (Hong et al. 2012).

In 90 % partial hepatectomy (PH, except the caudate lobe) of rats, omega-3 PUFA acid (ω-3 PUFA) was administered intravenously before PH surgery for liver regeneration. To analyze liver regeneration, survival rates, liver weights, liver weight/body weight ratios, nuclear associated antigen Ki-67, signal transduction, and other biotechnological assays were evaluated. Survival rates in the ω-3 PUFA-induced rats were remarkable as compared to death of all in the control group (Qiu et al. 2012).

CRP is often used to assess the status of postoperative infection. Anomalous CRP values might be perceived after surgery yet in the case of noninfection due to transfusion and muscle injury, which disrupted better perioperative management. The level of CRP was evaluated after spine surgery, which was shown to be noninfection. A dramatic decrease of CRP concentration was detected on postoperative day POD3 and POD7 in lumbar open discectomy (LOD) patients of noninfection. Thus CRP would be a more sensitive and effective parameter especially in LOD patients for early evaluation of infectious complications, and the typical prototype of CRP may assist to evaluate the early postoperative course (Choi et al. 2014).

4.10.6 Cardiovascular Disease (CVD)

Cardiovascular disease (CVD) or heart diseases affects the total cardiovascular system, the heart or blood vessels (like arteries, capillaries, and veins), vascular system of the brain and kidney, and peripheral arterial disease. The reasons of cardiovascular disease are miscellaneous ranging from hypertension to atherosclerosis, aging with other allied cardiovascular dysfunction and many others causes found in even healthy asymptomatic individuals. CVD is the leading reason of deaths worldwide, though, since the 1970s, cardiovascular mortality rates have reduced in many high-income countries. At the same time, cardiovascular disease and deaths have increased at a rapid rate in middle- and low-income countries. Although CVD usually influences older adults, the antecedents of CVD, remarkably atherosclerosis, begin in initial life, making primary anticipation efforts essential from childhood. Thus there is increased stress on preventing atherosclerosis by altering risk factors, like exercise, healthy eating, and avoidance of smoking.

Reports were found pointing out that hs-CRP acts both as risk factor in the general healthy population and as prognostic factor in those with CVD. CRP was the most useful biomarker in subjects with a history of CVD and intermediate risk of events at 10 years, where adding of hs-CRP to the classical models for event risk estimation improves the risk staging. There was actually no consensus on the clinical usefulness of CRP as a prognostic marker in subjects with acute or chronic disease (Brito et al. 2015).

In asthma and in some inflammatory diseases, cysteinyl leukotrienes (CysLT) behave as immunomodulating lipid mediators and a strong spasmogenic agent. Apoe$^{-/-}$ mice were fed with a hypercholesterolemic diet, and the expression of some key enzymes of the CysLT pathway and their related receptors (CysLT1/CysLT2) were analyzed in the myocardium (hypoxic and normal). Chronic inflammation with increased apoptosis, fibrosis, and leukotriene C4 synthase (LTC4S) and upregulation of expression of IL-6 and CysLT1 were demonstrated in the myocardial biopsies of Apoe$^{-/-}$ in comparison to biopsies from control C57BL/6 J mice. Acute bouts of hypoxia further induce the LTC4S and CysLT1 expression, increasing LTC4S enzyme activity, with associated increased expansion of hypoxic areas in the myocardium. In acute bouts of hypoxic stress, treatment with selective CysLT1 receptor antagonist (Montelukast) inhibits CysLT signaling pathway thus reducing myocardial hypoxic areas in Apoe$^{-/-}$ mice to nearly normoxic conditions. Even in human heart biopsies from patients with chronic coronary artery disease, the mRNA expression levels of LTC4S and CysLT1 were upregulated in chronic ischemic myocardium as compared to a nonischemic one. Thus, CysLT1 antagonists may have protective function on hypoxic heart by improving the oxygen supply to myocardial ischemia areas, for example, during episodes of sleep apnea (Nobili et al. 2012).

In CVD mainly atherosclerosis and atherothrombosis are mostly controlled by immune-mediated inflammation. In CVD and in various diseases associated with an elevated cardiovascular risk (acute coronary syndrome, systemic lupus erythematosus, rheumatoid arthritis, end-stage renal disease, and severe carotid stenosis), during humoral autoimmunity, IgG autoantibodies against apolipoprotein A-1 (apoA-1) might play an emerging cardiovascular risk. Anti-apolipoprotein A-1 antibodies (anti-apoA-1 IgG) may act against the active mediators of atherogenesis and as a probable diagnostic and prognostic biomarker of cardiovascular risk (Teixeira et al. 2012).

4.10.7 Autoimmune Disease

Autoimmune disease arises from autoimmunity, which is an imperfect immune response of the body against the body's own tissues and molecules. It may be organs (autoimmune thyroiditis) or a particular tissue (Goodpasture's disease). Its treatment involves typically immunosuppressive drugs. During an infectious disease, macrophages and other immune cells due to infection release cytokines (namely, IL-1β), which in turn activate NF-kβ-dependent transcriptional pathway and inflammatory-cell recruitment via adaptor protein, myeloid differentiation factor 88 (MYD88). But these cytokines may break the tissue architecture of endothelial and reduces its cell–cell interactions. Endothelial cell blocks the inflammatory cell infiltration and in other way also acts as an inflammatory mediator. The disruptive effects of IL-1β on endothelial stability in a human in vitro cell model is through binding of MYD88 to small GTPase ADP-ribosylation factor 6 (ARF6) signaling and its activator ARF nucleotide-binding site opener (ARNO; also known as CYTH2). It is independent of the NF-kβ pathway. SecinH3, an inhibitor of ARNO, increases vascular stability significantly in animal models of acute inflammation and inflammatory arthritis (Zhu et al. 2012).

CRP along with ESR was used as an inflammatory marker in juvenile arthritis disease (Mourão et al. 2014). Delivery-related CRP levels were significantly elevated during delivery among both multiple sclerosis (MS) pregnancy patients and in control pregnant women. CRP levels were higher only during pregnancy in both study groups than during the postpartum period. Delivery-related elevated CRP levels did not correlate with postpartum disease activity. MS patients having gestational diabetes had a significantly higher level of CRP in the beginning of pregnancy as compared to nondiabetic MS patients. MS patients having fatigue had significantly higher CRP levels throughout pregnancy as compared to patients without fatigue. Thus, higher CRP values were correlated with pregnancy-related comorbidities but not with the MS disease activity (Jalkanen et al. 2015).

In inflammatory and autoimmune disease and in CNS injury models, inflammatory mediators and DC migration are reduced by administration of cannabinoid receptor 2 (CB2R) agonists. CB2R signaling inhibits matrix metalloproteinase 9 (MMP-9) expressions and thus affects DC migration in the Matrigel migration assay (in vitro) and in draining lymph nodes (in vivo). CB2R-mediated MMP-9 expression resulted in reduced cAMP levels, decreased ERK activation, and lowered binding of c-Fos and c-Jun to MMP-9 promoter activator protein 1 sites and thus controls MMP-9-dependent DC migration to re-ensure homeostasis. In future, CB2R agonists might be targeted as potential therapeutic mediators for the treatment of different chronic inflammatory conditions activated by immune cells, including DCs (Adhikary et al. 2012).

Production of autoantibodies and breakdown of self-tolerance are the key players in SLE. Anti-dsDNA antibodies bind to cell surface receptor proteins of resident kidney cells, trigger downstream stimulation of signaling pathways, and release inflammatory mediators and fibrosis in the vascular, glomerular, and tubulointerstitial compartments of the kidney and associated acute or chronic renal failure during the pathogenesis of lupus nephritis (Yung and Chan 2012).

4.10.8 Joint Disease

Joint disease, affecting millions of people, is any disease or injury of the human joints. Arthritis and many others form the group of best-known joint diseases. Joint diseases may be acute or exceedingly chronic, causing uncomfortable, nagging, agonizing pain in one or many joints, and may affect many parts of the skeleton. Inflammatory joint diseases are represented by arthritis with inflammation of the joints, effusion of fluid into the joint cavity, stiffness, swelling, pain, and redness of the skin above the joint. In severe form, it destroys the joint cartilage and underlying bone with irreparable deformities. In ankylosis, the articulating members are adhered resulting in fusion with loss of mobility. In synovitis, inflammation is restricted to the synovial membrane/lining of the joint. Simple pains in the joints with no other accompanying evidence of arthritis are referred to as arthralgias. Rheumatism is a noninflammatory joint disease with unprecedented discomfort of the articular apparatus (joints with bursas, tendons, ligaments, and tendon sheaths). Spondylitis is the inflammation of the spine and joints.

"Spondyloarthritis" consists of an aggregation of a group of several diseases having similarities in clinical, radiological, and genetic content. Ankylosing spondylitis is the main representative of this spondyloarthritis group and is mainly characterized by a predominant axial involvement. For the diagnosis of ankylosing spondylitis, the presence of radiographic sacroiliitis is quite essential according to the newly modified New York criteria. The diagnosis of spondyloarthritis is often delayed as the occurrence of radiographic sacroiliitis takes nearly 8–11 years. In magnetic resonance imaging, the sacroiliac joint inflammation can be depicted before the appearance of radiographic damage thereby defining the new concept of "non-radiographic axial spondyloarthritis." Elevated levels of CRP at baseline have been used as biomarkers in predicting radiographic sacroiliitis progression. Thus CRP was helpful in definite diagnosis of ankylosing spondylitis (Tant et al. 2014).

The release of tenascin-C (TN-C) into knee synovial fluid after joint disease or in acute joint injury induces inflammatory mediators and matrix degradation in vitro in human articular cartilage. Human knee synovial fluid samples were collected from acute inflammatory arthritis (AIA), isolated knee meniscus injury, knee anterior cruciate ligament rupture, knee osteoarthritis (OA) with or without concomitant meniscus lesions, and knee healthy control groups for TN-C level and other cartilage markers. Same parameters were also undertaken in joints of dogs that undergone knee instability surgery. TN-C levels and its correlations with levels of matrix metalloproteinases 1 and 3 and aggrecanasedependent Ala-Arg-Gly-aggrecan (ARG-aggrecan) fragments were observed to be significantly higher in all the joint fluid of the human disease groups and in the dogs as compared to controls. Thus, TN-C being a marker of joint damage may also act as a stimulant of further joint degradation (Chockalingam et al. 2013).

4.10.9 Parasitic Infections

Any infectious disease caused or transmitted by a parasite is referred to as a parasitic disease. Parasitic diseases can affect all living organisms and any organ or tissue of the body. Some parasites can cause disease directly (malaria) but may also be through the toxins that they produce.

The etiology of parasitic diseases is protozoa (protozoan infection), helminths (helminthiasis), and bacteria (bacterial infection). Protozoa, helminths, and bacteria may be ectoparasites or endoparasites.

Malaria-associated acute respiratory distress syndrome (MA-ARDS) marked by pulmonary inflammation is a highly complicated disease with unknown pathophysiology. C57Bl/6 mice were infected with *Plasmodium berghei* NK65, *P. chabaudi*, and *P. berghei* ANKA, and the level of hemozoin in the lungs (in phagocytes, infected erythrocytes, and occasionally granulocytes) was correlated with the number of infiltrating inflammatory cells; alveolar edema; lung weight; increased pulmonary expression of cytokines, chemokines, and enzymes; and alveolar VEGF levels. Also, *P. falciparum*-derived hemozoin was injected intravenously in malaria-free mice. Consequently, hemozoin stimulated the pulmonary expression of cytokines (IL-6, IL-1β, TNF, IL-10, and TGF-β), pro-inflammatory chemokines (KC/CXCL1, IP-10/CXCL10, and MCP-1/CCL2), and other inflammatory mediators (iNOS, Hmox1, NOX2, and ICAM-1) (Deroost et al. 2013).

Patients infected with *P. vivax* uncomplicated and *P. falciparum* malaria in the endemic Brazilian Amazon were examined for hematological alterations and discharge of soluble mediators (CRP and NO) in acute (day 0) and convalescent phase (day 15). Laboratory inflammatory profiles reveal similar data in accordance with white blood cells, thrombocytopenia, and high band cell production during the acute phase of infection. CRP levels were higher in acute *P. vivax* infection as compared to acute *P. falciparum* infection, but higher NO was observed in acute and convalescent *P. falciparum* infections. Modifications in these mediators cannot envisage malaria infection and need further investigation (Lima-Junior et al. 2012).

Although CRP has been known to bind with varied nucleated cells, the direct binding of CRP molecule to erythrocytes in disease condition remains greatly unexplored. The binding of disease-associated CRP to erythrocytes of patients was demonstrated by flow cytometry, Western blotting, co-immunoprecipitation, ELISA, and surface plasmon resonance. A specific and strong RBC_{Mal}–CRP_{Mal} binding was observed. PC and calcium were found to be crucial for this interaction. Nearly a two- to threefold increase in RBC_{Mal}–CRP_{Mal} binding as compared to RBC_N confirmed disease specificity (Ansar et al. 2006). This binding altered the normal discoid shape of RBC_{Mal} with greater hydrophobicity and membrane fluidity. The effector function of CRP_{Mal} (10 μg/ml) has been exhibited by its greater potency to trigger the complement cascade as compared to RBC_N. Thus, these studies provide undeviating evidence for a significant phagocytic functional coaction of this acute-phase protein by activating the CRP complement cascade after binding with RBC_{Mal} (Ansar et al. 2006). Deficiencies of the complement-regulatory proteins (CD35, CD55, and CD59) on RBC_{Mal} of patients with *Plasmodium falciparum* were also reportedly known. The role of CRP_{Mal} in controlling the complement-regulatory proteins and downstream consequence on the complement cascade has been examined. In the presence of CRP, RBC_{Mal} demonstrated lowered complement-regulatory proteins with reduced affinity. These changes cause increased C3 deposition and more complement-mediated hemolysis, thus providing a novel mechanism of complement-fueled RBC_{Mal} destruction pathway refractory to erythrophagocytosis, and may account for pathogenesis of anemia (Ansar et al. 2009a).

The levels of CRP were elevated in pooled serum from *P. vivax*-, *P. knowlesi*-, and *P. falciparum*-infected patients. CRP might represent an important marker of infection of malaria, which could be utilized as a usual diagnostic tool for detecting *P. vivax*, *P. knowlesi*, and *P. falciparum* infections. However, the prospective of CRP as infection marker of malaria will need to be elucidated in a larger population of malaria-infected subjects (Mu et al. 2014).

4.11 Clinical History

In our study, three patients (designated as Patient 1, 2, and 3) were included whose clinical history and photos were included in Table 4.1. The patients were admitted in KPC Medical College and Hospital, Jadavpur, Kolkata, India. Informed consents were taken from these patients and their relatives, and the clinical history was given as per the consent of the institutional review board of the hospital. There were 2 patients (Patient 1 and 2), both female and diagnosed clinically as having rheumatoid arthritis. The first patient was admitted for her cataract surgery. The second patient was admitted for shortness of breath. Another patient (Patient 3) was suffering from perianal (or perineal) abscess and he came to the hospital to have the severely painful and tender abscess operated. The clinical history and photos of the patients (Patient 1 and 3) were collected during preanesthetic checkup and the history of Patient 2 was collected from the emergency ward during critical care management.

A 56-year-old woman (Patient 2) with shortness of breath and body ache was admitted to the ITU of the hospital at the onset. She was known to be diabetic (type II) and hypothyroid and had COPD. Her primary laboratory investigations showed high ESR, high CRP levels, very high *N*-terminal pro-BNP, a positive rheumatoid arthritis (RA) factor, and mildly deranged liver functions. Her echocardiography showed left ventricular hypertrophy (LVH) with high pulmonary artery pressure (PAP) and low ejection fraction (EF) (38 %) (Table1). Patient was managed conservatively by anti-failure management by giving a furosemide injection, nitroglycerine (GTN), and ramipril tablets and discharged. For her long-term rheumatoid arthritis, she is taking methotrexate, prednisolone, and folic acid by oral route. She had heart failure.

A 52-year-old male (Patient 3) was presented with fever, soft abdomen, and severe pain, redness, and tenderness in the anal region for 3 days. He had been diagnosed previously with diabetes (type II) with high BP. His laboratory findings revealed high TLC, DLC, and ESR and raised CRP levels. His blood serology report revealed negative for both malarial and dengue antigen. He had perineal abscess (as increased risk in diabetic patient). He had a palpable inguinal lymph node and an inflamed perineal region with fever. He had neutrophilic leukocytosis with high CRP and ESR. His procalcitonin level revealed absence of bacterial infection (Table 4.1). He was treated surgically by incision and drainage of perineal abscess under general anesthesia, and intravenous followed by oral antibiotics, tight control of sugar by regular insulin, and adequate analgesics and antihypertensives were given.

The level of inflammatory biomarkers like CRP was quite high in all the three patients.

4.12 Analysis of Biomarkers Seen in the Patients

It is quite evident from the clinical history from the hospital that no specific mediators were investigated during the routine investigations in the diagnosis process. Rather some broad-spectrum inflammatory markers were diagnosed which will point toward the process of inflammation and not to the pathways of activation of mediators, whereas in the laboratory people are developing inflammatory animal models to understand all possible molecules involved in inflammation. Very soon in hospitals, definite inflammatory markers for each disease will be investigated for routine diagnosis.

In this chapter, we have highlighted case studies of some patients from Kolkata, India, reveal-

Table 4.1 Clinical and laboratory parameters of patients

Parameters	Patients		
Sl no.	1	2	3
	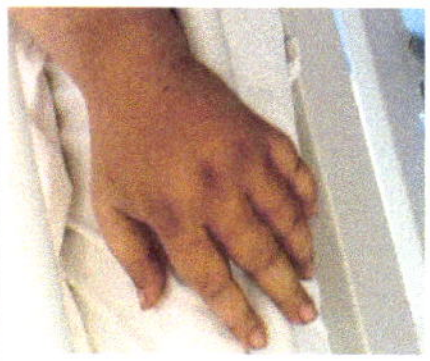	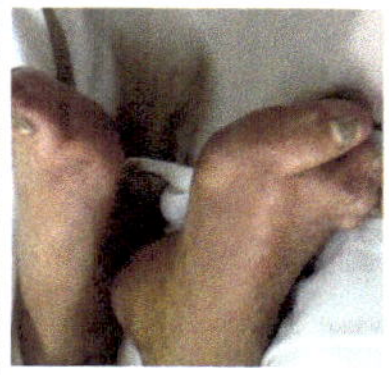	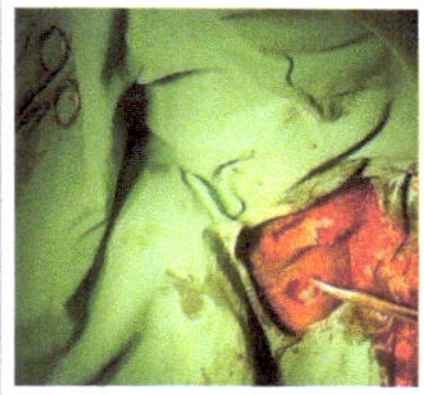
Physical parameters			
Age (years)	66	56	52
Sex	Female	Female	Male
Weight (Kg)	68	84 (obese)	73
Complaint			
1. Fever for 6 days 2. Decreased urine output for 1 day 3. Restless for the last 6 h		1. Shortness of breath for 2 days with exacerbation for the last 4 h 2. Body ache for 2 years 3. Swelling of knee joint 3 months back	1. Fever for 3 days 2. Pain and tenderness in the anal region and he can't sit for 5 days
Past medical history			
1. Diabetes mellitus type II for 4 years 2. Chronic kidney disease for 2 years		1. Diabetes mellitus type II for 2 years 2. Hypothyroid for 3 years 3. COPD	1. Hypertensive for 7 years 2. Diabetes mellitus type II for 4 years
Past surgical history			
1. Caesarian section—30 years back 2. Cholecystectomy—24 years back 3. Appendicectomy—18 years back		1. Hysterectomy and bilateral salpingo-oophorectomy—8 years back	1. Cholecystectomy—7 years back
Drug allergy			
Name not known		Not known till now	Sulfur drugs
On examination			
A. General examination			
Patient conscious, restless, and sometimes disoriented		Patient conscious, alert, and cooperative	Patient conscious, alert, and cooperative
GCS	13/15	15/15	15/15
Pulse/min	124	96	104
Temperature (°F)	101.4	98.1	101.4
Pallor	Mild	Positive	Nil
BP (mm of Hg)	136/90	140/82	152/90
JVP	Normal	Raised	Normal
Palpable lymph node	Not palpable	Not palpable	Inguinal lymph node palpable
Hydration status	Mild dehydration	No dehydration	Mild dehydration
Clubbing	Not present	Not present	Present

(continued)

Table 4.1 (continued)

Parameters	Patients		
Sl no.	1	2	3
B. Systemic examination			
Chest (air entry)	1. Adequate 2. Scattered rhonchi on both lungs 3. Maintaining SpO_2 is 94 % with 3 liters oxygen via nasal cannula with respiratory rate 29/min	1. Adequate 2. Rhonchi occasional, crepitation (++) more on left basal	Adequate
CVS (S1 and S2)	Audible, no gallop	Audible, gallop (++)	Audible
Abdomen	Soft	Soft and tender right hypochondrium, hepatojugular reflux positive	Soft
Intestinal peristaltic sound	Positive	Positive	Positive
C. Local examination			
1. No sore throat 2. No petechiae		1. *Lumbosacral spine*—tender (+) 2. *Knee joint*—tender and swollen	1. *Perineal abscess*—swollen (++), red (++), tender (++)
D. Laboratory investigations			
Hb (gm/dl)	9.8	12.3	13.4
Inflammatory and infective markers			
TLC (cells/cu.mm)	21,200	7900	22,400
DLC (%)	N (94), L (3), M (3), E (0)	N (62), L (34), M (1), E (3)	N (90), L (8), M (1) E (1)
ESR (mm/first hour)	87	96	74
CRP (mg/L)	342	126	260
Procalcitonin (ng/ml)	42	ND	12
Pro-BNP (pg/mL)	ND	35,624	ND
Malarial antigen	Negative	Negative	Negative
Dengue serology	Negative	Negative	Negative
Sugar level tests			
Blood sugar (mg/dl)	RBS-366	RBS -236	RBS -302
Glycated hemoglobin test (Hb A1c) (%)	8.6	7.6	9
Renal function tests			
Urea (mg/dl)	86	30	34
Creatinine (mg/dl)	2.8	1.3	1.2
Electrolytes			
Sodium (Na+) (mEq/L)	152	139	136
Potassium (K+) (mEq/L)	4.9	3.8	4.7
Calcium (serum) (Ca++) (mEq/dl)	8.6	9.3	8.7
Autoimmune markers			
Antinuclear antibody	ND	Negative	ND

(continued)

Table 4.1 (continued)

Parameters	Patients		
Sl no.	1	2	3
RA factor (above 20 IU/mL)	ND	Positive	ND
HLA (B-27)	ND	Negative	ND
Coagulation markers			
PTT (sec) (control—12 s)	26	13.7	14.2
APTT (sec) (control—26 s)	68	32	33
ABGA	Severe metabolic acidosis with increasing base deficit	Within normal ranges	ND
Chest X-ray	Bilateral lower zone opacities	Increased bronchovascular markings, sign of overload	Normal profile seen
Ultrasonography of the abdomen	1. *Kidney*—corticomedullary differentiation not maintained	Hepatomegaly	ND
	2. Liver mildly enlarged		
	3. Urinary bladder thickened, cystitis		
Cardiac function tests			
ECG	Sinus tachycardia with nonspecific ST changes, first-degree heart block	Sinus rhythm, rate 134/min with intermittent atrial fibrillation, nonspecific ST changes	Sinus rhythm (RBBB)
Echocardiography	Concentric LVH	LVH,dilated left atrium, PAP=56 mm of Hg	LVH, no RWMA
EF (%)	42	38	62
Bronchoscopy and bronchoalveolar lavage (BAL)/endotracheal (ET) suction			
	1. ET suction shows MDR *Pseudomonas aeruginosa*	ND	ND
	2. MDR *Acinetobacter baumannii*		
	3. MDR *Klebsiella pneumoniae*		
Urine routine examination and culture sensitivity			
	1. Pus cells plenty	Pus cells 2–3/HPF	ND
	2. Culture sensitivity shows MDR *Klebsiella*		

(continued)

Table 4.1 (continued)

Parameters	Patients		
Sl no.	1	2	3
Treatment summary			
1. Adequate calorie was ensured (by nasogastric tube/ parenteral nutrition) 2. Deep venous thrombosis prophylaxis was done 3. Tight control of blood glucose done by insulin (regular) infusion 4. Vasopressor (noradrenaline) infusion was started as patient was hypotensive in spite of adequate fluid management 5. Patient was on mechanical ventilator (pressure controlled) and then weaning off from ventilator 6. Injection polymyxin B (loading dose of 5 lac IU i.v.) followed by maintenance dose 7. Injection meropenem (loading dose 1 gm i.v. followed by 500 mg i.v.) 8. Nebulization with Duolin after every 8 h 9. Stress ulcer prophylaxis was done 10. Hemodialysis (as renal function deteriorated) was done 11. Electrolytes were replaced as per protocol as needed 12. Fresh frozen plasma (FFP) transfusion was given 13. Vitamin K injection (i.v.) was given		1. On sliding dose of human insulin (human Actrapid) subcutaneously 2. Furosemide injection (20 mg) i.v. thrice daily with bolus of 40 mg 3. Ramipril tablet (5 mg) orally at once daily 4. Injection GTN at the dose 0.9 ml/h (50/50) 5. Eltroxin tablet (25 μg) @orally per day 6. Inhaler with ipratropium bromide thrice daily 7. Methotrexate tablet (5 mg) orally at twice a week 8. Prednisolone tablet (10 mg) orally per day	1. Incision and drainage of perineal abscess done under general anesthesia 2. Blood sugar controlled by intravenous insulin infusion followed by subcutaneous insulin 3. Ramipril (5 mg orally/day) to control BP 4. **Antibiotic** piperacillin + tazobactam (4.5 gm for every 6 h) and linezolid (600 mg i.v. every 12 h) 5. *Analgesic injection* (a) Paracetamol (1gm) i.v. after every 8 h (b) Tramadol (100 mg) i.m./ SOS
Output			
Successfully discharged		Successfully discharged	Successfully discharged

COPD chronic obstructive pulmonary disease, *GCS* Glasgow coma scale, *°F* degree Fahrenheit, *BP* blood pressure, *JVP* jugular venous pressure, *CVS* cardiovascular system, S_1 first heart sound, S_2 second heart sound, *TLC* total leukocyte count, *DLC* differential leukocyte count, *ESR* erythrocyte sedimentation rate, *CRP* C-reactive protein, *N* neutrophil, *M* monocyte, *L* lymphocyte, *E* eosinophil, *ND* not determined, *FBS* fasting blood sugar, *RBS* random blood sugar, *RA* rheumatoid arthritis, *HLA* human leukocyte antigen, *PTT* prothrombin time test, *APTT* activated partial thrombin time, *ECG* electrocardiogram, *RBBB* right bundle branch block, *WNL* within normal limit, *LVH* left ventricular hypertrophy, *RWMA* resting wall motion abnormality, *PAP* pulmonary artery pressure, *EF* ejection fraction, *BNP* brain natriuretic peptide, *μg* microgram, *mg* milligram, *pg* pictogram, *i.v.* intravenous, *i.m.* intramuscular, *SOS* as and when required, *GTN* glyceryl trinitrate, *CBG* capillary blood glucose, *NA* not applicable, *MDR* multidrug resistance, *ABGA* arterial blood gas analysis, *HPF* high-power field

ing acute inflammation from disease together with their clinical history (Table 4.1). All the patients had some past medical or surgical history, representing some diverse inflammatory conditions. The question which we probe in here is the search of proper biomarkers in these represented acute inflammatory conditions. All the key players and biomarkers of inflammation change its role with the change in setup of disease and patients. Even in clinical diagnosis of an inflammatory patient, some broad-spectrum markers were analyzed without individual application of a universal biomarker.

4.13 Biomarkers of Inflammation in Different Subsets of Diseases

In this chapter, we had presented three case reports in search of proper inflammatory biomarkers. CRP is a diagnostic biomarker in a varied spectrum of diseases. This is discussed in other chapters also. The role of other inflammatory biomarkers in other subset of diseases is also discussed here.

Biomarkers can be defined as any alterations in the constituents of body or tissue fluids. These markers provide a medium for uniform classification of a disease with its risk factors and can be extended in understanding the basic underlying pathophysiology of disease. Biomarkers provide a powerful and dynamic tool to grasp the spectrum of inflammatory diseases with usage in observational and analytic epidemiology, clinical trials in populations, and screening with diagnosis and prognosis. Biomarkers can also reflect the entire steps of a disease from the earliest symptoms/screening to the terminal stages. Analytical assessment of the validity of biomarkers is required to correlate with respect to the stage of disease. Variability in the measurement of biomarkers ranges from individual error in laboratory technicians, machine dysfunction, improper storage of body fluid, and other bias and confounding issues.

Currently, there are no specific markers for inflammation; rather some broad-spectrum inflammatory markers were routinely investigated in hospitals. The clinical investigation revealed some broad-spectrum markers like C-reactive protein (CRP), erythrocyte sedimentation rate (ESR), total leukocyte count (TLC), differential leukocyte count (DLC), sodium–potassium level, urea–creatinine level, antinuclear antibody, pro-brain natriuretic peptide (pro-BNP), and human leukocyte antigen (HLA) which were accounted.

Nearly 7896 papers from PubMed and Google Scholar fit our criterion of research, of which 2641 were review articles, 151 were letters to the editors, and the rest were original articles. In this context, we reviewed some of the immunological mechanisms and the applicability of biomarkers that occur during acute phases with its remission/recovery period in some of the widely used experimental models of varied spectrum of diseases (Table 4.2).

Invasive tests sometimes are routine for the diagnosis and care of patients. Diagnosis is mostly based on clinical symptoms combined with radiological, endoscopic, and laboratory investigations. The employment of noninvasive biomarkers is always needed to avoid an invasive diagnostic test that causes discomfort and potential complications. The ability to determine the type, severity, prognosis, and response to therapy of a disease using definite biomarkers has long been a waited goal of clinical researchers. We describe the biomarkers assessed in this chapter, with special reference to acute-phase proteins and serologic markers, and thereafter, we describe the new biological markers (Table 4.2). The biological markers could be developed in the future based on serum markers of acute-phase response. The laboratory tests mostly used to measure the acute-phase proteins in clinical practice are the serum concentration of CRP and the ESR. Other biomarkers of inflammation include platelet count, leukocyte count, and serum albumin and serum orosomucoid concentrations and serologic markers like antibodies also. To detect specific pathologies, in the last decades, serological and immunologic biomarkers have been studied extensively in immunology and have been used in clinical practice. In different diseases, the presence of antibodies can aid as surrogate markers for the aberrant host immune response and also for future biomarkers. The field of the biomarker discovery have revolutionized by the progress of molecular biology tools (microarrays, proteomics, and nanotechnology). The advances in bioinformatics coupled with cross-disciplinary collaborations have highly enriched our ability to characterize, retrieve, and analyze large amounts of data generated by the technological advances. The present techniques available for biomarkers development are proteomics and genomics (single-nucleotide polymorphism genotyping, pharmacogenetics, and gene expression analyses). In the future days, the addition of new serological markers will add significant benefit to patients and clinicians. Our understanding of the pathophysiology of a disease is based on correlating serologic markers with clinical phenotypes and genotypes.

4.14 Discussions

Humans and animals are continuously exposed to microbial pathogens, trauma, stress, and injury that can pose a considerable relevance in our daily livelihood. After the initiation of the etiological agent, the organism elicits a set of highly organized immunological, physiological, metabolic, and behavioral responses to represent its

Table 4.2 Therapeutic and diagnostic applications of inflammatory biomarkers

Assessment of biomarkers	Research study observations/results	Therapeutic/diagnostic applications
Inflammatory responses after diesel exhaust (300 μg/m(3)) exposure lasted for 1 h (Xu et al. 2013)		
1. Symptom scores 2. PEF was assessed before exposure and at 15, 75, and 135 min of exposure 3. Monocyte and total leukocyte counts in peripheral blood 4. Serum IL-6 concentrations observed 20 h postexposure	1. Self-rated throat irritation in the upper airways was significantly higher 2. PEF decreased during diesel exhaust exposure 3. Monocyte and total leukocyte counts in peripheral blood were higher 4. Serum IL-6 concentrations were increased	Diesel exhaust has induced adverse acute effects on symptom scores, lung functions, and altered levels of inflammatory markers in healthy volunteers, from 75 min postexposure
Serum activin A and B levels predict outcome in patients with ARF (de Kretser et al. 2013)		
1. ARF patients require ventilator support for more than 6 h 2. The serum levels of activins A and B (members of the transforming growth factor-β family of proteins), with their binding protein, follistatin, have reported to be important regulators of inflammation, fibrosis, and ARF as compared to normal subjects.	1. Serum activin A and B were significantly elevated in most diagnostic patients. 2. Patients who had activin A and/or B concentrations above the reference maximum level were more likely to die in the 12 months following admission	1. The measurement of serum activin A and B levels in patients with ARF therapeutically predicts the risk of death 2. Modulating the activin A and B bioactivity should be explored as potential therapeutic agent
Diagnostic differences between severe sepsis patients and ARDS by biomarkers of lung epithelial injury and inflammation (Ware et al. 2013)		
1. The bio-clinical diagnostic discrimination of ARDS patients at risk from sepsis patients by 11 biomarkers of inflammation including plasma biomarkers (IL-8, IL-6) and lung epithelium generated biomarkers (SP-D, RAGE, CC-16) 2. Fibroblast activation, endothelial injury, proteolytic injury, and lung epithelial injury were also measured in early phases of ICU admission	Altered levels of five plasma biomarkers with three lung epithelium biomarkers help for discrimination and diagnosis of ARDS patients with severe sepsis patients	Plasma biomarkers may be useful as clinic–biological diagnostic tool of ARDS with sepsis patients
Prognostic value of plasma chitotriosidase activity in acute stroke patients (Bustamante et al. 2013)		
1. The elevated plasma activity of chitotriosidase, a component of innate immunity, reflects an inflammatory response 2. Chitotriosidase constitutes a sensitive parameter of macrophage activation 3. Plasma chitotriosidase activity was serially determined in 159 acute stroke patients and age-matched controls to assess its prognostic value in acute stroke patients 4. Additional predictive value of chitotriosidase was also tested	1. The baseline levels of chitotriosidase activity were increased in stroke patients compared to controls 2. Chitotriosidase activity was an independent predictor of neurological improvement at 48 h 3. Addition of plasma chitotriosidase activity showed a better prediction of improvement at 48 h	Acute stroke patients treated with tissue plasminogen activator with baseline chitotriosidase activity may constitute a prognostic predictor of short-term outcome

(continued)

Table 4.2 (continued)

Assessment of biomarkers	Research study observations/results	Therapeutic/diagnostic applications
New-onset AF after AMI is associated with admission biomarkers (Parashar et al. 2013)		
1. AF is an independent predictor of mortality after AMI 2. The biomarkers of myocardial stretch (NT-pro-brain natriuretic peptide [NT-pro-BNP]), myocardial damage (troponin-T [TnT]), and inflammation (hs-CRP) and new-onset AF during AMI were identified in patients at high risk for AF	1. New-onset AF was noticed in patients with AMI 2. Increase in NT-pro-BNP and hs-CRP is associated with increase in the rate of AF 3. TnT was not independently associated with new-onset AF	The markers of myocardial stretch and inflammation, but not the amount of myocardial necrosis, are important determinants of AF after AMI
Relationship between CSF biomarkers for inflammation, demyelination, and neurodegeneration in acute optic neuritis (Modvig et al. 2013)		
In patients with optic neuritis with demyelinating symptom and healthy subjects, CSF levels of CXCL13, CXCL10, MMP-9, CCL-2, osteopontin, chitinase-3-like-1, MBP, and NF-L were determined	1. Leukocyte infiltration biomarkers (CXCL13, MMP-9, and CXCL10) were strongly associated with MS-risk parameters 2. Osteopontin and chitinase-3-like-1 were associated ($p<0.0001$) and correlated with tissue damage markers (NF-L and MBP) to measure dissemination in space of white matter lesions	Leukocyte infiltration biomarkers (CXCL13, CXCL10, and MMP-9) strongly predict MS risk, and osteopontin and CHI3L1 suggest tissue damage-related inflammation
SAA, phospholipase A_2*-IIA, and CRP as inflammatory biomarkers for prostate diseases* (Menschikowski et al. 2013)		
Serum levels of inflammatory markers; SAA, $sPLA_2$-IIA, and CRP with PSA were determined in patients with localized PCa, BPH, and mPCa	Patients with BPH, PCa, and mPCa have elevated serum levels of SAA, $sPLA_2$-IIA, and CRP along with elevated levels of PSA as compared to healthy subjects	Significant differences in inflammatory circulating biomarkers were found between in PCa and mPca, suggesting their prognostic value during BPH development and PCa progression
Microalbuminuria and CRP as a predictor of CVD and as inflammatory biomarkers (Lazebnik et al. 2013; Bashir et al. 2014)		
1. Microalbuminuria, a risk factor and atherogenic marker for CVD, indicates the target organ damage 2. It is a valuable tool in screening and identification of patients with CVD 3. CRP is correlated with chest pain in CVD and increased levels of CRP in atherosclerosis and reflects inflammatory condition of vessel wall 4. CRP and microalbuminuria were estimated in patients of acute chest pain and type 2 diabetes along with healthy controls	It was found that microalbuminuria and CRP was much higher in CVD patients as compared to control groups	Sensitivity, specificity, and positive predictive value of CRP and microalbuminuria in patients as compared to normal can be used as important biomarkers in screening CVD

(continued)

Table 4.2 (continued)

Assessment of biomarkers	Research study observations/results	Therapeutic/diagnostic applications
Biological therapy by cell adhesion molecules in CD (Lazebnik et al. 2013)		
1. Diagnostic value of concentration of adhesion molecules (L-selectin, E-selectin, P-selectin, integrin-sVCAM-1) in blood serum for the assessment of the effectiveness of biotherapy in patients with CD and prognosis of the disease 2. Levels of leukocytes, ESR, and CRP were also analyzed before and after the biotherapy with infliximab and transplantation of MSC	1. Biological therapy along with transplantation of MSC decreases the levels of all adhesion molecules in all patients 2. Suppression of the synthesis of the main inflammatory cytokine TNF- α	Adhesion molecules are modern markers of inflammation used to assess the effectiveness of biological therapy in CD patients and to dictate the prognosis of the disease
Profile of circulating cytokines: impact on OSA, obesity, and acute cardiovascular events (Testelmans et al. 2013)		
1. OSA induces oxidative stress, systemic inflammation, and cardiovascular morbidity 2. The circulating cytokines profiles were measured in nonobese or obese patients with or without sleep apnea and with or without an acute cardiovascular event in a case-control study 3. Patients were assessed with sleep studies and inflammatory (hs-CRP, Leptin, RANTES, MCP1, IL-6, IL-8, TNF-α) and anti-inflammatory (adiponectin, IL-1Ra) cytokine profiles 4. Cardiovascular phenotyping was performed including carotid intima-media thickness, pulse wave velocity, and 24-h blood pressure monitoring	1.In comparison with patients without sleep apnea or without comorbidities, patients with the combination of an acute cardiovascular event 2. Patients with preexisting OSA showed a higher degree of systemic inflammation and significant increase in serum levels of hs-CRP, IL1-Ra, IL-8, IL-6, TNF-α, RANTES, and sICAM 3. Serum levels of different inflammatory markers were significantly higher in patients having OSA and an acute cardiovascular event	The diagnosis and treatment of OSA is potentially more important in patients after an acute cardiovascular event because these biomarkers could be associated with worsened cardiovascular outcome
Levels of hepcidin in cord blood: A biomarker for EONS (Cizmeci et al. 2014)		
1. Acute-phase reactant, hepcidin, has a critical role in inflammation and helps in host defense by interfering with microorganism's access to iron in EONS 2. Cord blood samples of infants born having EONS were collected and the level of cord blood hepcidin was determined	Hepcidin level was found to be significantly increased in newborns with EONS	1. Hepcidin behaves as an acute-phase reactant in the pathophysiology of EONS 2. Increased level of hepcidin in cord blood may be used as a reliable biological marker of EONS

(continued)

Table 4.2 (continued)

Assessment of biomarkers	Research study observations/results	Therapeutic/diagnostic applications
Acute-phase response in patients with AP (Kusnierz-Cabala et al. 2013)		
Concentrations of PTX3, SAA, CRP, HGF, PCT, PMN-elastase, IL-6, IL-18, and sTNFR-75 were measured in plasma of patients with severe and mild form of AP with control to age- and sex-matched healthy subjects	1. On each day of the study, significant correlations were found between PTX3 with SAA, IL-6, and PMN-elastase 2. The concentrations of these inflammatory markers were higher in severe AP patients as compared to those having the mild form of AP 3. The highest concentrations of PTX3 were noted on the early phase 3. The changes in PTX3 concentration in the early phase of AP is similar to that of IL-6 4. The peak levels of PTX3 are achieved earlier than another biomarker, CRP	1. PTX3 may be useful in early evaluation and prediction of the severity of AP 2. The relationship between the patterns of changes in PTX3 concentration with other inflammatory markers was evaluated in patients with AP at the early stage of the disease

CRP C-reactive protein, *MSC* mucosal stem cell, *CSF* cerebrospinal fluid, *ICU* intensive care unit, *AMI* acute myocardial infarction, *AF* atrial fibrillation, *IBD* inflammatory bowel diseases, *UC* ulcerative colitis, *CSF* cerebrospinal fluid, *CD* Crohn's disease, *ARF* acute respiratory failure, *CVD* cardiovascular disease, *OSA* obstructive sleep apnea, *EONS* early-onset neonatal sepsis, *SP-D* surfactant protein-D, *RAGE* receptor for advanced glycation end products, *IL* interleukin, *CC-16* club cell secretory protein, *ARDS* acute respiratory distress syndrome, *PEF* peak expiratory flow, *TNF* tumor necrosis factor, *hs-CRP* high-sensitivity C-reactive protein, *MMP* matrix metalloproteinase, *MBP* myelin basic protein, *NF-L* neurofilament light chain, *ESR* erythrocyte sedimentation rate, *AP* acute pancreatitis, *PTX3* pentraxin 3, *SAA* serum amyloid A, *HGF* hepatocyte growth factor, *PCT* procalcitonin, *PMN-elastase* polymorphonuclear elastase, *sTNFR75* soluble receptor for TNFalpha, A_2 = $sPLA_2$-IIA secreted group IIA phospholipase, *PSA* prostate-specific antigens, *PCa* prostate cancers, *BPH* benign prostatic hyperplasia, *mPCa* metastatic prostate cancers

strategy to fight the infection. Inflammation generated at the site of assault in the peripheral tissues communicates with the brain to modify its immune response and upgrade as necessary which aids in its ability to fight and eliminate the source. The groups of mediators, biomarkers, pathways, and key responders/effectors during inflammation are varied and change their *modus operandi* in a different disease in a different organism. Biomarkers' applicability lies in dissecting all the steps of a disease from diagnosis, screening, treatment, and drug formulation to prognostic consequences (Fig. 4.11).

As inflammation and its network of reactors form a juggling condition, thus to dissect all strings of its web, different animal models are developed in different diseases to mimic the original scenario. Inflammation is induced through varied stimulator molecules in these animals even for a single disease in different setups to decode the mechanism of inflammation and bring out probable diagnostic and therapeutic agents.

These studies have also provided evidence of systemically generated inflammatory biomarkers versus mediators both in acute and chronic inflammation. The process of inflammation is triggered by mononuclear phagocytes, which in turn alter the concentration of various biomarkers, the cascade of complement pathways, antibody production, and subsequent tissue repairing and resolution. Thus it traverses through both innate and adaptive immune system. While these responses are part of our normal homeostatic mechanisms, it is quite clear that systemic inflammation has a detrimental effect in humans and also in animals.

The inflammatory biomarkers responsible for the pathology of an inflammatory disease are of a heterogeneous nature (Table 4.2). In Table 4.2, we cited the potential therapeutic application of some biomarkers in varied acute inflammatory conditions.

The list is always never ending. Many discrepancies in the findings may at least in some sections be demonstrated by the specific methodologies used, different time points at which investigations were done, mice strain susceptibility, and/or concentration (or dose) of the administered inducer. Moreover, to the best of our collective knowledge, a very few studies aimed to clarify the kinetics of inflammatory cascades mediating acute and chronic phases of inflammatory disease.

In this pretext, the ongoing study aimed to evaluate cellular inflow and biomarkers of inflammation during the acute and chronic phases of inflammatory disease of humans induced in mice. Our review showed very striking differences among the induction phase and its recovery period which may be useful for future modeling of the experimental inflammatory diseases. This may particularly provide strong evidence for the detection of inflammatory biomarkers essential for the calculation of disease pathology and the designing of potential therapeutic outcome.

Disease subtypes are quite often described on the basis of medical history, physical findings, different serum markers, various imaging parameters, and histopathological and endoscopic characteristics of the related disease. Modern approaches in disease systematization are often invasive and/or generally related with a lack of specificity and sensitivity. From all the patients' history (Table 4.1), we can conclude that all of them had signs, symptoms, biomarkers, and expressions of inflammation and infection that are changed in human pathophysiology. We have documented those in our case histories. The varied physical biomarkers are inclusive of physical signs like local swelling, local edema, tenderness, local increased temperature, local redness, and local palpable lymph node in localized inflammation and infection. Other physical biomarkers are respiratory distress, fever (increased temperature), anuria, increased heart rate, low BP, deteriorating sensorium, and edema (due to increased vascular permeability of vessels). Radiologically, patient had diffuse opacities or consolidations in lungs, thickened urinary bladder and signs of kidney involvement (increased cortical echogenicity), and increased pleural reaction (revealed as pleural effusion). As per laboratory investigations, TLC with preponderance of neutrophil and its toxic granulation, high ESR, high CRP, increased procalcitonin level, high random blood sugar, rising urea–creatinine, increased sodium level (due to depletion of intravascular fluid due to increased vascular permeability), deranged coagulation profile (increased prothrombin time and activated partial thrombin time), deranged liver function test, metabolic acidosis (pH less than 7.5, low bicarbonate level, and increased base deficit), increased heart rate (documented by ECG),arrhythmia (due to electrolyte imbalance), and grossly hypokinesia of walls of the heart (due to sepsis) reversed after cure of disease (no hypokinesia and EF of heart was increased normally), ET suction or BAL fluid showed growth of different bacteria, and urine showed plenty of pus cells and growth of bacteria. A few laboratory investigations showed less than normal level in inflammation like low albumin, low pH, and low bicarbonate as we documented.

Many studies on biomarkers never achieve their goal because of the failure to stick to the same rules that would apply for the use of nonbiological variables. The development of any biomarker should rely on the standard design using epidemiological project or clinical trial. In formulating laboratory component, proper studies must be completed to determine reliability, accuracy, interpretability, feasibility, intra-individual or inter-individual variation, acquired or genetic susceptibility, and tissue localization, by setting the normal values with relation to variables like age, gender, and persistence of the biomarker (Sharma et al. 2013; Lazebnik et al. 2013; Testelmans et al. 2013; Cizmeci et al. 2014; Kusnierz-Cabala et al. 2013; Ansar et al. 2006, 2009a, b; Ansar and Ghosh 2013).

Advances in proteomics, genomics, metabolics, and molecular pathology have generated many potential candidate biomarkers with extensive clinical relevance. In the future, the integration of biomarkers and search for universal biomarkers, identified by high-throughput technologies, into medical science will be essential to achieve "personalization" of treatment/therapeutics and disease prevention.

The future of prevention protocol of inflammatory diseases and their detection and effective treatment will be highly influenced by the utilization of more effective markers of inflammation with superlative performance. Given the practical problem in collecting tissue samples or blood or fluids from patients in inflammatory diseases, biomarkers obtained from body fluids have a great prospect for the value-added patient management even through the drawbacks of the abovementioned hindrances or limitations. Since the immunosuppressive therapy of different inflammatory diseases currently rotates around long-term treatment with unwanted side effects, a paradigm shift from nonspecific cytotoxic drugs to specific and selective targeted therapeutic agents/drugs is an ardent medical need.

Perhaps the most regularly used nonspecific prognostic inflammatory biomarker is CRP. CRP levels are drastically elevated within 6 h after the initiation of inflammation. The final hike can sometimes be as much as 60-fold. Moreover, CRP is much more precise than some of the other generally used inflammatory biomarkers like ESR, total leukocyte count, etc. In various bacterial infections, CRP concentrations are unusually below 10 mg/L except in the case of neonates where 10-40 mg/L value typically represents a mild inflammation condition; levels between 40–200 mg/L represent significant bacterial infection or acute inflammation state. In serious bacterial infection or burns, concentration values may rise to 300 mg/L or even higher. The assessment of disease pathology, diagnostics, prognostics, and therapeutic applications of CRP in different diseases is already cited (Ansar et al. 2006, 2009a, b; Ansar and Ghosh 2013). Most common inflammatory markers include both inhibitors and mediators of inflammation as well as scavengers of prospective dangerous substances, namely, toxins. The following protein changes are observed during the inflammatory response: a rapid fall in serum prealbumin albumin and transferrin concentration as well as an elevated level of α1- and α2-globulin levels. The selected biomarkers with therapeutic applications of bone disease are CRP, calcitonin, collagen I, collagen I telopeptide, and osteocalcin; for cardiovascular disease, they are aldosterone, angiotensin, angiotensin-converting enzyme, antidiuretic hormone, atrial natriuretic peptide, etc; for diabetes, they are glucagon, insulin, and insulin-like growth factor; for gastrointestinal diseases, they are serotonin, gastrin, somatostatin, etc.; for women's health, they are estradiol, estrone, follicle-stimulating hormone, luteinizing hormone, etc.; there are many more biomarkers in other diseases (Sharma et al. 2013; Lazebnik et al. 2013; Testelmans et al. 2013; Cizmeci et al. 2014; Kusnierz-Cabala et al. 2013; Ansar et al. 2006, 2009a, b; Ansar and Ghosh 2013).

It is quite evident from the clinical history from the hospital that no specific biomarkers were investigated during the routine investigations in the diagnosis process (Table 4.1). Rather some broad-spectrum inflammatory markers (like CRP, ESR, procalcitonin, etc.) were diagnosed, whereas in the laboratory people are developing inflammatory animal models to understand all possible molecules involved in inflammation. Very soon in hospitals, definite inflammatory markers for each disease will be investigated for routine diagnosis.

Over the recent years, much clinical research is continuing to stress the significance of understanding the pathobiology of different inflammatory diseases for the procurement of efficient and safe pharmacological treatment modalities. In this pretext, we reviewed some of the immunological mechanisms/processes and the applicability of inflammatory biomarkers that occur during acute phases with its consequent remission/recovery period in some of the widely used experimental models of varied spectrum of diseases (Table 4.2). These diseases are quite common ailments and are posing quite a nuisance not only to global health but also in proving obstructions in its socioeconomic developments. Further research needs to be conducted in the area of genes and molecules involved in causing inflammatory disorders thereby improving the diagnosis and treatment of such diseases. The question we pose here is whether it is possible to screen the biomarkers of inflammation and their levels of expression in diseases associated with inflammation and correlate them with the clinical history. This would offer advantage both from the point of view of diagnosis, therapy, and prognosis of inflammatory disorders. The future scope of this paper remains in identifying such markers for inflammation and its correlation with disease condition.

Microheterogeneity of Proteins: Role in Diseases

5

Abstract

Protein posttranslational modifications largely influence their biological functions. The functions included effect on receptor binding, metabolism of the protein, tissue uptake, degradation, excretion, and protein–protein interactions. Modifications of proteins at and after translation also called as posttranslational modifications or PTM occur at a normal physiological concentration of the protein. However, in some diseases, modifications highly specific to the disease also are known to occur. Thus, disease-specific posttranslational modifications find importance as diagnostic markers in clinical medicine. Different approaches, most of which are analytical to study the posttranslational modifications in protein together their interactions, and their complex structure and function are illustrated by various researchers in immunology, biochemistry, biophysics, and clinical medicine and in proteomics. Immuno assay based affinity purification of protein by HPLC and chromatographic methods, and characterization of proteins by lectin binding assays, biophysical methods like MS or mass spectrometry, and surface-enhanced laser desorption/ionization method find importance as diagnostic techniques in clinical proteomic studies, in the detection of the disease-specific protein modification.

Keywords

Microheterogeneity • Disease • Posttranslational modifications • Clinical proteomics

Chapter Highlights

1. Variation in the chemical structure of a molecule (as the amino acid sequence of a protein) that does not produce a major change in its properties is called microheterogeneity.
2. Microheterogeneity gives rise to molecular variants having very minute differences in their molecular mass.

W. Ansar, S. Ghosh, *Biology of C Reactive Protein in Health and Disease*,
DOI 10.1007/978-81-322-2680-2_5

3. Microheterogeneity may be the outcome of posttranslational modifications.
4. Microheterogenity in proteins and protein complexes were purified by immuno-affinity methods, and their identification and characterization are carried out by various methods like ELISA, lectin binding studies, MALDI-TOF, MS SPR, Western blotting, and many other techniques.
5. Each technique has its advantages and also constraints. So a combination of traditional and modern approaches is required.
6. Newer approaches are coming in to establish these microheterogeneous variants. A rapid, highly reproducible, and cost-effective system for the determination of molecular variants of microheterogeneous proteins and protein complexes is always needed.
7. In proteomics and its interdisciplinary branches, the same protein needs extensive validation of its structure across a population of significantly large size under study. It forms the basis of accurate progress of diagnosis of disease in personalized medicine.
8. The application of proteomics in establishing disease-specific molecular variants as compared to normal individuals may be applied in the search of a potential disease biomarker.

5.1 Introduction

The proteome of a protein implies a static nature and was defined as the *protein complement expressed by its genome* in July 1995. But in reality, the proteome of a protein shows high dynamicity. Differential expression of different types of protein, their relative abundance, and their functional modification are aspects that are controlled by physiological or pathophysiological condition of the individual. Therefore, the proteome is a reflection of (1) the internal cellular machinery from where the protein is synthesized or functions or (2) the external conditions that are encountered by the individual. Therefore, proteomics is a dynamic process where different techniques were employed to enumerate the structurally diverse proteins and to correlate that diversity with the in vivo biological processes of proteins. Knowing the concept of vast protein diversity with a mere coding by only 30,000–40,000 genes, the interesting fact that occurs is the proteome is a much more complex process than the theoretical term itself implies. Alternative splicing and reactions occurring during post-translational modifications of proteins generate highly varied sets of proteins exceeding a million distinct protein molecular variants with functional variation (Haynes et al. 1998; Hochstrasser et al. 2002; Molloy and Witzmann 2002; Tyers and Mann 2003; Schweigert 2005).

It is not necessary that the normal physiological concentrations of specific proteins will change significantly, in many cases, but some posttranslational modifications are modulated for its proper functioning. The heterogeneity of protein can be caused by various sophisticated posttranslational modifications or by simple proteolysis (Haynes et al. 1998; Hochstrasser et al. 2002; Molloy and Witzmann 2002; Tyers and Mann 2003; Schweigert 2005).

The widespread range of posttranslational modifications includes (1) glycosylation, (2) ubiquitination, (3) phosphorylation, and (4) methylation. Less frequently observed modifications include (1) hydroxylation, (2) glutathionylation, (3) sulfation, and (4) transglutamination.

Studies reveal that glycosylations are observed in more than 50 % of all known proteins that modulate protein function. The minute differences between two protein species, like modification of a single amino acid, create two molecular variants, invariably exerting this minute changes to influence the protein function. The various physiological functions like the (1) the ability to be secreted, (2) the ability to be transported in plasma, (3) the potential of binding to receptor, (4) the property of degradation, and (5) their excretion and protein–protein complex formation may be influenced. Due to physiological or differential pathophysiological stimuli, protein posttranslational modifications can lead to the production of a dynamic repository of functionally diverse proteins that respond rapidly to these stimuli (Schweigert 2005; Baumann and Meri

2004; Davis 2004; Hancock 2002; Mann and Jensen 2003).

Plasma proteomics is the present trending field of research which is gaining interest in clinical medicine for the use of differential expression of different proteins in diseases. In proteomics, the protein–protein interaction process for intracellular communication is of outmost interest. These proteins exhibit a multi-protein complex which thereby affects the functions like receptor binding, plasma transport, degradation, and glomerular filtration in excretion (Figeys 2004; Russell et al. 2004; Grant and Husi 2001).

The two foremost tasks of clinical proteomics include the monitoring of PTM and the study of protein complexes to understand the potential impact of an altered protein structure and function in health and disease. Thus, new terminologies like "heteromics" and "interactomics" are described under the major domain of "functional proteomics." The aim of the functional proteomics is to characterize the protein modifications in the light of structural characteristics, functions, and also their potential interactions with other proteins. Functional proteomics contributes essentially to clinical diagnostics, especially in its application in personalized medicine. Crucial information gathered from functional proteomics studies has potential application in (1) using these proteins in early diagnoses of clinical conditions, (2) developing novel prognostic disease indices, and (3) generating novel therapies for diseases. Though the study of the role of protein complexes and their PTMs and their biological function is quite relevant, the vivid in-depth study suffers from the limitation of availability of suitable sensitive, reliable, and repeatable high biotechnological techniques (Schweigert 2005; Figeys et al. 2001; Nair et al. 2004; Arthur 2003; Hanash 2003).

5.2 Brief History of Microheterogeneity of Proteins

The human immune system has evolved in almost a million of years. This immune system has developed host defense systems highly specialized against pathogens. The human innate immune system lay its origin from the invertebrates. Innate immunity is a primitive host defense system wherein proteins encoded by germline recognize pathogens. Most of the innate immune cells are highly conserved in nature. This system bears pattern recognition receptors (PRRs) that recognize pathogens thereby triggering a variety of mechanisms to eliminate pathogens. The PRRs are host proteins that recognize pathogen-associated molecular patterns (PAMPs) and molecules with evolutionary primitive ancestors. Thus, recognition of pathogen molecules activates the effector molecules including (1) complement molecules, (2) cytokines, and (3) antimicrobial peptides. The major PRR families of proteins include plasma pentraxins and others group of proteins (Pepys and Baltz 1983; Tillett and Francis 1930; Li et al. 1994).

"Pentraxin" is the structure assigned for some acute-phase reactants like (1) CRP, (2) SAP, (3) female protein of Syrian hamster (HSAP), and many others. Pentraxin is a rare configuration due to their conserved pentameric arrangement of their protomers (Tillett and Francis 1930; Li et al. 1994).

Most of these proteins exhibit the common characteristics, but a few of them showed some variations in their structure and functions. CRP of horse shoe crab (*Limulus polyphemus*, an Arachnida), is phylogenetically the oldest member of this family, which differs from their vertebrate counterparts by being hexameric. The three major acute-phase proteins (CRP, SAP, and HSAP) differ significantly in their calcium-binding and calcium-dependent ligand-binding properties (Srinivasan et al. 1994).

5.3 Microheterogeneity of Proteins: Major Pathophysiology

Glycosylations can alter the structure and function of proteins. Such alterations in the acute-phase proteins (app) lead to microheterogeneity changes in the app thereby involved in the alteration of acute-phase responses (APR). The minute structural variation in APP has been defined as "microheterogeneity." The major cause of microheterogeneity is mostly contributed by the qualitative changes of some acute-phase glycoproteins. APPs involved in acute and chronic types of inflammation show differences in their microheterogeneity (Hansen et al. 1984 and Mackiewicz et al. 1987a, b; Kushner 1982; Kushner and Mackiewicz 1987). Such microheterogeneity differences in the APP is generated due to the differential activity of glycosylating enzymes thereby effecting alteration of arrangement and composition of their glycan side chains thus differing from their normal phenotype (Raynes 1982).

Microheterogeneity differences are prominent from comparative studies of pathophysiology of normal and disease states and also between acute and chronic disease conditions. Studies from malignant and nonmalignant and acute versus chronic diseases reveal microheterogeneity differences in α_1-acid glycoprotein (AGP) and α_1-antichymotrypsin (Hansen et al. 1984 and Mackiewicz et al. 1987a, b). Duration of inflammation and disease severity has also been reported to affect the glycosylation profile in AGP (Hrycaj et al. 1993; Havenaar et al. 1995). Such alterations are mostly accompanied by changes in glycoforms of proteins (distinct molecular variants) with increased or decreased branching and reactivity.

5.4 Microheterogeneity and Its Causes

Microheterogeneity has been reported to arise from mechanisms of proteolysis or due to different types of PTMs which could range from frequent ones like (1) glycosylation, (2) phosphorylation, (3) ubiquitination, and (4) methylation to the less frequent ones, such as (1) glutathionylation, (2) hydroxylation, (3) sulfation, and (4) transglutamination. This occurs by altered arrangement or composition of their glycan side chains due to differences in activity of various glycosylating enzymes. Although synthesis of the polypeptide chain is under direct genetic control, the attachment of carbohydrate moieties is controlled by posttranslational modification mechanisms (Spiro et al. 1986). Enzymatic treatment of native olive tree pollen with the specific enzyme glycosidase PNGase F is observed to reduce the allergen affinity toward specific IgE antibodies without modification of its secondary structure (Batanero et al. 1994). The electrophoretic heterogeneity of human prostatic acid phosphatase is not due to the oligosaccharide moiety of the glycoprotein or by the altered conformational states of the protein. The heterogeneity is due to the structural variations of the polypeptide itself at the C-terminus, partial deamidation, phosphorylation, or sulfation or may be attributed to other posttranslational modifications of the protein chain itself (Morris et al. 1989) (Table 5.1).

In different diseases or in inflammatory conditions, microheterogeneity of proteins, their causes of molecular variation, techniques employed in the detection of isoforms, and differential effects of these proteins were noted.

5.5 Immunological Characterization of PTM and Protein–Protein Complexes

The traditional method to detect and investigate the microheterogeneity of proteins was two-dimensional electrophoresis (2D-gel electrophoresis). Still today it is one of the key technologies used globally in the analysis of changes in posttranslational modifications. Two proteins having the same molecular mass differ in their isoelectric point (pI). The differences in the isoelectric point between proteins can be detected by two-dimensional electrophoresis. Changes during posttranslational modifications results in the minute molecular mass variations, and thus unique molecular variants are developed in differential clinical conditions. Moreover, by two-dimensional electrophoresis, the separation of glycoproteins based on their induction in glycosylation in different pathological conditions is also possible. The different isoforms or molecular variants are visible by protein spots. However,

Table 5.1 Microheterogeneity in glycoproteins and probable causes

Diseases/states	Proteins analyzed	Effects seen	Protocol/techniques followed	Causes	References
Pregnancy and in other disorders	Plasma proteins (alpha 1-protease inhibitor, alpha 1-antichymotrypsin, ceruloplasmin, alpha 1-acid glycoprotein, and alpha 2-HS glycoprotein)	Components of all five proteins were altered in the direction of less con A binding in diseased condition	In the first dimension lectin affinity crossed immunoelectrophoresis with concanavalin A and in the second dimension, electro-endosmotic elution with sugar was done	Differences in activity of various glycosylating enzymes but follow the same regulatory mechanism	Raynes 1982
Chronic inflammatory response to cancer and rheumatoid arthritis		No significant alteration in con A binding			
Patients having high blood estrogen level		General reduction of con A binding			
Childhood acute lymphoblastic leukemia (ALL)	Sialate-*O*-acetyltransferase enzyme in ALL cell lines and lymphoblasts from bone marrow of children diagnosed with B- and T-ALL and 9-*O*-acetylated sialoglycoproteins (Neu5,9Ac(2)-GPs) on lymphoblasts	Positive correlation between the enhanced sialate-*O*-acetyltransferase activity and the enhanced expression of Neu5,9Ac(2)-GPs in these lymphoblasts. Sialate-*O*-acetyltransferase activity increased at the diagnosis of leukemia, decreased with clinical remission, and sharply increased in relapsed patients	Enzyme activity seen by radio-thin-layer chromatography, radiometric assay, newly developed nonradioactive ELISA, and fluorometrically coupled radio-high-performance liquid chromatography	Differences in activity of various glycosylating enzymes	Mandal et al. 2009
Lung cancer, benign lung inflammation patients, and healthy individuals	Acute-phase proteins (orosomucoid, ceruloplasmin, antitrypsin, and haptoglobin)	Correlation was found between tumor size and acute-phase protein level. No difference in acute-phase protein concentration in benign and malignant diseases. Patterns of APPs in the patients with benign inflammation and malignant disease were different	Crossed immunoelectrophoresis without lectin and with lectin. Glycan-dependent microheterogeneity of APP profile in isolating high-risk patients and in monitoring radically treated cancer patients for relapse	Nonspecific inflammation in the tissues surrounding the tumor	Hansen et al. 1987

(continued)

Table 5.1 (continued)

Diseases/states	Proteins analyzed	Effects seen	Protocol/techniques followed	Causes	References
Altered glycosylation seen in plasma in some inflammatory states using human hepatoma cell line Hep 3B and keratocarcinoma cell line COLO-16	APPs (alpha 1-protease inhibitor and ceruloplasmin)	Altered glycosylation patterns of these proteins seen in human serum of various inflammatory states	Agarose affinity electrophoresis with free Con A as a ligand	Altered glycosylation seen in plasma proteins is regulated by the effects of monokines on hepatocytes. Gene expression and glycosylation of these APPs may be regulated by different mechanisms	Mackiewicz et al. 1987a, b.
Early rheumatoid arthritis (RA) patients without clinical features of intercurrent infection and patients with acute bacterial infections	Alpha 1-acid glycoprotein (AGP)	High values of AGP reactivity coefficients were found in acute patients which was similar to those found in acute bacterial infections patients	Affinity immunoelectrophoresis with concanavalin A	Microheterogeneity of AGP in early and long-standing RA is controlled by differences in cytokine action at different stages of the disease	Hrycaj et al. 1993
Sera obtained from patients with hyperimmunoglobulinemia D and periodic fever syndrome	AGP	Increased concentration of AGP as compared to control values were found during attacks and in remissions. However, in febrile attacks, as compared to healthy controls, the presence of diantennary glycan-containing glycoforms of AGP also increased. No changes were found during remissions	Crossed affino-immunoelectrophoresis using concanavalin A and *Aleuria aurantia* lectin as diantennary glycan- and fucose-specific affino-components	Disease severity and persistent inflammation	Havenaar et al. 1995

Healthy elderly and younger individuals living independently in a community	CRP, AGP, and SAA	Concentrations of CRP in the older persons were significantly higher than the younger group. Concentrations of SAA and AGP were similar in the two groups. AGP glycosylation forms observed in chronic inflammatory states in the elderly individuals showed reduced binding affinity for concanavalin A	Binding affinity with concanavalin A	Quantitative and qualitative alterations of acute-phase proteins occur with physiological aging in humans	Ballou et al. 1996
Normal healthy individual	Human prostatic acid phosphatase	Neuraminidase retarded the anodic electrophoretic mobility of acidic forms of human prostatic acid phosphatase	Polyacrylamide gel electrophoresis	Different numbers of sialic acid residues to a single enzyme protein, variation in electrophoresis due to posttranslational modifications, and organ-specific variations of human acid phosphatases	Smith and Whitby, 1968; Beckman and Beckman, 1967
Sera were collected from 19 patients with primary squamous cell carcinoma of the lung, from 16 patients with an inflammatory lung disease, and from 17 persons with normal health	Alpha 1-acid glycoprotein (AGP, orosomucoid)	Distribution of AGP into varied microheterogeneity forms. Microheterogeneity patterns of AGP in the three diseases were significantly different from each other. The concentration of AGP in the patients was significantly different from the concentration in the healthy group	Crossed immuno-affinity electrophoresis with addition of Con A in the first dimension and sugar in the second dimension	Microheterogeneity of AGP in different inflammatory conditions. Microheterogeneity was also observed both among patient and normal individuals	Hansen et al., 1986

these denaturing separating methods fail to characterize the protein completely. At present, antibodies are used for the characterization of both PTM and protein complexes (Clack et al. 2003; Schweigert 2005; Packer and Harrison 1998).

High-affinity antibodies capture both (1) proteins of interest and (2) related protein complexes. Capturing proteins and their interacting proteins or protein complexes by high-affinity antibodies is the most promising analytical approach in proteomics. As antibodies can specifically identify an epitope, they find application in protein-detection array system (Schweigert 2005; Lal et al. 2002).

Traditionally techniques like immunohistochemistry, Western blots, and immunoprecipitation have been used to detect protein antibody interactions. These techniques were used to determine and characterize gene products—such as (1) the tissue distribution and cellular localization of proteins, (2) their posttranslational modifications, and (3) the level of protein expression and its interaction to form complexes (Fig. 5.1). One-dimensional gel electrophoresis followed by Western blot is used to identify minute differences in protein modification. Using the property of specificity of antigen–antibody binding, antibody chips are being designed to understand the expression of desired proteins. In a surface plasmon resonance (SPR) protein–protein interaction, the protein antibody interaction and protein cell interaction were mostly investigated. For the same purposes, radioimmunoassay (RIA), enzyme-linked immunosorbent assays (ELISA) and Western blotting were used (Schweigert 2005; Burgess and Thompson 2002; Hage 1999).

Thus, to monitor posttranslational modifications in different assays, it is important to know the different modifications on one hand, and on the other it is important to know their specific interacting antibodies (Fig. 5.1). In mass spectrometry (MS), specific affinity interactions can

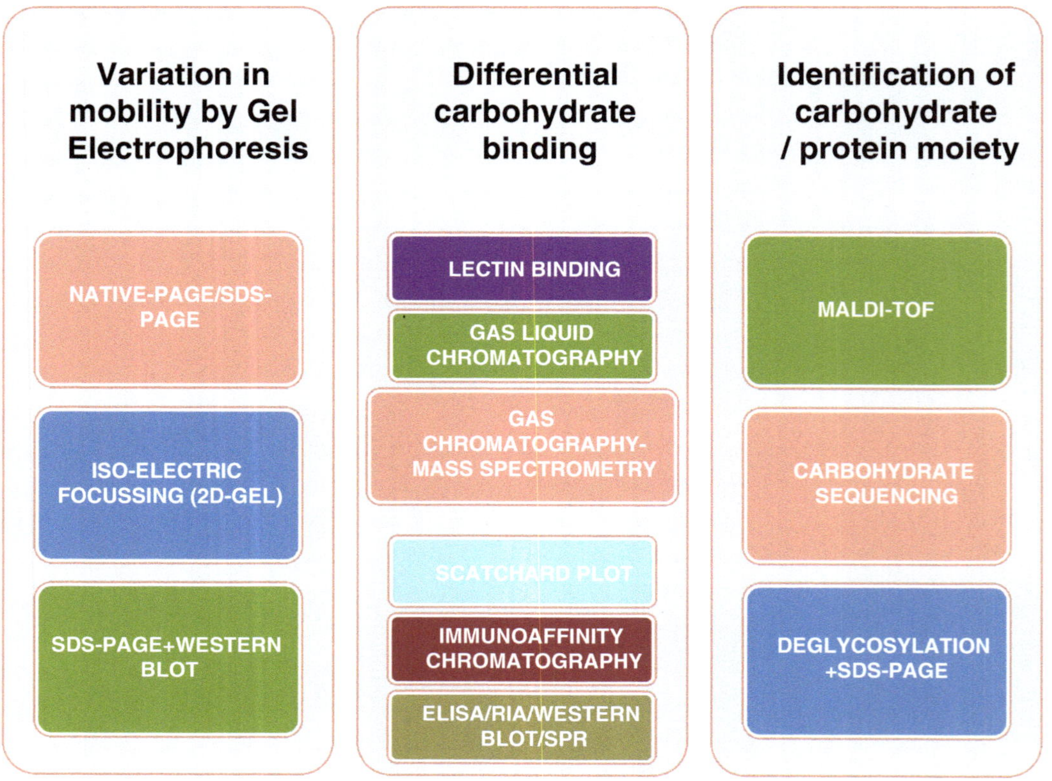

Fig. 5.1 *Techniques employed in the detection of microheterogeneity*. The small but distinct differences in the molecular variants during electrophoresis, the differential lectin binding of each isoform, and finally carbohydrate sequencing and molecular modeling give a clear idea about microheterogeneity

be used to identify protein isoforms and complexes. In specific affinity interaction processes, traditional approaches of antigen separation with an immobilized immuno-affinity column or immunoprecipitation are also noteworthy (Schweigert 2005).

In MS analysis the column-packing material can be performed directly or can be done after the analyte is eluted from the column. Affinity chromatography is quite often employed together with various other detection methods. All these techniques and methods have been successfully used in different approaches of proteomics to characterize the microheterogeneity of varied proteins and plasma peptides. These proteins and plasma peptides in healthy and diseased states showed distinct molecular variants. These plasma proteins showing microheterogeneity are CRP, amyloid-β peptides, β2-microglobulin, serum amyloid-α, transthyretin (TTR), and many others. For the development of an advanced immunoassay, the surface-enhanced laser desorption/ionization (SELDI) platform is adopted. In SELDI, antibodies are used as an affinity capture tool, for qualitative and quantitative estimation of various proteins in differential biological mixtures by MS. In the SELDI-based immuno-affinity assay, proteins are directly detected without tags. Moreover, in sandwich ELISA technology, a protein of interest or an antigen is sandwiched by specific antibodies which bind against that antigen only. Thus by using SELDI, sandwich ELISA, SPR, SELDI-time of flight MS, Westernblotting; different isoforms of a protein in differential clinical condition was detected using antibodies as bait. In surface plasmon resonance-biomolecular interaction analysis (SPR-BIA), to characterize proteins and protein–protein interactions, the Biacore instrument was coupled with the mass spectrometer (Tolson et al. 2004; Nelson et al. 2000a, b; Schweigert 2005; Clarke et al. 2001; Kiernan et al. 2003; Lim et al. 2003; Nedelkov et al. 2004; Tubbs et al. 2001; Sydor et al. 2003; Lin et al. 2003; Nelson et al. 2000a, b; Davies et al. 1999).

MS enables detection of minute mass differences clearly which forms a major limitation in traditional methods of one-dimensional electrophoresis and immunological detection. Two-dimensional electrophoresis followed by Western blot offers better resolution of the microheterogeneous protein variants, but this method suffers from the limitation by similar minute mass accuracy. One-dimensional electrophoresis followed by Western blot analysis and MS analysis is also one of the approaches for mass resolution in isoforms of proteins.

In on-chip immunoassay, the antibody in the protein chip must be an active state as well as in high density. Therefore, the substrate on the solid surface of the chip has to be modified in order to achieve maximum binding capacity and activity. Thus, the reactive surfaces on the chip can covalently cross-link with the proteins. These are the basic consideration while developing protein chips. Generally, polyclonal antibodies are capable of binding different isoforms of protein. Polyclonal antibodies can bind different molecular variants of a protein regardless of their differences and can therefore be often used to analyze microheterogeneous forms. Moreover, further characterization of the molecular variants by other immunoassays is highly needed. But the greatest disadvantage of using polyclonal antibodies over monoclonal antibodies is the more pronounced cross-reactivity with other proteins which is highly nonspecific. This is more observed while studying protein complexes (Schweigert 2005).

Other factors controlling the usage of antibodies, either monoclonal or polyclonal, as tools for analysis are stability, affinity, and specificity of the antibodies. However, extensive experimentation needs to be carried out before their use. In microarray-based studies, a very small percentage of the commercial antibodies are suitable for application.

5.6 Identifying Different Posttranslationally Modified Proteins

Antibodies specific to posttranslational modification may be developed and studied for a particular protein. Specific definite antibodies are developed that are recognizing only specific proteins that have tyrosine residues that are phosphorylated or specific to other modifications.

For glycosylation of proteins, or its sugar modifications, certain lectins are employed that bind sugars of the glycoproteins. The most common method to determine protein posttranslational modification is to allow the complex mixture of proteins to run in 2D-gel electrophoresis.

Recently, SDS-PAGE is combined with shotgun proteomics, a newer approach known as PROTOMAP, which enables the detection of minute changes in migration of proteins in gel electrophoresis, mostly due to proteolysis or posttranslational modification (Schweigert 2005).

Carbohydrate chains are known to respond to disease, injury, and environmental changes. To know how and why all this happens, the carbohydrate structure needs to be first evaluated. Mass spectrometry may be employed for this. Mass spectrometers break the carbohydrate molecule and convert them into separate ions or charged particles. These ions are then categorized and analyzed depending on their mass-to-charge ratio. These ratios are analyzed for sequencing of the entire glycoprotein molecule.

5.7 Detection of Proteins in Complex Mixtures

Antibodies are the most common tools used to detect particular proteins and their modified forms. ELISA techniques can be used for quantitative estimation of protein expression.

For rapid determination of protein mixtures, proteomic study techniques like MALDI and electrospray ionization (ESI) are being used. More recently, in novel proteomic assay, the fast parallel proteolysis (FASTpp) thermal protease resistance was exploited to detect specific proteins in lysates of *E. coli*. FASTpp may be applied in the future to detect mechanistic effects of point mutations or in proteomic perturbations in cancer (Schweigert 2005).

5.8 To Identify Interaction of Protein with Protein

As most proteins are functional only in concoction with other proteins, the major goal of proteomics is to identify the interacting proteins. This finds application in determining protein members involved in cell signaling pathways.

The traditional method used to detect protein–protein interactions involves the yeast two-hybrid analysis. Several recent techniques like SPR, protein microarrays, dual polarization interferometry, immuno-affinity chromatography followed by MALDI-TOF, phage display, and many computational methods are used. Brief descriptions of these methods used are discussed above section (Schweigert 2005).

5.9 Detection of Microheterogeneity

Affinity electrophoresis with a lectin, Con A, as a ligand finds a major application in determining microheterogeneity of acute-phase glycoproteins. Many proteins in the plasma exist as multiprotein complexes. They participate in functions like (1) plasma transport, (2) binding to receptor, and (3) subsequent degradation and excretion. Thus, PTM and proteins in complexes are foremost processes with potential impact on health and disease. Several techniques are being employed in the detection on microheterogeneity in proteins. Some techniques used to study microheterogeneity in APP and CRP are described below.

5.10 Lectin Binding Studies to Identify Microheterogeneity

Lectins are proteins that bind to carbohydrates with particular specificity. Despite specificity of lectin, it can bind to one carbohydrate moiety with higher affinity. The major specificity toward carbohydrates of some common lectins used is described in Table 5.2. Lectins find application in detecting the nature and type of glycosylation of glycoprotein by different technologies like affinity chromatography, blotting, and electrophoresis (Goodarzi and Turner, 1996; Fig. 5.2). All such techniques need skilled technicians, a large quantity of lectins, and a low number of specimens and moreover give a semiquantitative result. Lectins can be used in sandwich ELISA in a multi-well system. In sandwich ELISA, lectin is used as one partner and antibody as the other partner to detect the protein. Another method to identify glycoprotein using lectins is the steroid hapten, digoxigenin (DIG)–glycan detection assay. The purified glycoprotein is allowed to be absorbed on a plastic surface of a microtiter plate. After washing the unbound proteins and blocking the uncoated site, a DIG- or biotin-labeled lectin is used to interact with the bound protein in the microtiter. Unbound lectin is washed out, and the bound lectin is detected by using an anti-DIG antibody or a streptavidin conjugate enzyme. Streptavidin shows high affinity toward biotin (Goodarzi and Turner 1996). Another method includes the binding of Iodinated glycoproteins to Sepharose or agarose-bound lectins with different sugar linkages and maintaining specificity at 4 °C overnight. Extensive washing is done to eliminate unbound radioactivity. Bound radiolabeled proteins were monitored in a gamma counter (Ansar et al. 2009b).

5.11 Microheterogeneity in CRP in Animals

Microheterogeneity of CRP is an interesting part of study of CRP. Such microheterogeneity differences are recorded as early as in the horse shoe crabs in the evolutionary history of animals. Microheterogeneity of CRP has been reported from different species like horseshoe crabs, fish, and humans.

Table 5.2 Differential carbohydrate specificity of some lectins

Lectin	Abbreviation	Specificity
Concanavalin A	Con A	Mannose α 1–3 or mannose α 1–6
Datura stramonium agglutinin	DSA	Galactose β 1–4 *N*-acetyl glucosamine
Snowdrop lectin	GNA	α 1–3 and α 1–6 linked high mannose structures
Aleuria aurantia agglutinin	AAA	Fucose α 1–6 *N*- acetyl glucosamine
Lens culinaris agglutinin	LCA	Mannose α 1–3 or mannose α 1–6 or fucose α 1–6 Glc NAc
Lotus tetragonolobus agglutinin	LTA	Fucose α 1–2 galactose β 1–4 or fucose α 1–3 *N*- acetyl glucosamine
Lentil lectin	LCH	Fucosylated core region of bi- and triantennary complex type N-Glycans
Sambucus niagra agglutinin	SNA	*N*-acetyl neuraminic acid or α 2–6 galactose
Maackia amurensis agglutinin	MAA	*N*-acetyl neuraminic acid or α 2–3 galactose
Peanut agglutinin	PNA	Galactose β 1–4 *N*-acetyl glucosamine on *O*-linked chains
Wheat Germ agglutinin	WGA	GlcNAcβ1-4GlcNAcβ1-4GlcNAc, Neu5Ac (sialic acid)
Ricinus communis agglutinin	RCA	Galactose β 1–4 *N*-acetyl galactosamine
Jacalin	AIL	(Sia)Galβ1-3GalNAcα1-Ser/Thr (T-antigen)
Maackia amurensis leukoagglutinin	MAL	Neu5Ac/Gcα2,3Galβ1,4Glc(NAc)
Maackia amurensis hemoagglutinin	MAH	Neu5Ac/Gcα2,3Galβ1,3(Neu5Acα2,6)GalNac
Ulex europaeus agglutinin	UEA	Fucose α1-2Gal-R

Gal, galactose; *N-Ac Gal NAc*, *N*-acetylgalactosamine; *N-Ac Glc NAc*, *N*-acetylglucosamine; *Neu*, neuraminic acid

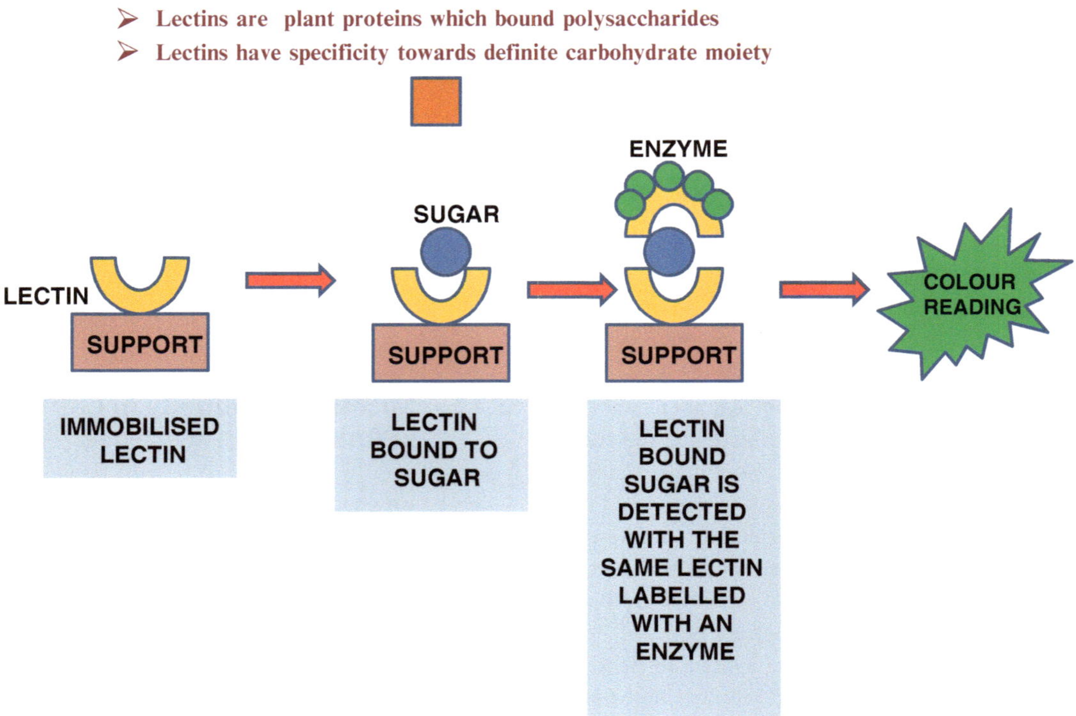

Fig. 5.2 Lectins and interactions

Lectins specific to *N*-acetyl-D-galactosamine was isolated from an *Epiphragmophora trenquelleonis* snail (ETL) and it was partially characterized. ETL1 showed two protein subunits, while three subunits were revealed in ETL2. Under reducing conditions, results of SDS-PAGE analysis of both lectins revealed four protein subunits, and *N*-linked oligosaccharides were detected in some subunits of both ETL1 and ETL2. Moreover, *O*-linked oligosaccharides were detected in some subunit of ETL2 and ETL1. Both lectins exhibited microheterogeneity in IEF with varied pI values. Posttranslational modifications lead to observed differences between ETL1 and ETL2 due to differences in their protein subunits (Castagna et al. 1996).

A sialic acid-specific lectin has been isolated from the albumin glands of the *Cepaea hortensis*, a garden snail, by performing affinity chromatographic purification on a fetuin-Sepharose column followed by gel filtration on Superdex 200. Three serological identical bands were detected, of which only two were found to be glycosylated (*N*- and partially *O*-glycosidic bound). Three sugars, namely, mannose, galactose, and *N*-acetylglucosamine, could be detected in these isoforms. IEF of the purified lectin revealed a differential pattern with bands in the pH range of 4.3–5.0. The heterogeneity was also revealed by amino acid sequencing of internal tryptic peptides (Gerlach et al. 2002).

5.11.1 Microheterogeneity in Horseshoe Crabs

In the American horseshoe crab (*Limulus polyphemus*), the protein limulin, in the hemolymph plasma, is present. Limulin binds to sialic acid and phosphorylethanolamine-binding hemagglutinin. It is also a hemolytic C-reactive protein (Armstrong et al. 1996). Three types of C-reactive protein in the plasma of the Japanese horseshoe crab (*Tachypleus tridentatus*) were identified, based on their differential affinities toward fetuin–agarose and phosphorylethanolamine–agarose. The affinities were determined by an artificial phosphorylethanolamine–protein conjugate and a fetuin in the quantitative precipitin assays. The amino acid sequences of the isolated CRP revealed different homologous proteins termed as *Tachypleus tridentatus* CRP-1 (tCRP-1), tCRP-2, and tCRP-3 (Fig. 5.3). All these three possibly constitute isoprotein mixtures. These three types of CRPs are based on different affinities against different ligands, designated as tCRP-1, tCRP-2, and tCRP-3, and exhibited varied calcium-dependent hemolytic and sialic acid-binding and hemagglutinating activities (Iwaki et al. 1999). Only tCRP-2 and tCRP-3, but not tCRP-1, showed agglutination of RBC when treated on mammalian RBC. The most abundant C-reactive protein in the plasma, tCRP-1, exhibited the highest affinity to the phosphorylethanolamine–protein conjugate. But tCRP-1 lacks the property of both sialic acid-binding and hemolytic activities. The tCRP-2 showed limulin-like properties as it bound to both phosphorylethanolamine–agarose and fetuin–agarose and also exhibited sialic acid-binding activities and calcium-dependent hemolytic property. Moreover, tCRP-2 showed a higher affinity to colominic acid, a polysialic acid from bacteria. On the other hand, tCRP-3 shows stronger sialic acid-binding, hemolytic, and hemagglutinating activities than tCRP-2. But tCRP-3 has no affinity to phosphorylethanolamine–protein conjugate, to phosphorylethanolamine–agarose, and also to colominic acid. Thus, tCRP-3 is basically a novel hemolytic C-reactive protein lacking the common property of phosphorylethanolamine–agarose binding affinity. Nearly 22 clones of tCRPs exit both in their amino acid sequences and in *N*-linked glycosylation and indicate the possibility of isolectins. The structural and functional diversities of all the tCRPs provide a novel model for studying the functioning of innate immune mechanism in invertebrate

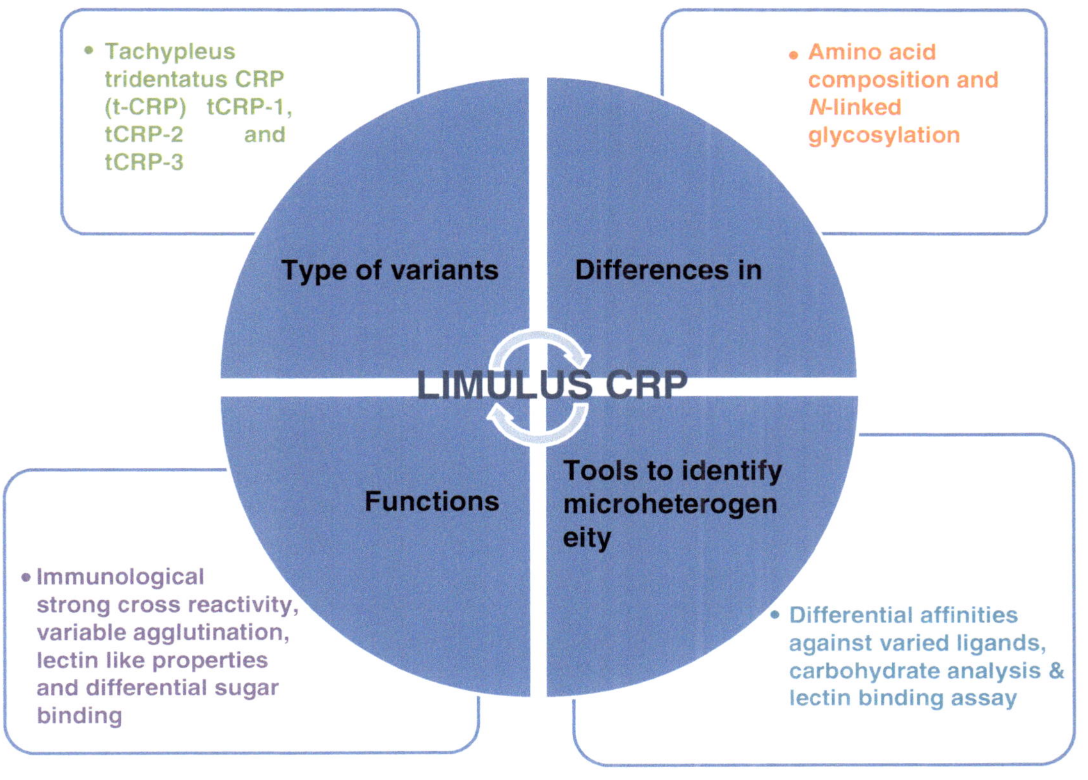

Fig. 5.3 *Microheterogeneity in Limulus*. The different molecular variants of *Limulus*; their differences and functions are noted

organisms, which survive without the thorough benefit of adaptive immunity.

5.11.2 Microheterogeneity in Fishes

Studies from the vertebrate freshwater fishes including *Labeo rohita* and *Catla catla* revealed that alterations in the environmental condition could affect the CRP levels. Differential toxic aquatic conditions could cause an increase in the CRP levels and also gives rise to different molecular variants of CRP showing differences in amino acid and carbohydrate compositions. They showed differential minute alterations in their amino acid and carbohydrate compositions (Sinha and Mandal 1996; Sinha et al. 1999a, b, c, d, 2001; Mandal et al. 1999; Paul et al. 1998, 2001). Such findings reveal a prominent role of CRP as an indicator of environmental pollution and aquatic toxicity.

Fish CRP was purified by using calcium-dependent affinity chromatography on a phosphorylcholine (PC)–Sepharose column. Different forms of fish CRP was obtained from the sera of *Labeo rohita* from either freshwater (CRP_N) fishes or fishes in water polluted with sublethal doses of toxic molecules like (1) mercury (CRP_{Hg}), (2) cadmium (CRP_{Cd}), (3) phenol (CRP_{Ph}), and (4) hexachlorocyclohexane (CRP_{Hx}). These CRPs purified from polluted water have nearly three- to fivefold elevated serum concentration than CRP_N. When these induced forms of CRP proteins are run on native polyacrylamide gel electrophoresis (PAGE), they show remarkable differences in their molecular mass, charge, and/or shape as revealed by their differences in electrophoteric mobility. At the peak of induction, the pollutant-specific molecular variant has been reported to replace the normal form as revealed by the kinetic studies. These pollutant-specific CRPs differ significantly from each other as studied by their amino acid and carbohydrate compositions and showed strong immunological cross-reactivity, variable agglutination properties with RBC from different sources in presence of calcium ions, isoelectric focusing, lectin binding assay, and binding to PC and pneumococcal C-polysaccharide (CPS). Electron microscopy revealed the secondary structures of these purified CRPs, including pentraxin structure (Sinha and Mandal 1996; Sinha et al. 2001; Mandal et al. 1999).

Similarly, pollutant-specific glycosylated molecular variants of CRP were purified from the sera of the major carp, *Catla catla*. Fish CRP was purified from fishes confined in either freshwater (CRP_N) or in water polluted with sublethal doses of mercury (CRP_{Hg}), cadmium (CRP_{Cd}), phenol (CRP_{Ph}), and hexachlorocyclohexane (CRP_{Hx}). CRPs in polluted fish are induced than CRPs purified from freshwater as shown by differences in electrophoretic mobility. The carbohydrate content of these CRPs ranges from 20 to 50 %. The binding constants of these CRPs with CPS were varied. The molecular variants varied in their amino acid and carbohydrate compositions, secondary structures, kinetic studies of metal intoxication, binding to PC and lectins, and differential electrophoretic mobility in native gel and in isoelectric focusing. Interestingly, their electrophoretic mobilities become identical on desialylation and deglycosylation implying that these molecular variants differ in their glycan component. These CRPs share common properties of a CRP, including the pentraxin structure as showed by electron microscopy and on examination of their immunological cross-reactivity. Thus, different molecular variants were induced in a toxic stressful environment as compared to CRPs purified from freshwater. Moreover, these variant coexists with normal CRPs, and they show similar overall molecular topology but show differences in their quantitative extent of binding (Sinha and Mandal 1996; Paul et al. 1998, 2001; Mandal et al. 1999).

An acute-phase protein isolated from channel catfish (*Ictalurus punctatus*) serum can precipitate pneumococcal C-polysaccharide (CPS) in a calcium-dependent manner. Similarly, a non-glycosylated, phosphorylcholine (PC)-reactive protein (PRP) of molecular weight 100 kDa was also isolated by affinity chromatography from sera of catfish. The planar, pentagonal symmetry of PRP was confirmed by electron microscopy.

Interestingly, purified PRP has gamma mobility, whereas in serum samples PRP has gamma-beta mobility. The channel catfish PRP is similar in structure function to human, dogfish (*Mustelus canis*), and rainbow trout (*Oncorhynchus mykiss*) CRP (Szalai et al. 1992).

5.11.3 Microheterogeneity in CRP as Marker for Pollution in Freshwater Fishes

The levels of CRP were reported to be elevated in the serum of fishes maintained in water contaminated with cadmium and mercury as compared to those isolated from the fishes maintained in normal freshwater. Both the total carbohydrate and sialic acid content in CRP have been reported to vary in the order of CRP isolated from fish maintained in water contaminated with mercury and of CRP isolated from fish maintained in water contaminated with cadmium as compared to that of CRP isolated from fish maintained in freshwater. These studies in *Labeo rohita* revealed differences in the content of both charged and uncharged sugars; a probable cause of CRP microheterogeneity is induced by metal pollutants (Sinha and Mandal 1996) thus indicative of a possible application of CRP as a biomarker to environmental pollution.

5.11.4 Microheterogeneity in Rats

In rats the role of glycosylation in the microheterogeneity of corticosteroid-binding globulin (CBG) was revealed by electrophoretic migrations. The glycoprotein, CBG, migrates as doublet bands in PAGE and as numerous bands in IEF (isoelectric focusing) technique. Desialylation with neuraminidase does not remove this doublet in either PAGE or SDS-PAGE, indicating that the doublet in rat CBG does not arise due to sialic acid residue differences. But desialylated CBG migrates as two bands in IEF. As migration in IEF is thoroughly based on charge and as only sialic acid molecules are charged in *N*-linked glycosylation, any heterogeneity seen by the desialylated glycoprotein might reside within the protein moiety only. In rat CBG, there is also variation in the protein moiety as indicated by sodium dodecyl sulfate (SDS) treatment. The differences in the number, type, or location of sugars attached to each of the variants were differential as detected by the treatment of the separated upper and lower variants of native CBG with *N*-glycosidase F (PNGase F). The upper variant shows a quicker deglycosylation over time as compared to the lower variant. Blocking the *N*-terminal end could not account for any variation, as both the upper and lower variants were sequenced separately resulting in identical sequences of the first 13 amino acids. Moreover, sugars could not be detected by wheat germ agglutinin (WGA) lectin or by digoxigenin glycan detection system. *O*-Glycosidase enzyme treatment showed the presence of *O*-glycosylation. Thus, the differences in the protein moiety as well as the sugar entities contribute to the heterogeneity of CBG (Ali and Bassett 1995).

5.12 Microheterogeneity of Human Acute-Phase Proteins

Human APP has been divided broadly into two groups such as positive or negative acute-phase proteins. Human APP may be divided into three groups based on the usual magnitude of the plasma changes observed (Fig. 5.4). Of at least 30 acute-phase proteins, CRP and SAA are the two proteins whose concentration arise more than 1000 fold and even up to 3000-fold within 48 h following injury (Morley and Kushner 1982).

In glycosylated proteins, i.e., proteins that have covalently attached carbohydrate chains to their polypeptide backbone. These are posttranslational modifications, which are mostly carried out by cytoplasmic enzymes. These modifications confer minute changes to the structure and function of a protein, and their composition is very sensitive to many environmental conditions. As glycosylation affects the reactivity of a protein, an increasing interest exists to understand

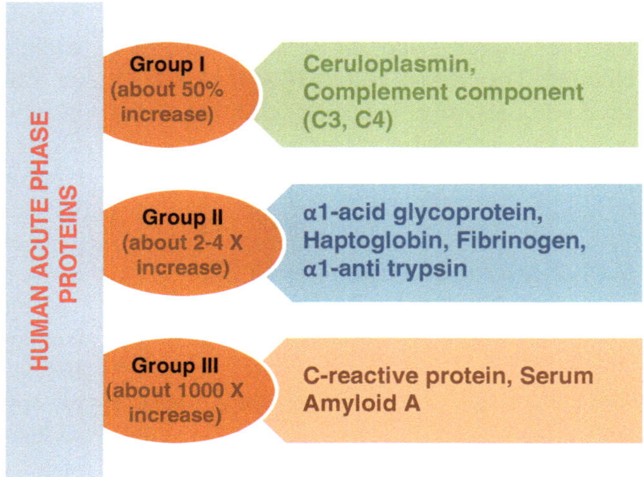

Fig. 5.4 *Classification of human microheterogeneous acute-phase proteins.* Classification on the basis of usual change in plasma concentration (Morley and Kushner 1982)

the glycosylation of a protein. In different pathological conditions, carbohydrate structure changes are observed. So glycosylation therefore is important in understanding disease. But, the protein glycosylation involves a complex process with variations. Variations exists in (1) glycosylation sites, (2) the composition of the attached chains, (3) linkages in each chain, and (4) the amino acid types—carbohydrate bonding and the definite carbohydrate sequences in different glycosylation of different proteins. However, within any group of protein molecules, there is considerable heterogeneity in the carbohydrate structures termed as glycoforms. These glycoforms are synthesized at any one time. Glycoforms are typified by some molecules showing reduced chain length, increased branching, and also further addition of a single carbohydrate moiety to the internal backbone of chain.

5.12.1 Role of Microheterogeneity of APP in Different Diseases

Different APPs (acute-phase proteins) and their associated differences in microheterogeneity are correlated to different disease conditions. While some do not show any differences in the disease or normal individuals, others show marked differences in disease vs. normal states. Such microheterogeneity differences in the glycosylation pattern of the proteins have been attributed to alterations in the intra-hepatocellular processes associated with the glycosylation pattern, induced by cytokines (Mackiewicz et al. 1989). The glycosylation patterns of transferrin isolated from patient sera suffering from ulcerative colitis are highly branched and sialylated and were directly related to the status of inflammation, as compared to the healthy individuals. The microheterogeneity patterns of AGP and ACT do not show any difference with that of the normal individual. Recent studies in ulcerative colitis, showed that patients had significantly higher ACT and AGP concentrations and very low transferrin level as compared to healthy individuals. Thus a direct correlation between concentrations of ACT and AGP, the glycosylation patterns of transferrin, changes in standard protein electrophoresis or count of blood cell were prominently observed (Grzymisławski et al. 2006). Microheterogeneity in alpha 1-acid glycoprotein (AGP) on the other hand has been reported to be as a useful marker for the clinical stages of inflammation based on serum CRP and cytokine IL-6 levels in patients suffering from esophageal or stomach carcinoma before and after operation (Iijima et al. 2000). In systemic lupus erythematosus (SLE), microheterogeneity of AGP has been reported to play a prominent role in the detection of intercurrent infection (Mackiewicz et al. 1987a, b). Serum CRP levels and microheterogeneity of AGP have been observed to show significant correlation with disease status in patients suffering from rheumatoid arthritis (Pawłowski et al. 1986). In patients suffering from ankylosing spondylitis (AS), microheterogeneity differences in AGP were reported together with increased levels of alpha 1-antitrypsin, AGP, CRP, haptoglobin, IgA, and gastrointestinal inflammation (Mackiewicz et al. 1989). Pulmonary sarcoidosis shows altered glycosylation pattern of acute-phase glycoproteins in tune with other chronic inflammatory diseases (Hrycaj et al. 1996). Although transthyretin (TTR), a traditional biomarker for nutritional and inflammatory status, is recently reported as a potential biomarker for the diagnosis of ovarian cancer, the observed microheterogeneity itself of TTR in serum and ascite fluid does not show any correlation with the disease status in ovarian cancer patients (Gericke et al. 2005).

5.13 Microheterogeneity of Human CRP

CRP is a trace component of the normal human serum. It has a molecular weight of 105,000. It is structurally composed of five apparently identical subunits with a cyclic symmetry. The serum CRP concentration rises rapidly in response to acute inflammatory or necrotic processes. There are only a few extensive studies on human CRP in various aspects and in various diseases. The exact cause of microheterogeneity of human CRP has not been reported yet. But their altered synthesis, synthesis outside the liver, structural alterations, cleavage or complication of the circulating CRP may play diverging physiological roles like alternation of its binding properties (Lasson and Göransson 1999). The microheterogeneity of human CRP has been reported in nearly sixteen different pathological conditions (Das et al. 2003, 2004a, b). CRPs were purified from sera of patients, and by different immunological techniques, different molecular variants were established (Fig. 5.5). Furthermore different molecular variants were noted in malaria, in visceral leishmaniasis, and in tuberculosis. The differences in glycosylation of these CRPs purified from patients in all these three diseases were reflected in their function to bind erythrocytes (diseased and normal), to activate the complement cascade and lyse the damaged erythrocytes (Ansar et al. 2006, 2009a, b). It was shown that molecular variants of human CRP are induced in different disease conditions as compared to normal (Fig. 5.6). Depending on the nature of inflammation in diseased condition, these variants are differentially glycosylated. These glycoforms showed differential binding affinities toward different lectins and plasma proteins. Microheterogeneity is not only exhibited in the glycosylation motif but also in amino acid sequencing as revealed by MALDI (Das et al. 2003; Ansar et al. 2009a, b).

CRP was treated with high concentrations of urea and in the absence of calcium (designated F-CRP). F-CRP had alpha-electrophoretic mobility and a pI of 5.4 as compared to native CRP having a pI of 6.4 and gamma mobility. Altered F-CRP had significant antigenic, electrophoretic, and binding site modifications, a new antigenic reactivity than native molecules. F-CRP retained some identical characteristics of native CRPs as revealed by an affinity chromatography process where the protein was separated with an apparent molecular weight of 75,000 with no evidence for free CRP subunits. But F-CRP has lost the capacity of calcium-dependent binding to the C-polysaccharide. Moreover, F-CRP showed the persistence of calcium-independent binding reactivity for polycations. Thus, F-CRP has increased electrophoretic mobility and exposure of a new antigenic reactivity but has alterations in the phosphocholine-binding sites. The antigenic reactivity of F-CRP needs to be evaluated (Potempa et al. 1983).

The three structural forms of CRP are a) the pentameric ringlike structure which is formed on a membrane in a calcium-dependent ligand reaction, b) the small globulin-like form which is formed on a negatively charged membrane in absence of calcium, and the c) fibril-like structure which is formed by face-to-face stacking of a number of pentameric CRPs. The freshly purified CRPs form short single-strand fibrils, while those long and bundled fibril forms are observed when

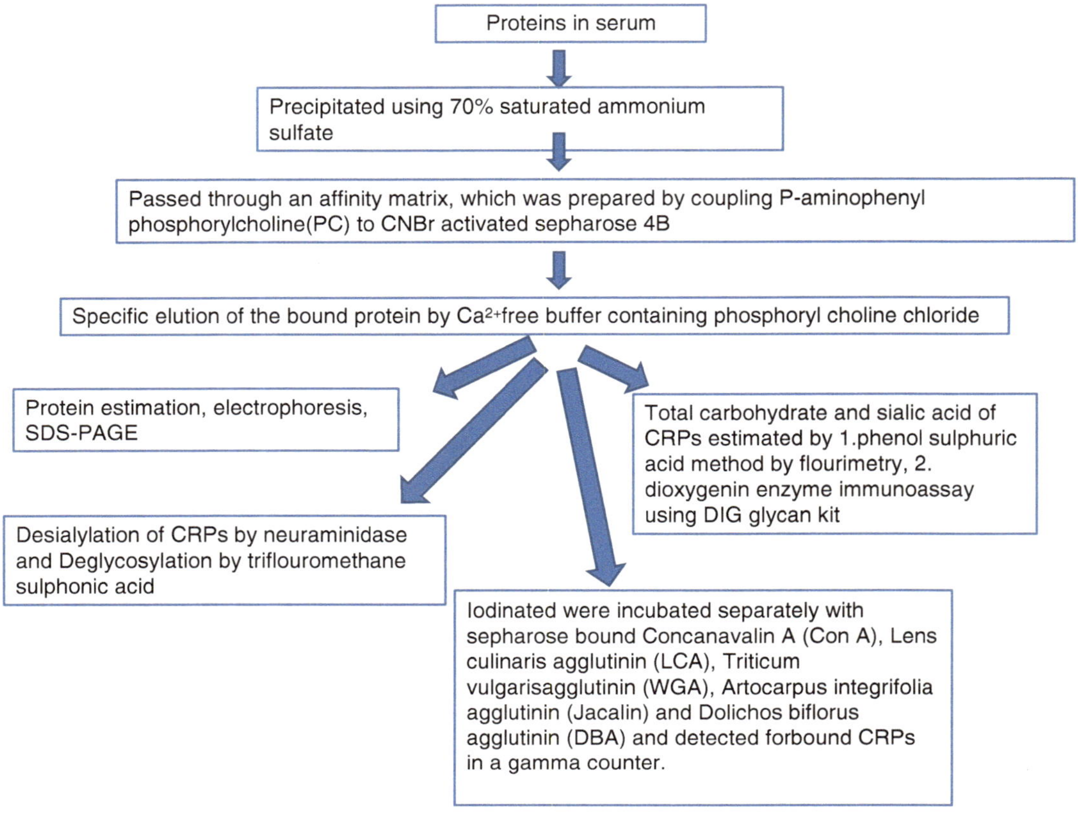

Fig. 5.5 *Purification and characterization of human CRP*. Human CRP was purified by affinity chromatography, and then the variation in carbohydrate moiety was characterized by fluorometry, gel electrophoresis after deglycosylation of CRPs, lectin binding assay with iodinated CRPs, or differential lectin binding of CRPs with various lectins by using different lectin binding kits (Das et al. 2003)

CRP is stored for a longer period of time. These structural forms are interconvertible with changes in the conditions, to perform multiple functions of CRPs (Wang et al. 2002).

By affinity chromatography, two distinct pentraxin proteins were isolated from bovine serum in a calcium-dependent manner. As proteins cross-reacted specifically with certain rabbit anti-human CRP antisera and with a sheep antihuman SAP antiserum, respectively, they are designated as bovine CRP and bovine SAP. Electron microscopy revealed their identity as pentraxin proteins. Non-glycosylated bovine CRP was composed of a single subunit, while glycosylated bovine SAP contained two major types of subunits and a minor polypeptide. Interestingly, neither protein behaved as an acute-phase reactant nor are they reactive to phosphoryl choline in the presence of calcium (Maudsley et al. 1987).

5.13.1 Variation in Carbohydrate Composition of CRP Purified from Malaria, Visceral Leishmaniasis, and Tuberculosis Patients

CRP is differentially glycosylated in different pathological conditions and CRP purified from normal individuals has no glycosylation attached to it (Das et al. 2003; De Beer and Pepys 1982). In 2009, Ansar et al. purified CRPs from the sera of malaria (MAL), visceral leishmaniasis (VL), and tuberculosis (TB) patients. Large variations

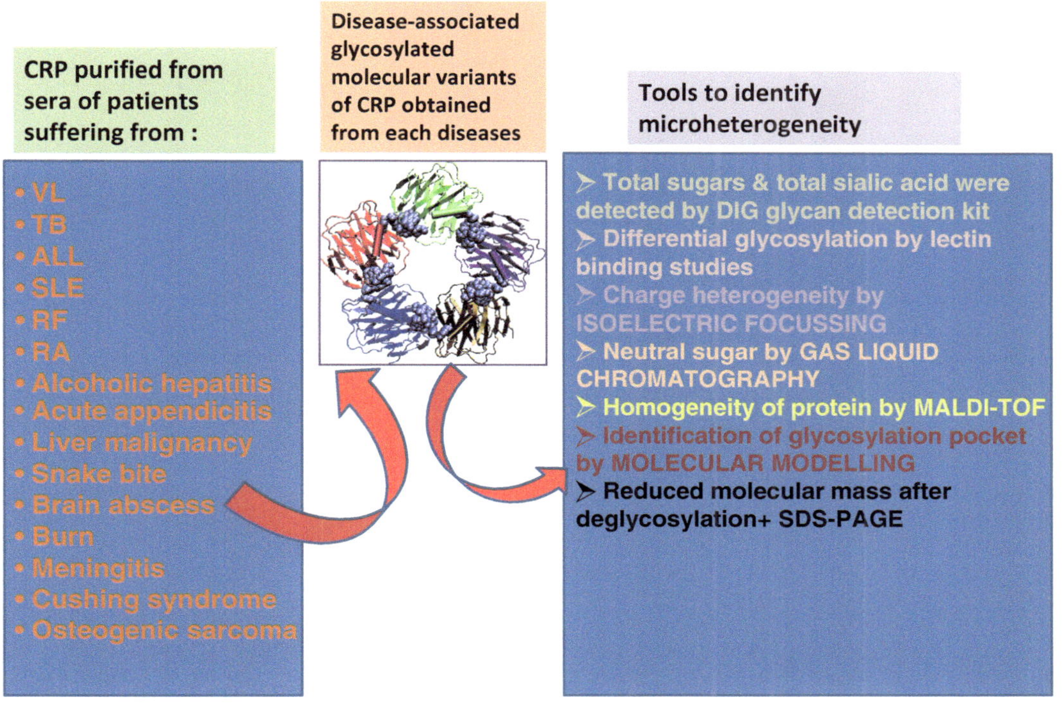

Fig. 5.6 *Human glycosylated CRP and disease correlation*. Different glycosylated CRPs were induced in some disease conditions, whereas CRP present in normal healthy individual is non-glycosylated

among CRP isolated from tuberculosis, malaria, and visceral leishmaniasis patients were observed in their carbohydrate moieties demonstrated by their property of differential binding toward different lectins (Fig. 5.7) (Ansar et al. 2009b).

A glycosylation and sialylation profile was detected by DIG enzyme immunoassay techniques using purified CRPs (1.0 μg/spot). Densitometric studies of CRPs from patients were taken. Linkage specificity of CRP carbohydrate chains was done by a DIG–glycan differentiation kit using lectin binding study. Fetuin (MAA, SNA, and DSA positive), carboxypeptidase Y (GNA positive), and asialofetuin (PNA and DSA positive) were used as positive controls.

The *carbohydrate composition* for the presence of three hexoses (mannose, glucose, and galactose) in CRPs (100 μg) was analyzed by *GLC and GLC-MS to understand the* abundance of hexoses against m/z. The occurrence of sialic acid in CRPs (100 mg) was analyzed by ion chromatography. The peak was compared with that of standard sialic acids. Analysis of sialic acid of CRPs was done by MALDI-TOF also. Intensity (%) was plotted against mass (m/z) and the peak was compared with that of standard sialic acids. Mass spectra were recorded between 400 and 560 nm obtained by averaging 50–200 individual laser shots. Glycosylated CRPs were digested separately with deglycosylating enzymes at 37 °C overnight and analyzed by SDS-PAGE (10 %). Protein bands were visualized by Coomassie staining. Isoelectric focusing of purified CRPs (1.5 mg) was applied to ampholine polyacrylamide tube gel (4 %) in a pH gradient (3.5–10.0) and stained with silver nitrate. Glycosylated CRPs before and after enzyme treatment were visualized. Iodinated CRPs were incubated with Sepharose-bound lectins (20 μl) with different linkages like Con A, UEA, WGA, LPA, DBA, SNA, and MAA and binding (%) was determined.

Commercially available plant lectins like SNA, PNA, DSA, GNA, and MAA were used.

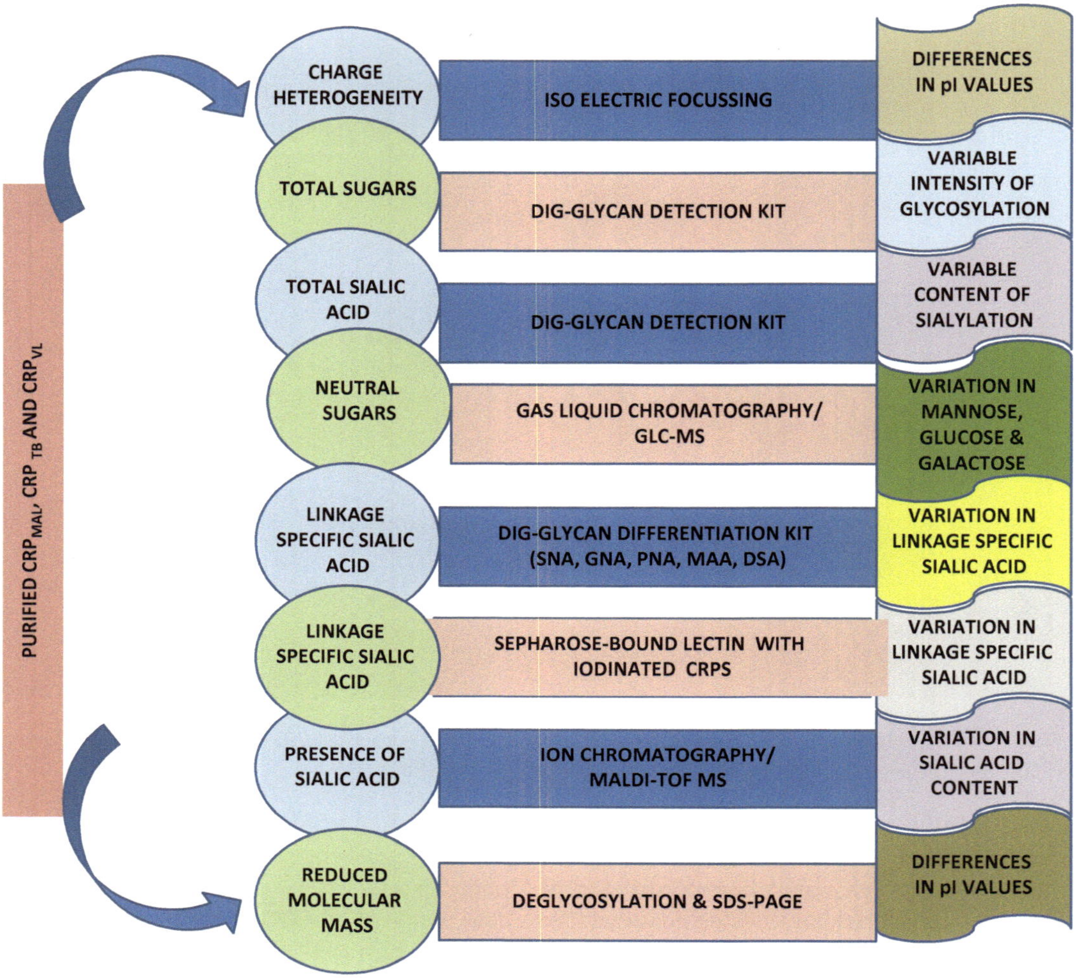

Fig. 5.7 *Glycosylation status and profile of CRP_{TB}, CRP_{MAL}, and CRP_{VL}.* Glycosylated crp_{tb}, crp_{mal}, and crp_{vl} were purified from serum of patients with tuberculosis, malaria, and visceral leishmaniasis (Das et al. 2003; Ansar et al. 2009a, b)

These lectins are known to recognize different linkages of sialic acids (Table 5.2). The lectins applied are conjugated with the DIG. DIG selectively recognizes the terminal sugars like sialic acid and enables immunological detection of the bound lectins. Total sugars and sialic acids were initially qualitatively detected with the help of DIG–glycan detection kit. Equal amounts of CRPs, along with glycoprotein fetuin (positive control) and creatinase (negative control), were absorbed onto nitrocellulose paper and were detected by digoxygenin (DIG) enzyme immunoassay. The proteins were oxidized by NaIO4 for detection of neutral sugars and sialic acid. DIG-labeled glycoconjugates were subsequently detected in an enzyme immunoassay. A specific antibody conjugated with alkaline phosphatase was used in this assay. Densitometric scanning of DIG dot blot analysis of CRPs was also performed (Ansar et al. 2009a, b).

The dot blot of three CRPs showed positive reactivity in DIG sialic acid detection, and differential reactivity kits where sialic acid-binding lectins were used. Moreover, non-glycosylated CRPs showed no reactivity with lectins. The densitometric scanning of the dot blot indicated the presence of varied amounts of these linkage-specific sialic acids in CRPs. CRPs purified from three pathological conditions are glycosylated. The variable amount of glycosylation was reflected in the intensities of the spots that prominently differed from each other (Ansar et al. 2009a, b).

The carbohydrate compositions were further determined by gas–liquid chromatographic (GLC) and mass spectroscopic analysis (GLC-MS). Neutral sugars in CRPs were detected as alditol acetates by GLC (Sloneker 1972). The CRPs were hydrolyzed with trifluoroacetic acid (TFA, 2.0 N) to liberate monosaccharides. The liberated monosaccharides were reduced with sodium borohydride. The resulting alditol was dried in vacuum over P_2O_5 and was acetylated with distilled pyridine and acetic anhydride at room temperature. The alditol acetates were then extracted with distilled chloroform and analyzed by GLC and GLC-MS using suitable columns. The constituent sugars were identified from the retention time of authentic sugars and/or from their mass fragmentation pattern in EI mode. The GLC chromatogram peaks of three CRPs not only revealed the presence of three hexoses, namely, galactose, mannose, and glucose, but also showed striking differences in their compositions. However, the ratio of the hexoses differed significantly in the CRPs from different diseases. High variation in the mannose content in three diseases is also seen (Ansar et al. 2009a, b).

To investigate the charge heterogeneity, IEF studies was performed with the CRPs. IEF was carried out in capillary tubes on ampholine polyacrylamide gel (4 %) with pH ranging from 3.5 to 10.0 under native condition. The samples were focused at a constant voltage of 400 V for 6 h, and then the focused tube gels were stained with silver nitrate. The isoelectric point (pI) of the individual protein was determined by measuring their migration from the cathode. All three CRPs separately showed a single band in 2D-gel electrophoresis confirming their purity. However, slight differences in their pI values were observed in these three CRPs (Ansar et al. 2009a, b).

In view of differential mobility, the possibility of the variation of negatively charged sugars, sialic acid, and its commonest derivatives, namely, (8) 9-*O*-acetylated sialic acid (*O*-AcSA), was examined by quantitative fluorometric analysis as described in Shukla and Schauer 1982. All three different CRPs revealed the presence of these derivatives confirming again the sialylation of these three CRPs. Differences in the sialic acid content and (8) 9 *O*-AcSA probably contribute toward the variability observed in CRP induced in these pathological conditions. Sialic acid could not be detected in non-glycosylated control CRPs. Equal concentration of iodinated CRPs when incubated separately with Sepharose/agarose binds lectins of different sugar linkages and specificities, namely, Con A, WGA, DBA, LPA, and UEA. A wide variation was observed in the lectin binding properties of CRPs. CRP from malaria patients showed the highest binding with Con A indicating the presence of more glucose and mannose. By Scatchard plot, the binding constant was also determined (Ansar et al. 2009a, b).

CRPs (100 μg/500 μl) were hydrolyzed separately with 0.05 M of trifluoroacetic acid for 1 h to release the sialic acid for ion chromatography. The hydrolysate was evaporated under reduced pressure. The hydrolyzed sample was dissolved and analyzed by the ion chromatography system, and the peaks were analyzed by the software. The *N*-acetyl neuraminic acid was identified by comparing it against a sialic acid standard. Ion chromatography revealed the presence of sialic acid in these CRPs. In non-glycosylated CRPs, no detectable peak was observed (Ansar et al. 2009a, b).

The presence of terminal sialic acid was determined by MALDI-TOF-MS. CRPs (50 μg) were subjected to acid hydrolysis. Treatment with propionic acid (4 M) for 4 h was done to release sialic acids. Then it was purified through Dowex 50 W × 8 (100–200 mesh) cation and Dowex 2×8 (200–400 mesh) anion exchange columns. The purified sialic acids of CRPs, derivatized with DMB, were analyzed by MALDI-TOF-MS. Sialic acids released from bovine submaxillary mucin (BSM) were used as control for comparison. All mass spectra were recorded in the positive ion mode using the reflector. The acquired spectra were the average of 1000 laser shots. Mass spectrometry of CRP_{TB} showed three well-resolved peaks which coincide with that of Neu5Ac, Neu5Gc, and Neu5,7Ac2 and one small peak co-migrating with Neu5Gc9Ac. CRP_{VL} showed that two well-resolved peaks coincided with Neu5Ac and Neu5Gc. Three small peaks corresponding to Neu5,9Ac2, Neu5Gc9Ac, and Neu5,7(8),9Ac3 were also seen (Ansar et al. 2009a, b).

Glycosylation was further shown by deglycosylation of CRP and analysis of deglycosylated CRPs in SDS-PAGE and IEF for possible reduction in their molecular weight and isoelectric point (pI). CRPs were deglycosylated separately using *Arthrobacter ureafaciens* neuraminidase in denaturation buffer. Reaction mixture was heated and centrifuged and then *N*-glycosidase F or *O*-glycosidase or combination of *N*- and *O*-glycosidase treatment was done. Deglycosylated CRPs were analyzed by SDS-PAGE. Deglycosylated CRPs were passed on a PC–Sepharose column to remove enzymes present in the reaction mixture and subsequently analyzed by 10 % SDS-PAGE. There is a decrease in the CRPs' molecular mass after the enzyme treatment. Non-glycosylated CRP even after enzyme treatment moved to the same location (Ansar et al. 2009a, b).

5.13.2 Microheterogeneity of Human CRP: Role in Different Diseases

Human CRP is clinically important. From different pathological conditions, CRPs were purified from the sera collected from the patients where their concentration ranges from 22 to 342 μg/ml. Differences in molecular mass, charge, and/or shape between CRP samples were noted as significant variations in electrophoretic mobilities on native PAGE and SDS-PAGE. CRPs purified from multiple individuals or pooled sera of a particular disease had a single mass reflected in SDS-PAGE. Isoelectric focusing also demonstrated that the purified CRPs differed from each other. The presence of sialic acid, glucose, mannose, and galactose was revealed by gas–liquid chromatography (GLC), fluorometric analysis, and mass spectroscopy. The differential glycosylated motif in these purified CRPs was demonstrated by digoxigenin kits, by neuraminidase treatment, and by lectin binding assays. Protein homogeneity was confirmed by MALDI analysis. Molecular modeling suggested that the calcium-dependent PC binding and the potential glycosylation sites were different. Thus, CRP is differentially glycosylated in some pathological conditions.

5.14 Discussion: Physiological Impact of Microheterogeneity

The impact of posttranslational modification of proteins is widespread. It may influence the interactions or binding with other proteins and have an effect on different metabolic activities and receptor binding and in protein–protein complexes, tissue uptake, degradation of proteins, and also many other aspects.

Most posttranslational modifications happen in physiological conditions, but some disease-specific changes are also noteworthy. These modifications are immensely important as are of diagnostic importance in clinical proteomics and clinical medicine. There are different analytical methods used to study the posttranslational modifications which are discussed in details in this chapter.

The microheterogeneity exerts major influences on the function of the protein like (1) secretion, (2) plasma transport, (3) receptor binding, (4) degradation, (5) excretion, and (6) protein–protein complex formation. Microheterogeneity variants differ with regard to their immunomodulatory properties. The Con A nonreactive variant of AGP is more effective in modulation of lymphocyte proliferation than Con A reactive AGP serum variants (Pos et al. 1990). The concentration and microheterogeneity of APPs can be used in early diagnosis, management, and prognosis of chronic inflammatory stages.

Thus, posttranslational modification generates a repository of molecules that rapidly respond to stimuli (Baumann and Meri 2004; Davis 2004;

Hancock 2002; Mann and Jensen 2003). Microheterogeneity in APP has been primarily reported to affect the process of physiological aging in humans (Ballou et al. 1996).

Using various reproducible and cost-effective biotechnologies, the molecular variants can be determined for microheterogeneous proteins and the protein complexes. Thus, in proteomics, thorough characterization of protein is necessary to validate the phenotype of the protein in large population-based studies. This provides the basis for the diagnostic and prognostic relevance of disease-specific isoforms of proteins with regard to personalized medicine. The search for disease-specific biomarker will be easily formulated by the study of microheterogeneity of proteins.

Pathobiological Role of CRP in Diseases: Clinical Medical Applications of CRP

6

Abstract

Although C-reactive protein has been designated as a marker for inflammation and recently as a marker in assessment of risk in cardiovascular disorders from a number of studies across the globe, the role of CRP in other diseases have also been studied. Some of the major diseases in which CRP levels have been correlated with disease states are detailed in this chapter. Bearing a hepatic origin, C-reactive protein is a protein found in the plasma and belongs to the pentraxin family participating in the inflammatory reaction. The CRP concentration in response to injury or infection becomes escalated about 1000-fold over and above normally expressed levels. CRP is known for its function as a mediator in the classical innate immunity pathway and constitutes a part of both humoral and cellular effector systems of inflammation. Although it is well known as a marker for acute-phase response in injury of the tissue, during the infection of the host by pathogenic agents like viruses, bacteria, parasites, and inflammations, it has recently gained the status of a disease marker in cardiovascular disorders and also has been reported for its clinical and pathological significance. In this chapter we highlight the status of C-reactive protein (CRP) in different disorders and pathological conditions.

Keywords

C-reactive protein • Systemic Lupus Erythromatosus • Thyroid diseases • Ovarian diseases • Rheumatoid arthritis

Chapter Highlights

1. Human CRP, since its discovery is a subject of great clinical interest.
2. CRP is a trace plasma protein that is expressed dramatically and rapidly during acute phase response to infection, injury and other conditions of inflammation. CRP is an inflammatory biomarker.
3. The phagocytic and complement activation property of CRP was quite significant. Although CRP was used as a marker of disease

W. Ansar, S. Ghosh, *Biology of C Reactive Protein in Health and Disease*,
DOI 10.1007/978-81-322-2680-2_6

activity, its primary function in many diseases is still unclear.

4. In spite of the phosphocholine dependant ligands of CRP, it also binds with histones, chromatin under non-physiological conditions and to small nuclear ribonuclear particles under physiological conditions.
5. The binding properties of CRP in *in vivo* and *in vitro* conditions are a matter of debate.
6. The present chapter focuses the clinical significance of CRP in some diseases.

6.1 Introduction

Human CRP, discovered in 1930, belongs to the pentraxin family which is a phylogenetically ancient, highly conserved, major acute-phase plasma protein (Czarnywojtek et al. 2014). Macleod and Avery in 1941 isolated a C-reactive material and identified it as biochemically protein in nature. It was found to react with C-polysaccharide (CPS) only in the presence of calcium ions (Ca^{2+}). Such proteins were termed as "acute phase" (Nagasaki et al. 2007) initially, while the term CRP was coined in 1941 to this serum protein responsible for precipitin with CPS (Yu et al. 2013). A rapid increase in its serum concentration up to 3000-fold as a consequence to infection or after an injury in the tissue (Tuzcu et al. 2005) is reported. Liver hepatocytes synthesize it and secrete it in plasma and its regulation is mainly controlled by cytokines interleukin-6 (IL-6). CRP forms a key molecule in the host innate immunity (Nebiker et al. 2013) and has been reported of its property of identifying pathogens by initially binding to their surface components (phosphorylcholine, PC) thereby leading to activation of the complement cascade, leading to downstream effects causing opsonization of the pathogen and culminating in its phagocytosis and eventual removal. Target cell elimination is initiated by the interaction of CRP with the host immune system including both humoral and cellular immunological pathways of inflammation (Bilgir et al. 2014) and is therefore known as a marker for inflammation (Piciu et al. 2013). The dependency of CRP is on calcium to be able to bind to ligands including nuclear constituents like basic histone proteins, small nuclear ribonucleoproteins (snRNPs) and chromatin (Martocchia et al. 2010- Jublanc et al. 2004). CRP has been implicated in processing, scavenging, and eventual removal of nuclear antigens that may be contributory to autoimmune disorders therefore preventing host autoimmune responses to nuclear antigens.

6.2 Biological Role of CRP

Forming the first line of defense of pathogens, CRP, although it has structural differences with immunoglobulin (Ig) molecule, shows similarity in function with the Igs, like the ability to promote agglutination, complement fixation, bacterial capsular swelling, phagocytosis, and precipitation of polycationic and polyanionic compounds (Oliveira et al. 1979). CRP has been implicated of its role in the clearance of bacteria and of dying and altered cells and has been reported to contribute to complex immunomodulatory functions. Quantitative estimation of CRP level in serum is in widespread clinical application as a sensitive inflammatory biomarker.

6.3 CRP and Clinical Significance

CRP concentration in the plasma or serum increases markedly in various conditions like bacterial infection, abscess, Crohn's disease, connective tissue disorders (except SLE), neoplasia (except leukemia), trauma, and necrosis, whereas it showed only slight elevation in viral infection, steroids and estrogen therapy, ulcerative colitis, and SLE diseases. The relevance of CRP as a diagnostic tool (45, 80) is summarized in Fig. 6.1.

Elevated levels of CRP is very early, sensitive, and an indicator of most forms of microbial infection particularly in cases of bacterial infections in newborn babies, especially preterm delivered babies suffering from sepsis, pneumonia, and meningitis, who fail to show any other known symptoms on infections like increased

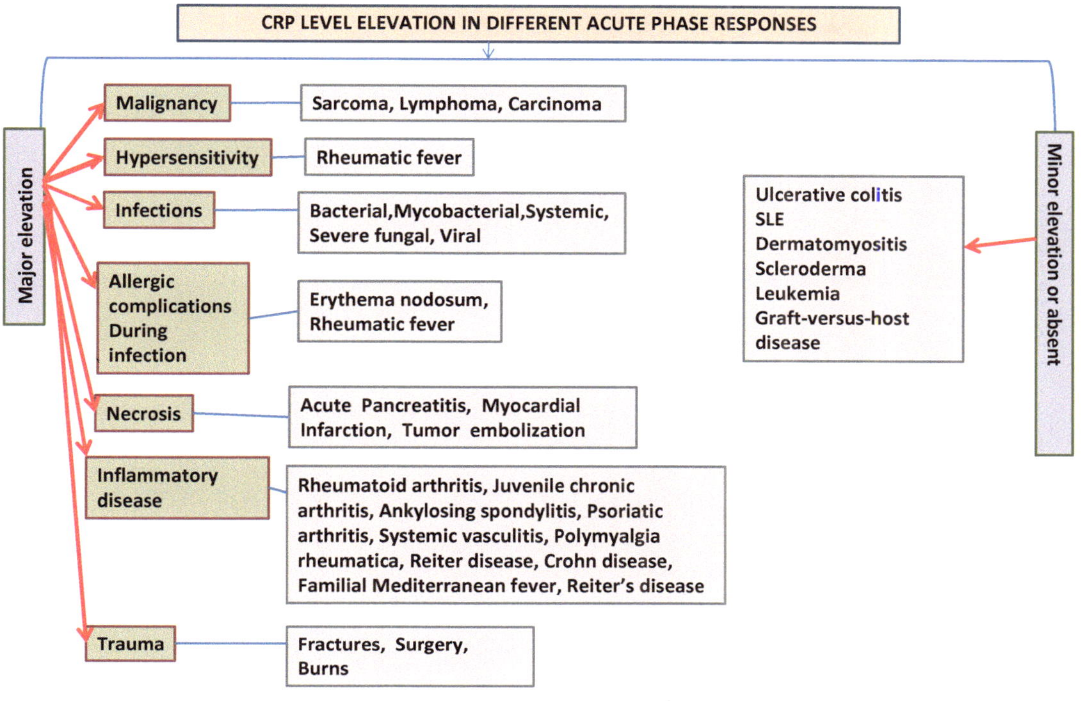

Fig. 6.1 CRP in disease, diagnosis, and prognosis

body temperature and white blood cell (WBC) count. Therefore C-reactive protein (CRP) finds application as a useful diagnostic marker of neonatal septicemia and meningitis. CRP is also elevated in most cases of acute osteomyelitis when the usual symptoms like leukocytosis and plain radiographic abnormalities are also not seen and therefore are an early indicator to the disease. It also indicates intercurrent infection, i.e., an infection occurring with another infection in systemic lupus erythematosus (SLE) and leukemia. Postsurgeries, i.e., within 2–6-h surge in the CRP level, have been reported, but continuous rise of CRP level after 7–10 days or secondary rise is an indicator of persistent postoperative infection or thromboembolic complications, i.e., a block in the blood flow associated with the surgery. C-reactive protein (CRP) also finds profound importance as the only useful marker in diagnosing acute pancreatitis.

CRP has also shown promises in evaluating clinical course and response to treatment in connective tissue disorders. It also finds importance in the prognosis of acute myocardial infarction in relation to its long-term outcome (Pepys and Hirschfield 2003; Hirschfield and Pepys 2003).

Elevated CRP level in treated visceral leishmaniasis (VL) patients has been reported to predict higher future occurrence of post-kala-azar dermal leishmaniasis (PKDL) (Gasim et al. 2000; Pepys and Hirschfield 2003; Ansar et al. 2009a) which is indicative of its predictive potential. At the high-normal range, CRP is a predictive marker for a long-term risk of both stable and unstable angina, myocardial infarction, and even that of death. While CRP level at a high-normal baseline is reported to be indicative of an elevated risk of coronary disorders even in individuals with normolipidemic conditions, elevated CRP and adiposity relations are suggestive of its predictive value in the development of metabolic syndrome.

CRP finds importance as a diagnostic tool in diseases like neonatal septicemia and meningitis, postoperative infection, cases of postoperative thromboembolic complication, intercurrent infection of SLE, intercurrent infection in leukemia, acute pancreatitis, acute appendicitis, acute osteomyelitis, acute rheumatic fever, and insulin resistance. CRP has been reported to find importance as a prognostic marker in diseases like rheumatoid arthritis both in adult and juvenile arthritis, rheumatic fever vasculitic syndrome, familial Mediterranean fever, ankylosing spondylitis, Reiter's syndrome, psoriatic arthropathy, Crohn's disease, myocardial infarction, stable and unstable angina, and visceral leishmaniasis. In diseases like myocardial infarction; angina; cardiovascular mortality morbidity, and mortality in the elderly; metabolic syndrome; and PKDL, CRP finds importance in risk assessment.

Normal individuals have the average plasma CRP concentrations of 2 mg/L or<2 mg/L. The CRP level in a normal individual can also be as high as 10 mg/L which may be contributed by limited immunological stimulation by minimally apparent or unapparent low-grade process (Pepys and Hirschfield 2003). Therefore, the 10 mg/L is regarded as a clinically unimportant parameter. However, recent findings highlight that even a slightly elevated level of CRP within the normal range could be predictive of future coronary disorders. The differential responses of CRP in most similar inflammatory disorders are not well understood. However, differential CRP response may reflect the progressing tissue damage both on its nature and extent of damage or it may be different due to genetic variability in individuals affecting production of inflammatory mediators, exhibiting differential responses to mediators and/or differential, variable production of CRP molecules itself.

Although its diagnostic specificity is not well spelled out, CRP is reported for its clinical importance as changes in their serum or plasma level reflect the ongoing process of inflammation and its intensity of reactions and enable differentiation of inflammatory from noninflammatory effects thereby signaling a disease condition (Ansar et al. 2009a, 2006), need for treatment, assessing prognosis, and predicting future risk. Moreover, the CRP level cannot be regulated and controlled by the effect of anti-inflammatory or immunosuppressive drugs without controlling the underlying root cause of the disease. Therefore, recent researches have strongly established that CRP is a more important biomarker of acute-phase reactions over other prevalent ones like ESR.

6.4 CRP in Health and Different Diseases

In normal healthy subjects CRP is reported to be in trace plasma protein, usually lower than 10 mg/L (Das et al. 2003; Clyne and Olshaker 1999), while in response to severe infections mediated by bacterial, parasitic agents and due to burn (>200 mg/L), infection caused by viruses (10–40 mg/L), trauma, injury, wound, necrosis of tissue, aging, different pathophysiological inflammatory disease conditions, and malignant neoplasia serum level of CRP has been observed to escalate rapidly about 1000-fold by 48 h after an acute event (Clyne and Olshaker 1999). Recent research has focused on the use of high-sensitivity C-reactive protein (hs-CRP), an inflammatory biomarker, in the detection of patients with elevated risk for cardiovascular disorders. Currently, hs-CRP allows precision in the very low ranges (0.2 mg/L) needed for this purpose, and studies reveal that CRP levels<1 mg/L, 1–3 mg/L, and>3 mg/L indicate low, intermediate, and higher vascular risk groups (30). In patients suffering from chronic infections and inflammatory disorders, the CRP levels which may show high persistence, together with the disease, show a sharp fall to normal levels in consequence to cure inflammation pathology either spontaneously or in response to treatment by drugs or anti-inflammatory agents.

Viral infections that are localized reveal to be less potent in stimulating CRP production as compared to that of infections mediated by bacterial agents. Certain disorders which reveal a normal or modestly elevated CRP level despite exhibiting severe, active, tissue-damaging disease pathology include scleroderma, dermatomyositis, ulcerative colitis, Sjögren's syndrome, and leukemia. In active systemic lupus erythematosus (SLE), a complex heterogeneous disease, patients having high plasma concentrations of CRP are accompanied with some infection probably bacterial infections (Ter Borg et al. 1990), whereas low CRP elevations in SLE may be due to anti-CRP antibodies related to SLE. Nowadays, obesity has been considered to be a systemic inflammatory disorder of low-grade inflammation. Obese and overweight adults and children have high serum CRP levels (Das 2001) and caloric-restriction-induced loss of weight together with decrease in the plasma CRP level (Tchernof et al. 2002). Mean concentration of CRP is substantially higher in smokers than in nonsmokers (Haverkate et al. 1997). Clearance of CRP from the system is monoexponential and is not dependent on the serum CRP level or pathobiology. Therefore, quantitative estimation of CRP finds importance as an excellent biomarker in activity in disease biology.

CRP and its status in different diseases are summarized in Fig. 6.2. The activity of CRP in different diseases includes:

6.5 C-Reactive Protein (CRP) and Systemic Lupus Erythematosus (SLE)

Unlike other diseases, systemic lupus erythematosus or SLE is not a single disease; it is rather a heterogeneous group of diseases with consistent CRP response mostly mute in SLE. Although the level of CRP has been shown to correlate directly with the disease pathophysiology of many inflammatory diseases, on the contrary, CRPs at insufficient concentration are known to directly correlate in the development and progression of chronic inflammatory and autoimmune disorders like SLE (Vogt et al. 2007).

The role played by CRP in SLE as an inflammatory biomarker is complex. SLE patients are

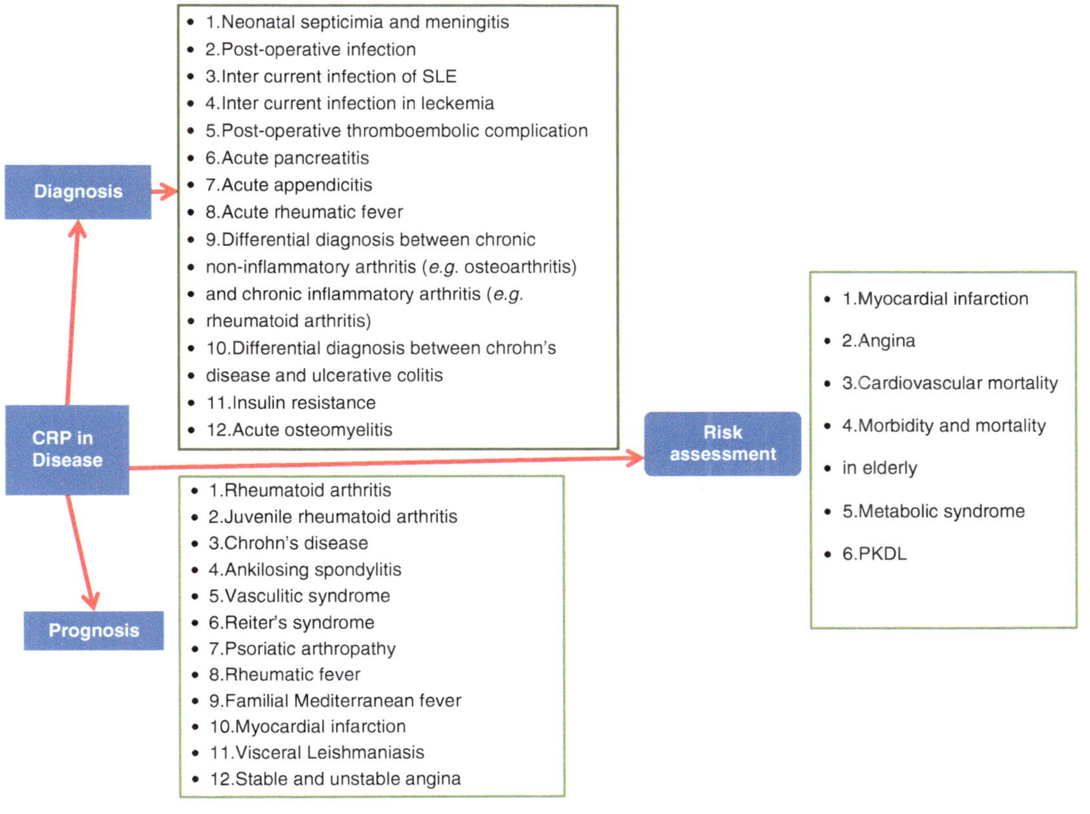

Fig. 6.2 Level of CRP in different acute-phase responses

designated into groups of inactive, mild, active, and confirmed infection based on their CRP levels in serum. Patients with active SLE display only modest elevated or even normal level of CRP during an active disease condition. A marked elevation of CRP with SLE indicates accompanying some intercurrent infections (Ter Borg et al. 1990).

However, lupus serositis patients exhibit marked CRP values (50–60 mg/L) (Rezaieyazdi et al. 2011). Recently, hs-CRP finds application in detecting micro-inflammation, and serum concentration of hs-CRP bears direct correlation with the disease pathophysiology in lupus. Though hs-CRP serum levels reveal significant elevation in patients with SLE in comparison to healthy individuals used as controls in the study, it is indicative of the fact that CRP is not a sensitive biomarker as an indicator for disease activity (Rezaieyazdi et al. 2011). The reason behind the reduced CRP response in SLE may be attributed to the modest production of CRP rather than their enhanced rapid clearance from the plasma mediated by CRP autoantibodies found in blood and may not be due to deficiencies in the IL-6 and IL-1 cytokines responsible for the induction of CRP synthesis. Recent studies have revealed that highly expressed IFN type I in SLE inhibits CRP expression. Therefore, low CRP levels may also be a major contributory factor to the pathophysiology and biology in this disease (Gaitonde et al. 2008).

6.6 Thyroid Diseases

In many different thyroid disorders, the concentration of CRP has been reported to be indicative of the inflammatory conditions. hs-CRP has been observed to be elevated in various types of hypothyroidism and shows positive correlation with the stage of the disease and the length of duration of that stage (Czarnywojtek et al. 2014) together with a manifestation of stiffness in the arterial wall of the common carotid artery (Nagasaki et al. 2007). Elevated hs-CRP correlated positively with serum TSH level is elevated in subclinical hypothyroidism (SCH, Yu et al. 2013), correlating with low-grade inflammation (Tuzcu et al. 2005). Case reports on painful swelling of the left thyroid lobe together with clinical manifestation of increased C-reactive protein (CRP) above 100 mg/L has been reported (Nebiker et al. 2013). Treatment with 50-mg tablets of propylthiouracil three times daily to patients suffering from hyperthyroidism revealed no difference that bore a statistical significance in the concentration of hs-CRP before or after treatment and control group of individuals (Bilgir et al. 2014). Posttreatment monitoring of disease-free patients who had undergone treatment for thyroid carcinoma after undergoing total thyroidectomy, radioiodine ablation, and thyroid hormone replacement therapy revealed an elevated level of hs-CRP leading to increased cardiovascular risk in such patients (Piciu et al. 2013). Elevated CRP finds significant importance as an important biomarker in detecting patients suffering from nonthyroidal illness syndrome (NTIS) showing low T3 thyroid hormone levels due to reduced conversion from T4 thyroid hormones (Martocchia, et al. 2010). hs-CRP level elevation is found to bear positive correlation with low levels of thyroxine in hyperlipidemic euthyroid individuals (Jublanc et al. 2004). Elevated levels of CRP, in SCH, remained unaffected by L-thyroxine therapy adding to the cardiovascular risk associated with hypothyroidism (Christ-Crain et al. 2003).

6.7 CRP and Diabetes

Inflammation is observed in the pathogenesis of both atherosclerosis and diabetes. CRP level elevation is associated with an increased risk of insulin resistance by impairment of delivery of glucose to the skeletal muscle (Tanigaki et al. 2012). Type 2 diabetes mellitus exhibiting conditions of inflammatory atherothrombosis condition shows more prevalence of a cardiovascular disorder. Type 2 diabetic patients showing symptoms of low-level inflammation show elevation in the plasma level of CRP. Marginally low elevation in CRP level predicts the probability of development of cardiovascular diseases in individuals who are diabetics as well as nondiabetic individuals. Inflammatory and metabolic factors like high blood glucose, adipokines, i.e., cytokines synthesized and released from adipocytes, modified lipoproteins, and plasma FFA or free fatty acids associated with diabetes, have also been reported to trigger endothelial cells, smooth muscle cells and monocytes, and macrophage immune cells to synthesize and release CRP in circulation. Local CRP production has been reported to accelerate the genesis of vascular disorders in type 2 diabetic individuals (Mugabo et al. 2010). Increased CRP levels are reported to predict diabetes (Mugabo et al. 2010) at the onset. A study on an American Indian population including highly diabetic individuals revealed that elevated CRP concentration strongly correlated with symptoms of cardiovascular disorders in nondiabetic women but not in diabetic women or in men, irrespective of glycemic status indicating that CRP has an important role as a variable predictive biomarker in population subsets (King et al. 2003).

6.8 Ovarian Disease

The inflammation marker CRP in its elevated plasma level concentration is reported to be indicative of increased risk in patients suffering from ovarian cancer (Trabert et al. 2014; Poole et al. 2013; Toriola et al. 2011). Although in EOC patients elevated levels of preoperative serum concentration of interleukin-6 and interleukin-8, together with CRP levels as compared to the healthy individual, enable diagnosis of patients with epithelial ovarian cancer (EOC), CRP has been correlated with poor prognosis and therefore a poor prognostic factor and indicative of poor overall survival and disease-free survival posttreatment in such patients (Dobrzycka et al. 2013). In women suffering from polycystic ovarian syndrome (PCOS), elevated serum levels of PEDF and hs-CRP levels (Cheng et al. 2013) together with high-soluble CD40 (El-Mesallamy et al. 2013) as compared to normal healthy women are reported (El-Mesallamy et al. 2013). The elevated level of CRP has been correlated with chronic inflammation observed in PCOS patients (Deligeoroglou et al. 2012) which may be directly related with risks of cardiovascular disorders and high concentration of CRP together with the chemokine; monocyte chemoattractant protein-1 (MCP-1) could lead to the development of atherosclerosis (Sasidevi et al. 2013). Elevated plasma level of CRP and homocysteine (Hcy, an α amino acid, synthesized from methionine) is reported to be associated with an abnormal recovery in heart rate in patients suffering from polycystic ovary syndrome (PCOS) (Kaya et al. 2010).

Reports from studies conducted in families with a family history of patients suffering from PCOS indicated that CRP levels are controlled genetically and are a heritable trait occurring in PCOS women and their siblings when both parents express elevated CRP levels (Sasidevi et al. 2013). Metformin treatment in PCOS individuals have revealed significant reduction of serum hs-CRP levels, and reported to effectively improve insulin resistance in obese and overweight women suffering from PCOS (Esfahanian et al. 2013). A multivariate analysis of the biomarkers to detect cancer of the ovary revealed that elevated plasma concentrations of CRP, SAA, interleukin-6 (IL-6), and interleukin-8 (IL-8) indicated the role of these markers in the diagnosis of cancer of the female ovary (Autelitano et al. 2012). Together with other clinical markers like body weight, abdominal circumference, and ovarian size, CRP is also found to be a potential candidate in indicating the severity of ovarian hyperstimulation syndrome (OHSS) (Nowicka et al. 2010).

6.9 Infections like Mumps

Elevated levels of CRP have been recorded from studies in infectious diseases like mumps orchitis (Niizuma et al. 2004) and other forms of mumps infection (Strati et al. 1993). A preliminary report suggested the probable role of CRP in the development of acute infections in mumps (Strati et al. 1987, 1993).

6.10 Testis

C-reactive protein (CRP) has also been studied for its predictive potential for cardiovascular diseases (CVD) in testicular cancer survivors (TCSs, Wethal et al. 2010). Biomarkers in circulation including both interleukin-1 receptor antagonist and overexpressed C-reactive protein have been reported from patients who are suffering from testicular cancer and are long-term survivors of the disease (Orre et al. 2009). Of all the different acute-phase proteins including haptoglobin, α1-acid glycoprotein and transferrin, and fibrinogen, CRP has been reported most efficient in predicting epididymitis as compared to other noninflammatory conditions of the testis (Doehn et al. 2001).

6.11 Prostate Diseases

An elevated plasma level of CRP (>=8.6mgl-1) in prostate cancer finds application as an indicative biomarker in the disease as a poor prognostic indicator factor in both the overall survival of the patient and disease-free survival (DFS) in prostate cancer patients undergoing radiotherapy (Thurner et al. 2015). Although CRP has been associated with prostate cancer, CRP genetic single-nucleotide polymorphisms (SNPs) have not been found to associate with the risk of overall or lethal prostate cancer (Markt et al. 2014). Both the markers C-reactive protein (CRP) and erythrocyte sedimentation rate (ESR) are able to identify cases of epididymitis and different causes of an acute scrotum (Asgari et al. 2006).

6.12 Pathological Relevance of CRP

Serum CRP levels most commonly are related to infections mediated by microbial agents like bacteria; ischemic tissue injury (de Beer et al. 1982a); cardiovascular disorders like myocardial infarction or pulmonary embolism; tissue necrosis; physical or traumatic injury, such as fracture of the bones, surgery, wounds, or burns; and many inflammatory pathological disorders, like rheumatic fever, rheumatoid arthritis, vasculitis, chronic active hepatitis, malignant neoplasms, or ulcerative colitis. Clinical data reveal that the level of tissue damage together with the type of inflammatory stimulus and the organ or tissue where the process of inflammation sets in and develops bears direct correlation with the concentration of plasma CRP. CRP levels are indicative of the length or depth of a surgical innervation process. Although CRP responses are observed as a consequence to infection by microbial agents, the response is observed to be more pronounced in case of a bacterial infection rather than that of a viral infection. Limited concentration of CRP expression are observed as a consequence to infections by viruses like (1) Influenza A and B virus, respiratory syncytial virus (RSV), and human parainfluenza viruses causing viral pneumonitis; (2) coxsackievirus B, echoviruses, Epstein–Barr virus, rubella virus, and HIV causing viral myocarditis; (3) enterovirus infection causing aseptic meningitis; and (4) hepatitis virus A causing infectious hepatitis. CRP expression and its response at low levels may bear significance in differential diagnosis of infection by microbial agents. Infections, mediated by bacterial agents, lead to the induction of a high level of CRP responses. A high level of CRP after the third and fourth day of postsurgical procedure of an individual undergoing surgery is indicative of persistent bacterial infection in the host wound. Thus, monitoring the level of expression of CRP after surgical procedures finds extreme importance in clinical medication. CRP elevated levels reported from aged individuals and in neonates especially the pretermed delivery cases are indicative of occult bacterial infection.

The response of synthesis and release of CRP and its serum levels may enable monitoring of the effects of administration of anti-inflammatory drugs. The CRP response and its serum levels may permit monitoring of the anti-inflammatory effect of drugs. Thus, immunomodulatory, or anti-inflammatory molecules including salicylates, steroids, nonsteroidal anti-inflammatory drugs (NSAIDS), penicillamine, gold injections, and cytotoxic agents often induce a fall in CRP as well as ESR, presumably by suppressing underlying inflammatory mechanism(s) and inducing hemolysis.

6.13 Discussions: CRP, a New Biomarker in Disease Targeting

The correlation of circulating CRP in the serum or plasma with the extent of severity and disease progression in diverse pathological conditions, together with its significance as a prognostic marker, pronounces the role of CRP both as a biomarker in diseases and also a marker contributing to pathogenesis in disease. Novel drugs are being used that plays a role in blocking CRP binding with specificity, thereby disrupting its pro-inflammatory role in vivo (Pepys and Hirschfield 2003; Pepys et al. 2012) which would serve as a powerful tool for determining the role of CRP not only in the detection of atherosclerosis but also elevated CRP levels playing an important role in the pathogenesis of atherosclerosis and its associated complications, together with providing a cardioprotective role in acute myocardial infarction.

The structural and functional role of C-reactive protein composed of its individual three-dimensional (3D) structure and/or in the form of complex with ligands (Thompson, et al. 1999)—coupled with the knowledge of development of an inhibitor targeted toward the serum amyloid protein (SAP) (Allin et al. 2011; Ansar et al. 2006)—is being employed as a strategy in drug designing.

7 C-Reactive Protein: A Clinical Marker in Cardiovascular Disease

Abstract

Atherosclerotic vascular disease reveals the process of inflammation including several key mediators and markers of inflammation as major manifestations of its pathobiology. Different inflammation markers observed in atherosclerotic disease have been implicated for their prediction of risk in vascular disease or vascular events. Among them, extensively studied is C-reactive protein (CRP), a protein categorized under the class of acute-phase response (APR) proteins. CRP has been reported for its most consistent relationship with cardiovascular disease (CVD) and is capable of predicting risk in the future. Recent epidemiological studies in population have confirmed positive associations between CRP and CVD in populations across the globe. The role of CRP in binding with lipoprotein and activation of the complement cascade reactions and its localization in atherosclerotic vessels indicate strongly the involvement of CRP in the pathophysiology of heart diseases. Initially identified for its major role as a marker in acute processes like tissue injury, infection by pathogen, and inflammation, recently it has earned the disease biomarker status in different CVDs. The role of hs-CRP in detection of risk of cardiovascular diseases is highly accomplished. Use of hs-CRP has well-known clinical and pathological significance. Among hundreds of putative cardiovascular risk factors together with factors including behavioral, biochemical, environmental, and genetic risk markers, CRP remains a unique risk predictor globally. The future prospect of C-reactive protein as a disease biomarker in CVD will help add to the knowledge of the pathobiology and to follow-up studies in patients suffering from CVDs.

Keywords

Cardiovascular diseases (CVDs) • C-reactive protein • Cardiovascular risk factors • Atherosclerosis • Inflammation • hs-CRP

W. Ansar, S. Ghosh, *Biology of C Reactive Protein in Health and Disease*,
DOI 10.1007/978-81-322-2680-2_7

Chapter Highlights

1. The process of inflammation and its role in the pathogenesis of atherosclerosis and in triggering acute coronary syndromes.
2. Summarized studies confirming the relationship of inflammatory biomarkers and mediators with C-reactive protein.
3. hs-CRP and its role in predicting risk to CVD and categorizing patients as low-, intermediate-, and high-risk individuals.
4. Prediction of coronary events, stroke, and peripheral disease and its progression by levels of C-reactive protein.
5. Association of CRP and other systemic inflammatory process markers in CVD.
6. CRP binding selectively to LDL, deposited atherosclerotic plaques, and/or complex ligand leads to activation of the complement cascade and can be pro-inflammatory in nature.
7. Found to co-deposit with activated complement components; in lesions of all acute myocardial infarction, human CRP and complement increase the final size of myocardial infarction in animal models in experimentation. Thus, CRP finds importance as a target for therapy in myocardial infarction.
8. The role of CRP in different cardiovascular diseases and its applicability in assessment of cardiovascular risk globally.
9. CRP plays a role in the pathogenesis of atherosclerosis.
10. Future directions of research based on current research.

7.1 Introduction

Plasma concentration of cholesterol and low-density lipoproteins (LDLs) is reported to have association with an elevated risk of heart attack and death from heart disease. Inflammation together with LDL is a major contributor to heart attacks. LDLs that are deposited from the plasma in the atherosclerotic plaques stick on the walls of blood vessels thereby obstructing the continuous flow of blood in the vessel.

The inflammatory process and gradual deposition of LDL on atherosclerotic plaques lead to complete damage of the plaques. Subsequently small fragments of the plaques on the blood vessel surface break off and are released into the bloodstream. These tiny fragments of atherosclerotic plaques then enter into the bloodstream and are carried away by the blood and sometimes get deposited in the heart or brain, leading to a heart (cerebral) stroke or (cardiac) attack, respectively.

The pathophysiology of atherosclerosis is understood to be that of a chronic inflammatory disorder. Damage of the vasculature leads to its initiation. The induction factors of atherosclerotic plaques include (1) oxidized LDL, (2) ROS or reactive oxygen species, (3) diseases like diabetes, (4) different infections, and many others. In atherosclerosis, the immune system shows the following: (1) macrophage activation, (2) activation of immune cells like B and T cells, (3) activation of endothelial cells, (4) alteration in the Th1/Th2 response, and also (5) elevation of inflammatory cytokines.

The key process in the inflammatory response involves:

1. The recognition of the immune cells in the circulation
2. The recruitment of the blood cells to the inflamed blood vessel by interacting with various adhesion molecules and chemokines (Yeh and Willerson 2003)

In patients suffering from cardiovascular disorders, the quantitative estimation of adhesion molecule expression on endothelial cells is difficult. Among the complete list of lipid and inflammation biomarkers used to predict risk of cardiovascular disorders, C-reactive protein (CRP) remains at the zenith of all other clinical biomarkers employed in predictions of risk of CVD.

The properties of CRP that make it a distinct biomarker in predicting CVD include (1) functional stability exhibited even after long-term storage, (2) half-life that is long, (3) its less variation day to day, and (4) independency with age and sex. CRP is a member of the innate immune system and can activate the classical complement cascade after aggregation or after ligand binding. CRP binding to phospholipids of (1) damaged cells and or oxidized LDL subsequently activates the complement system and leads to uptake of these cells by cells like macrophages (Fig. 7.1) (Yeh and Willerson 2003) eventually leading to their removal.

The pro-atherogenic property of CRP is contributed by its robust disease status predictive property. CRP also stimulates the endothelial cells leading to the expression of various adhesion molecules like intercellular adhesion molecule-1 (ICAM-1), vascular cell adhesion molecule-1 (VCAM-1), selectins, and chemokine-like monocyte chemotactic protein-1 (MCP-1). In the host body, CRP plays a biological role by (1) stimulating cytokine secretion like interleukin-6 (IL-6) and endothelin-1, (2) reducing the bioavailability and expression of nitric oxide synthase (NOS) in endothelial cells, and (3) activating macrophages thereby leading to expression of cytokines and tissue factor and thereby enhancing LDL by cells.

CRP also leads to amplification of the pro-inflammatory effects of mediators including endotoxin. In different in vitro experiments, CRP at a concentration in excess of 5 μg/mL could elicit these pro-inflammatory responses. However, this CRP concentration is much higher as compared to that of the serum CRP concentration of 1 to 3 μg/mL, which is in turn quite correlated with risk in cardiovascular disorders. Therefore, serum CRP levels are often not the true indicator of the

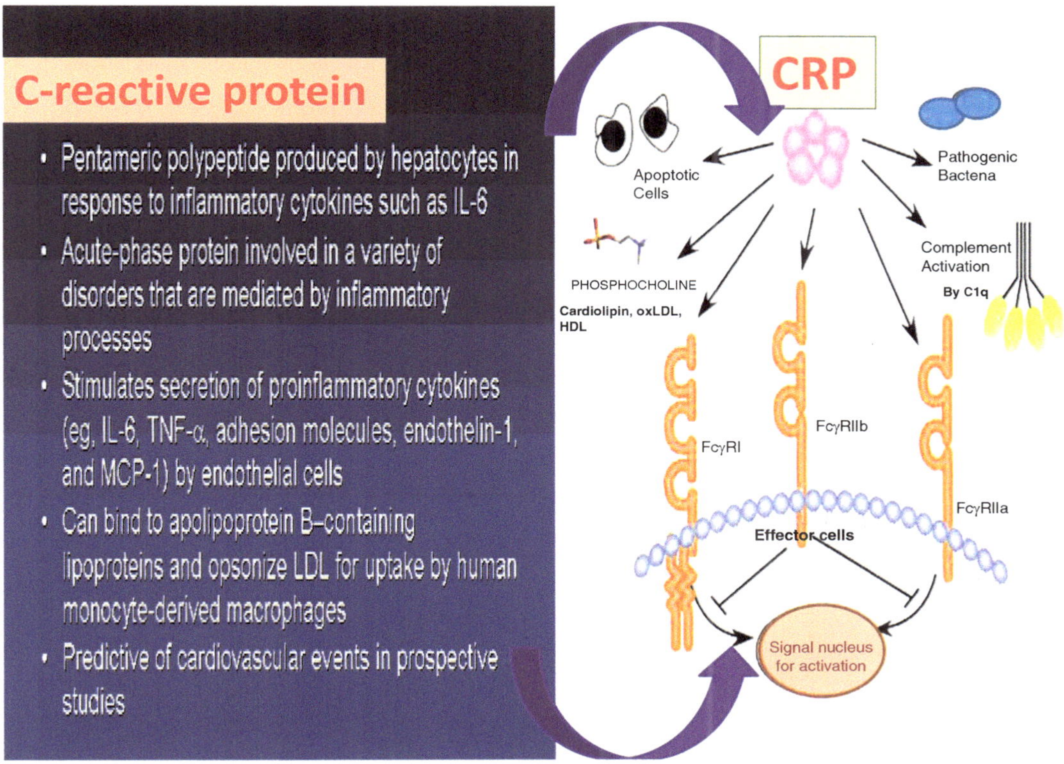

Fig. 7.1 *The ligand-dependent innate immune functions of CRP*. PC, the prototypical ligand of CRP, is present on bacterial cell walls and damaged cell membranes. CRP opsonizes microbes for phagocytosis and activates the complement. CRP binds with FcγR receptors for phagocytosis and release of pro-inflammatory cytokines. *PC* phosphorylcholine

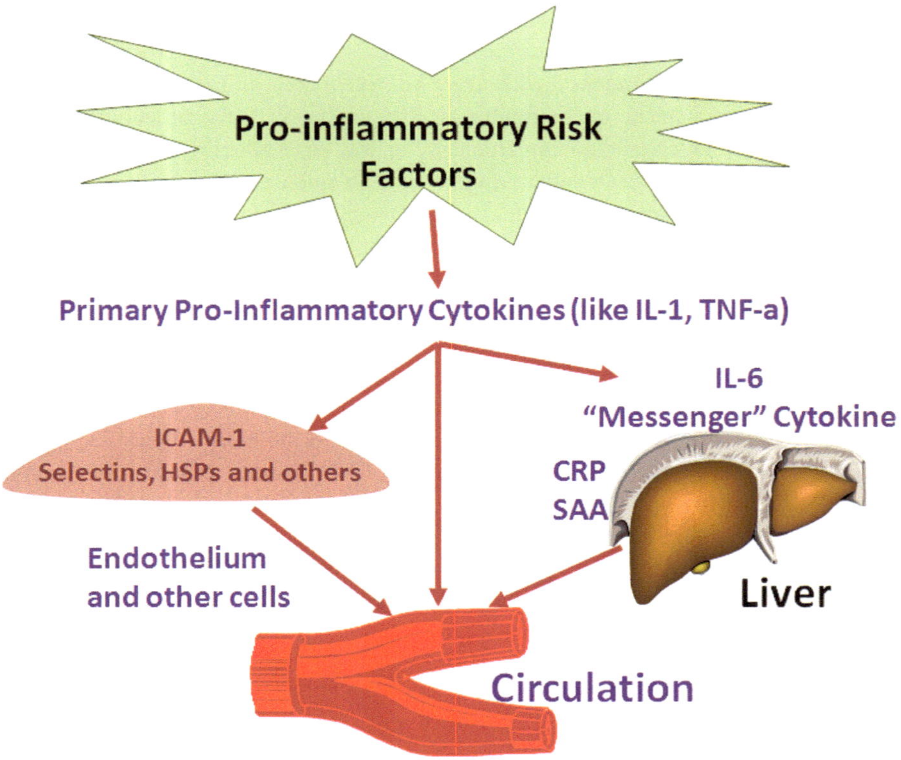

Fig. 7.2 *Inflammatory pathways in atherogenesis*. The pro-inflammatory cytokines lead to secretion of acute-phase protein molecules (CRP, SAP) and adhesion molecules (ICAM-1) to the circulation

tissue CRP levels since CRP has been reported to be deposited in arterial walls as atherosclerotic plaques. The presence of concentrated CRP locally may promote the development of atherosclerosis. The process of atherogenesis involves the synergetic effects caused by the liver that secretes acute-phase proteins together with the secretion of selectin and other molecules from the endothelium to the blood circulation (Yeh and Willerson 2003) (Fig. 7.2).

Researchers have shown that an elevated CRP level is indicative of heart disorders. Elevated blood CRP level is indicative of inflammation and may be as important as cholesterol as cardiovascular disease marker. CRP helps in determining the development of atherosclerosis which involves the hardening of the arterial wall and causing heart disease. People with elevated CRP levels are at twofold more risk of a heart attack as compared to individuals with lower CRP levels. Although CRP forms an important marker of inflammation, the reason behind elevated CRP levels in inflammatory states remains largely unknown. Chronic infection mediated by certain bacteria or viruses may lead to rise in the CRP level.

In normal healthy individuals, CRP is reported to be a trace plasma protein; usually its concentration is lower than 10 mg/L (Das et al. 2003; Clyne and Olshaker 1999). In response to various infections and burn (>200 mg/L), trauma, necrosis of different tissues, aging, many forms of inflammation, and malignant neoplasia, the serum concentration of CRP increases rapidly to nearly 1000-fold by 48 h even after an acute event (Clyne and Olshaker 1999). CRP is the important and most studied among all other acute-phase proteins whose concentration elevates with systemic inflammation (Fig. 7.1). In individuals lacking inflammation, the CRP levels is observed to remain less than 1 μg/mL; however, in patients suffering from various microbial particularly bacterial infections, autoimmune dis-

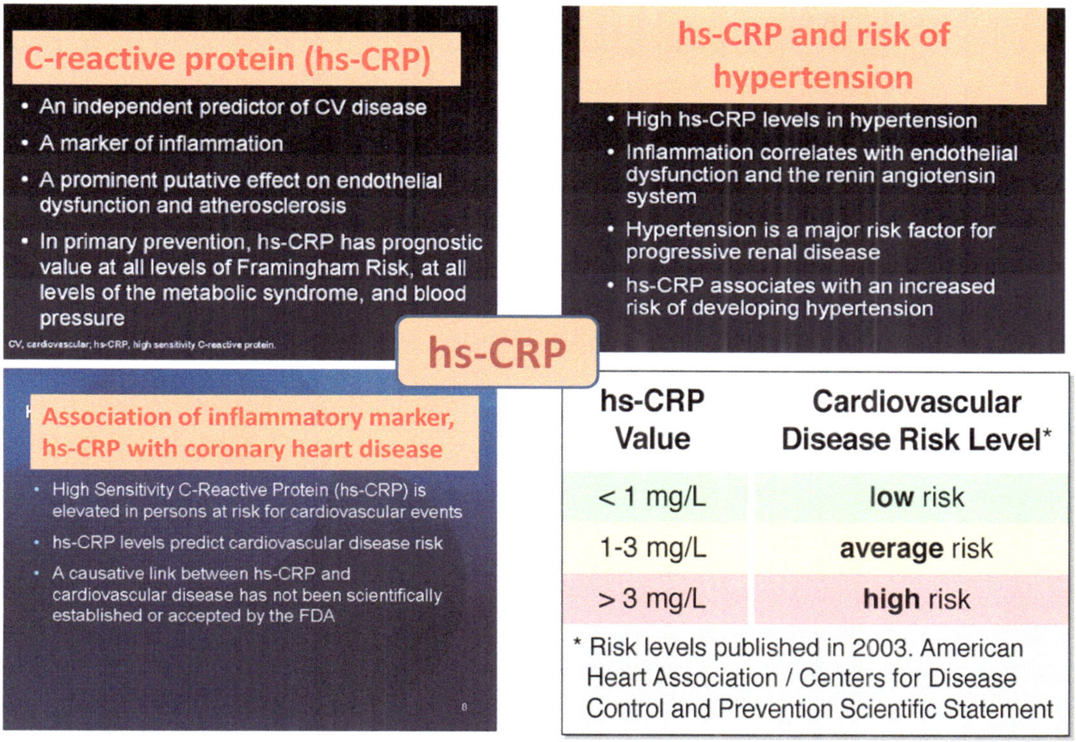

Fig. 7.3 Clinical application of hs-CRP for cardiovascular risk prediction (Ridker 2003)

orders, and cancer, the CRP concentration has been observed to be around 100 μg/mL or higher. Therefore it is understood that a simply elevated concentration of CRP in the serum bears no indication of the specificity of the disorder occurring in the individual and cannot provide an insight on the different disease conditions one from the other. However, CRP lacks specificity; it finds importance as a predictive biomarker in cardiovascular diseases and predicting risks associated with CVD.

Moreover, remarkably the predictive property of CRP falls within the range of 1 to 5 μg/mL. In the time preceding to the discovery and applicability of high-sensitivity CRP (hs-CRP) test, this concentration of CRP was regarded to be normal. In fact, when tests showing greater than 10 μg/mL serum CRP levels in apparently healthy men or women were used, it used to be repeated so as to exclude chances of occult infection or any other systemic inflammatory processes.

In order to understand the gradual transition of biological role of CRP molecule from being just one of the members of an acute-phase protein family to the elucidation of its biological role as one the major sensitive biomarkers in predicting future risk of cardiovascular events, it is necessary to understand the biological role played by CRP in the pathogenesis of atherosclerosis (Yeh and Willerson 2003).

However, in recent researches, much focus has been given on the use of hs-CRP or high-sensitivity C-reactive protein. hs-CRP is an inflammatory biomarker which is being employed in the diagnosis of increased risk for cardiovascular diseases in patients (Fig. 7.3). Currently, the precision of hs-CRP lies in the very low ranges (0.2 mg/L) in detecting patients with high risk of cardiovascular diseases. CRP levels ranging from < 1 mg/L and 1 to 3 mg/L to > 3 mg/L that are used to designate low-, intermediate-, and higher-risk vascular groups (Yeh and Willerson 2003) are currently employed.

7.2 Epidemiology of Cardiovascular Disorders

The dominance of people suffering from cardiovascular diseases (CVDs) is an alarming global concern. Among the different global causes of death, CVDs rank first in causing the most global deaths. Gradually an increasing number of individuals annually show mortality due to causes of CVDs as compared to any cause of death. According to current reports from the World Health Organization (WHO), an estimated population consisting of about 17.5 million individuals died in 2012 suffering from different CVDs. This is contributed to an overall mortality of 31 % globally. Among this mortality rate of 31 %, around 7.4 million individuals succumbed to mortality caused by coronary heart disorders and around 6.7 million died due to stroke. Eighty percent and above of the mortality due to CVD globally is reported from countries lying within the cover of low and middle income. It is speculated that in the year 2030 more than 23 million individuals will undergo death annually due to different CVDs (http://www.who.int/mediacentre/factsheets/fs317/en/). Nearly 16.7 million globally died of CVD in 2002, of which 7.2 million people succumbed to death suffering from coronary heart disorders, while stroke claimed the lives of 6.0 million people, 0.9 million people died due to hypertensive heart disorders, inflammatory heart disease claimed 0.4 million lives, rheumatic heart disease claimed 0.3 million individuals, and other CVDs were the root cause for the another 1.9 million lives (http://www.who.int/mediacentre/factsheets/fs317/en/).

The epidemiology of CVD is studied under the headings of:

Descriptive epidemiology wherein the distribution of CVD is measured by studying individuals in a population within the parameters of age, gender, ethnicity, and time and place of study. Analysis and understanding of the relationships between CVD and risk factors for the disease in a population, risk model, and multicausal developments are done under *analytical epidemiology* while *experimental epidemiology includes* experiments, researches, and strategies designed to reduce/prevent CVD at both individual and community levels.

7.2.1 Epidemiology of CVD across the Globe

Epidemiological studies from low- and middle-income countries indicate that dietary changes together with nutritional transition and with the increased intake of energy-rich diets including unhealthy fats, oils, sodium, and sugars increase the influence of CVD incidence (Hu 2008).

Adult lifestyle patterns usually starting from childhood and youth like smoking, dietary habits, sporting behavior, and others are the presumed causative factors of CVD. In the age group of 30–44 years, the risk of CVD morbidity and mortality increases. Premature death below the age of 64 years (i.e., 25–64 years range of age) or death in the elderly population is more difficult to interpret as with increase in age and varied other health causes succumbing the individual. The risk factors in men includes smoking and high triglyceride levels while in women certain risk factors are diabetes mellitus, depression, oral contraceptives, hormone replacement therapy, and polycystic ovary syndrome. The varied risk factors of CVD are tabulated in Table 7.1. The common risk factors of CVD include hypertension,early lesions in blood vessels, atherosclerotic plaques, modern lifestyle (increased sympathetic activity, psychosocial stress, leading position in job), dietary intake (a high intake of salt, processed food, low levels of water hardness, high tyramine content of food, alcohol use), positive family history, abnormal blood lipids, higher low-density lipoprotein (LDL) levels, obesity, and diabetes mellitus. These are the common traditional cardiovascular risk factors to name quite a few. These risk factors can be avoided or restricted or stopped to reduce cardiovascular morbidity and mortality (Castelli 1996).

Table 7.1 Steps in cardiovascular epidemiology

(1) Locating sites and plotting distribution areas of occurrence of CVD, (2) monitoring, 3) surveillance, and (4) monitoring trends of changes
Reporting and studying the natural history of CVD
Testing association of risk factors with occurrence and prediction of CVD
Designing of awareness and implementation of prevention programs for CVD
Assessment of effectiveness of prevention programs
Analysis and risk factor types
Important risk factors that can be controlled: High blood pressure, abnormal concentration of lipids in blood, use of tobacco, lack of physical activity, obesity, unhealthy and fatty food intake, diabetes
Risk factors that cannot be controlled: Age, family history, gender of the individual, race to which the individual belongs *Novel risk factors*: Elevated level of blood homocysteine, elevated CRP, fibrinogen levels, abnormal blood clotting

7.3 Cardiovascular Disorders

Cardiac disease forms the major root cause of deaths globally and also forms the root cause of major disabilities associated with CVD. The different cardiovascular diseases are listed in Fig. 7.4. The different cardiovascular conditions include:

1. *Rheumatic heart disease*

Mostly occurring in childhood, this condition is manifested by one or several attacks of rheumatic fever that damages the valves of the heart. Development: rheumatic fever is a disease that initiates after an untreated beta-hemolytic streptococcal throat infection in children. As a consequence, the heart valves are permanently damaged or the valves are scarred, with improper opening and closing, cardiac muscle weakening, and cardiac sac damage. This may progress to heart failure.

2. *Valvular heart disease*

Under normal conditions, the cardiac valves control and regulate the direction of blood flow through the heart. But valvular damage may be caused by (a) narrow valves termed as stenosis, (b) leakage, and (c) valves not closed properly termed as prolapse.

3. *Hypertensive heart disease*

High blood pressure termed as hypertension (primary or secondary) caused by certain conditions, including diseases or infections like adrenal gland tumor, kidney disorder, and disorder in the blood flow through the blood vessels, may lead to blood volume irregularity in the heart and blood vessels and to fatal consequences.

4. *Aneurysm*

An aneurysm is diagnosed as a bulge or weakness arising in the arterial wall caused by high blood pressure or a weak spot in a wall of the blood vessel which may enlarge over time and if ruptured is life threatening. Although it can affect any blood vessel, the most common sites affected include the walls of the abdominal aorta and the brain base arteries.

5. *Atherosclerosis*

Atherosclerosis is a slowly progressive disease. This condition is manifested by thick and stiff artery walls mediated by fat deposits termed as plaques restricting the blood flow. It can occur anywhere in the circulatory system. When it occurs in the cardiac arteries, this condition is termed as disease of the coronary artery, and when it occurs in the periphery like the legs, it is termed as peripheral arterial disease (PAD). Atherosclerosis develops gradually over a long time and leads to fatal consequences causing heart attack and stroke. Atherosclerosis is characterized by the deposition of fatty substances like LDL, cholesterol, cellular metabolic waste products, calcium, and fibrin on the inner arterial wall lining. Hyperlipidemia in the blood enables buildup of plaque in the arteries. PAD develops due to atherosclerosis, leading to the narrowing and/or complete blocking of the blood vessels in the legs. It is manifested by leg pain while walking. People suffering from PAD are at an

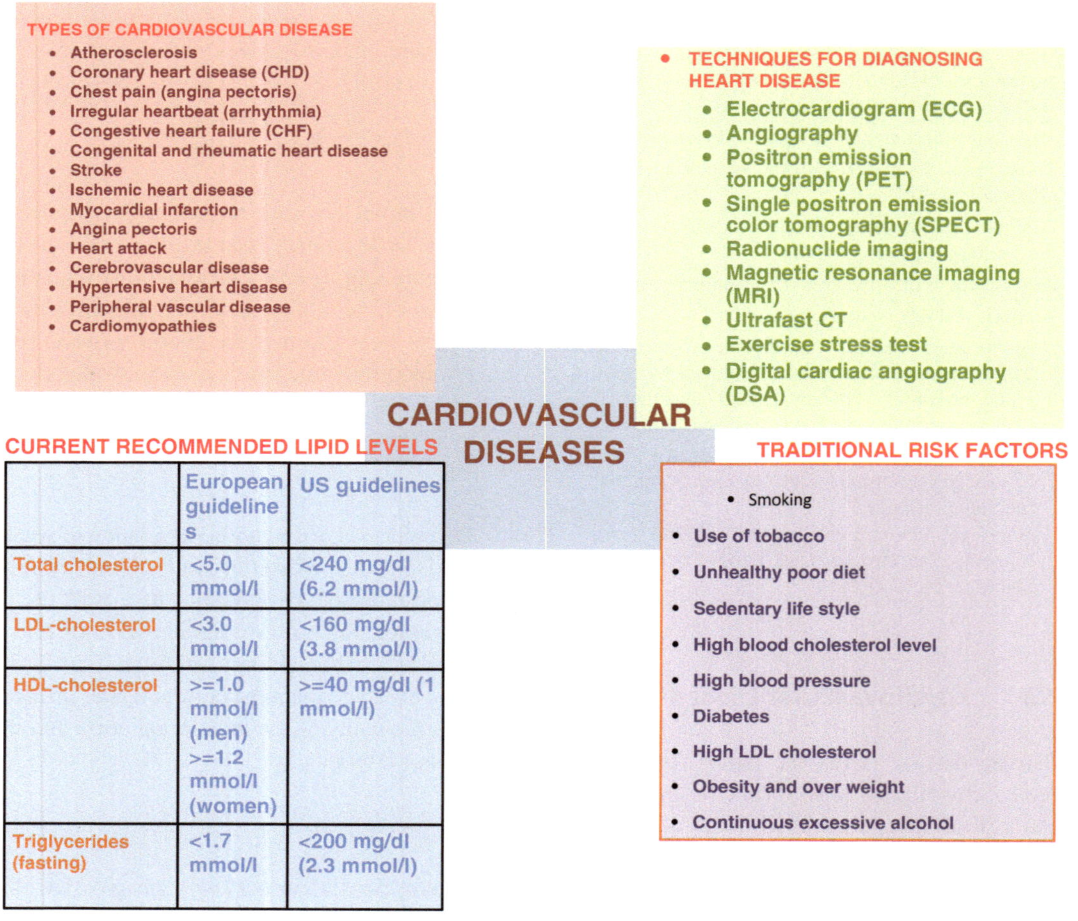

Fig. 7.4 Cardiovascular diseases: Different types of cardiovascular diseases, their diagnostic weapons used in clinical medicine, and the recommended lipid levels for adults to avoid cardiovascular diseases are represented

enhanced risk of later developing infection from gangrene in legs. Atherosclerosis is often seen to be initiated very early in childhood and manifests completely when the individual reaches his/her middle ages or even later toward his/her old age manifested by a cardiac event like cardiac stroke or peripheral vascular disease (Olson 2000). In atherosclerosis, the inflammatory processes are reflected by the gradual progression of plaque instability indicating various underlying atherothrombotic events.

6. *Ischemic heart disease*

It develops by narrowing or blockage of the coronary arteries thereby reducing blood supply to the heart.

7. *Angina pectoris*

It is manifested by pain in the chest due to reduced blood and oxygen supply to the cardiac tissues or ischemia. The extent of severity of the pain is directly correlated with the level of oxygen deprivation. Angina occurs due to atherosclerosis that is the narrowing and/or blockage of the arteries that supply the heart. The chest pain of angina sometimes may spread to the arm, shoulders, and jaw. Symptoms of angina also include shortness of breath and profuse sweating. Women suffering from angina experience severe intense pain in the shoulder, middle of the back area, throat, neck, and jaw as compared to men. Nitroglycerine is the drug used to exert prophylactic effect thereby leading to relaxation of the veins while beta-blocker class of compounds is given to control the overactivity of the cardiac muscle.

8. *Coronary artery disease (CAD)*

It is also known as ischemic heart disease. It is caused by mechanisms leading to atherosclerosis. It is initiated by the gradual narrowing of arteries and blood vessels and finally by complete blockage of the arteries that normally function to bring blood to the heart thereby completely depriving the heart from fresh blood supply. Most common types of heart disease are the main root cause contributing to heart attacks and angina.

9. *Heart attack*

It is also termed as myocardial infarction. It is initiated due to stoppage of blood supply to the heart. Symptoms include chest pain, which sometimes spread to the left arm, shoulder, or jaw, exhibiting severe shortness of breath and profuse sweating, and some individuals may also faint. Women experience difficulty breathing, tiredness, shoulder pain, jaw pain, or upper back pain. This sometimes leads to sudden death due to complete nonfunction of the heart. Abnormality in the rhythm of the heart is manifested.

10. *Cerebrovascular disease*

It is mediated by disorder in the blood vessels in the brain. A cerebrovascular stroke is caused by irregularity or stoppage of blood supply to the brain.

11. *Cerebrovascular disease*

It is caused by atherosclerosis.

12. *Stroke*

It initiates due to interruption of blood supply to the brain. It occurs due to blockage or burst of blood vessels in the brain or neck leaving the brain deprived of oxygen thereby causing permanent damage to parts of the brain. Aftereffects of stroke include impairment in speech or vision, weakness, or paralysis.

13. *Transient ischemic attacks*

TIAs occur due to brief blockage of blood supply affecting the brain function. It shows symptoms like transient weakness on one side of the body with loss of body balance, blindness, double vision affecting one or both eyes, difficulty in speech, and severe headache. However, the damages are temporary. A TIA is an indication for an occurrence of stroke in the individual.

14. *Cardiomyopathy*

It is caused by defect in the function of cardiac muscle. While some are hereditary, others may be a result of infection or other conditions. An enlarged heart is observed in idiopathic dilated cardiomyopathy. Other symptoms include complete loss of cardiomyocytes, a dilated and enlarged heart, hypertrophy, and thickening of cardiac muscles.

15. *Pericardial disease*

The disorder results from the abnormal functioning of the pericardium, i.e., saclike structure, that encapsulates the heart. Inflammation (pericarditis), fluid accumulation (pericardial effusion), and stiffness (constrictive pericarditis) may affect the pericardium.

16. *Congenital heart disease*

Congenital heart diseases are those that occur at birth due to malformations of the heart's structures. They may occur due to defective genes or adverse exposure of fetus in the womb to certain elements, such as some medicines or too much alcohol. Holes in the heart, abnormal function of valves, and abnormal heart chambers may lead to this condition. Hereditary factors, maternal diseases, chemicals like alcohol, or drug intake during fetal development may be some of the lead causes.

17. *Arrhythmia*

It is manifested by an abnormal rhythm of the heart that can dictate the function of the heart. In arrhythmia, the heartbeat may be too slow, too fast, or irregular. When the heart rate is less than 60 beats/min, it is termed as *brachycardia*, and when the heart rate is more than 100 beats/min, the condition is termed as *tachycardia*. Under this condition, the heart fails to pump enough blood to cater and to meet body requirements; due to fast beating, it may become exhausted and fatigued. Fibrillation develops when heartbeat is sporadic in a quivering pattern.

7.4 Traditional Cardiovascular Risk Factors and Their Limitations

Despite growing awareness of different CVDs, atherosclerosis forms the major cause of death. From epidemiological studies, a series of common cardiovascular (CVD) risk factors categorized as traditional include factors like hypercholesterolemia, hypertension, smoking habit, diabetes, obesity, and family history of early coronary artery disease (Table 7.1). The Framingham study is a well-documented and most referred report in this context of epidemiological studies in detecting CVD risk factors, among other studies (Castelli 1996).

Some of these factors are reported to have led to mortality and morbidity but can be avoided or treated. The Framingham risk score takes into account different parameters including age, sex, concentration of HDL or high-density lipoprotein cholesterol in blood, blood pressure, habit of smoking, and other diseases like diabetes in predicting risk of cardiovascular disorder. The Framingham risk score is widely recommended in algorithm risk assessment (Lloyd-Jones et al. 2006; Castelli 1996).

A series of cardiovascular risk factors including (1) family history of early coronary artery disease, (2) hypertension, (3) hypercholesterolemia in blood, (4) smoking habits, (5) obesity, and (6) diabetes are now accepted as the most common traditional risk factors for various cardiovascular diseases. Very few and common risk factors are noted here (Table 7.1). With respect to time, these factors can be avoided or restricted or treated, with marked reduction on different types of cardiovascular mortality and morbidity. With modernization and health consciousness, people are changing their diet and lifestyle to refrain themselves from various CVDs. Moreover the action of statins on lipid levels and on other physiological processes has made quite an advancement in controlling CVDs (Castelli 1996).

Interestingly, atherosclerotic patients like acute coronary syndromes do not reveal traditional risk factors (Fig. 7.4). From the Framingham study, it was established that nearly 35 % of the patients suffering from coronary disease showed normal cholesterol levels. Thus in this present scenario, it is pertinent therefore to search for new biomarkers of CVD to help improve prediction of its risk associated with cardiovascular disorders. However, the role of CRP in the use of statin is quite accomplished (Fig. 7.5). So CRP can be used as cardiovascular risk factors (Castelli 1996; Gomes 2002).

HMG-CoA reductase inhibitors (statins), independent of their effects on lipid profiles, reproducibly reduce CRP values. It is quite unclear whether statins directly affect the hepatocytes, leading to anti-inflammatory effects within or outside the atherosclerotic plaques. However, recent researches suggest that statins are capable of reducing risk of malfunctions of the cardiovascular system both in patients without normal level of LDL but with a baseline CRP level above the median in the blood and those with elevated level of LDL cholesterol in the blood (46). Thus the measurement of CRP concentration finds importance in detecting and applying in prophylactic anti-atherosclerotic therapy even in individuals falling in the category of being low risk (Ridker et al. 1999; 2001a, b; Ridker 2001). CRP binds with LDL and is present in the atherosclerotic plaques, and thus it may be considered as a causative molecule in coronary heart diseases. Statins also potentially affect LDL cholesterol concentration, and probably due to its relevant anti-inflammatory actions, it may be assumed that the

CRP used to target statin therapy in primary prevention: AFCAPS/texcaps

STUDY GROUP	STATIN DOSE (mg/dL)	PLACEBO (mg/dL)	NNT
Low LDL-C/low CRP	0.025	0.022	----
Low LDL-C/high CRP	0.029	0.051	48
High LDL-C/low CRP	0.020	0.050	33
High LDL-C/high CRP	0.038	0.055	58

Median LDL-C= 149 mg/dL
Median CRP= 0.16 mg/dL

CRP, C-reactive protein; AFCAPS/Tex Caps, AirForce Texas Coronary Athersclerosis Prevention Study; NNT, number needed to treat; LDL-C, low-density lipoprotein cholesterol.
Adapted from Ridker PM, et al. N Engl J Med. 2001;344:1959-1965.

Fig. 7.5 *Application of CRP is determining the dose of statin.* AFCAPS/TexCAPS, Air Force Texas Coronary Atherosclerosis Prevention Study; *NNT* number needed to treat, *LDL-C* low-density lipoprotein cholesterol (Ridker et al. 2001d)

cardioprotective benefits of statins find greater application in people with elevated baseline CRP levels than in those individuals with lower CRP levels. From observational epidemiology, various human genetic studies, different experimental and animal models, and randomized clinical trials of statins reveal that CRP itself remains a direct causal mediator of coronary heart diseases and thus further investigation is required in this field (Danesh and Pepys 2009).

From a study conducted over a population of 200,000 individuals of which around 47,000 suffered from coronary heart disorders, it was shown that conditions of elevated CRP concentration of CRP controlled genetically are unrelated to conventional risk factors and risk of coronary heart disease. Thus CRP concentration alone may not be the causative factor for coronary heart disorders (CCGC et al. 2011).

From recent researches, it is revealed that elevated CRP levels find importance in indicating long-term risk of sudden cardiac death (SCD) in apparently normal individuals. CRP may also be used in the assessment of risk of cardiovascular disorder in healthy individuals and which could directly impact on the reduction of mortality from SCD. Nearly more than 50 % of SCDs occur together with symptoms of ischemic heart disease and poor survival rates of out-of-hospital cardiac arrest. Therefore it is crucial to identify these risk factors even before the event. So, in modern clinical diagnosis, CRP along with other cardiac risk factors is necessary to be assessed in prediction of risk among apparently normal individuals. Homocysteine (Hcy) is a sulfur-containing amino acid and is a risk factor in atherosclerosis and thrombosis and also a predictive marker for risk of development of myocardial infarction and coronary heart diseases (Albert et al. 2002).

Acute plaque rupture in blood vessels together with associated thrombosis in deep veins is the most common manifestation in SCD individuals, and atherosclerosis remains the underlying cause of most cases of SCD found in men. CRP alone or synergistically with elevated concentration of homocysteine may lead to prediction in individuals at risk for SCD apparently even in healthy populations. Even switching to a healthy diet and/or the use of drugs can affect the lipid level and lower it in healthy individuals; the increased homocysteine concentration was reported to show strong association with myocardial infarction over the short span of time but not when the follow-up periods were longer. Thus the association among these plasma markers (CRP and homocysteine) and the plasma lipid levels with the risk of developing SCD later in apparently normal individuals helps researchers to detect effects ranging from small to moderate in SCD, which needs to be tested in a larger population (Albert et al. 2002).

In a JUPITER clinical trial of 17,802, apparently individuals having low LDL cholesterol levels (<130 mg/deciliter) and elevated concentration of hs-CRP are benefitted from the use of statin (rosuvastatin dose 20 mg/day or higher) therapy. The individuals were diagnosed for the occurrence and risk of stroke, arterial revascularization, myocardial infarction, unstable angina, or death from any other cardiovascular causes (Emerging Risk Factors Collaborations et al. 2012).

Rosuvastatin therapy therefore reduces cholesterol primarily LDL and hs-CRP by nearly 50 % and 37 % approximately. The results of the studies on stroke, myocardial infarction, unstable angina, death from cardiovascular causes, and death from any cause when analyzed reveal that effects of statin therapy in all these subgroups of patients were quite consistent (Ridker et al. 2008). The independent predictive value of hs-CRP (hazard ratio of 1.20) for future vascular events was actually greater than that of total cholesterol (hazard ratio of 1.17) (Emerging Risk Factors Collaborations et al. 2012).

7.5 The Search for New Risk Markers

The search of new predictive risk factors for CVD continues. Markers like homocysteine, plasminogen activator inhibitor, lipoprotein (a), and von Willebrand factor, when tested for their predictive capability of CVD, revealed that some of these predictive risk factors are capable in predicting early coronary disease. Recently, various inflammatory markers were counted in predicting CVDs. The risk factors of atherosclerosis may be classic, uremia related, or due to dialysis (Fig. 7.6). These inflammatory markers including leukocyte count, fibrinogen, soluble CAM or cell adhesion molecules, and cytokines like interleukin-6 (IL-6), TNF-α, and CRP are being investigated as risk factors. The work of pathologist Russell Ross, published in *New England Journal of Medicine*, in the month of January 1999 is most cited in this context. It is therefore important at this juncture to understand the role of inflammation and the development of atherosclerosis (Ross 1976, 1999).

7.6 The Role of Inflammation in Atherosclerotic Processes

From recent studies probing underlying reactions in understanding the biochemical responses to injury, Russell Ross and later Ross and Peter Libby showed that the initiation of aggression against the vascular endothelium is mediated by several factors like (1) LDL lipoproteins which undergo subsequent oxidization, (2) increased vascular wall stress factors like hypertension, (3) hyperglycemia in case of diabetic individuals, (4) free oxygen radicals in smokers, and many other risk factors for atherosclerotic disease. Atherosclerosis involves simultaneous events of (1) migration of vascular smooth muscle cell, (2) formation of foam cell, (3) cell-mediated immunological responses initiated by T-cell activation, (4) platelet adherence and aggregation, and (5) leukocyte infiltration adherence (Ross 1976; 1999). The role of CRP in inducing atherosclerosis through an inflammatory cascade is repre-

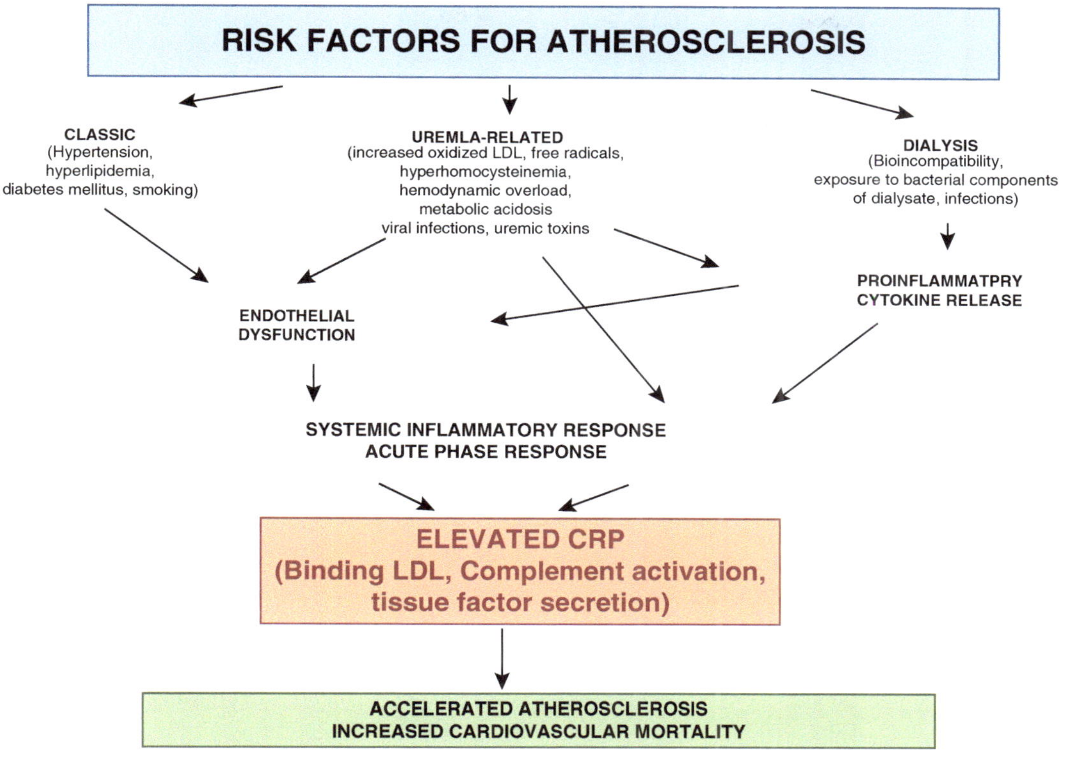

Fig. 7.6 The risk factors of atherosclerosis

sented in Fig. 7.7. It is studied under the following headings:

7.6.1 Endothelial Dysfunction

This process leads to endothelial dysfunction initiated by activation of endothelial cells that reduces their NO production (Ross 1976; Farzaneh-Far et al. 2001) which plays a major role in anti-atherogenesis and acts by increasing the production of (1) chemokines and (2) adhesion molecules (Patel et al. 2001; Libby 1995). Chemokines, including MCP-1 (monocyte chemotactic protein-1), interleukins like IL-8, growth factors like MCF (macrophage colony-stimulating factor), and macrophage antigen-1, in turn attract mononuclear cells, T cells, and macrophages derived from monocytes to the blood vessel wall (Krishnaswamy et al. 1998; Gonzalez-Amaro et al. 1998; Valente et al. 1992; Wang et al. 1991).

Then the adhesion molecules including VCAM-1 and other selectins aid in the mobility of the mononuclear cells to the subendothelium. This forms the earliest events in the atherogenic process and has been detected by electron microscopy (Blake and Ridker 2001; Price and Loscalzo 1999; Cybulsky and Gimbrone 1991).

7.6.2 Subendothelium Processes

In the next steps, here the mononuclear immune cells produce pro-inflammatory cytokines IL-1, IL-6, and TNF, which, in turn, function as chemoattractants that activate the adhesion cells on one hand and express matrix metalloproteinases on the other thereby leading to promotion of the plaque growth. Adhesion cells express metalloproteinases that lead to cell migration and initiate proliferation of smooth vascular myocytes of the tunica media. They also capture LDLs and convert macrophages into foam cells (Blake and Ridker 2001).

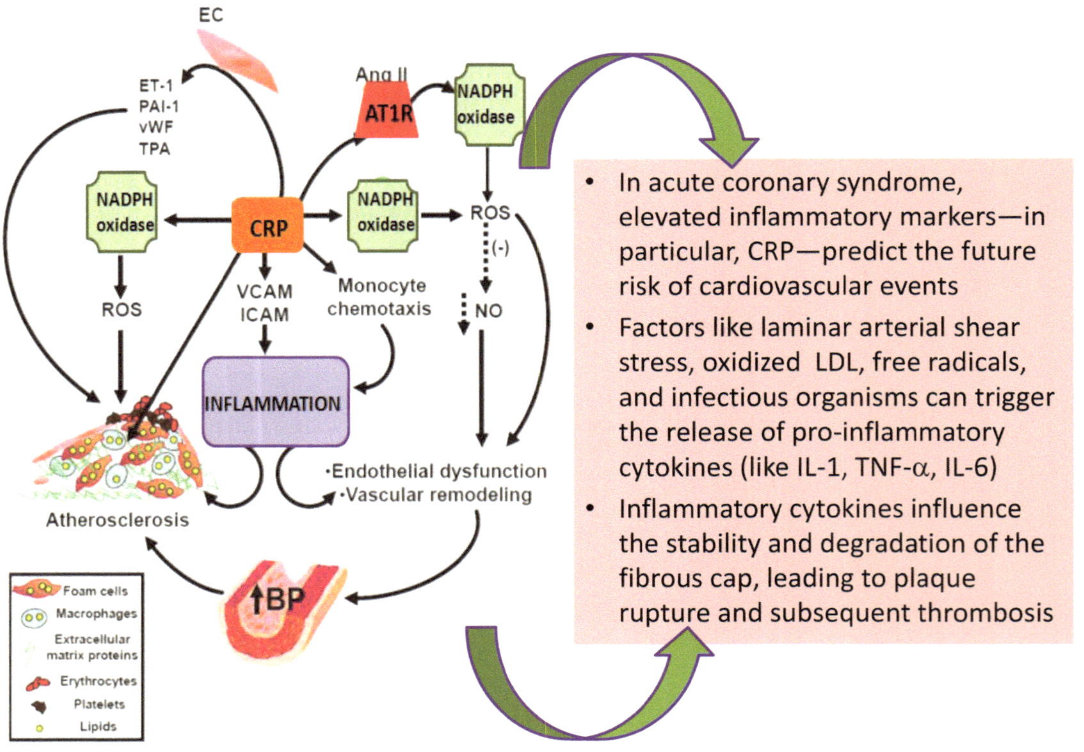

Fig. 7.7 *The process of the inflammatory disorder atherosclerosis*. The role of CRP in inducing atherosclerosis through an inflammatory cascade is represented. *BP* blood pressure, *Ang II* angiotensin II, *AT1R* angiotensin type 1 receptor, *CRP* C-reactive protein, *ET-1* endothelin-1, *ICAM* intercellular adhesion molecule, *NAD(P)H* oxidase, nicotinamide adenine dinucleotide phosphate oxidase, *NO* nitric oxide, *PAI-1* plasminogen activator inhibitor-1, *ROS* reactive oxygen species, *TPA* tissue plasminogen activator, *VCAM* vascular cell adhesion molecule, *vWF* von Willebrand factor

Endothelial injury together with the inflammatory process thus enables a continuous destruction process progressively leading to atherogenesis (Farzaneh-Far et al. 2001). With the progression, the lesion forms an atheroma plaque which matures to develop a collagen fibrous cap and inner agglomerate formed by dead macrophage foam cells incorporated into a growing necrotic lipid center. The stability of the atheroma plaque, or its tendency to rupture, is controlled by the balance between both synthetic factors and proteolytic factors (Patel et al. 2001; Libby 1995). Synthetic factors include certain inflammatory and growth factors, like FGF (fibroblast growth factor) and PGF (platelet growth factor), which promote proliferation of smooth muscle cells and subsequent segregation of collagen. Collagen helps in the formation of stable fibrous cap. But many inflammatory cytokines, due to their proteolytic activity, weaken the fibrous cap. TNF-α, expressed by endothelial cells, macrophages, and smooth muscle cells in blood vessels, causes collagen degradation by activating the matrix metalloproteinases. IL-1 also enables degradation of the collagen like TNF-α. IL-1 also stimulates IL-6 production, expressed in atheromatous lesions, and it also stimulates the production of acute-phase reactants from the liver. Acute-phase reactants are delivered in the blood. Acute-phase reactants like CRP, SAA, and fibrinogen play an important role in the process. In this group of deleterious cytokines, the interferon-γ which has effects on plaque stability and reduces collagen production is also included. These pro-inflammatory cytokines lead to alteration of consistency and thickness of the fibrous cap thus enabling it to rupture (Gomes 2002).

7.6.3 Rupture of Plaque

Various morphological characteristics determine the stability of the atheroma plaques. The fibrous caps are ruptured when they are thin or with macrophages with ratio higher as compared to smooth muscle cells; more than 50 % of the volume of the plaque is formed of the necrotic lipid center. Moreover, sometimes the plaques are heavily infiltrated by macrophages and less infiltrated by smooth muscle cells causing them to rupture. The plaque ruptures to expose the thrombogenic lipid center releasing thromboxane A2 and collagen. This leads to rapid accumulation of platelets. Coronary thrombosis is initiated that triggers acute coronary syndromes. Different coagulation molecules affect thrombus development. However, inflammatory cytokines aggravate the development of the thrombus formation by inducing P-selectin expression on platelet cell surface. P-selectin causes platelet adhesion to other platelets and leukocytes and to the endothelium. Inflammatory cascade plays a crucial role in destabilizing the plaque and its consequent rupture. Thus a complete circle of platelet activation followed by thrombus formation continues to add to the disease pathology (Patel et al. 2001; Libby 1995; Gomes 2002).

7.7 Diagnosis of a Heart Disease

Any individual experiencing cardiovascular problems will show the symptoms of undue fatigue, palpitations (with a skip of a beat or excessive rapid beat), dyspnea (difficult or labored breath), chest pain, angina pectoris (stable angina), unstable angina, and vertigo or sudden unconsciousness. The diagnosis of coronary heart disease (CHD) depends on family medical histories of the patient, blood pressure, associated risk factors, physical examinations, and the results obtained from clinical or hematological tests and diagnostic procedures (Fig. 7.4). Listening to the heart for heart murmurs (swishing or whooshing sounds) provides important clues about heart trouble. The doctor sometimes asks the patient to fill out a questionnaire called the Duke activity status index which measures the ease with which the patient is doing the routine tasks. It also provides information about the blood flow in the coronary arteries. No single test can diagnose CHD. So, if a patient is suspected of having CHD, then the clinician will recommend the suspected individual with one or more of the following tests.

7.7.1 Electrocardiogram (EKG/ECG)

A test of the heart is performed to detect and record the cardiac electrical activity. The test shows and records the beat and rhythm of the heart whether steady or irregular. An EKG records the strength and timing of electrical signals when it passes through the heart. Electrodes are attached to the arms, legs, and chest for recording that are connected to a recorder that records the electrical signals generated from every heartbeat. EKG indicates CHD-induced heart damage signs from a current or previous heart attack.

7.7.2 Echocardiography (Echo)

Echocardiography is a biophysical technique that sends sound waves to capture a picture or the activity of the heart. The physiology, structure, functions, shape, and size of chambers and valves of the heart can be known. Echo reveals areas where the blood flow is poor, thickness of the cardiac muscle, muscle not showing normal contraction, and movement of each valve and also shows sign and sites of previous injury (if any) too.

During an echocardiogram, a lubricating gel is rubbed on the skin to allow smooth movement of a small mechanical device called transducer. Transducer will be allowed to move smoothly over the chest with a continuous contact between the sensor and the skin with a final notation of heart activity and images.

7.7.3 Chest X-Ray

A chest X-ray uses the radiography method to visualize all the organs and structures inside the rib cage, including ribs, the heart, lungs, and blood vessels. It can reveal signs of heart failure, lung disorders, chest infection, osteoporosis, malformed blood vessels of lungs, and other symptoms.

7.7.4 Stress Testing

During stress testing, patients are advised to perform exercises so as to make the heart pump and beat fast during the tests. In order to work and beat fast, the heart requires an increased supply of blood and oxygen. Coronary arteries that are narrowed by deposited plaques are not able to supply sufficient oxygenated blood. Therefore, this stress test shows the blood flow in the heart. Signs and symptoms of coronary heart disease, like abnormal function of the heart, abnormal blood pressure in suspected individuals, shorter breath, chest pain, and abnormal heart rhythm or activity, could be known from this test.

In stress tests, while the patient is at rest or is performing exercise, pictures of the heart are taken.

7.7.5 Blood Tests

Blood tests are biochemical tests performed in order to check the concentration of fats, cholesterol, sugar, and proteins in the body of the individual suspected or at risk of CVD. Over and above normal levels may be indicative of the disease state or his/her risk for coronary heart disease. In a heart attack, cardiac muscle cells undergo death with the discharge of proteins in the bloodstream which are quantitatively estimated during blood tests. Elevated levels of these molecules indicate a recent or ongoing heart attack. Cardiac enzyme tests show recent damage to the cardiac muscle.

7.7.6 Coronary Angiography and Cardiac Catheterization

A coronary angiography test employs dye in a procedure called cardiac catheterization. This is followed by imaging the coronary arteries by special X-rays. It reveals information about cardiac functioning and pressure inside heart chambers. A thin, flexible catheter tube is inserted into the blood vessel in the arm, upper thigh, or neck. This is then led to the coronary arteries. The dye is then released into the blood. Special X-rays taken during dye flow in the blood through the coronary arteries detect any blockages in the large coronary arteries. But in coronary microvascular disease, there are no blockages in the large coronary arteries and thus are not detected by coronary angiography. A coronary angiogram is a relatively safe method with rare serious complications. However, the procedure itself could cause death accounting for 1 or 2 in every 1000 lives.

7.7.7 MRI or Magnetic Resonance Imaging

An MRI scan uses magnetic fields and radio waves to show the structure and function of the heart with each beat and can be visualized in a computer. An MRI scan is performed with the patients lying inside a scanner with surrounding magnetic field. The images are produced by magnetic field and radio waves. Both still images and video of the heart at work together with all major blood vessels can be seen in an MRI scan.

7.7.8 A CT (Computerized Tomography) Scan

A CT scan is used as a diagnostic test. It uses modernized X-rays and a computer to produce detailed images of the bone, blood, and soft tissue of the patient body taken from several angles. Here, the patient is stationary and made to lie on a flat bed, and an X-ray tube is made to rotate around his/her body.

7.7.9 Radionuclide Tests

This test involves the injection of radioactive isotopes like thallium or cardiolytes into the vein of the patient and imaged through a special camera kept close to the chest, both during resting condition and while performing exercise. The images show areas of poor blood supply. It is more informative as compared to ECG test.

7.8 Markers of Inflammation in Various Coronary Diseases

Throughout the last decade, the search for new inflammatory markers was made in patients suffering from acute coronary syndromes. The markers of clinical importance are discussed:

1. Elevated IL-6 and TNF cytokines together with the biological role in the process leading to the development of atheromatous condition have already been outlined (Biasucci et al. 1999; Ridker et al. 2000).
2. ICAM-1 has been reported to be positively associated with increased risk of myocardial infarction (32).
3. Systemic markers like acute-phase proteins are easier to measure and easily obtained inflammatory indicators. Thus they find importance as clinical markers of future prediction of cardiovascular risk (Pepys 1996; Davies 1996; Rus et al. 1996; Heinrich et al. 1990).
4. Elevated levels of other acute-phase reactants like SAA and fibrinogen in patients with acute coronary syndromes have been shown to be associated with greater cardiovascular risk (Morrow et al. 2000; Liuzzo et al. 1994; Ridker et al. 2001a, b; Ridker 2001).
5. Extensive study on CRP has been conducted than the others in this context.

However, the various inflammatory markers are interrelated (Patel et al. 2001; Libby 1995) and different inflammation components contribute, independently or through complex feedback mechanisms, leading to hepatocyte-produced acute-phase reactant released in the blood (Pepys 1996; Heinrich et al. 1990).

A study reported in *Fortune*, an American magazine, in the year 2002 revealed that the causative agent behind most myocardial infarctions is initiated by arterial inflammation rather than by coagulation (Ristisch 2002).

7.9 CRP and Cardiovascular Diseases: Is There Any Association?

For decades, CRP was known as a nonspecific marker in inflammatory disorders, like rheumatoid arthritis. Recently, it has gained the status of predicting risk for CVD and predicting mortality as a consequence to CVD globally. In coronary syndrome patients undergoing percutaneous or surgical coronary revascularization, CRP was assumed as an important risk factor. In the recent years, the association of CRP with CVDs has aroused much interest among scientists. This has contributed to the paradigm shift in our understanding of the underlying biological events in the process of inflammation leading to acute coronary syndromes and also in the formation, progression, rupture, and thrombosis of the atheroma plaque (Gomes, 2002) in atherosclerotic disorders.

Therefore, in order to apply CRP for predicting of cardiovascular risk in an individual, true baseline CRP value is to be set out first so that serious undercurrent pathologies can be understood. A persistently high CRP level above 10 mg/L is indicative of significant acute-phase response occurring in the individual which calls in full the case and family history of patient with relevant investigations to determine the ongoing cause. Simultaneously, the problem and its cause should be sorted and alleviated by treatment.

However, in conditions of chronic inflammation and diseases like rheumatoid arthritis and patients undergoing hemodialysis with end-stage renal failure, there is a characteristic persistent elevated serum CRP concentration. These individuals are sometimes reported to show an association with an early CVD (Patel et al. 2001; Libby 1995; Gomes, 2002).

7.10 CRP: The Nonspecific Inflammation Marker

CRP, produced by liver cells on stimulation with IL-6, is recently termed as the "golden marker" among all other inflammatory markers in CVD and related pathologies. Its quantitative estimation in different stages of coronary artery disease has strongly established it as a highly reliable predictor of cardiovascular risk (Pepys 1996; Heinrich et al. 1990). These properties include:

1. It does not degrade in the time interval between blood sample collecting and its biochemical testing.
2. Detectable in the blood and property of being produced under stimulation. Healthy individuals reveal traces of the values less than 0.3 mg/cc.

Although patients suffering from acute coronary syndromes with elevated CRP reveal poor outcomes, sometimes the measurements lacked sufficient sensitivity. In the 1990s, with the discovery of highly sensitive methods for measuring CRP (hs-CRP) (39–42) limits, in normal individuals, limits were restricted to 0.0–0.5 mg/cc. This normal baseline fixation of CRP concentration allowed assessment of underlying mechanism of systemic inflammation in both apparently healthy individuals and those with coronary disease (De Beer and Pepys 1982; De Beer et al. 1982b; Macy et al. 1997; Ladue et al. 1997; Ridker 1999, 2001) accurately.

From the Adult Treatment Panel III guidelines (including a predicted risk of ≥20 % together with other risk factors like diabetes), for people at intermediate risk for a cardiovascular event, the assessment of inflammation biomarkers like CRP or fibrinogen may help in the prevention of another additional CVD event for a period of 10 years. This was seen among every 400–500 people screened in a very recent study from the Emerging Risk Factors Collaboration meta-analysis. CRP or fibrinogen information increases the C-index (a measure of risk discrimination) by 0.0039 and 0.0027, respectively, and predicted 10-year risk categories of "low" (<10 %), "intermediate" (10 % to <20 %), and "high" (≥20 %). Moreover, information on conventional risk factors (like HDL and total cholesterol, age, sex, smoking habit, blood pressure, family history of diabetes) increases the C-index by 0.0050 only. In this study of 246,669 individuals with no prior CVD history, statin was administered after testing their CRP or fibrinogen level. This study that led to the understanding of the role of CRP as a prognostic model for CVD was developed based on these informations (Emerging Risk Factors Collaborations et al. 2012).

7.11 CRP in Cardiovascular Diseases

However, in order to assess the risk of cardiovascular disorder, it is necessary to use CRP concentration critically considering true baseline CRP values. It should be taken into account that the baseline CRP values should not be perturbed by other underlying events. A persistently elevated CRP level above its normal concentration, i.e., above 10 mg/L, is indicative of acute-phase reaction. In this case, a full scrutiny of the patient's medical history along with relevant hematological and cardiac investigations is required to diagnose the cause clinically. In some chronic inflammatory conditions, individuals show persistently elevated serum CRP levels and patients may be exposed to the risk of premature cardiovascular diseases (Gomes, 2002).

Tissue necrosis stimulates acute-phase responses. Post-myocardial infarction, a major CRP upsurge is observed in patients and the extent of the response reflects the extent of myocardial necrosis. CRP levels peak at around 48 h after the onset of the process, and this observation indicates strongly the outcome after a myocardial infarction (De Beer and Pepys 1982; De Beer et al. 1982a). Moreover, CRP has been observed to be co-deposited with activated complement molecules in all acute myocardial infarcts (Kushner et al. 1963; Kushner and Kaplan 1967; Kushner 2002). Thus there is compelling strong experimental evidence suggesting that the CRP response reflects (1) tissue damage and (2) necrosis and (3) aggravates ischemic myocardial injury severity (Griselli et al. 1999).

CRP is the only acute-phase reactant that has been reported to be an inflammation biomarker and helps in predicting the risk of CVD and other metabolic syndromes (Pepys and Hirschfield 2003). Diet patterns influence CVD by (1) elevating risk factors and (2) role of CRP on inflammation pathology. Recent researches show that CRP has been able to predict diseases like (1) CVD, (2) diabetes type 2, and (3) cerebrovascular diseases (Lee et al. 2014). Upregulated CRP levels have been reported to correlate with atherogenic dyslipidemia and insulin resistance (Sung et al. 2012). An increased CRP level revealed an increased risk for coronary heart disease in comparison to lower CRP levels (Buckley et al. 2009). Of the risk factors for CVDs such as lifestyles, obesity, and fatty diets (Fig. 7.4), CRP is found to be an important determinant (Basu et al. 2006). High intake of *trans*-fatty acids has shown direct effects on increased inflammatory responses (Lopez-Garcia et al. 2005) while the anti-inflammatory properties of honey reduced CRP levels which indicated their role in inflammation (Al-Waili 2004; Lee et al. 2014). Several studies reveal the association of CRP with hypertension and effect of BP on CRP. BP is the most important predictor of future hypertension currently in use. Studies have shown direct correlation of CRP concentration with that of diet (Al-Waili 2004).

Debates related to the importance of assessing the role of CRP and other inflammation biomarker concentrations in the prediction of initial cardiovascular events remained. In a study in 2013, conducted by the British Heart Foundation and others on normal individuals, who had no history of CVD, it was reported that assessment of the CRP or fibrinogen level in people at intermediate risk for a cardiovascular event could help to prevent one additional event over a period of 10 years for every 400 to 500 people screened (Emerging Risk Factors Collaborations et al. 2012). High-sensitivity CRP has been reported as a potential marker for risk associated with CVD diseases in diabetes mellitus (Pfützner and Forst 2006; Ridker 2003) patients.

Although CRP has shown to have no predictive value in atherosclerosis, elevated CRP levels are reported to be a good indicator of elevated risk of CVD and aid in early prediction in HIV-infected individuals (Westhorpe et al. 2014; Markowicz et al. 2014). An increase in serum CRP levels in patients suffering from cardiovascular and cerebrovascular disease in psoriasis which is an infectious skin disease (Takahashi et al. 2014) is reported. An elevated level of CRP at baseline is reported to predict higher high-frequency heart rate variability (HF-HRV) at follow-up in a recent study in 2014 (Jarczok et al. 2014). CRP and goblet plate reactivity (G-HPR) are reported to be important markers in predicting high death risk in ST segment elevation myocardial infarction (STEMI) patients (Ganie et al. 2014). CRP together with high leptin has been reported to increase the risk associated with CVD (Amrock and Weitzman 2014).

CRP reflects the degree of inflammatory response and also has been used as a marker of immune injury to tissues. CRP sometimes could directly participate in amplifying the immune responses and henceforth increase tissue damage. CRP is known to be present in atherosclerotic plaques but they are not present in the normal vessel wall. CRP may deposits in early atherosclerotic lesions before the deposition of monocytes. CRP is also known to induce the cause of chemokine and adhesion molecule expression in the endothelial cells of humans. During the activation of endothelial cells, CRP acts synergistically with lipopolysaccharide to induce monocytes to produce tissue factor. Thus, CRP acts not only as a marker of inflammation but also as an amplifier of inflammatory responses (Yeh et al. 2001).

In atherosclerosis, the inflammatory processes involve the gradual progression of plaque instability and an undercurrent atherothrombotic event. Increased production of CRP indicates atherothrombotic events and different inflammatory responses (Fig. 7.7) (Pereira and Borba 2008). The extent of atherosclerosis is highly individualistic and depends on the individual's exposure to the type of low-grade acute-phase intercurrent stimuli. But "CRP responder" individuals have higher CRP serum concentrations which may be through acquired or transmitted through genetic processes. These individuals are more susceptible to varied atherosclerotic progression and complications regardless of other relationships (Pepys and Hirschfield 2003).

Coronary vascular disease is a high-prevalence disease, wherein about 40–50 % of individuals diagnosed reveal cholesterol levels at normal or mild increase. After an acute coronary event, there is an increase in CRP level which can be clearly associated with individuals presenting coronary vascular disease or in those apparently healthy individuals. Patients with higher peak CRP values above the normal level may develop larger infarct thus giving rise to acute myocardial infarction. In patients with higher CRP concentration (>3.0 mg/L), continuous rise in CRP during the hospitalization and after the discharge showed the worst prognosis in the case of unstable angina. Also, patients with higher CRP values and having stable angina may be exposed to greater coronary vascular disease risk. Even healthy men and women having higher CRP values have the risk of greater coronary vascular disease or stroke. The association and connection between CRP and traditional coronary vascular disease risk factors (like smoking, hypertension, cholesterol, and obesity) (Fig. 7.4) have been well studied in many researches. The effect of various medications on CRP concentrations used in patients having coronary vascular disease is also well reported (Fig. 7.5) (de Ferranti and Rifai 2002).

CRP has also been reported to be associated with the risk of coronary vascular disease in patients having a family history of acute coronary disease and stable angina and even in those individuals who have never been diagnosed with coronary vascular disease The risk imparted by CRP is independent of hyperlipidemia (Pietilla et al. 1996). Thus in clinical research and in diagnosis, CRP finds profound importance in predicting both short-term and long-term cardiovascular outcomes and in the screening of coronary vascular disease. In the future prospect, CRP being a biomarker may modify treatment and preventive therapies (Haverkate et al. 1997; Thompson et al. 1995).

7.12 hs-CRP

High-sensitivity CRP or hs-CRP aids in accurate measurement of CRP to determine a person's risk for cardiovascular disease (Fig. 7.3). Although hs-CRP is reported to be a marker in CVD and diabetes, elevated CRP levels show no correlation to Indian adolescent women suffering from polycystic ovarian syndrome (PCOS) (Marcucci et al. 2014). Interrelations of increased hs-CRP level and oxidized LDL (OxLDL) in the blood perform a dominant role in cardiovascular diseases (Obradovic et al. 2015). Elevated levels of hs-CRP, circulating neurohormones including N-terminal pro-B-type natriuretic peptide (NT pro-BNP), and big endothelin (big-ET) have been shown to be increased in heart failure and to contribute to both hemodynamic deterioration, cardiovascular remodeling and predict mortality in hospitalized patients with dilated cardiomyopathy (DCM) (Li et al. 2014).

7.13 Biochemical Role of C-Reactive Protein in CVD

CRP overexpression is observed to contribute to instability in the atherosclerotic plaque and the formation of immature microvessels. CRP has been reported to upregulate the expression of signaling molecule Notch-3 that play an active role in angiogenesis and development and maturation of vasculature. Monomeric CRP (mCRP) together with Notch-3-mediated signaling reactions enables formation of thicker sprouts in angiogenesis which leads to upregulated expression of N-cadherin on one hand and downregula-

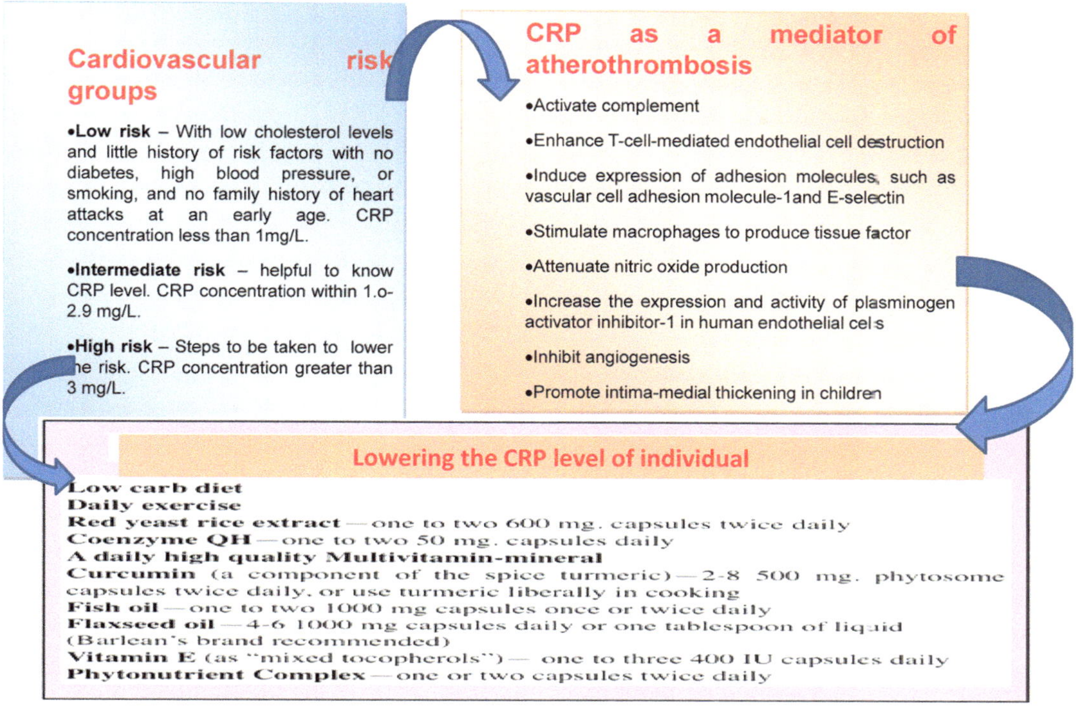

Fig. 7.8 *CRP as atherothrombosis mediator and predicting cardiovascular risks*. CRP's direct pathogenic role in arterial disease helps in predicting future cardiovascular risks. The level of CRP can be reduced by certain paradigm

tion of expression of VE-cadherin on the other, thereby promoting angiogenesis through the PI3K/Akt signaling pathway (Boras et al. 2014). CRP while mediating atherothrombosis plays a direct pathogenic biological role in arterial disease (Fig. 7.8).

Elevated CRP levels can predict cardiovascular risks (Fig. 7.8) and heart conditions including myocardial infarction, stroke, PAD, SCD, recurrent ischemia, and death in patients with stable and unstable angina and coronary intervention and acute myocardial infarction in individuals with no history of cardiovascular disease and are independent of age, smoking, cholesterol, diabetes, and other major cardiac risk factors (Ridker 2003). The prognostic potential of CRP is therefore very strong.

7.14 Clinical Significance of CRP in Cardiovascular Diseases

CRP has been shown to predict cardiovascular disorders such as myocardial infarction, stroke, PAD, and SCD and risk of both recurrent ischemia and death in individuals with stable and unstable angina, those undergoing percutaneous angioplasty, and patients presenting acute coronary syndromes.

CRP finds clinical relevance as a biomarker that has a strong predictive potential of cardiovascular events over and above the traditional marker, LDL cholesterol, and adds to the information on prognosis at all levels of metabolic syndrome calculated by the Framingham risk (Gomes, 2002).

Based on high-sensitivity assays and CRP levels obtained as <1, 1 to 3, and >3 mg/L, individuals are grouped as (1) low-, (2) moderate-, and (3) high-risk groups for future cardiovascular diseases (Fig. 7.8). LDL cholesterol below 130 mg/dL in patients but with CRP levels >3 mg/L is categorized in the high-risk group (Ridker 2003; Ridker et al. 1997). This has a tremendous importance in clinical studies.

7.15 Quantitative Estimation of CRP

CRP/hs-CRP is measured with a simple blood test. Nonspecific elevation of CRP is indicative of inflammation in the body while hs-CRP—high-sensitivity CRP—enables a measure of inflammation in blood vessels (Fig. 7.3). This is the test needed to help measure heart disease risk. Levels of CRP can determine people with low, intermediate, and high risks to cardiovascular diseases.

Table 7.2 Prevention of heart diseases: There are several levels to prevent heart diseases

Primordial: Social, legal, and other nonmedical activities lowering the risk factors (e.g., socioeconomic development, smoke-free environment)
Primary: Controlling risk factors contributing to CVD including health education programs, anti-smoking campaign, sports programs, nutrition counseling, and regular checkup of blood pressure and blood parameters including cholesterol, blood lipids, glucose.
Secondary: Screening and treatment of symptomatic patients and setting up of personal risk profile
Tertiary: Cardiovascular rehabilitation and prevention of recurrence of CVD (new heart attack: 5–7 times higher risk among CVD patients)
Individual:
No smoking
Maintenance of a healthy weight
Intake of a healthy diet
Reduced saturated fat in diet
Regular exercises for about half an hour every day
Prevention or effective management of other health complications like diabetes, hypertension, high blood pressure, etc.
Limiting salt intake

7.16 Increased CRP and Conventional Treatment

Conventional treatment to lower CRP levels (Fig. 7.8) includes prescription of statin drugs used to lower LDL cholesterol. Change of lifestyle, like exercise and weight loss, not smoking, drinking less alcohol, following a heart-healthy diet, reducing the other risk factors for CVD, and proper medication and control of other diseases like diabetes, blood pressure, etc., helps to lower CRP. Several methods are employed to prevent heart diseases (Table 7.2).

7.17 Discussions

From various researches, it is clear that CRP finds importance as a clinical tool for screening and predicting risk in cardiovascular diseases. Results from epidemiologic studies show that elevated CRP levels show positive associations with the cardiovascular disorders. CRP, selectively binding to LDL or to "damaged" LDL, is found to deposit on atherosclerotic plaques, and this CRP–ligand complex then in turn activates the complement cascade which is pro-inflammatory in nature. In experimental animal models, both CRP and complement have been shown to increase the probability of myocardial infarction. Therefore CRP may find application as a therapeutic target in cardiovascular diseases.

CRP–LDL binding is biologically beneficial to the individual due to downstream effects of (1) prevention of macrophage foam cell formation, (2) attenuation of inflammatory effects of LDL, (3) inhibition of LDL oxidation, (4) and reduction of various pro-atherogenic macrophage effects.

In experimental animal models, nonnative CRP shows atheroprotective effects. Moreover, temporary inhibition of CRP and phosphocholine binding together with localized nonnative pentameric CRP may show promises in the treatment of atherosclerosis and myocardial infarction. Thus the total biosynthesis of CRP in the individual may not be required to stop. Thus, CRP does not cause a disease, but in case of myocardial infarction, it actually worsens the condition.

CRP is a more accountable and predictable risk factor than traditional cardiovascular risk factors like high BP, obesity, and increased plasma lipid content. However, true baseline values of CRP and the characteristic associations between these risk factors should be assessed to understand the importance and applicability in prognostic or diagnostic tests. Additionally CRP contributes mostly to the predictive value above traditional risk factors for CVD in risk estimation. Quantitative estimation of CRP may be more informative in elimination of risk associated even in individuals with intermediate risk or also in apparently healthy individuals. Globally, in the diagnosis of cardiovascular events, CRP finds importance in the prediction of cardiovascular risks, and universal screening of individuals in patients is highly recommended and gaining importance globally.

Role of CRP in Leishmaniasis

8

Abstract

Different types of leishmaniasis are differentiated by range and complexity of clinical expressions ranging from asymptomatic infection to life-terrorizing illness. Visceral leishmaniasis (VL) is caused by an obligate intercellular parasite of the mononuclear phagocyte, *Leishmania donovani,* which causes a life-threatening disease. *Leishmania* during its stay in the human host have adapted to survive and proliferate in the host's macrophages. The survival and proliferation of *L. donovani* in macrophages are largely due to the protection conferred by some family of glycosylinositol phospholipids or phosphoglycans on the cell surface or some secreted/expressed molecules of the host. C-reactive protein (CRP) is a prominent acute-phase protein of man. The serum concentration of CRP increases dramatically to nearly 10–1000-fold during inflammation following activation of hepatocytes by inflammatory cytokines. However, the function of CRP in inflammatory conditions and resistance to different infections is still less understood. CRP, a pattern recognition molecule, is present in host circulation. CRP binds to phosphorylcholine (PC) and some phosphorylated carbohydrates found on the surface of a number of microbes during their first entry into the mammalian host. Previously it was reported that CRP binds to the surface of *L. donovani* through their lipophosphoglycan (LPG) component and it increases the uptake of the parasite into host macrophages. *Leishmania* uses CRP to increase its infection without inducing any detrimental macrophage activation. The pathophysiology of different kinds of leishmaniasis was also abridged. CRP, being a phylogenetically conserved innate immune system recognition molecule, recognizes microbial determinants and components of damaged cells as an opsonin. CRP plays its effector function by activating the complement cascade and phagocytosis. A complete definition of the varied ligands used by CRP in recognizing the parasite is essential to understand its role in homeostasis and host defense. The main endeavor of this chapter is to unwind the functional significance of CRP in *Leishmania*

W. Ansar, S. Ghosh, *Biology of C Reactive Protein in Health and Disease*,
DOI 10.1007/978-81-322-2680-2_8

infection, perpetuation, and survival in response to diverse host immune responses in the pathophysiology of its homeostatic mechanisms.

Keywords
Leishmania • Lipophosphoglycan • Phosphorylcholine • Post-kala-azar dermal leishmaniasis • Kala-azar

Chapter Highlights

1. CRP binds with the promastigote of *Leishmania donovani* in a calcium-dependent manner to activate the complement cascade. CRP binds to lipophosphoglycan (LPG), the chief constituent glycocalyx coat of the parasite. LPG is another ligand of CRP apart from phosphorylcholine (PC).
2. CRP also causes opsonization of the parasite in a complement-independent manner into the macrophages. CRP also activates the complement cascade in lysis of the parasite.
3. *Leishmania* uses CRP to enter into the host macrophages at a normal plasma concentration of CRP.
4. CRP was also reported to precipitate an excretory factor found in the medium of cultured parasites.
5. LPG and other surface glycoproteins like gp63 affect deposition of complement-mediated opsonic proteins on the outer surface of the parasite. This causes increased uptake and enhanced survival of the parasite.
6. Human CRP mediates the final transformation to amastigote-like forms from *Leishmania mexicana* metacyclic promastigotes.
7. The parasites survive in the macrophages even in the presence of IFN-γ and nitric oxide like effector molecules. The excessive secretion of IFN-γ and NO was related with the development of the disease.

8.1 Introduction

CRP has been reported to bind to numerous surface molecules of a number of different organisms or microbes like bacteria, parasite, fungi, and many others. Any microbe after entering into the vertebrate host immediately encounters two of the most ancient yet effective protective immune responses: (1) destruction by phagocytosis and (2) lysis by complement. Whether a microbe would be successful or unsuccessful in initiation of its infection depends on how they deal with these two obstacles in different immune mechanisms (Mosser and Brittingham 1997; Mosser and Edelson 1984). The first identification of a role for CRP in defense against infection in humans was related to its capacity to bind to phosphorylcholine (PC) present in the surface membranes of microorganisms (Gotschlich and Edelman 1967). PC is a widespread molecule present on the surface of a number of microbes. CRP binds with the PC present on the surface of microbes and helps in the destruction of these microbes by initiating either phagocytosis or complement-opsonized phagocytocis or complement-mediated lysis. In case of *Streptococcus pneumoniae*, CRP protects the host organism from the fatal pneumonia due to its binding with the cell wall polysaccharide by activating the complement-opsonized phagocytosis (Szalai et al. 1995). The protective effects of CRP are not limited to bacteria but also to other microbes. It has been reported to bind with fungus like *Aspergillus fumigatus*, the yeast *Candida albicans*, and sporozoites of *Plasmodium yoelii* and *P. falciparum* and promote their phagocytosis by different human leukocytes. In contrast, in some cases, CRP causes a complement-independent opsonization of some microbes or parasite and helps in their entry into macrophages. Inside the macrophage, CRP does not induce detrimental macrophage activation but helps in the uptake and survival of the parasite inside the macrophage. Thus CRP indirectly helps the parasite to enhance its infection or the parasite is using the host immune molecule, CRP, for its infection and survival (Szalai et al. 1995; Bodman-smith et al. 2002a). Once within the macrophage, the parasite weakens and deteriorates normal signaling of the cells to restrain or reduce the bang of microbicidal systems. For this purpose, the parasite has developed unique molecular mechanisms, and one such strategy involves multiple receptor-mediated interactions in which various macrophage receptors, different parasite ligands, and host opsonins are involved. Such strategy involves parasite ligands (LPG and gp63), host opsonins (CRP, IgG, and C1q), and multiple receptor-mediated interactions (e.g., macrophage receptors (FcγR1, MAC-1) and other receptors specific for C1q, CR1, CR3, and fibronectin). These molecules are used by parasites for its uptake, infection, and spread in the host body (Mosser and Rosenthal 1997; Bodman-smith et al. 2002a; Wilson and Pearson 1986; Wilson and Pearson 1988; Rivzi et al. 1988; Mosser and Edelson 1984; Chakrabarty et al. 1996). Thereby, they not only avoid the detrimental effects of complement components but also exploit the beneficial effects of opsonization by host opsonins for establishment of disease. Although the main function of CRP is to opsonize foreign intruders and force them to enter into macrophages for their clearance from the system, the parasite exploits its membrane structure for this entry and thus avoids the deleterious effects of CRP (Mosser and Rosenthal 1997; Wilson and Pearson 1988; Bodman-Smith et al. 2002a; Mosser and Edelson 1984).

Leishmaniasis is exemplified by a range and complexity of clinico-medical manifestations ranging from various asymptomatic infections to life-terrorizing expression of illness. Experimental research studies and clinical evidences indicate multidimensional role of varied factors/components leading to parasite endurance and multiplication. In early phase of infection, the production of reactive oxygen species and nitrogen intermediates plays a noteworthy role in restraining the parasitic propagation. In later phase, hepatic resistance is conveyed dominantly by the nitric oxide synthase (NOS)-2 gene regulation and the synthesis of inhibitors of NOS-2 gene expression. Both the correlated roles played by transforming growth factor-beta (TGFβ) and interleukin-10 (IL-10) with reduced parasite killing are also worth mentioning. The hepatic infection of the parasite is highly self-restraining due to synthesis of diverse cytokine responses including reasonable secretion of tumor necrosis factor (TNF), whereas in the spleen surplus TNF production mediates vicious pathology. In *Leishmania* pathology, $CD8^{+}$T cells play various roles comprising both cytotoxic activity and synthesis of various cytokines and chemokines. The capacity to stimulate Th1cytokines is related with subclinical self-healing or asymptomatic infection. However, in symptomatic subjects, Th1 cytokine secretion is not lowered but there appears some unresponsiveness to the stimuli of these Th1 cytokines (Malla and Mahajan 2006).

Previously, it has already been reported that the involvement of CRP helps *Leishmania* to enter the macrophage by phagocytosis at normal plasma level of CRP (Culley et al. 1996) and probably through Fc receptor rather than indirect complement-mediated uptake (Bodman-smith et al. 2002a). The binding of CRP with LPG, a glycocalyx coat of the parasites, has been well established (Culley et al. 1996). The transformation of promastigotes to amastigotes of the same parasite is believed to be mediated by CRP binding (Bee et al. 2001). Direct opsonization of the parasite through CRP did however pilot to more proficient uptake in vitro than CRP-unopsonized cells. During pneumococcal infection, the effects of CRP on cytokine responses have been reported. In human pneumococcal pneumonia and in mice infected intranasally with *S. pneumoniae*, a strong local and systemic inflammatory cytokine response occurs. This first response is related with greater survival of the parasite (Thomas-Rudolph et al. 2007). CRP interacts with FcγR, protects mice from pneumococcal infection, and helps in *S. pneumoniae* removal from the bloodstream (Suresh et al. 2007; Mold and Du Clos 2006; Szalai and McCrory 2002; Szalai et al. 1995).

CRP, being a pattern recognition molecule, is used by the parasite for its safe entry and establishment in the host. Also, the potential role of CRP, specifically induced in different leishmaniasis patients, was explored by searching its immune functions. A different complement-dependent or complement-independent role of CRP was elucidated in the survival of the parasite in blood, bone marrow, liver, or other tissues. Such interaction elucidated different approaches adopted by parasites for invasion which enhance infectivity and survival in macrophage by altering the host's cellular reaction suggesting its possible implications in the pathogenesis of leishmaniasis.

Based on this information, it is worth searching for the mechanism of immune response, which helps in the survival and development of the organism with respect to the well-documented protective/destructive role of CRP. CRP is a prominent acute-phase protein of man, with serum levels increasing strikingly following activation of hepatocytes by inflammatory cytokines. However, the function of CRP in inflammation and resistance to *Leishmania* infection is still less understood.

8.2 Epidemiology

WHO, 2010, reported the prevalence of leishmaniasis to nearly 12 million cases and about 350 million people are at the risk of this disease (Fig. 8.1). Leishmaniasis has a digenetic life cycle encompassing a promastigote and an amastigote stage (Fig. 8.2). Promastigotes are motile insect stage found in the midgut of sand fly. Amastigotes are non-motile, intracellular, mammalian stage. The organism is propagated by the bite of multiple species of bloodsucking sand flies (*Phlebotomus*) which carry the promastigote in the pharynx and anterior gut. It gains entry to mononuclear phagocytes where they modify into amastigotes and multiply until the infected cell ruptures. The discharged organisms infect many other cells. The sand fly attains the organisms during the blood meal period; the amastigotes transform into flagellate promastigotes and divide in the gut till the pharynx and anterior gut are loaded. Dogs and rodents are some common reservoirs. The mammalian host includes rodents, gerbils, hyraxes, bats, porcupines, opossums, and many others.

Visceral leishmaniasis (VL) or kala-azar or black fever or dumdum fever is a vector-borne chronic disease in man infected by the intracellular protozoan parasite, *Leishmania donovani.* The disease is mainly endemic in the tropical and subtropical regions and the Mediterranean basin. VL cases mostly (nearly 90 %) found in six countries, namely, India, Nepal, Bangladesh, Brazil, Ethiopia, and Sudan. The death toll of VL is comparable to another parasitic disease, malaria. The disease is highly diverse and complex. The disease is perpetuated by nearly 20 leishmanial species, and it is transmitted to humans by nearly 30 phlebotomine sand fly vectors. Reservoir hosts are dogs, foxes, and cats, and the vector is the *Lutzomyia* sand fly. *Leishmania* belongs to the phylum Sarcomastigophora, order Kinetoplastida, family Trypanosomatidae, and genus *Leishmania* (Ready 2014; Chappuis et al. 2007; www.who.int/mediacentre/factsheets/fs375/en/).

About 200 million people are at risk and 500,000 people suffer from this disease (website http://www.whoindia.org). It perpetuates the infection by converting promastigotes to the obligate intracellular amastigotes within the mononuclear phagocytes. There are four different forms of leishmaniasis present, namely, (1) visceral leishmaniasis (VL/kala-azar/dumdum fever), (2) cutaneous leishmaniasis (oriental sore), (3) mucocutaneous leishmaniasis (espun-

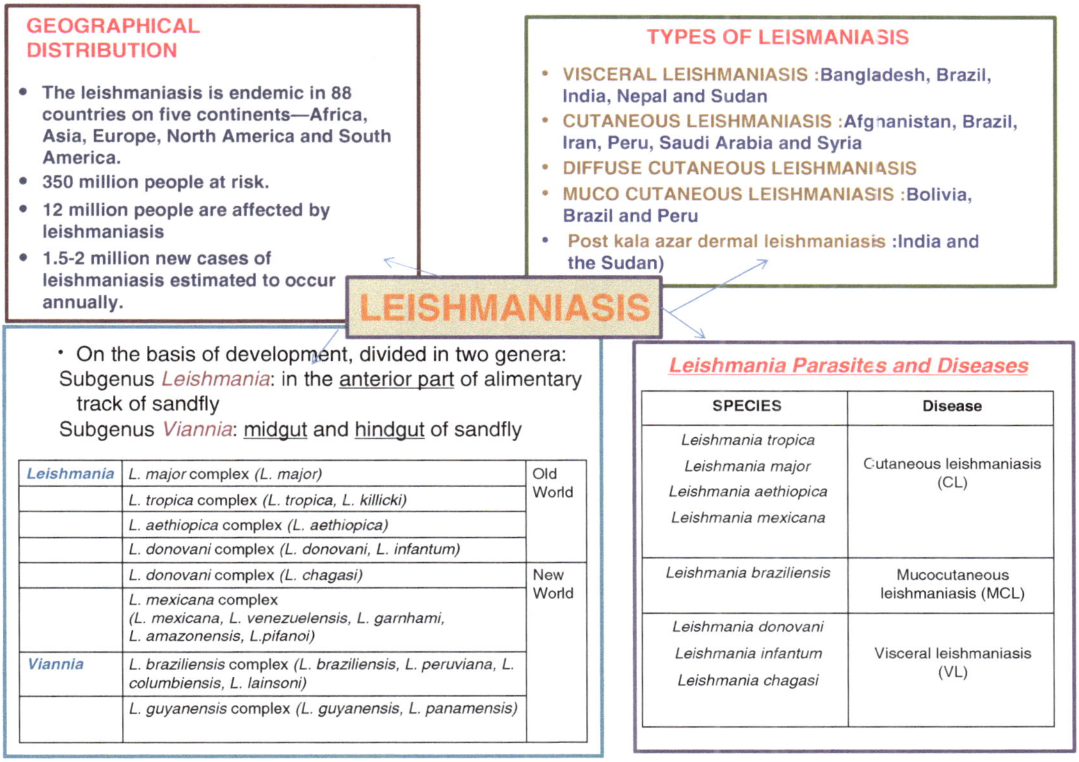

Fig. 8.1 The geographical distribution types of leishmaniasis insect species and their related parasitic diseases are noted

dia), and (4) post-kala-azar dermal leishmaniasis. VL is also known as black sickness or kala-azar in Asia and it is characterized by some physical and clinical features (Fig. 8.1). India carries nearly 40–50 % of the global burden of leishmaniasis. The direct evidence of the demonstration of VL includes splenic aspirate and biopsy, liver biopsy, direct agglutination tests, fast agglutination screening tests, bone marrow aspirate and culture, presence of blood buffy coat, presence of *Leishmania* IgG titer in serum, observation of *Leishmania* amastigote form in bone marrow smear, and hepatosplenomegaly in the abdominal ultrasound and computer tomography. Immunological tests includes aldehyde test, antimony tests, and complement fixation tests with WKK antigens (Figs. 8.3 and 8.4) (Ready 2014; Leishman 2006; Chappuis et al. 2007; Sundar and Rai 2002; http://www.who.int/leishmaniasis/clinical_forms_leishmaniases/en/index2.html).

Cutaneous leishmaniasis (CL) is the mainly prevalent form of the disease. It is characterized by one or more sores, papules, or nodules on the skin. In due course of time, the sores can change their shape and appearance. The sore appears as a volcano with elevated edge and central crater. Generally the sores are painless but turn painful if infected secondarily. The sore is commonly known as Baghdad boil. Swollen lymph nodes may be present near the sores. It creates ulcers on the exposed parts of the body, leading to permanent scars, disfigurement, stigma, and disability in some cases. According to WHO, 2010, nearly 90 % of CL occurs in Afghanistan, Brazil, Iran, Peru, Saudi Arabia, and Syria. WHO, 2010, also reported the annual incidence of 1.0–1.5 million cases worldwide. The urban CL is thought to be an anthroponosis while the rural CL is zoonosis with human infections occurring only sporadically. *Phlebotomus* sand fly is the vector for the

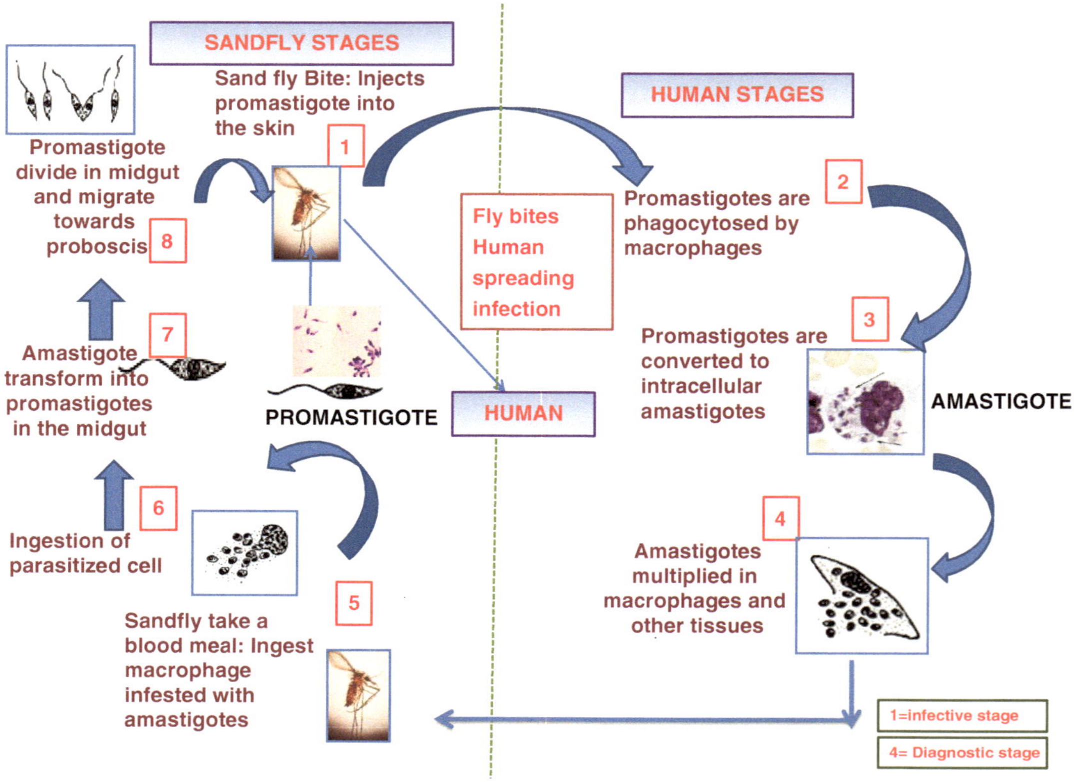

Fig. 8.2 *Life cycle of leishmaniasis*. (1) Leishmaniasis is transmitted by the bite of female phlebotomine sand flies. The sand flies inject the infective stage, promastigotes, during blood meals. (2) Promastigotes that reach the puncture wound are phagocytized by macrophages. (3) They transform into amastigotes. (4) Amastigotes multiply in infected cells and affect different tissues. (5) Sand flies become infected during blood meals on an infected host when they ingest macrophages infected with amastigotes. (6) In the sand fly's midgut, the parasites differentiate into promastigotes. (7) They multiply and migrate to the proboscis

Old World CL. In patients of the Ethiopian CL, a similar kind of lesion was also seen, but it may also give rise to diffuse cutaneous leishmaniasis (DCL). DCL patients had least cell-mediated immunity against the parasite. Thus the patients had disfiguring nodules over the surface of their body. In CL, the diagnosis is confirmed by clinical manifestation with parasitological tests. VL is the harshest form of the disease and turned out to be fatal if left untreated. The disease is systemic and thus it affects the vital organs of the body. VL is characterized by irregular bouts of fever, progressive anemia, pancytopenia, hypergammaglobulinemia, substantial weight loss, and hepatosplenomegaly and is complicated by serious infections. If VL is left untreated in developing countries, the fatality rate can reach as high as 100 % within 2 years. In VL, hypersplenism; reduced bone marrow activity; lymphadenopathy; hyperglobulinemia; high levels of liver transaminases, CRP, and ESR; and cellular distraction in the spleen cause leucopenia, anemia, and thrombocytopenia with concurrent secondary infections. Leishmaniasis is a curable and treatable disease. VL is mostly diagnosed by detecting a combination of clinical signs with serological or parasitological tests. Mucocutaneous leishmaniasis causes total or partial damage of mucous membranes of the nose, throat, and mouth. Mucocutaneous leishmaniasis is caused by *Leishmania braziliensis.* Mucocutaneous leishmaniasis is caused by *L. tropica* in the Middle East. Mucocutaneous leishmaniasis is mostly predominant in Central and

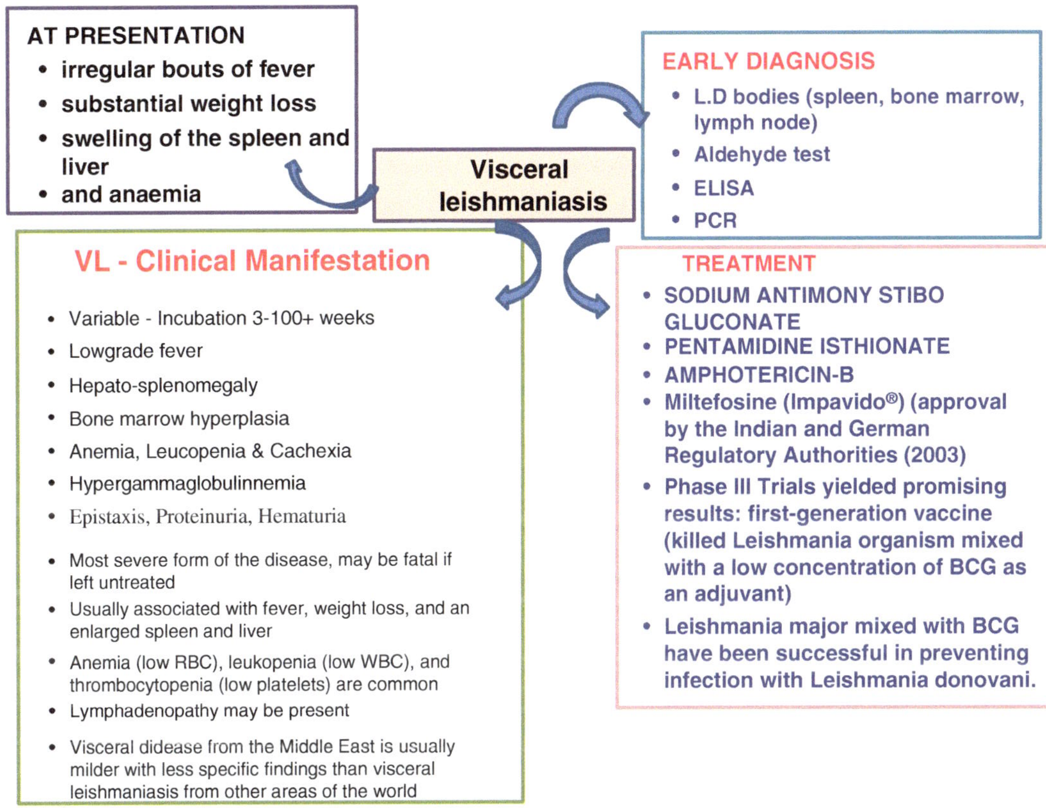

Fig. 8.3 The initial presentation, the early diagnosis, clinical features, and treatment of VL are demonstrated

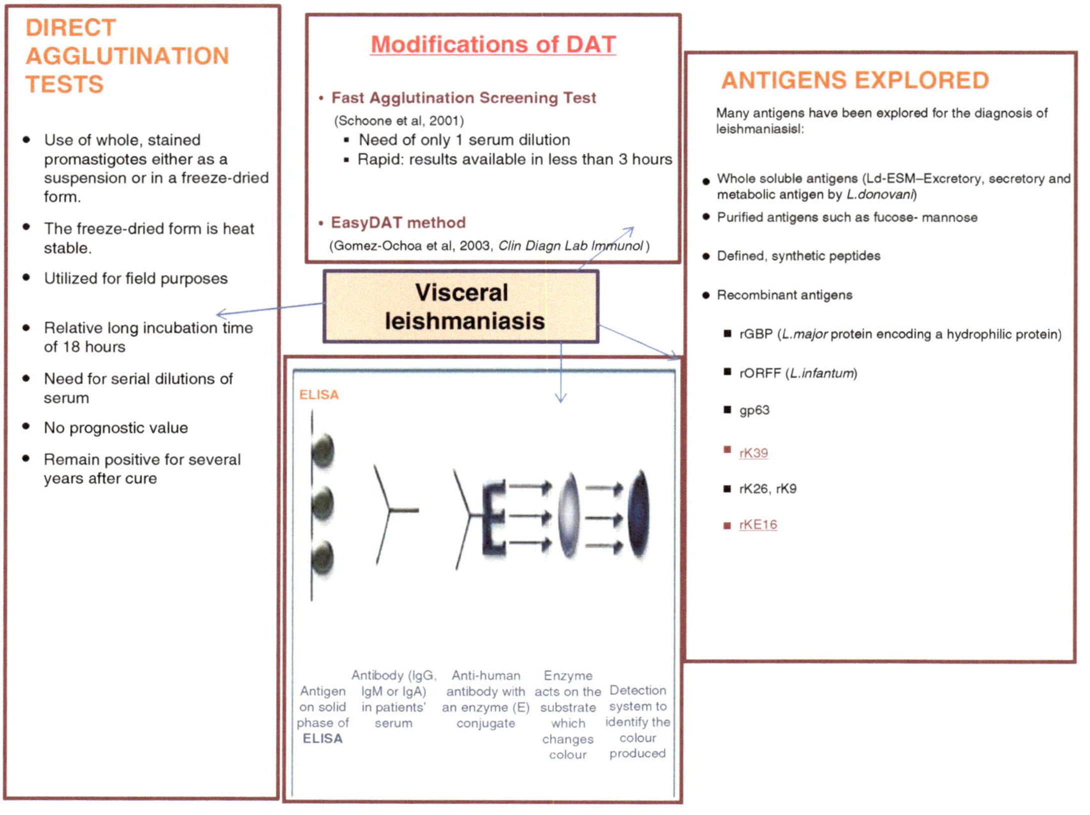

Fig. 8.4 Clinical analysis of VL. The basis of direct agglutination test and its modifications, the ELISA-based detection, and the different antigens exploited for the diagnosis of VL is represented

South America and the Yucatan peninsula. According to WHO, 2010, nearly 90 % of cases of mucocutaneous leishmaniasis occur in Bolivia, Brazil, and Peru. The vector of mucocutaneous leishmaniasis is the *Lutzomyia* sand fly. This is the most destructive form of leishmaniasis (Ready 2014; www.who.int/mediacentre/factsheets/fs375/en/; Chappuis et al. 2007; http://www.who.int/leishmaniasis/about_disease/en/).

Though *Leishmania* infections are distributed worldwide, they are mostly restricted in five continents. In nearly 88 countries, this disease is endemic in the tropical and subtropical regions. Nearly 1.5–2 million new cases occur every year worldwide. Fifty to seventy-five percent of all new cases are actually the cutaneous form, the most common form. Approximately 1–1.5 million cases of cutaneous leishmaniasis were recorded per year (WHO 1995, 2010; Desjeux 2001; Chappuis et al. 2007; Sundar and Rai 2002; http://www.who.int/leishmaniasis/about_disease/en/).

VL is typically caused by three species, namely, *L. donovani*, *Leishmania infantum*, and *Leishmania chagasi*. Each parasitic species has a unique epidemiological pattern and different species-specific clinical features. The clinico-medical manifestations of all types of VL change now and then. In the Indian subcontinent, leishmaniasis is almost uniquely caused by *L. donovani*. *L. donovani* affects individuals of all age groups. *L. infantum* causes VL in children and immunosuppressed individuals in the Mediterranean basin, whereas *L. chagasi* causes VL disease in children in Latin America. *L. tropica* is reported to cause visceral disease in nonimmune persons (Ready 2014; Herwaldt 2001; Chappuis et al. 2007; Sundar and Rai 2002; Marsden and Jones 1985).

The zoogeographical distribution pattern of leishmaniasis is restricted to the nearby localities of natural distribution of its vector, the sand fly (*Phlebotomus argentipes*). The spread of the sand fly vector as well other reservoir systems of *Leishmania* is mainly contributed toward economic progression, urbanization, deforestation, procurement of newer human habitats, migration from rural areas to urban ones, and many others. Moreover, with the increase in the populations of immunodeficient HIV-infected patients, mostly in Africa and southern Europe, the number of new host populations was also increasing (WHO 1995; Ready 2014; Desjeux 1999; Alvar et al. 1997; Chappuis et al. 2007; Sundar and Rai 2002).

HIV and *Leishmania* coinfection leads to full swing of clinical disease, with high mortality and relapse rates. The better choice in treatment of such coinfection is the antiretroviral treatment which lessens the development of the disease. *Leishmania* and HIV coinfection is quite a common emerging disease in Southern Europe, where nearly 25–70 % of adults with VL have AIDS. Thus leishmaniasis acts as an opportunistic infection, and it has been included as an AIDS-defining illness. In HIV-infected patients, the presence of the *Leishmania* parasite in the peripheral bloodstream makes these subjects a convenient reservoir and source of disease infection for the vectors. Even, due to high parasite load in peripheral blood, the transmission rate of leishmaniasis among intravenous drug consumers by use of shared common syringes is quite high. The emergence and resurgence of leishmaniasis in newer hosts and in newer geographical areas forwarded a newer challenge in the research of *Leishmania* including areas of diagnosis, treatment strategies, and disease control. Due to its coinfection with leishmaniasis, it is spreading to other non-endemic areas of the world (WHO 1995; Desjeux 1999; Alvar et al. 1997; Chappuis et al. 2007; Sundar and Rai 2002).

Nearly 90 % of the authenticated worldwide *Leishmania* cases occur in India, Bangladesh, Sudan, and Nepal. In India, an estimated number of 165.4 individuals are at the risk of VL in four states. WHO, 2010, reported 15538 cases and 47 deaths of VL in India. In India, it is a fatal setback in Bihar, Jharkhand, Eastern Uttar Pradesh, and West Bengal where both kala-azar and PKDL occur even in women and also in children within 0–9 years of age. PKDL mostly occurs within 2 years of recovery from VL. The mortality rate of untreated cases of kala-azar is up to 90 %; with treatment it decreases to nearly 15 % and up to 20 % subclinical infection. In India also, there are reports that the sand fly vector is developing resistance or that there is a possibility of a new variant or a vector of the disease. Additionally, there were reports of the development of resistance in patients toward sodium antimony gluconate (SAG), the chief drug for *Leishmania* treatment. The widespread *Leishmania*–malaria coexistence in Bihar caused a serious hindrance in the diagnosis and treatment of the disease. Moreover, clinical trials in India with amphotericin B, the third-line medicine by the National Malaria Eradication Program, have shown encouraging results. The phase III trials with a first-generation vaccine using killed and dead *Leishmania* organism added with BCG adjuvant have also shown promising results. In India, as no effective vaccine therapy is available, so the best therapeutic approach includes sand fly vector control and treatment of patients before it turn out to be drug resistant (Ready 2014; Bora 1999; Thakur 2000; WHO 1995).

In Central America and Asia and Kenya and other parts of Africa, VL remains a prominent health problem. In these endemic regions, splenic aspirate smear and its culture are the standard routine methods for monitoring therapy and further relapse (Wasunna et al. 1995). The epidemiology of kala-azar and asymptomatic leishmanial infection in a Bangladeshi community was measured by serological and leishmanin skin test (Bern et al. 2007). VL was reported to be a major health problem in Kenya and also in other parts of Africa, Central America, and Asia. Apart from the currently standard methods of monitoring therapy and relapse of VL by splenic aspirate smear and culture, less invasive techniques include the usefulness of measuring the serum concentrations of acute-phase reactant markers, alpha-1-acid glycoprotein (AGP), CRP, and serum amyloid A protein (SAA) (Wasunna et al. 1995).

VL is prevalent in most countries of the Mediterranean areas, including Turkey. Mostly, in Turkey, *Leishmania infantum* is responsible for VL. From January 2000 to December 2003, nearly 19 children were reported with VL, and from December 2003 to January 2008, almost 13 consecutive cases of VL were also reported among children. The laboratory and clinical findings at diagnosis include elevated erythrocyte sedimentation rate (ESR) and CRP, high fever, splenomegaly, hepatomegaly, and anemia in patients. All the patients were positive for *Leishmania* antibodies and *Leishmania* amastigotes were detected in 69.2 % cases. Patients were cured with liposomal amphotericin B after initial treatment failure by using meglumine antimoniate (Tanir et al. 2006; Arik Yilmaz et al. 2009).

An adult VL case (31-year-old male patient) was also reported from Zonguldak (a province located at the western Black Sea region) from Turkey. VL was confirmed in the patient by the diagnosis, by the detection of *Leishmania* IgG titer, by the observation of *Leishmania* amastigote in the bone marrow smears, and by the presence of *Leishmania* promastigote in the bone marrow culture. The patient was treated with pentavalent antimony (Oztoprak et al. 2010).

The known sequel to visceral leishmaniasis is post-kala-azar dermal leishmaniasis (PKDL). PKDL is mostly restricted to skin. PKDL is quite common in East Africa, Sudan, and India. In Sudan, nearly 50 % of the patients suffering from kala-azar develop PKDL (Gasim et al. 2000). In Sudan, PKDL may occur concurrently with or immediately after an episode of VL. Thus, it is largely constricted to countries or regions where *Leishmania donovani* is the prominent causative parasite. In interepidemic periods of VL, PKDL acts as a reservoir for parasites and thus probably has an important function in transmission of VL. The patients were healed spontaneously in a time period of 6 months to several years after the VL episode. Though the diagnosis of PKDL is mainly clinico–medical, parasites can also be seen by microscopic smears with restricted sensitivity. In more than 80 % of cases, parasites may be detected by PCR and monoclonal antibody techniques. The leishmanin skin test and serological tests are of limited value. In Sudan, most cases will self-cure, but treatment strategies are required in severe and chronic cases. In Indian PKDL, treatment is also needed. In India, sodium stibogluconate is given to patients at 20 mg/kg for 4 months and for 2 months in Sudan. Liposomal amphotericin B also seems to be effectual. Some new compounds of major potential interest like miltefosine can also be administered orally or topically. Although modern research has brought diverse new insights in the pathobiology and management of PKDL, various issues in exacting in relation to control measures remain unsolved and warrant serious attention (Sundar et al. 2015; Mukhopadhyay et al. 2014; Singh et al. 2011; Mondal and Khan 2011; Ganguly et al. 2010; Zijlstra et al. 2003).

Patients of PKDL have papular, macular, or nodular rash mostly found on face, trunks, upper arms, and also in other parts of the body. Depending on the severity, the rash usually first initiates around the mouth from where it gradually stretches to other parts of the body. In a population, PKDL cases serve as a reservoir of *Leishmania* infection. Thus PKDL patients should be diagnosed and treated effectively. PKDL patients were characterized by the presence of macular, papular, or nodular lesions or a mixture of them on their faces. In India, as compared to Sudan, nearly the outcome of <10 % of VL cases is PKDL. PKDL occurs after 6 months to 3 years of the cure of VL (Zijlstra et al. 2003; Sundar et al. 2015; Mukhopadhyay et al. 2014; Singh et al. 2011; Mondal and Khan 2011; Ganguly et al. 2010). Miltefosine, the first-line therapy, recommended by WHO for the treatment of VL in the Indian subcontinent, has shown its efficacy also in the treatment of PKDL (Sundar et al. 2015).

Thirty-three PKDL patients from Muzaffarpur, Bihar, India, were enrolled between July 2009 and June 2010 at the Kala-Azar Medical Research Center, Muzaffarpur, at the field site of the Institute of Medical Sciences, Banaras Hindu University. The patients were followed up for

1 year after effective treatment with miltefosine (Sundar et al. 2015).

8.3 VL and its Pathogenesis

The life cycle of *L. donovani* revolves around two distinct forms, the flagellar promastigote form developed in the alimentary tract of the sand fly arthropod vector and the amastigote form which perpetuates in the mammalian host. Only the female vector flies carry the disease during their blood meal, by inoculating the promastigote into the skin of the vertebrate host. The promastigotes are engulfed by macrophages and dendritic cells in the dermis of the skin, and by losing their body flagella, they gradually differentiate and transform into amastigotes. Through a multifaceted host–parasite interaction, the amastigotes multiply, thrive, and survive in the phagolysosomal vacuoles. The parasites spread through the vascular and lymphatic systems and infect other immune cells like macrophages and monocytes in the reticuloendothelial system and finally infiltrate the spleen, bone marrow, liver, and different lymph nodes (Rittig and Bogdan 2000; Lodge et al. 2006; Chappuis et al. 2007).

Leishmaniasis is a fatal and systemic disease. Rate of infection does not always show a relationship with clinico-medical illness. Many cases of asymptomatic infections along with symptomatic clinical cases are quite prominent in endemic areas (Chappuis et al. 2007). Even in many areas, different species overlap with different disease forms of leishmaniasis. Thus it is quite relevant to understand the disease pathogenesis in leishmaniasis to intervene in proper therapeutic approaches. The causative organism not only subverts the detrimental effect of the host adaptive and innate responses but can also resist the microbicidal activity of various immune cells like macrophages and monocytes. The pathobiology of the *Leishmania* parasite is highly critical, and most of the information is culminated from different experimental murine or mice models (Bodman-Smith et al. 2002a).

The main thrust of *Leishmania* research is to search the molecular mechanisms that allow these parasites to enter into host macrophage, survive, and thrive in the hostile intracellular environments. Lipophosphoglycan (LPG), a novel surface glycoconjugate, is a multifunctional virulence determinant which is essential for perpetuation of *Leishmania* infections in the host. Another major surface protease, a 63-kDa molecule, the gp63, serves as a ligand toward host cells after transmission (Mosser and Rosenthal 1997; Bodman-Smith et al. 2002a).

In the differentiation of less infectious *L. major* promastigotes to the highly infectious "metacyclic" form, it propagates from the logarithmic growth phase to the stationary growth phase and gradually the structure of the LPG is modified. The LPG molecule is composed of linear chain arrangements of phosphorylated oligosaccharide repeat units attached to the membrane by glycosyl-phosphatidylinositol glycolipid anchor. The average number of repeat units per molecule of LPG is doubled during metacyclogenesis. Other structural changes of LPG also occur. The structural modification during developmental phase of the LPG molecule is important in modulating the binding of the promastigotes to different receptors in the midgut of the sand fly and to the human macrophages. The LPG is modified so that the metacyclic promastigotes turned out to be resistant toward complement-mediated lysis (McConville et al. 1992).

During metacyclogenesis of *Leishmania* parasite inside the insect vector, the promastigotes morphologically change in size and shape. Both LPG and gp63 undergo varied qualitative and quantitative modifications as the promastigotes differentiated from procyclic to infective metacyclic forms. During the metacyclic stage, many proteins like Meta-1, SHERP, and HASP were also upregulated (Muskus and Marín 2002).

LPG and gp63 molecules are significantly involved in the identification of virulent strains by the macrophages. Various other carbohydrate receptors are also involved in the recognition of avirulent (strain UR6) and virulent (strains AG83 and GE1) strains of *Leishmania* parasites into the host macrophages. Avirulent *Leishmania* promastigotes probably use the mannosyl fucosyl receptors (MFR) for their initial attachment and consequent internalization into the host macrophages. The virulent *Leishmania* parasites restricted the use of this receptor (Chakraborty et al. 1998).

The expression of a greater number of LPG and gp63 in virulent strains of Indian *L. donovani* (AG83 and GE-I) than the avirulent strain (UR6) perpetuates much faster internalization of the virulent one inside the macrophages. Both LPG and gp63 synergistically help the faster entry of virulent strain in macrophages. Moreover avirulent strain (UR6) remains unaffected by either LPG or gp63. The transfection of avirulent UR6 strain with the gp63 gene cloned from *L. amazonensis* helps in faster entry of the transfected UR6 like the virulent strain as compared to the non-transfected UR6. Thus, the expression of the gp63 gene is significantly involved in the recognition and internalization of *Leishmania* parasite into the macrophages (Chakraborty et al. 1998).

The combating mechanism of the parasite starts as it enters the mammalian host as a promastigote. Though the promastigotes are engulfed by the macrophages, they are resistant toward proteolysis and varied degrading mechanisms inside the phagosome. Gradually the promastigotes reside and transform to amastigotes inside the neutrophils, dendritic cells, monocytes, and macrophages. The C3b complement-opsonized parasites are also internalized by phagocytic cells. Once inside the phagosome, the parasite is resistant toward lytic clearance and degradation. The surface glycoprotein, gp63, converts the C3b to iC3b and helps in faster phagocytic uptake. Phagocytic engulfment favors the survival of the parasite as it can subvert the phagosomal environment (Malla and Mahajan 2006; Silverstein 1977; Channon et al. 1984; Remaley et al. 1984; Kausalya et al. 1996).

The macrophage applies various strategies to combat the parasitic intervention. These strategies involved the production of oxidative burst, attack of hydroxyl and superoxide radical on the parasite in the phagosome, and also production of degrading acidic enzymes in the phagolysosomes. Various enzymes secreted by the phagosome and the lysosome are flooded over the parasite in the phagolysosome. In the phagosome, the *Leishmania* parasite resists this attack by coating itself with acid phosphatase enzymes. The acidic environment maintained in the phagolysosome is resisted by the parasite by proton pump to maintain its intracellular pH close to neutrality. The surface molecule LPG plays a crucial role in inhibiting the various lysosomal enzymes. The parasite evolves various immune evasion mechanisms to attenuate the macrophage microbicidal machinery. The responsiveness of the macrophage toward lipopolysaccharide in the induction of IL-10 production is reduced. Inside the macrophage, the parasite gradually subverts the humoral immune responses (Malla and Mahajan 2006; Silverstein 1977; Channon et al. 1984; Remaley et al. 1984; Kausalya et al. 1996).

Presence of anti-leishmanial antibodies in the sera of the patients due to humoral immunity depends on the species of *Leishmania* infecting the human. In many forms of this disease, the presences of anti-leishmanial antibodies were critically evaluated for the diagnosis and prognosis of the disease. In many cases of VL, a strong titer of anti-leishmanial antibody was found in the patient sera. However, the role of anti-leishmanial antibodies in the pathogenesis of this disease is quite unclear. The presence of high antibody titer during the active phase of VL is used in various serodiagnostic applications (Kaul et al. 2000; Chakraborty et al. 1998; Sunder et al. 2002).

The differential antibody (IgG of different subclasses) intensity in kala-azar and PKDL in pediatric patients was studied as a marker of leishmaniasis toward differential disease susceptibility (Ansari et al. 2008).

Sialic acids are a group of nine-carbon sugars that featured outstandingly at terminal positions of surface-exposed glycoconjugates adding important dimensions in functioning of the cell surface (Bandyopadhyay and Mandal 2008; Mukhopadhyay and Mandal 2006; Shukla and Schauer 1982; Varki 2008). We have demonstrated the presence and role of sialic acids on *Leishmania* by many of our researches (Chatterjee et al. 2003; Chava et al. 2005). Although biological roles of the *O*-acetylated sialic acids have been attributed in other systems (Chava et al.

2004; Kelm and Schauer 1997; Sharma et al. 1998), their role in causing enhanced infectivity has also been demonstrated.

The presence of 9-*O*-acetylated sialoglycans (9-*O*-AcSGs) on different hematopoietic cells from VL patients had an enhanced presence of antibodies directed against them was demonstrated in VL. The antibodies, namely, anti-*O*-AcSGs, even at normal physiological concentrations, triggered the classical complement-mediated deposition of C3 molecules on promastigotes and subsequent lysis of the parasite. Other complement pathways like alternate pathways and CRP-mediated pathways play a negligible role. Thus anti-*O*-AcSGs act as a significant source of classical complement pathway activation at normal physiological conditions, conferring that they play a function in protecting the host from the parasite infection (Bandyopadhyay et al. 2004).

Several species of *Leishmania* promastigotes can fix the complement components by activating the alternative complement cascade where C3 is deposited on the surface of the parasite. This resulted in complement-mediated lysis of parasites or accelerated uptake of the parasite into macrophages. Thus the in vivo infective functioning of the parasite is balanced between the extracellular lysis and enhanced internalization into phagocytic cells (Mosser and Edelson 1984). The *Leishmania* parasite gains entry into macrophages through receptor-mediated endocytosis, and the binding determinants include at least sialyl, glucosyl, mannosyl, and N-acetylglucosaminyl terminal residues (Chang 1981).

Cell-mediated immune response of the host species mainly plays a major role in combating and curtailing the infection. The critical pathophysiology revolves mainly around the T-cell unresponsiveness and production of circulatory interleukin-10 (IL-10) of the host individual toward different *L. donovani* antigens. The Th1/Th2 paradigm of the host responses in resisting or being susceptible toward the *Leishmania* infection is the topic of research. Th1 induces the production of IFN-γ to activate the cell-mediated immune responses. Th2 stimulates the production of IL-4 cytokines. IL-10 stimulates the production of CD25 − Foxp3 − T cells, whose role was currently demonstrated in the pathobiology of human VL in India. In cases of malnutrition and young age, the gene encoding for the solute carrier family 11 A1 (SLC11A1; formerly NRAMP1), the family which controls macrophage activation, decreased synthesis of interferon-γ and regulates polymorphisms in the promoter of the tumor necrosis factor gene or concomitant immunosuppressive diseases, like HIV infection; the increased threat of developing clinical illness is amplified, and as such, the crucial role of the cell-mediated immune responses cannot be neglected (Sharma and Singh. 2009; Chakraborty et al. 1998; Malla and Mahajan 2006; Bodman-Smith et al. 2002a).

The ability of CRP to stimulate pro-inflammatory cytokines (Ballou and Lozanski 1992; Galve-de Rochemonteix et al. 1993) and to regulate the production of nitric oxide and superoxide has been previously established by several investigators (Pal et al. 2000). Although toxic oxygen may have a role in killing, nitric oxide has been more evidently associated with parasite killing. *L. donovani* is known to tip the cytokine balance toward the less advantageous (to the host) Th2-type response (as measured by decreased IL-4 production) much away from the Th1 response. CRP has also been known to modify macrophage cytokine production, and therefore relations of promastigotes with CRP premature in infection might be supposed to lead to a change in the stability of the cytokine profile synthesized. However, the parasite dynamically tries to organize the macrophage response through molecules like LPG present on the outer surface of the infective promastigotes. For obvious reason, LPG is known to behave as a potent macrophage suppressor and is able to resist macrophage functions like nitric oxide synthesis, interleukin (IL)-1 production, and phagolysosomal fusion (Sharma and Singh 2009; Chakraborty et al. 1998; Malla and Mahajan 2006; Mortensen and Zhong 2000; Ghalib et al. 1995a; Ghalib et al. 1995b; Wilson et al. 1996).

Before the adaptive immune response comes into play, CRP in a pattern recognition manner may afford a more rapid and forceful response by the host, laying the groundwork for the full reaction by the immune system. CRP binding to the organism, at concentrations much below those, initiates during the total acute-phase response to trigger FcγR-bearing cells to produce different protective, inflammatory cytokines. Studies also showed that, once the acute-phase stages of CRP are attained, it plays a central role in the inflammatory response (Das et al. 2003, 2004a; Ansar et al. 2009a).

Different effector molecules like nitric oxide (NO), interferon-gamma (IFN-γ), CRP, and IL-6 are the key molecules of the host defense mechanism against *Leishmania* infection in kala-azar (KA) and post-kala-azar dermal leishmaniasis (PKDL) as compared to healthy individuals. NO and CRP are both involved in the pathogenesis of *Leishmania*. The circulating concentration of IFN-gamma, NO, CRP, and IL-6 was showed to be elevated in pretreatment KA patients as compared to PKDL or healthy individuals, while, in posttreatment KA patients, CRP and IL-6 remained above the control level and IFN-γ and NO levels significantly decreased. Presence of neutralizing cytokines along with the possibility of unresponsiveness to Th1-type stimuli helps in detrimental pathogenesis in *Leishmania* (Sharma and Singh 2009; Chakraborty et al. 1998; Malla and Mahajan 2006; Mortensen and Zhong 2000; Ghalib et al. 1995a, b; Wilson et al. 1996; Ansari et al 2007).

The parasite survives even in the presence of high levels of NO and IFN-γ as there always remains a counteracting interaction between these molecules. Hence, the study implies that due to immunosuppression of the human host during visceral leishmaniasis, the appearance of various effector molecules against the parasite is not suppressed. A balance should be maintained between related cytokines and effector molecules (IFN-γ and NO) that enhance leishmanicidal affect. The excessive synthesis of IFN-γ and NO could be related with the succession of the disease. CRP, as reported in other papers, could be a useful as a noninvasive inflammatory biomarker for monitoring KA, particularly in response to therapeutic protocols and relapse (Ansari et al. 2007).

The leishmanicidal activity is highly dampened with difficulties like toxicity or drug resistance. The incidence rate of leishmaniasis is on the hike for children, and the reasons are yet to be defined (Ansari et al. 2008).

8.4 Relation between CRP and *Leishmania*

CRP, a member of the pentraxin family of proteins, is a highly ancient and conserved molecule. It is synthesized in the liver and generally shows the characteristics of calcium-dependent ligand-binding property (Bodman-Smith et al. 2002a). The disease induced glycosylated molecular variants of human CRP emerged in certain pathological conditions and showed the differential binding interaction with several biological macromolecules (Ansar and Ghosh 2013; Das et al. 2003, 2004a). CRP acts as an acute-phase protein in humans and plays an important player in innate immune system. Similar to the immunoglobulin G (IgG) molecule, it activates the complement cascade, binds to Fc receptors, and operates as an opsonin for multiple pathogens like bacteria, parasite, etc. Unlike IgG, which prominently recognizes distinct antigenic epitopes of pathogen, CRP recognizes only altered self and foreign antigenic molecules based on varied pattern recognition processes. CRP provides early defense and leads to a pro-inflammatory signal and stimulation of the humoral branch of adaptive immune system (Mortensen et al. 1976; Marnell et al. 1995; Oliveira et al. 1979; Bodman-Smith et al. 2002a; Ansar et al. 2009a).

In contrast to other immune systems, where CRP opsonization results in macrophage stimulation, CRP causes a complement-independent uptake of the parasite into macrophages and recovers infection without stirring detrimental macrophage activation. It has already been reported that CRP is a protein which *Leishmania* can utilize to go into the macrophage at normal plasma levels of CRP. Evidence from researches was presented that CRP bound to LPG molecule constitutes the glycocalyx coat of the parasites. LPG and gp63, two major surface glycoproteins of *Leishmania,* are both involved in the parasite virulence, as well as resistance of parasites to complement-mediated lysis. Parasites actively control the macrophage response through LPG since LPG is recognized to act as an effective macrophage suppressor and restrain macrophage functions. Thus, CRP by binding with parasite promastigotes via LPG helps in the initiation of parasitic infection in the host (Bodman-Smith et al. 2002a; Culley et al. 1996).

8.5 Role of CRP in *Leishmania Infected* Animals

CRP shares many functional attributes in common with human IgG, including the ability to stimulate the classical complement cascade, the capacity to relate with human FcγR, and the ability to bind to diverse cell lines and blood cells. The interaction of CRP with human FcγR mediates various functions that are equivalent to those of IgG, including diverse cytokine secretion, opsonized uptake of bacteria, and binding to altered or exposed self-molecules on damaged cells. Later during infective phases, phagocytosis of amastigotes may initiate through immunoglobulin and Fc receptors (FcR). Antibody may also be required for perpetuation of infection. CRP binds to PC-substituted carbohydrate moieties on the surface of *S. pneumoniae* and LPG on the surface of metacyclic *L. donovani*. In addition, CRP binding to *L. mexicana* metacyclic promastigotes activates transformation to a prominent amastigote-like stage. Although transformation of *L. mexicana* promastigotes to amastigote-like forms is known to be mediated by binding of CRP, no information is available regarding the role of CRP inside the macrophage. It may be envisaged that like *L. mexicana,* P-FcR on the *L. donovani* surface is the contributory factor for the transformation of promastigotes to amastigotes in in vivo condition (Marnell et al. 2005; Volanakis 2001; Bodman-Smith et al. 2002a; Bharadwaj et al. 1999, 2001; Mold and Du Clos 2006; Szalai et al. 1995; Marnell et al. 1995; Bee et al. 2001).

In dogs naturally infected with *Leishmania infantum* parasite, the measurement of urinary CRP (uCRP) could be used as a device to detect and calculate the possible renal damage associated with leishmaniasis. The presence of azotemia and proteinuria also affects the urinary (uCRP) and serum CRP values in dogs suffering from leishmaniasis. *Leishmania*-infected dogs with proteinuria and elevated serum creatinine (sCr) had high levels of uCRP and had more renal damage since uCRP/creatinine ratio was prominently elevated. The uCRP/creatinine ratio has a noteworthy increase in dogs with proteinuria having elevated sCr values (Martínez-Subiela et al. 2013).

Rossi in 2014 reported the importance of a negative acute-phase reactant, an antioxidant enzyme, paraoxonase (PON1), that works as a better predictor than CRP and electrophoretic fractions for the clinical recovery of leishmaniotic dogs undergoing standard treatments. PON1 activity had abnormal values at admission in severely sick leishmaniotic dogs. PON1 activity decreased in dogs with systemic inflammation but not in those dogs with mild leishmaniasis. During the course of treatment, the PON1 activity normalized earlier than other markers like CRP due to oxidative phenomena. So, PON1 activity should therefore be tested on admission, and for low values of PON1, severe inflammation may be suspected. PON1 measurement may be done repeatedly during treatment for early identification of responsive dogs (Rossi et al. 2014).

The activity of PON1 in leishmaniotic dogs was also monitored by other researchers along with other markers. The changes in the concentration of high-density lipoproteins (HDLs) before and after treatment were used to monitor the extent of oxidation-related inflammation in leishmaniotic dogs. The serum concentration of CRP, antioxidant enzyme paraoxonase (PON1), and HDL cholesterol (HDL-chol and HDL %) were measured in leishmaniotic and control dogs. At admission, the levels of HDL-chol and PON1 were low in leishmaniotic dogs and gradually their value increases after treatment. CRP is negatively correlated with HDL-chol and HDL%, whereas PON1 is positively related. As HDLs decrease through an oxidative mechanism, so it may be used as a marker of inflammation, along with CRP in leishmaniotic dogs (Ibba et al. 2014).

The kinetics of different acute-phase proteins (APPs) in dogs experimentally infected with the parasite *L. infantum*, during the time course of therapy against canine leishmaniasis, was reported from Spain in 2011. Seven months postinfection, the concentrations of CRP, serum haptoglobin, and serum amyloid A from infected beagles were quite high. Experimental infection with *L. infantum* amastigote forms induces a hike in the concentration of all these APPs. After clinical recovery, the levels of all APPs decreased significantly. Thus this study pointed the relevance of APPs as premature markers for canine VL as well as for scrutinizing the response of the organism toward treatment in canine leishmaniasis (Martinez-Subiela et al. 2011).

8.6 Role of CRP in *Leishmania Infected* Humans

It was previously reported that human and rabbit CRP can precipitate excreted factors from *Leishmania tropica* and *Leishmania donovani* in a calcium-dependent manner. The reported reaction appears to occur in a similar fashion to that reported to occur between CRP and the various galactans (Pritchard and Volanakis 1985).

In 2011, recombinant CRP was purified from *Leishmania tarentolae* growth medium, and the binding specificity of recombinant CRP with PC was also verified (Dortay et al. 2011).

But the binding of CRP to *Leishmania* parasites was first reported in 1994 by Raynes et al. (1994). During the infection of *Leishmania donovani* in human host, the parasite turns out to be promastigotes at the infectious metacyclic stage of development. The promastigote is actually an obligate intracellular parasite of the mononuclear phagocytes of human. Lipophosphoglycan (LPG) molecule present on the promastigote cell surface helps in both uptake and survival of these parasites within the macrophage. CRP specifically binds to the surface LPG of the promastigote of *Leishmania donovani*, at concentrations found in normal human serum (10 μg/ml). CRP even in its normal concentration is sufficient enough to bind specifically with *Leishmania* parasites. The infectious metacyclic promastigotes after CRP-opsonization were increasingly engulfed and uptake by human monocyte-derived macrophages. Thus CRP is actually helping the parasites to enter the macrophages to avert the host immune response. Apart from phosphorylcholine (PC), the characteristic ligand of CRP is LPG. LPG has a repeating phosphorylated disaccharide unit molecule as its backbone. LPG also forms a unique ligand for CRP. Protein A remains another ligand for CRP apart from PC (Das et al. 2004a). This is the first report showing the binding of CRP to another novel ligand LPG by using mAb against LPG with its known ligand specificities (Culley et al. 1996). In the year 2000, the repeating phosphorylated disaccharide units of LPG was characterized by various experimental assays by the author FJ Culley and his group itself. The avidity of binding CRP with this phosphorylated disaccharide unit was also determined. CRP showed highly specific binding for a unique chemically synthesized carbohydrate ligand (phospho-oligosaccharides and synthetic oligosaccharides of diverse conformations and lengths) representing the repeating disaccharide of LPG. Increasing the number of ligand phosphorylated disaccharides does not elevate the avidity for CRP, but the number of CRP binding sites increases. The binding interaction of CRP with phosphorylated disaccharide units was evaluated by competitive binding. CRP was competed to bind to whole *Leishmania* parasites with unsubstituted monosaccharides and/or sulfated, amino, and phosphorylated monosaccharides. Of all these parasite surface molecules, only phosphorylated monosaccharides were able to slow down during binding. The avidity for CRP was influenced by both the position of phosphorylation and the carbohydrate molecule. The optimum structure of the phosphorylated disaccharide units needed for CRP recognition is a lone phosphate group placed in between two monosaccharide pyranose rings, encircled within a linear molecule. Both the phosphate group and the carbohydrate are required for the interaction with CRP. CRP may bind with large collections of phosphorylated carbohydrates. These carbohydrates are characteristic of the surface of microorganisms and help to carry over the host innate immune functions through receptor–ligand interactions (Culley et al. 2000).

The metacyclic infective promastigote forms of *Leishmania mexicana* transformed into the amastigote form into their human hosts. The kinetics of this transformation was maximum at 32 °C/pH 5.5 and intermediate response at 26 °C/pH 5.5 and little transformation occurred at 26 °C/pH 7.2. All these transformations occurred in vitro by stimulation in presence of normal human serum. In complement heat-inactivated serum (previously serum was incubated at 56 ° C for 1 h), the transformation was markedly reduced. This activating effect could be repeated by exposure to a purified human CRP. As mentioned earlier by other authors (Culley et al. 1996), CRP binds to the whole outer surface of *L. mexicana* metacyclic promastigotes in a dose-dependent manner within its normal serum concentration (10 μg/ml). The activating effect of CRP in the transformation was highly specific as there is no transformation in CRP-depleted serum. However, the stimulatory effect of whole serum was restored by addition of purified CRP. The effect of purified CRP in transformation occurred at its normal serum concentrations. This is also validated by FJ Culley's study on *L. donovani* (Culley et al. 1996) where CRP even in its normal concentration is sufficient enough to bind specifically with *Leishmania* parasites (Bee et al. 2001).

Previously it was reported that in human host, CRP-opsonized *Leishmania* parasites were quickly uptake by macrophages (Culley et al. 1996). But in the presence of surplus CRP in the fluid phase, CRP-opsonized uptake of *L. donovani* was inhibitable. CRP could not alter the parasite survival over a span of 72 h within the peripheral blood-derived macrophages (PBMs) or in differentiated U937 cells. Thus CRP helps the parasite not only in its uptake but also in its survival within the macrophages. By altering the macrophage function, CRP showed its meaningful significance in CRP-opsonized uptake of *L. donovani* and its endurance within human macrophages and the subsequent macrophage cell responses to the parasite infection. However, CRP in in vitro condition enhances macrophage responses to erythrocytes coated with phosphorylcholine. CRP un-opsonized or opsonized promastigotes of *L. donovani* do not stimulate the production of interleukin (IL)-10, tumor necrosis factor-alpha, or IL-12 from peripheral blood mononuclear cells. Interleukin (IL)-10 or IL-12 and tumor necrosis factor-alpha help in the destruction of microbes within the macrophages. Thus CRP behaves differentially in *Leishmania* infection in altering the macrophage responses. *Leishmania* parasite is actually using CRP to improve infection without stimulating detrimental macrophage responses (Bodman-Smith et al. 2002a).

Different effector molecules like CRP, interferon-gamma (IFN-γ), nitric oxide (NO), and IL-6 are the key molecules of the host defense against *Leishmania* infection in kala-azar (KA) and post-kala-azar dermal leishmaniasis (PKDL) as compared to healthy individuals. CRP and NO are both involved in the pathogenesis of *Leishmania*. The circulating concentration of IFN-gamma, NO, CRP, and IL-6 were increased in pretreated KA patients as compared to PKDL or control individuals. In contrast , in posttreated KA patients, CRP and IL-6 level remained high and IFN-γ and NO levels low as compared to healthy individuals. However, in sodium antimony gluconate (SAG)-unresponsive patients, significantly increased levels of CRP, NO, and IFN-gamma were observed as compared to responsive patients. Interestingly, in PKDL patients, NO levels were significantly higher while other parameters were comparable to normal individuals. Presence of counteracting cytokines along with the possibility of unresponsiveness to Th1-type stimuli helps in detrimental pathogenesis in *Leishmania*. The parasite survives even in the presence of high levels of NO and IFN-γ as there always remains a counteracting interaction between these molecules. Hence, the study implies that due to immunosuppression of the human host during visceral leishmaniasis, the expression of different effector molecules is not repressed. A balance should be maintained between correlated cytokines and effector molecules (IFN-γ and NO) that enhance the leishmanicidal effect. The excessive synthesis of IFN-γ and NO could be associated with the advancement of the disease. CRP, as reported in other papers, could be a useful noninvasive inflammatory marker for monitoring KA, particularly in response to therapy strategies and relapse (Ansari et al. 2007).

Also, in another study, the CRP levels were shown to be higher in KA and asymptomatic infected patients in a Bangladeshi community (Bern et al. 2007). CRP is a widely accepted inflammatory marker and its immune responses vary differentially in different stages of inflammation. In different eukaryotic hosts, its heterologous expression was seen. The binding specificity of recombinant intact CRP pentamer from the protozoan *Leishmania tarentolae* (2 μg/mL culture medium) was for PC in a calcium-dependent manner. The binding was not through LPG molecule as reported earlier (Dortay et al. 2011).

The incidence rate of leishmaniasis is on the hike for children, and the reasons are until now to be defined. The children had shown a significantly elevated level of IL-10 as compared to adult patients with kala-azar. Though the level of anti-leishmanial antibody (IgG4 and IgG3) was prominently raised in PKDL and children in comparison to KA patients, the total concentrations of IgG, IgG1, IgG2, and CRP were significantly lower. In polymorphic or macular cases of PKDL, no noteworthy difference in the level of anti-leishmanial antibody level was seen. The varied antibody concentration and the elevated level of circulating blood IL-10 and CRP could be considered as biomarkers of various differential disease susceptibilities in KA and PKDL cases in adults or pediatric patients (Ansari et al. 2008).

A case report from Shush in Iran of a 26-year-old male patient suffering from visceral leishmaniasis (VL) was also reported in 2011. The patient had fever with cutaneous lesions, was negative for Leishman–Donovan bodies as screened by indirect fluorescent antibody (IFA), had pancytopenia, and had splenomegaly. The enlarged spleen was filled by *Leishmania* protozoa as revealed by diagnostic splenectomy. The patient had high levels of CRP, lactate dehydrogenase (LDH), and erythrocyte sedimentation rate (ESR) (Khorvash et al. 2011).

Human CRP affinity-purified from VL patients differs from CRP purified from TB patients in having differential glycosylation linkages in their sugar composition. Glycosylated CRPs, as a mediator of innate immune system, cleared damaged erythrocytes by activating the CRP–complement cascade. The specific binding of glycosylated CRPs (from VL and TB patients) with erythrocytes greatly increases the hydrophobicity and fragility and decreases the rigidity of diseased and damaged erythrocytes. The CRP-bound erythrocytes are hemolyzed by the activation of CRP–complement pathway, even at physiological normal concentration of CRP (10 μg/ml). CRP in its normal physiological condition (10 μg/ml) is also known to increase the deposition of C3 complement component on *Leishmania donovani* promastigotes present in human serum (Bandyopadhyay et al. 2004; Culley et al. 1997). Interestingly, deglycosylated CRPs revealed a much decreased binding with diseased erythrocytes validating the role of glycosylated molecules on CRP. Thus, the authors postulated the protective role of CRP in the clearance of damaged erythrocytes in VL and tuberculosis. As anemia is a common symptom in VL and tuberculosis, the significant role of CRP in influencing hemolysis by activating the CRP–complement cascade was compared with healthy control individuals (Ansar et al. 2009a).

The invasive recommended protocol to aspirate bone marrow/spleen for treatment of kala-azar is to monitor the duration and response of therapeutic drugs as well as to monitor drug resistance phenomenon. The less invasive diagnostic tools include the evaluation of the role of CRP in VL patients at different steps of the disease. This includes the relationship, if any, between disease activity and CRP levels during the response to strategic therapy. In 2–12-year age group children, the plasma level of CRP was estimated on admission, each 5th day during therapy, and a continuous follow-up to 2 and 6 months. Before treatment, the mean serum

CRP value in VL affected children was significantly higher in comparison to the healthy subjects. After treatment protocol started, there is a simultaneous reduction in serum CRP level. Also, patients having an early low serum CRP level (<60 mg/l) had faster parasitic clearance from the spleen in evaluation with patients with high CRP levels (>60 mg/l). Late responders in treatment had significantly higher mean serum CRP levels than in early responders. Thus the researchers pointed a promising diagnostic role of CRP and its estimation every 5–10 days during kala-azar responsive or resistance therapy (Singh et al. 1999; Singh 2014).

In 1995, researchers again pointed the noninvasive therapeutics and monitoring of VL during and after therapy using acute-phase reactant markers like alpha-1-acid glycoprotein (AGP), CRP, and serum amyloid A protein (SAA). At admission, VL patients had significantly elevated levels of CRP, SAA, and AGP, and the serum concentrations gradually decrease with effective strategic therapy to reach the normal levels by 3-month follow-up (CRP) or by the end stage of therapy (SAA and AGP). Patients having more serum acute-phase protein concentrations showed significantly time-consuming parasite clearance. Thus the specificity and sensitivity of these acute-phase proteins were used as noninvasive markers of monitoring disease activity and predictors of parasite clearance. The monitoring of the acute-phase proteins might be useful for the response of VL patients to different therapies and also relapse after cure from VL (Wasunna et al. 1995).

In the diagnosis of VL in children, in some cases, when it is difficult to find the *Leishmania* parasites, then protein profiling of ten proteins may act as supplementary help. Due to inflamed condition in VL, the concentration of CRP, alpha-1-antitrypsin, and orosomucoid increases while concentration of albumin, prealbumin, and transferrin decreases. Due to hemolytic mechanisms in VL, the concentration of haptoglobin decreases. Thus, in this study also, CRP was taken as an inflammatory marker of inflammation (Bouree et al. 2000).

Post-kala-azar dermal leishmaniasis (PKDL) is the consequence to VL in East Africa and India. In Sudan, almost 50 % of the VL patients develop consecutively PKDL. As reported earlier, the presence of elevated levels of IL-10 in plasma and in keratinocytes during VL foretells the succeeding development of PKDL. Patients, who developed PKDL during the first 6 months after kala-azar disease, had higher plasma CRP levels at diagnosis and even after 30 days of follow-up period. Patients who did not develop subsequent PKDL had low plasma levels of CRP at diagnosis. Thus visceral leishmaniasis patients, who have a high threat of developing PKDL even after VL treatment therapies, can be diagnosed at an early stage by measuring plasma CRP (Gasim et al 2000).

The role of elevated levels of serum immunoglobulins in the resistance or progression of VL is highly unexplored. The effector role of antibodies depends on their interactions with Fc receptors (type I and II) and also relies on the patterns of antibody Fc N-glycosylation. Thus, IgG Fc N-glycopeptide profiling of 187 kala-azar patients along with 116 endemic controls, 177 asymptomatic individuals, and 43 non-endemic controls individuals was done. Patients with VL had altered IgG Fc N-glycan profiles as compared to asymptomatic or healthy uninfected individuals. The altered glycosylation profiles correlated with concentrations of serum cytokines and CRP. The specific Fc N-glycosylation and its relation with levels of different serum cytokines and CRP are prominently correlated with the development of harsh clinico-medical symptoms in VL, and, clearly, the Fc glycosylation changes after therapeutic treatment. Thus altered N-glycosylation of IgG Fc of VL patients regulates the effector immune function of antibodies toward *Leishmania* parasites (Gardinassi et al. 2014).

Leishmania infantum interferes with the oxidative metabolism of phagocytes in dogs. To assess this, the levels of derivatives of reactive oxygen metabolites (d-ROMs) were measured in *Leishmania* seropositive dogs or asymptomatic control and dogs with other diseases. Dogs, having higher concentrations of d-ROMs, also had higher serum CRP levels. Thus inflammation may mask the decreased d-ROMs values as induced by *Leishmania* infection (Paltrinieri et al. 2010).

The increased concentration of CRP along with high levels of orosomucoid and alpha-1-antitrypsin and decreased levels of albumin, prealbumin, and transferrin was used in the diagnosis of visceral leishmaniasis, in some cases when it is difficult to find the parasites. These experiments were done to profile ten proteins in 7 children infected with visceral leishmaniasis. Thus protein profiling diagram gives a supplementary data in the diagnosis of visceral leishmaniasis (Bouree et al. 2000).

8.7 Discussion

It is thought that an understanding of the process by which the parasite invades into the host cells holds promise for the prevention or modification of infectious diseases. This may be possible with the proper insight into the machinery that the parasite utilizes to enter into the host cells. The interface of promastigotes with macrophages has been an extensively studied aspect of *Leishmania* research (Mosser and Rosenthal 1997) although the entire mechanism of parasite invasion into macrophages still remains to be elucidated because it is evidenced that a thriving pathogen possesses more than one mechanism to divert destruction by the body's first line of defense such as the complement system (Mosser and Rosenthal 1997). In the case of *Leishmania* promastigotes, this divulgence has been taken to an elevated level. Not only are the damaging effects of complement activation evaded by the parasite, but also the useful effects of opsonization are exploited. It has been showed earlier that promastigote binding and phagocytosis are mainly the receptor-mediated immune reactions (Mosser and Rosenthal 1997).

Multiple macrophage receptors, host opsonins, and parasite ligands have been associated in the binding of promastigotes to macrophages. Internalization of the *Leishmania* promastigote is considered to occur through one of several receptor systems. This receptor system includes receptors for complement, fibronectin and mannose, immunoglobulin (Ig)G Fc receptors, and many others.

CRP shares many functional properties in common with human IgG, including their capability to activate the classical complement cascade (Kaplan and Volanakis 1974), bind to various cells (Marnell et al. 2005), and interact with human Fcγ receptor (FcγR) (Marnell et al. 1995; Bharadwaj et al. 1999). Phagocytosis of amastigotes may occur through Fc receptors (FcR) and immunoglobulin also (Bodman-Smith et al. 2002a; Kima et al. 2000). CRP binds to cells through PC-substituted carbohydrates on the outer surface of *S. pneumoniae* (Culley et al. 2000; Szalai et al. 2005a). It also binds with LPG of metacyclic *L. donovani* (Culley et al. 1996; Guy and Bolosevic 1993) and activates transformation to a more or less amastigote-like stage in *L. mexicana* (Bee et al. 2001).

CRP is well known to bind FcγRI with high avidity and this binding has been shown to cause opsonized uptake of PC-coated erythrocytes in FcγRI-transfected COS-7 cells. Subsequently, during infection, phagocytosis of the amastigote parasite may occur through Fc receptors, and an immunoglobulin or antibody may also be required for continuance of infection (Guy and Bolosevic 1993; Kima et al. 2000; Bodman-Smith et al. 2002a; Peters et al. 1995; Wilson and Pearson 1988).

It has been suggested that LPG and gp63 cause deposition of some complement components as opsonic factors on the outer surface of the parasite. Consequently this causes higher uptake of the parasite, a less hostile immune response by the macrophage, and an improved endurance of the parasite. However, there appears to be little substantiation that endurance of parasites taken up by the Fc receptors is not less than those opsonized by the complement receptors. Though nitric oxide is more strongly linked to *Leishmania* parasite killing, toxic oxygen may have a role in killing the parasites. Moreover, the parasite actively tries to regulate the macrophage reaction through different molecules like LPG present on the external surface of infective promastigotes. LPG is a well-known potent macrophage suppressor as it acts to inhibit macrophage functions through the regulation of nitric oxide synthesis, interleukin (IL)-1 production, and phagolysosomal fusion. At later phases, a fully differentiated amastigote has developed increased ability to resist pH balance and lysosomal enzymes and to tip over free radicals (Ballou and Lozanski 1992; Beverley and Turco 1998; Mold and Du Clos 2006; Bredt and Synder 1994; Du Clos 2000; Galve-de Rochemonteix et al. 1993; Kima et al. 2000; Peters et al. 1995; Proudfoot et al. 1995; Ratnam and Mookerjea 1998).

CRP binds with phagocytic cells directly through FcγRI and FcγRII and mostly binds with macrophages, monocytes, and neutrophils through FcγR receptors (Marnell et al. 2005). CRP also binds to human FcγR on U-937 cells, on transfected COS-7 cells, and on human leukocytes and myeloid cells (Bharadwaj et al. 1999; Marnell et al. 1995). CRP exhibits multiple functional similarities to IgG including recognition of ligands, activation of complement via the classical pathway, binding to receptors on phagocytic cells, induction of cytokine synthesis by triggering cellular responses, and enhancement of phagocytosis. These functional similarities are explained by their (CRP and IgG) shared ability to interact with complement component C1q and with Fcγ receptors. FcγRs are well characterized as important to antibody-mediated immunity and, when cross-linked by bound IgG antibodies, trigger cellular responses (Marnell et al. 2005). Direct uptake of CRP opsonized cells is a much more efficient uptake process in vitro than CRP-unopsonized uptake of cells. Attributes of CRP on cytokine responses to pneumococcal infection have also been reported. In mice infected intranasally with *S. pneumoniae* and in human pneumococcal pneumonia, a strong local developed systemic inflammatory cytokine response prevails, and this early response is correlated with increased endurance and persistence (Thomas-Rudolph et al. 2007). CRP binds with FcγR, protects mice from pneumococcal infection, and assists *S. pneumoniae* removal from the bloodstream of the host (Suresh et al. 2007; Mold and Du Clos, 2006).

CRP shares many functional attributes in common with human IgG, including the ability to activate the classical complement cascade (Kaplan and Volanakis 1974), the capacity to interact with human FcγR (Bharadwaj et al. 1999; Marnell et al. 1995), and the ability to bind to various cell lines and blood cells (Bodman-smith et al. 2002a). The interaction of CRP with human FcγR intervenes several functions that are common to those of IgG, including cytokine synthesis and opsonized uptake of bacteria and binding with exposed or altered self-antigens on damaged cells (Marnell et al. 2005). Later during infection, phagocytosis of amastigote parasites may occur through immunoglobulin molecules, Fc receptors (FcR), and antibody proteins for the maintenance of infection (Bodman-smith et al. 2002a; Kima et al. 2000; Bharadwaj et al. 2001; Guy and Bolosevic 1993). CRP binds to PC-substituted carbohydrate on the surface of *S. pneumoniae* and LPG on the surface of metacyclic *L. donovani* (Culley et al. 1996, 2000). In addition, CRP binding to *L. mexicana* metacyclic promastigotes also stimulates the transformation to a new amastigote-like stage (Bee et al. 2001).

CRP possesses at least one function with the Fc-region of Ig, and it initiates complement sequence upon reacting with its substrates (Mortensen et al. 1976). It shares many functions with IgG, like complement activation, opsonization, and phagocytosis (Marnell et al. 1995), and also has similar amino acid composition, suggesting a common evolutionary origin (Oliveira et al. 1979).

Accordingly, this chapter will deal with the binding of CRP with leishmanial parasite. The main goal of this chapter is to investigate the modulation of the binding of CRPs with *Leishmania* and its detection and application in immunological diagnosis or pathophysiology in different forms of leishmaniasis.

9 CRP and Diabetes: Sugar Is Not So Sweet

Abstract

The complications and burden of the metabolic syndrome of diabetes is increasing worldwide. This metabolic disease is represented with inappropriate hyperglycemia either due to deficiency of insulin secretion or reduction in the biologic effectiveness of insulin. The pathogenesis of the multifactorial diabetes is very complicated. To control this pandemic disease, the pathophysiology of the disease and stipulated drug targeting covering definite areas are needed. During the progression of a prediabetic patient to diabetic, inflammation plays a key role, including insulin resistance and decreased beta cell secretory capacity. Insulin resistance plays a prominent role in the pathophysiology of various macrovascular complications. Drugs targeting through different inflammatory pathways represent a newer approach in the therapeutics of diabetes and its related complications.

Keywords

Diabetes • Glycemic index • Cardiovascular risk • Insulin resistance • Inflammation • Atherosclerotic plaque • Atherosclerosis • Metabolic syndrome • Glycemic load • Cardiovascular disease • Type 2 diabetes

Chapter Highlights

1. The burden of diabetes and its complications is increasing worldwide.
2. Inflammation plays a key role in the natural history of diabetes. This includes the progression from prediabetes to diabetes, including the decreased beta cell secretory capacity and insulin resistance. Insulin resistance is an important part of the metabolic syndrome and plays a role in the pathogenesis of various macrovascular complications.
3. Inflammation has been long known to increase insulin resistance and decrease beta cell insulin secretion. Circulating cytokines moreover affect the beta cell function directly causing increased apoptosis and secretory dysfunction. These cytokines can also indirectly influence the beta cell role by increasing the adipocyte-mediated inflammation.

W. Ansar, S. Ghosh, *Biology of C Reactive Protein in Health and Disease*,
DOI 10.1007/978-81-322-2680-2_9

4. Inflammation may be one of the underlying mechanisms of cardiovascular disease in subjects with diabetes and/or obesity. Thus, anti-inflammatory drugs are the best option to combat diabetes.
5. In type 2 diabetes mellitus (DM), in adults, CRP acts as a pro-inflammatory marker. CRP is associated with the development of insulin resistance. CRP is a precise predictor for insulin resistance and hyperglycemia in diabetic patients. CRP levels of diabetic subjects were higher than normal people.
6. Type 1 and type 2 DM patients exhibit higher levels of hs-CRP as compared to healthy subjects which suggest endothelial damage. In type 2 DM, as a predictor of macrovascular complications, hs-CRP might be used as a key predictor molecule. Patients with metabolic syndrome have elevated hs-CRP level.
7. The main inflammatory markers that are used and have been proved to provide prognostic information on the outcome and progression of the disease in diabetic patients are CRP, TNF-α, IL-6, intercellular adhesion molecule (ICAM)-1, and VCAM-1.
8. CRP and IL-6 during the acute phase has a causative role in diabetes mellitus (DM) and a function associated with the risk of DM complications. More specifically, increased CRP levels are independent predictors of type 2 diabetes mellitus.
9. High levels of ferritin, hepcidin gene expression, hs-CRP, and TBARS were related with the risk of developing type 2 diabetes mellitus, and these factors are strongly associated to the nutritional and insulin resistance status.

9.1 Introduction

Metabolic syndrome is a collection of risk factors that greatly increases the risk of diabetes, coronary artery disease, and stroke. Metabolic syndrome is actually defined as the presence of three or more of the risk factors. These risk factors include hypertension, insulin resistance, increased abdominal obesity, higher hypertriglyceridemia, and low high-density lipoprotein cholesterol (HDL-C) (Grundy et al. 2005; Voils and Cooper-DeHoff 2014; Ryo et al. 2004; Mojiminiyi et al. 2007; Lee et al. 2009). Many researchers by observational studies and confirmatory studies had pointed the association between metabolic syndrome with the history of myocardial infarction and stroke prevalence (Ninomiya et al. 2004; Koren-Morag et al. 2005; Kurl et al. 2006).

Diabetes is a chronic disease condition where the body cannot properly use the insulin it produces or when the pancreas does not produce an enough amount of insulin as required by the body. Hyperglycemia, insulin resistance, hyperinsulinemia, hyperlipidemia, and hyperhomocysteinemia are the important pathophysiological components of diabetes mellitus that succumb to endothelial/vascular dysfunction through several underlying physiological processes (Festa et al. 2000).

It is well established that the endothelium is not just composed of a single layer of cells but rather it acts as a regulator exerting various significant autocrine, paracrine, and endocrine actions and thus it affects several cell types. In addition, the endothelium regulates and maintains the vascular tone through several vasoactive mediators and primarily by nitric oxide. However, all these functions are altered in diabetes mellitus (Fig. 9.1, Table 9.1). Hyperglycemia, insulin resistance, and elevated free fatty acids (FFAs) stimulate systemic inflammatory processes and impair nitric oxide bioavailability, leading to impairment of endothelial function (Tousoulis et al. 2013). Changes in the expression of pro-inflammatory molecules and adhesion molecules with their regulated alterations exist in diabetes mellitus (Table 9.1). The endothelium after activation expresses adhesion molecules and different causative factors that participate in the inflammatory process and, subsequently, in the pathophysiology of atherosclerosis in diabetes mellitus. Endothelial cells are stimulated by pro-inflammatory molecules like TNF-α and CRP to promote atherogenic processes. The major

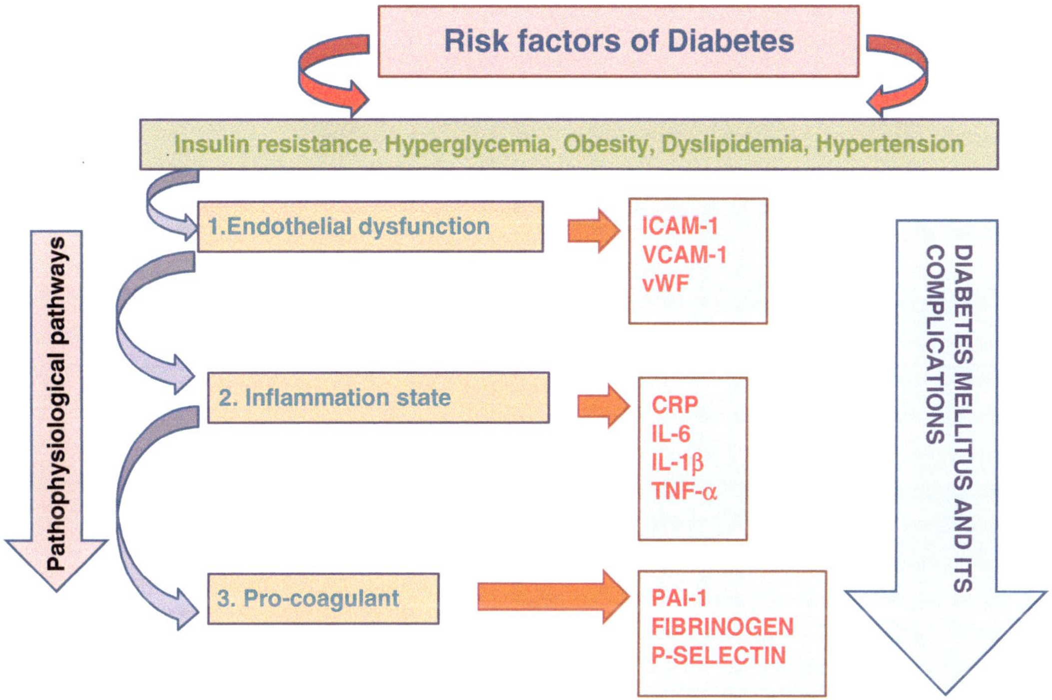

Fig. 9.1 *Different factors and related complications of diabetes*. Insulin resistance, hyperglycemia, obesity, dyslipidemia, and hypertension are the hallmark of diabetes mellitus. These factors through various complex pathophysiological pathways and mechanisms impair the endothelial function, increase several inflammatory processes, and alter the procoagulant state (Tousoulis et al. 2013).
CRP C-reactive protein, *ICAM* intercellular adhesion molecule, *IL* interleukin, *PAI* plasminogen activator inhibitor, *TNF* tumor necrosis factor, *VCAM* vascular cell adhesion molecule, *vWF* von Willebrand factor

Table 9.1 Diabetes mellitus-related endothelial dysfunction: mechanisms and related markers[a]

Biomarkers	Status (UR/DR)	Mechanisms	Manifestation
Insulin resistance	UR	–	–
Glucose levels	UR	–	Increased arterial stiffness
HbA1c	UR	–	–
Homocysteine	UR	–	–
TNF-α	UR	Insulin resistance	–
MCP-1	UR	Abnormal cluster of hyperglycemia	–
Selectins (E-selectin)	UR	Increased oxidative stress/ reduced nitric oxide production	–
ICAM-1	UR	–	–
VCAM-1	UR	–	–
IL-1, IL-6	UR	–	–
VEGF (GF)	UR	Increased inflammatory status	Decreased vasodilation
ET-1	UR	Elevated free fatty acids	Increased vasoconstriction
PAI-1	UR	Glycosylated end products	Increased atherogenesis
Fibrinogen	UR	Vascular smooth muscle cell dysfunction	–
vWF	UR	Endothelial dysfunction	Impaired arterial remodeling
EPC	DR	–	–
ADMA	UR	–	–

[a]*ADMA* asymmetrical dimethylarginine; *EPC* endothelial progenitor cell; *ET* enthothiline; *HbA1c* glycosylated hemoglobin; *ICAM* intercellular cell adhesion molecule; *IL* interleukin; *MCP* monocyte chemoattractant protein; *PAI* plasminogen activator inhibitor; *TNF* tumor necrosis factor; *VCAM* vascular cell adhesion molecule; *GF* growth factors; *VEGF* vascular endothelial growth factors; *vWF* von Willebrand factor; *UR* increased, upregulated; *DR* decreased, downregulated (Tousoulis et al. 2013)

inflammatory markers that are used as a prognostic factor on the consequences and progression of the disease in diabetic subjects are CRP, TNF-α, IL-6, intercellular adhesion molecule (ICAM)-1, and VCAM-1 (Tousoulis et al. 2013; Uysal et al. 1997; Frohlich et al. 2000; Schultz and Arnold 1990; Su et al. 2010).

Hyperglycemia is the most common effect of uncontrolled diabetes, and over time it leads to severe harm to many systemic influences; especially the blood vessels and nerves are badly affected. In 2014, 9 % of the adults above 18 years had diabetes. In 2012 nearly 1.5 million deaths were directly caused by diabetes (Global status report on noncommunicable diseases 2014 (2012)).

The global burden of diabetes is predicted to increase to nearly 380 million by 2025. Diabetes would be a major health problem in near future. However, both type 1 and type 2 diabetes are the major problem to challenge health services worldwide (Kaul et al. 2010; CDC 2006). The incidence of diabetes is escalating in an alarming rate worldwide and is nearing epidemic proportions. Type 2 diabetes almost accounts for nearly 90 % of all diabetes cases and is related with allied cardiovascular morbidity and mortality (Blaschke et al. 2006). The prevalence of type 2 diabetes has also been rapidly rising worldwide. In type 2 diabetes, either due to reduction in the biologic effectiveness of insulin or deficiency of insulin secretion there is inappropriate hyperglycemia. The relation of obesity with type 2 diabetes patients was observed with altered adipokine concentrations and elevated inflammatory markers. The underlying pathophysiological mechanism in type 2 diabetic patients is impaired insulin secretion and insulin resistance, but other specific determinants or key players of main abnormalities of these metabolic defects remain uncertain yet (Hansen et al. 2010; Bergman 1989; Dunmore and Brown 2013; Tsui et al. 2011; Lamers et al. 2011; Wei et al. 1998; Turner et al. 1998).

Type 2 diabetes mellitus is a heterogeneous disorder characterized mainly by two interassociated metabolic defects like impaired β-cell function and insulin resistance. In type 2 diabetic patients with impaired glucose tolerance (IGT), numerous factors like genetic, some host-related, and many environmental conditions contribute mostly to the progression of insulin resistance in this disease. The risk of budding type 2 diabetes is mainly contributed by obesity with visceral adipose tissue accumulation and its endocrine factors and changes in adipose tissue function (Schmidt et al. 1999; Al-Hamodi et al. 2014; Stumvoll et al. 2005; Kahn 2008; Jensen 2008; Bastard et al. 2006; Rasouli and Kern 2008).

Type 2 diabetes mellitus is mainly a chronic inflammatory condition mainly characterized by the presence of high or low insulin resistance with associated systemic low-grade inflammation. The increased risk of cardiovascular disease is a prominent alliance of type 2 diabetes mellitus. Presence of inflammation has been suggested as a key factor in the progression and development of insulin resistance along with cardiovascular disease comorbidities like atherosclerosis (Schmidt et al. 1999; Festa et al. 2000; Dandona et al. 2004; Goldberg 2000).

The indicators of obesity are elevated circulating levels of acute-phase proteins, secretion of pro-inflammatory cytokines like TNFμ and IL-6, and tissues exposed to an excess of nutrients. The increased circulating saturated fatty acids are a constant stimulus for the secretion of pro-inflammatory cytokines. These factors at the terminal end of a signaling cascade inhibit the insulin receptor. Micronutrients, especially a systemic iron overload, could contribute to abnormal glucose metabolism and the development of diabetes. Ferritin is an iron status marker but it reflects the amount of body iron stored in healthy individuals. Ferritin is also an acute-phase reactant, and thus in inflammatory conditions or by the induction of cytokines like IL-6 and IL-1, its synthesis is upregulated. In type 2 diabetes mellitus, the concentration of ferritin is increased (Dungan et al. 2015; Tuomi et al. 1993; Schmidt et al. 1999; Dasu and Jialal 2010; Guzmán and Olguín 2014).

9.2 Inflammation and Diabetes

CRP is a critical component of the innate immune system. It is a complex protein that our body synthesizes from the liver excessively when faced with major problems like infection or trauma. CRP was discovered almost more than 70 years back by scientists exploring different human inflammatory responses or mechanisms. The key role of CRP in different heart diseases, however, has only recently been addressed and uncovered (Abdelmouttaleb et al. 1999; Barzilay et al. 2001; Grau et al. 1996).

CRP is synthesized in different amounts and functions differentially depending on a variety of genetical factors as well as many lifestyle habits. On an average, high levels of CRP are attributed to individuals who smoke regularly, have high blood pressure, are overweight or obese, and fail to exercise quite often. Indeed, thin athletic individuals tend to have much lower levels of CRP. Nonetheless, nearly half of the differential variation in CRP plasma levels between different study subjects is mostly inherited. This property chiefly correlates with the fundamental function of CRP in inflammatory responses. Inflammation are critical for survival of organisms during wound healing, for warding off microbes like bacteria and viruses and in many other homeostasis disruptions. Studies over the past decade have revealed that too much inflammation in some conditions can have unfavorable effects, mostly on the blood vessels that carry nutrients and oxygen to different tissues of the body. Scientists now comprehended that atherosclerosis (where cholesterol accumulates in the arteries) is in several manners an inflammatory disorder of the blood vessels, just like arthritis which is an inflammatory disorder of the joints and bones (Ridker et al. 1998a, b,

2001a, b; Kervinen et al. 2001; Pradhan et al. 2001).

Many researchers have revealed that blood markers that reproduce the inflammatory process are raised among subjects at high risk for upcoming heart diseases. Inflammation is a critical factor in all stages of heart disease, including the quite early initiation of atherosclerotic plaques within the wall of arteries, till the acute rupturing of these atherosclerotic plaques that upshoot in heart attack and, all too often, with sudden death also. Until latest studies, existing markers of inflammation were not appropriate for use. By contrast, CRP has been publicized to be very steady and quite easy to compute and measure (Ridker et al. 1998a, b).

CRP is a potent predictor of risk, especially when combined with cholesterol assessment. Some physicians choose to compute CRP alongside with a panel of other "novel" risk factors like homocysteine and lipoprotein-a. Other clinicians may select to measure plasma CRP in addition to some more expensive tests that actually measure specific cholesterol subfractions. However, in all direct assessments, the predictive value for CRP has been significantly greater than those practical values for these alternative "novel" markers of health risk. Moreover, only CRP has confirmed to add valid prognostic relevance to that already available data from standard cholesterol screening protocols (Nix et al. 2015; Onat et al. 2014; Wu et al. 2002; Rodriguez-Moran and Guerrero-Romero 1999; Yudkin et al. 1999).

In some cases, different imaging techniques like "whole-body scans" that distinguish calcification in the aorta and the heart arteries have been promoted as valid screening techniques. Moreover, the presence of calcification in arteries does also increase cardiovascular risk. Indeed, such imaging scans are currently very expensive and are not recommended by the American Heart Association. An additional concern for all these imaging procedures is that end results are often misinterpreted by patients and clinicians. Thus, it can lead to unnecessary coronary interventions, like angioplasty and bypass surgery. While CRP plasma levels also have been revealed to add prognostic values at all levels of coronary calcified calcium, this information should be validated primarily to encourage at-risk individuals to adopt further heart-healthy lifestyle habits, as a precautionary measure, and not to seek insistent interventional cardiac procedures (Mekonnen et al. 2014; Hayashino et al. 2014a, b; Fichtlscherer et al. 2000).

Unlike LDL cholesterol, CRP predicts not only the risk of heart disease but also the risk of upcoming type 2 diabetes. Patients with CRP levels higher than 3 mg/L have a risk of having diabetes 4 to 6 times higher than subjects with lower values of CRP. Part of the connection between diabetes and heart disease is due to inflammation, and in other parts, for many patients, inflammation is the outcome of obesity, particularly the "central obesity," or the gradual tendency to put on weight in and around the stomach. This is due the fact that adipocytes (fat cells) produce some messenger proteins that switch on the production of CRP itself from the liver (Xu and Whitmer 2006; Hayashino et al. 2014a, b).

The metabolic syndrome is a known condition to predispose individuals to different heart diseases and diabetes. Clinicians classify patients as having the metabolic syndrome depending on the presence of three factors of the following 5 conditions: (i) low HDL cholesterol, (ii) central obesity, (iii) high triglycerides, (iv) increased blood sugar levels, and (v) high blood pressure. However, the metabolic syndromes also bring about a number of less frequently measured abnormalities like insulin resistance and the problems with blood clotting. CRP levels increase as the number of components of the metabolic syndrome increase. Even among individuals known to have the metabolic syndrome, CRP levels add important prognostic information on risks. Thus, many physicians now also measure CRP as part of the process of defining the metabolic syndrome. This practice is increasingly common among endocrinologists and other physicians interested in the prevention of diabetes as well as heart disease (Pfützner et al. 2010).

There is limited clinical data addressing that inflammation is hypothesized to play a role in the development of type 2 diabetes mellitus (DM). CRP is a known inflammatory biomarker produced and synthesized by the liver under the activation of interleukin (IL)-1, IL-6, and tumor necrosis factor-α cytokines (Cheung et al. 1998; Pradhan et al. 2001). There were many reported cases of elevated levels of both IL-6 and CRP among subjects featuring insulin resistance syndrome and clinically overt type 2 diabetes (Pickup et al. 1997). In patients with chronic kidney disease with type 2 diabetes, CRP and tumor necrosis factor-α (TNF-α) were demonstrated as independent risk factors (Yeo et al. 2010; Cheung et al. 1998; Ventre et al. 1997). CRP as an inducer of pro-inflammatory cytokine was significantly raised in diabetic patients as compared to normal individuals. In these patients CRP level was also highly correlated with insulin resistance. Thus, the role of CRP as a precise predictor for insulin resistance in diabetic patients was confirmed. In patients with obese type 2 diabetes, elevated levels of different inflammatory markers and altered adipokine concentrations have been reported. The underlying pathophysiological anomalies in type 2 diabetes include determinants like impaired insulin secretion and insulin resistance, yet some specific determinants are coming up. Inflammation played a crucial mediator role in the pathogenesis of type 2 diabetes, thereby connecting diabetes with a number of commonly coexisting inflammatory conditions. CRP, a sensitive subclinical inflammatory biomarker, is associated with hyperglycemia, level of glycosylated hemoglobin (HbA1c) and insulin resistance in overt type 2 diabetic individuals (Hansen et al. 2010; Pradhan et al. 2001; Hayashino et al. 2014a, b).

The levels of CRP, triglyceride, and leptin are considerably elevated in obese diabetes patients as compared to the healthy normoglycemic controls. Elevated levels of CRP and IL-6 independently predict the development of type 2 diabetes. Thus, CRP played a possible role in the inflammatory diabetogenesis (Hopps et al. 2011; Hansen et al. 2010; Cheung et al. 1998; Ventre et al. 1997; Karbowska et al. 2009).

Moreover, inflammation plays a pivotal role in the pathobiology of several glucose disorders in adults. This concept was substantiated by a subset of nonobese adults rapidly developing incident diabetes but without antecedent glucose abnormalities. In these nonobese adults, the presence autoimmune inflammatory markers like antibodies against the β-cell (islet cell antibodies and antibodies to glutamic acid decarboxylase) were included (Tuomi et al. 1993; Tuomilehto et al. 1994; Pietropaolo et al. 2000); elevated markers of inflammation in diabetic and nondiabetic patients (Tuomi et al. 1993; Tuomilehto et al. 1994; Pietropaolo et al. 2000), prediction of incident diabetes in nondiabetic individuals with baseline levels of certain inflammatory markers (Pickup et al. 1997; Pickup and Crook 1998); helping of inflammatory mediators (tumor necrosis factor α, decrease insulin sensitivity) to precipitate diabetes (Pickup and Crook 1998; Cheung et al. 1998); and the implication of inflammation as a part of the insulin resistance syndrome (Pickup and Crook 1998; Duncan et al. 1999 ;Cheung et al. 1998; Ventre et al. 1997).

9.3 Biomarkers of Diabetes Mellitus

Diabetic patients exhibit pathologically upregulated or downregulated expressions of various biomarkers. The biomarkers of endothelial function, like vascular cell adhesion molecule (VCAM)-1 and von Willebrand factor (vWF), were enhanced in diabetic subjects. Moreover, markers of systemic inflammation like CRP and tumor necrosis factor (TNF)-α were included in diabetic patients. Biomarkers of diabetes mellitus includes inflammatory biomarkers, coagulation-related biomarkers, stress-related biomarkers, and some recently highlighted biomarkers like micro-ribonucleic acids and endothelial microparticles (Tousoulis et al. 2013) (Table 9.2).

Additionally, the injured endothelium finally results in vasoconstriction, smooth cell proliferation, some coagulation disorders, leukocyte aggregation, thrombosis, and vascular inflammation predisposing a condition of atherosclerosis. The endothelial integrity was altered through various inflammatory and oxidative processes (Fig. 9.2). The process of systemic inflammation results in increased secretion of cytokines and

Table 9.2 Important biomarkers in diabetes

Important biomarker	Analytical methods	Diabetes-related distinctiveness
CRP (hs-CRP)	1. Ultrasensitive solid-phase ELISAs 2. Immunoturbidimetric CRP assays 3. Immunonephelometry (laser nephelometry)	1. Increases rapidly, has long rising periods, quite stable in plasma 2. Attenuates NO production 3. Decreases endothelial NO synthase 4. Upregulates the expression of different adhesion molecules in the wall of endothelial cells 5. Triggers the oxidation of molecules like low-density lipoprotein cholesterol Induces the PAI-1expression 6. Activates the macrophages to secrete tissue factor 7. Stimulates the release of matrix metalloproteinase-1
IL-6	ELISA	1. Affects insulin sensitivity through adenosine monophosphate activated protein kinase 2. Alters glucose homeostasis and metabolism both directly and indirectly by the functioning of cells like skeletal muscle cells, adipocytes, hepatocytes, and neuroendocrine cells 3. Affects extracellular matrix dynamics at both at the mesangial and podocyte levels 4. Stimulates proliferation of mesangial cells 5. Increases expression of fibronectin 6. Enhances permeability on endothelial cells 7. Induces the production of adrenocorticotropin
TNF-α	ELISAs	1. Self-regulatory properties 2. Growth-stimulating properties 3. Growth inhibitory processes 4. Cytokine-triggering effect 5. Inflammation and apoptosis 6. Expressed in cells like: B cells, T cells, macrophages, monocytes, mast cells, neutrophils, and adipocytes
Fibrinogen	1. Immunological assays (ELISA or nephelometry) 2. Automated immunoassays of total fibrinogen 3. The Clauss fibrinogen assay (on the basis of the thrombin clotting time) 4. Automated clotting rate assays	1. Interacts with ICAM-1 2. Lowers platelet inhibition

(continued)

Table 9.2 (continued)

Important biomarker	Analytical methods	Diabetes-related distinctiveness
ICAM-1	1. Flow cytometry 2. ELISAs	1. Leads to changes through the protein kinase C pathway, cAMP, phospholipase A2, Ca^{+2}, and proteosomes (extracellular signal transportation) 2. Promotes the recruitment of mononuclear cells in diabetic glomeruli 3. Role in glomerular hyperfiltration 4. Interactions of lymphocyte function-associated antigen-1 5. Binds with fibrinogen gamma chain (at the sites 117–133)
VCAM-1	1. Flow cytometry 2. ELISAs	1. Interacts with secreted protein and it is rich in cysteine 2. Interacts with VLA-4
Endothelial microparticles	1. Flow cytometry	1. Factor XI-dependent procoagulant properties 2. Impaired NO release 3. Apoptosis 4. Thrombosis, cell inflammation, angiogenesis, and cell-to-cell communication
miRs	1. Microarrays 2. Real-time qPCR 3. Northern blot 4. RNA sequencing 5. In situ hybridization	1. Interaction with TGF-beta 2. Affect insulin secretion 3. Affect pancreatic β cells and insulin-target tissues

CRP C-reactive protein, *hs-CRP* high-sensitivity C-reactive protein, Ca^{2+} calcium doubly charged positive ion, *cAMP* cyclic adenosine monophosphate, *ELISA* enzyme-linked immunosorbent assay, *ICAM* intercellular cell adhesion molecule, *IL* interleukin, *miR* micro-ribonucleic acid, *NO* nitric oxide, *PAI* plasminogen activator inhibitor, *qPCR* quantitative polymerase chain reaction, *RNA* ribonucleic acid, *TGF* transforming growth factor, *VCAM* vascular cell adhesion molecule, *VLA* very late antigen (Tousoulis et al. 2013)

migration of leukocytes into the vessel wall. Elevated levels of circulating cytokines cause enhanced oxLDL, formation of foam cell, and scavenging activity of macrophages especially in diabetes mellitus in the critical atherosclerotic process (Tousoulis et al. 2013).

Inflammatory biomarkers mostly involve the changes in the expression of various adhesion molecules, pro-inflammatory molecules, and alterations and regulations of these molecules. The endothelium can be stimulated by different pro-inflammatory molecules like TNF-α and CRP to promote an exclusive atherogenic phenotype. The main inflammatory biomarkers having prognostic information in diabetic subjects are CRP, IL-6, intercellular adhesion molecule (ICAM)-1, VCAM-1, and TNF-α (Kampoli et al. 2011).

C-reactive protein (CRP) is one of the commonly used inflammatory biomarkers (Table 9.2) having diagnostic and prognostic value. CRP may exert modulatory effects as it is present in atherosclerotic plaques. The expression of CRP is mainly controlled by IL-6 during the acute phase. Increased CRP levels are independent predictors of type 2 diabetes mellitus in apparently healthy individuals, supporting the notion that subclinical inflammation is an underlying causative factor in the pathology of type 2 diabetes mellitus. CRP also predicts the outcomes in diabetic patients with cardiovascular disease. High-sensitivity C-reactive protein (hs-CRP) has been proven to be useful in predicting adverse cardiac

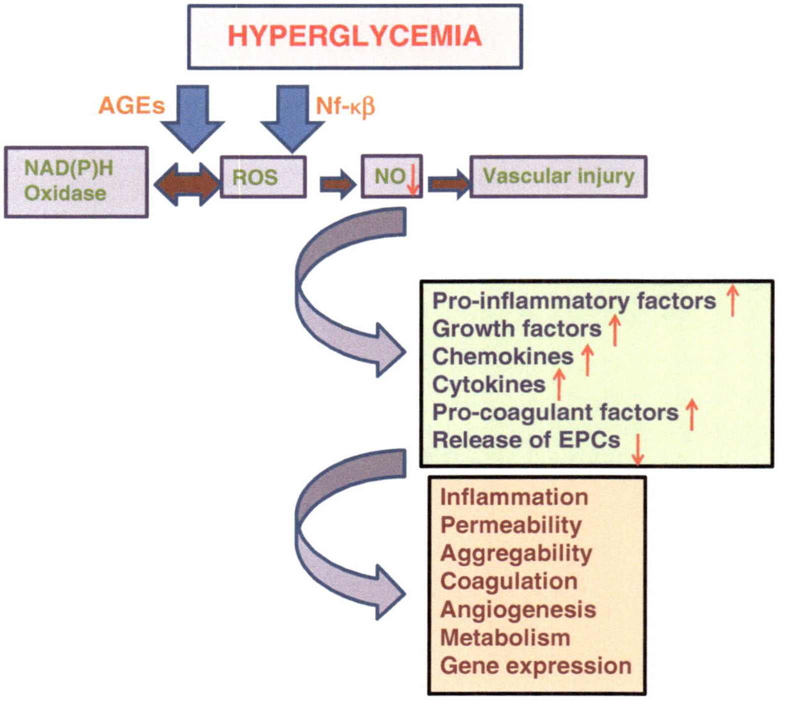

Fig. 9.2 *Role of hyperglycemia in atherogenesis*. Hyperglycemia plays a pivotal role in states of diabetes mellitus. Different pathways are triggered by advanced glycated end products and mediated by the nuclear factor (Nf)-κB. Thus, it leads to increased oxidative status and impaired nitric oxide (NO) production/bioavailability. So, levels of growth factors, cytokines, procoagulant factors, and others are increased and consequently induce and promote atherogenesis. *AGE* advanced glycated end product, *NAD(P)H* nicotinamide adenine dinucleotide phosphate, *ROS* reactive oxygen species, *EPC* endothelial progenitor cell, *BH4* tetrahydrobiopterin (Tousoulis et al. 2013)

outcomes in diabetic subjects (Paiva et al. 2008; Biasucci et al. 2009) (Table 9.2).

Several pro-inflammatory and/or anti-inflammatory cytokines have been recognized as important inflammatory biomarkers in diabetes mellitus. IL-6 and TNF-α (Table 9.2) are important cytokines, which are critically participating in atherogenesis and mainly associated with the risk of diabetes mellitus complications. Their levels increased in type 2 diabetes mellitus. The increased levels of CRP, fibrinogen, IL-6, IL-1, and TNF-α regulated various procoagulant activities, leukocyte adhesions to the endothelium by important mediators like VCAM-1 and ICAM-1 (adhesion molecules), and alterations of vasoregulatory responses. These responses subsequently elevated the risk of diabetes mellitus complications. Diabetes mellitus is nothing but a unique state of balance between pro-thrombotic and anti-fibrinolytic conditions. The markers of endothelial functions are mostly plasminogen activator inhibitor (PAI)-1, tissue factor, and vWF. Both endothelial cells and macrophages attribute toward the generation of altered coagulation processes by the increased expression of PAI-1 and tissue factor and through platelet activation. Simultaneously acute-phase reactions also increase the level of coagulation factors like fibrinogen. PAI-1, vWF, and fibrinogen are prothrombotic molecules that are produced under cytokine stimulation and thus are the end products of the acute-phase responses (Spranger et al. 2003; Meigs et al. 2004).

The increased oxidative stress through superoxide production and impairment of NO bioavailability in the vascular wall are the major contributors of endothelial dysfunction in the vasculature and associated complications of diabetes mellitus. Also, in humans with diabetes, obesity, and insulin resistance, different circulating markers of oxidative stress (Table 9.2) (F2 isoprostanes, antibodies against oxLDL) are increased. Micro-ribonucleic acids (miRs) and microparticles (MPs) are the two new biomarkers emerging as potential challengers of the classical biomarkers of diabetes (Meigs et al. 2007).

Micro-ribonucleic acids (miRs), a class of nearly 22 nucleotide noncoding ribonucleic acids, have significant contributions toward atherogenesis and act as modulators in endothelial dysfunction by either expressing or suppressing the effect of eNOS activity. However, overexpression of miR21 enhances NO production but downgrades endothelial cell apoptosis (Bartel 2004 ;Tousoulis et al. 2013) (Table 9.2).

MPs (membrane particles of <1 mm in diameter) constitute a heterogeneous population of varied cellular origins, numbers, sizes, and antigenic symphonies which are budded into the circulation from various blood cells (platelets, leukocytes, and erythrocytes) and endothelial cells (EMPs). MP generation regulates apoptosis, mechanical injury, markers of endothelial damage, and platelet activation and cellular activation through cytokines. Microparticles are found in blood circulation of healthy individuals, but their number is increased in various cardiovascular diseases (Shantsila et al. 2010; Tousoulis et al. 2013) (Table 9.2).

9.3.1 Role of CRP in Diabetes

CRP, the known marker of systemic inflammation, is gradually outgrowing as an independent risk factor for different cardiovascular diseases (Schmidt et al. 1999; Abdelmouttaleb et al. 1999; Ridker et al. 2001a). Elevated CRP levels have been associated to an increased risk of cardiovascular events like myocardial infarction (Ridker et al. 1998a, b) and have also been associated to an elevated risk of development of later diabetes (Kervinen et al. 2001). Interestingly, CRP levels may be higher in nondiabetic people as compared to a diabetic individual (Ridker et al. 2001a, b). The relation between the level of glycemic control with concentration of CRP in people with diabetes is quite obscured. CRP is associated with HbA1c levels in some studies (Ridker et al. 2001a, b; Hayashino et al. 2014a, b).

The role of inflammation in patients with a mixture of cardiovascular disease and diabetes was explicated by the association between the level of glycemic control and the related inflammation. The purpose of the study was to investigate the relation. The presence of higher HbA1c is considerably linked with higher CRP plasma levels among adults suffering from diabetes. The relation between CRP and HbA1c in adults with diabetes was significant in comparison with people with only a high level of CRP with a minimum value of HbA1c This logistic regression model showing the relationship was formulated in subjects with the same age, race, and sex, having smoking habit, similar BMI values and insulin level, and also a more or less similar length of time with diabetes (Ridker et al. 2001a, b; Hayashino et al. 2014a, b).

CRP, being an inflammatory marker, has been related nowadays to the development of insulin resistance and also with type 2 diabetes. Earlier research study has also established that CRP levels are correlated with HbA1c in individuals without diabetes, whereas CRP concentration is higher in diabetic people. It is also reported that people having concentration of CRP >0.30 mg/dl has also successively higher levels of HbA1c as seen established diabetic patients. Thus, apart from the established role of inflammation in diabetes, the implication of ongoing levels of hyperglycemia in the development of diabetes was also established (Ridker et al. 2001a, b; Pradhan et al. 2001; Hayashino et al. 2014a, b; Barzilay et al. 2001).

Recent research evidence supports a link between hyperglycemia and inflammation. CRP concentration is known to be higher in people with frank diabetes and impaired glucose tolerance. Also, an increased CRP level has been found to be a risk factor for the later development of diabetes. The link between CRP and insulin resistance was also reported. By demonstrating simultaneous inflammatory reactions, oxidative stress causing endothelial dysfunction, and insulin resistance at the physiologic level, the relation of hyperglycemia to inflammation has been stated previously in adult patients also (Hayashino et al. 2014a, b).

Baseline levels of CRP and IL-6 were significantly higher among type 2 DM cases than among controls. The highest relative risks of future type 2 DM for women is strongly correlated with the lowest quartile of these inflammatory markers (CRP and IL-6). The relative risk values were deducted after adjustment for body mass index, family history of diabetes, smoking habits, exercise, use of alcohol, and various hormone replacement therapies, baseline hemoglobin A (1c), and fasting insulin level. Thus, elevated levels of CRP and IL-6 predict the development of type 2 DM and they have a possible role during inflammation in diabetogenesis (Ford 1999; Pradhan et al. 2001; Wang et al. 2013; Pickup et al. 1997; Hayashino et al. 2014a, b).

Endothelial dysfunction and significantly increased inflammatory markers are the hallmarks of type 2 diabetes. In type 2 diabetes, low-grade systemic inflammation is regulated by various pro-inflammatory cytokines, acute-phase proteins like CRP, cell adhesion molecules, interleukins (IL), and tumor necrosis factor alpha (TNF-α). Physical exercise reduces the concentration of inflammatory biomarkers by decreasing cytokine secretion from immune cells, skeletal muscles, and other endothelial cells and also by decreased adipocytokine production. Thus, in type 2 diabetes, aerobic and rigorous physical training has differential effects on cytokine secretion levels and also depends on the different modes/forms of exercise (intensity, duration, and type of exercise). But combined exercises has higher anti-inflammatory outcomes than different aerobic or resistance exercises in lowering the CRP, cytokines (IL-6, IL-1β, TNF-α), leptin, and resistin level and simultaneously highly increasing anti-inflammatory cytokines (IL-4, IL-10) and adiponectin also (Hopps et al. 2011; Cheung et al. 1998; Ventre et al. 1997; Mirza et al. 2012).

The impact of glycemic load (GL) and also glycemic index (GI) on cardiovascular outcomes has been assessed by a number of meta-analyses of cohort studies. In women, the consumption of high-GL/GI diets and increased cardiovascular disease (CVD) risk has been significantly associated. This association is much stronger in women with greater adiposity and diabetics. Women, whose CRP levels are higher, may be reduced, as an emerging CVD risk factor as compared to men. CVD risk factors like LDL-C may be reduced on low-GI diets, whereas increases in dietary GL, another CVD risk factor equally in women and men, have been associated with increased risk of diabetes. The relative risks of CVD in relation to GL and GI, with corresponding confidence intervals, are needed to further elucidated in men (Mirrahimi et al. 2014). Hyperglycemia is a related factor to the increased serum CRP concentration in noncontrolled type 2 diabetic people with and without infectious diseases as compared to nondiabetic subjects (Rodriguez-Moran and Guerrero-Romero 1999).

In diabetes, glycemic control is of great importance, as the classification of glycemic index of foods has been employed as a device to evaluate possible prevention and clinical treatment strategies for this disease. Low-GI diets have been known to perk up the serum lipid profile, lowering serum CRP levels and assisting in weight control. Furthermore, low-GI or glycemic load diets (mean GI multiplied by total carbohydrate) have been correlated with greater concentration of high-density lipoprotein cholesterol (HDL-C) and with lowered CRP concentrations. Thus, it is associated with reduced risk of developing diabetes and further cardiovascular disease. Additionally, several positive relations among dietary GI and threat of various cancers (like colon, breast, and prostate cancer) have also been revealed. Thus, not only in diabetes but glycemic index might have a pivotal role in the clinical treatment and prevention of some chronic diseases (Esfahani et al. 2009; Hayashino et al. 2014a, b).

High-density lipoprotein and lipoprotein (a) are both independent risk factors for cardiovascular disease. Niacin, the most potent agent to increase high-density lipoprotein and reduce lipoprotein (a), has also been found to reduce inflammatory markers with anti-atherosclerotic properties like CRP, lipoprotein-associated phospholipase-A2 (Lp-PLA2), and small-dense LDL and to increase large-particle LDL. Niacin reduces inflammation and oxidative stress, improves endothelial function, and also exerts various pleiotropic effects. Niacin remains high in the list of lipid-modulating agents to be used in clinical practice, and, second after statins, the most relevant medications for lipid management (Gouni-Berthold and Berthold 2013; Hack et al.1997).

Both type 1 and type 2 diabetes increase the threat of developing microvascular and also macrovascular complication and thus have a devastating blow on excellence of life of the patients. The role of inflammation and hyperglycemia is the major causative feature for the progress of these complications. The microvascular complication affects the small blood vessels causing foot amputation (diabetic neuropathy), blindness (diabetic retinopathy), and end-stage renal diseases like diabetic nephropathy in type 1 and type 2 diabetes patients. Also, low-grade inflammation has pathogenic link in type 1 and Type 2 diabetes patients with the common retinal–renal nerve. Circulating and locally produced markers of inflammation, including the cell adhesion molecules (vascular adhesion cell molecule-1, VCAM-1; intracellular adhesion molecule-1, ICAM-1), pro-inflammatory cytokines (IL-6; TNF-α), and CRP are highly related with the progression and succession of diabetic microvascular complications (Kaul et al. 2010).

In vascular atherosclerosis, type 2 diabetes mellitus and, in insulin resistance, systemic inflammatory activities take part in an important pathogenic role. High-sensitivity C-reactive protein (hs-CRP), an inflammatory biomarker, is used constantly to monitor insulin resistance and various cardiovascular risks in both nondiabetic and diabetic individuals. It is known that hs-CRP straightway participates in the progression of atherogenesis. Antidiabetic drugs, like thiazolidinediones (pioglitazone and rosiglitazone), specifically targeting insulin resistance could also reduce inflammation, process of atherogenesis, and thus consequently cardiovascular risk. Thiazolidinediones act as selective ligands for the nuclear transcription factor (peroxisome proliferator-activated receptor gamma, PPAR γ). Thiazolidinedione agent, pioglitazone was altered by hs-CRP by its pronounced insulin-sensitizing (in lowering the glucose level) and anti-inflammatory properties in diabetic and nondiabetic individuals. Coadministration of pioglitazone with anti-lipidemic statin therapy altered low-grade inflammation. Thus, the early insulin resistance treatment strategy to resist systemic vascular inflammation and cardiometabolic syndrome in subjects with higher levels of hs-CRP has an underscore benefit (Pfützner et al. 2010; Schmidt et al. 1999).

Thiazolidinediones improve insulin sensitivity and endothelial dysfunction and are used in the treatment of diabetes. Thiazolidinediones exert beneficial effects on the lipid profile by activating the peroxisome proliferator-activated receptor gamma (PPAR-gamma). PPAR-gamma agonists in the endothelial vessel wall have effects on CRP, matrix metalloproteinase-9 (MMP-9), ATP-binding cassette transporter A1 (ABCA1), plasminogen activator inhibitor type-1 (PAI-1), and adiponectin. Thus, PPAR-gamma agonists have their implications in the treatment protocols of advanced stages of atherosclerosis process, particularly in type 2 diabetic individuals (Blaschke et al. 2006).

Among apparently healthy subjects with low-to-normal lipid levels, the increased risk for myocardial infarction and stroke was clinically evaluated by levels of hs-CRP >3 mg/L. Varied epidemiological and laboratory data showed hs-CRP in correlated with inflammation, to degree of insulin resistance, to impaired insulin sensitivity and also to development of different dysglycemic conditions (like cardiometabolic syndrome and incident type 2 diabetes). The role of hs-CRP measurement in the current definition of the cardiometabolic syndrome to improve valid detection of risk for both diabetes and cardiovascular processes in patients is noteworthy. Furthermore, multiple clinical studies are now ongoing to evaluate whether traditional agents used to perk up glycemic control may also prominently reduce hs-CRP level (Ndumele et al. 2006).

Insulin resistance in type 2 diabetes mellitus inclines patients to develop cardiovascular disease with related risk factors for atherosclerosis like inflammation, dyslipidemia, hypertension, and altered hemostasis. CRP has a straight pro-atherogenic effect in upregulating angiotensin II type 1 receptors through the activation of some pro-inflammatory factors. Higher CRP concentrations in subjects with type 2 DM than nondiabetic subjects predisposes a greater role of inflammation in the hastening atherosclerosis process of these patients (Dandona 2008; Schmidt et al. 1999).

In apparently healthy individuals, higher levels of CRP predict the danger of developing type diabetes. CRP is a predictor of cardiovascular risk and an important participant in atherogenesis. Monocytes/macrophages in the inflammatory vessel wall secrete a significant amount of CRP. CRP is expressed in vascular cells and human atherosclerotic plaques. Various inflammatory and metabolic factors are related with diabetes, like high glucose level, modified lipoproteins, adipokines, and free fatty acids. These factors stimulate CRP secretion from smooth muscle cells, endothelial cells, and from monocytes/macrophages. Thus, the local CRP production and level in diabetic atherosclerotic plaques could be elevated than in nondiabetic people. Furthermore, the possible association between local CRP secretion and the amount of severity of coronary artery disease may attribute to the accelerated progression of vascular disease in type 2 diabetic patients (Mugabo et al. 2010; Zimmermann et al. 2014).

Diabetes has no appreciable relationship with different inflammatory markers like white blood cell and platelet counts, factor VIIIc levels, and concentration of albumin and fibrinogen. But diabetic individuals had greater median levels of CRP concentration at baseline than the normoglycemic individuals. Also individuals with elevated CRP levels were further likely to develop diabetes on follow-up. Thus, inflammation, as measured by CRP concentration, plays a significant role in the pathobiology in adults with glucose disorders like diabetes. This was confirmed by normal fasting glucose (FG) values of older individuals at baseline, and understanding the pathophysiology may lead to better organization and treatment protocol of glucose disorders among elderly (Ridker et al. 1998a, b; Hayashino et al. 2014a, b; Duncan et al. 1999).

High incidence of obesity, related insulin resistance, and type 2 diabetes mellitus are related with cardiovascular diseases. Adipose tissue produces multiple cytokines (TNF-alpha, IL-6), CRP, and molecules like PAI-1, angiotensinogen, leptin, adiponectin, visfatin, apelin, and resistin. These cytokines decrease insulin sensitivity endothelial dysfunction and atherosclerosis by inducing inflammatory processes. Resistin has a role in the procurement of inflammatory processes and endothelial dysfunction and is thus shown to be a predictive component of coronary artery disease and cardiovascular mortality (Karbowska et al. 2009).

9.4 Outline of Clinical Details of Patients

In our study two patients (designated as Patient 1 and 2) were included whose clinical history, laboratory investigations data, and photos were included in Table 9.3. The patients were admitted in MB Nursing Home, Park Circus, Kolkata, India. Patients 1 and 2 were admitted during their operation of perineal abscess (Patient 1) and laparoscopic cholecystectomy (Patient 2) in the hospital. Informed consents were taken from these patients and their relatives, and the clinical history was given as per the consent of the institutional review board of the hospital. Patient 1 was diagnosed with a case of perianal (or perineal) abscess in known diabetic individuals. He came to the hospital to have the severely painful and tender abscess operated. Patient 2 was admitted for laparoscopic cholecystectomy in a subject with known diabetes with very high sugar level. Both Patient 1 and Patient 2 were suffering from type 2 diabetes mellitus. Patient 2 has a cataract problem, a common complication of type 2 diabetes mellitus. So here, in the short span of this book, we focused on some biomarkers in different acute inflammatory conditions with their screening, diagnostic, and therapeutic applications in diabetes subjects.

Currently, there are no specific markers for inflammation; rather some broad spectrum inflammatory markers were routinely investigated in hospitals. The clinical investigation accounted the presence of some broad spectrum markers like

Table 9.3 Clinical history and laboratory parameters of diabetic patients

	Patient 1	Patient 2
Parameters	Physical parameters	
Age (years)/sex (M/F)/weight (kg)	52/male/73	64/male/62
Chief complaint	1. Fever for 3 days 2. Pain and tenderness in the anal region and he can't sit for 5 days	1. Intermittent pain abdomen for the last 3 months 2. Pain is increased after taking fatty meal
Past medical history	1. Hypertensive for 7 years 2. Type 2 diabetes mellitus for 4 years	1. Type 2 diabetes mellitus for 24 years
Past surgical history	1. Cholecystectomy, 7 years back	3. Appendicectomy, 18 years back
Birth history and vaccination	NA	NA
Drug allergy	Sulfur drugs	Not known till now
	Patient conscious, alert, cooperative	Patient conscious, alert, cooperative
GCS	15/15	15/15
Pulse/min	104	112
Temperature (°F)	101.5	98.1
Pallor	Nil	Nil
BP (mm of Hg)	152/91	130/80
JVP	Normal	Normal
Palpable lymph node	Inguinal lymph node palpable	No palpable lymph node
Hydration status	Mild dehydration	No dehydration
Clubbing	Not present	Not present
Chest (air entry)	Adequate	1. Bilateral equal good air entry
CVS (S1 and S2)	Audible	Audible
Abdomen	Soft	Abdomen soft, epigastrium and right hypochondrium tender on deep palpation
Intestinal peristaltic sound	Present	Present
	1. Perineal abscess, swollen (++), redness (++), tender (++)	1. Cataract in right eye
Hb (gm/dl)	13.4	11.2
TLC (cells/cu.mm)	22,400	13,200
DLC (%)	N (90), L (8), M (1), E (1)	N (86), L (10), M (2), E (2)
ESR (mm/first hour)	74	56
CRP (mg/L)	260	54
Procalcitonin (ng/ml)	12	ND
Blood culture (aerobic)	ND	ND
Pro-BNP (pg/mL)	ND	ND
Malarial antigen	Negative	ND
Dengue serology	Negative	ND
Blood sugar (mg/dl)	RBS 302	Glucometer revealed high RBS RBS 662
Glycosylated hemoglobin test (Hb A1c) (%)	9	12.1
Urea (mg/dl)	34	39
Creatinine (mg/dl)	1.2	1.3
Sodium (Na+) (mEq/L)	136	134
Potassium (K+) (mEq/L)	4.7	3.94
Calcium (serum) (Ca++) mEq/dl)	8.7	ND
Antinuclear antibody	ND	ND

(continued)

Table 9.3 (continued)

	Patient 1	Patient 2
Parameters	Physical parameters	
RA factor (above 20 IU/mL)	ND	ND
HLA (B-27)	ND	ND
PTT (sec) (Control-12 s)	14.2	13.1
APTT (sec) (Control-26 s)	33	27
Liver function test 1. Albumin (gm/dl) 2. Other liver enzymes	Within normal range	Within normal range
ABGA	ND	ND
D-7. Chest X-ray	Normal profile seen	Normal profile seen
D-8. USG of abdomen	ND	1. A big solitary stone (size 4 cm diameter) 2. Common bile duct not dilated, no stone is there 3. Pancreas normal in shape and texture
D-9. Cardiac function tests		
ECG	Sinus rhythm (RBBB)	Sinus rhythm
Echocardiography	LVH, no RWMA	No RWMA
EF (%)	62	64
D-10. Bronchoscopy and BAL/ET suction	ND	ND
D-11. Urine routine examination and culture sensitivity	ND	1. Pus cell 1–2/HPF 2. Culture not done 3. Ketone bodies positive
E. Treatment summary	1. Incision and drainage of perineal abscess done under general anesthesia 2. Blood sugar controlled by intravenous insulin infusion followed by subcutaneous insulin 3. Ramipril (5 mg orally/day) to control BP 4. Antibiotic piperacillin + tazobactam (4.5 g) for every 6 h and linezolid (600 mg i.v. every 12 h) 5. Analgesic injection, (a) paracetamol (1 g) i.v. after every 8 h, (b) tramadol (100 mg) i.m./SOS	1. Intravenous fluid by normal saline. 2. Sugar controlled by intravenous regular insulin infusion, and hourly CBG done, and titrated by regular insulin 3. Analgesic anti-emetic drugs and antibiotics given 4. After stabilizing the patient laparoscopic cholecystectomy done under general anesthesia 5. On discharge, patient was advised to take oral hypoglycemic agent and mixtard insulin and also advised to attend an eye specialist
F. Output	Successfully discharged	Successfully discharged

GCS Glasgow coma scale, *min* minutes, *°F* degrees Farenheit, *BP* blood pressure, *JVP* jugular venous pressure, *CVS* cardiovascular system, *TLC* total leukocyte count, *DLC* differential leucocyte count, *ESR* erythrocyte sedimentation rate, *CRP* C-reactive protein, *N* neutrophil, *M* monocyte, *L* lymphocyte, *E* eosinophil, *ND* not determined, *FBS* fasting blood sugar, *RBS* random blood sugar, *RA* rheumatoid arthritis, *HLA* human leukocyte antigen, *PTT* prothrombin time test, *APTT* activated partial thrombin time, *ECG* electrocardiogram, *RBBB* right bundle branch block, *WNL* within normal limit, *LVH* left ventricular hypertrophy, *RWMA* resting wall motion abnormality, *PAP* pulmonary artery pressure, *EF* ejection fraction, *BNP* brain natriuretic peptide, *μg* microgram, *mg* milligram, *pg* picogram, *i.v.* intravenous, *i.m.* intramuscular, *SOS* as and when required, *GTN* Glyceryl trinitrate, *CBG* capillary blood glucose, *NA* not applicable, *MDR* multidrug resistance, *ABGA* arterial blood gas analysis, *HPF* high-power field, *ET* endotracheal, *CBG capillary blood glucose*, *mEq* milliequivalents, *dl* deciliter, PaO_2 partial pressure of arterial oxygen, *USG* ultrasonography, *ST* the ST segment which represents the period when the ventricles are depolarized in ECG, *BAL* bronchoalveolar lavage, *DVT* deep venous thrombosis, *GFR* glomerular filtration rate, *CRRT* continuous renal replacement therapy, *CVP* central venous pressure

CRP, erythrocyte sedimentation rate (ESR), total leukocyte count (TLC), differential leukocyte count (DLC), sodium–potassium level, urea–creatinine level, human leukocyte antigen (HLA), pro-brain natriuretic peptide (pro-BNP), and antinuclear antibody for investigation.

A 52-year-old male (Patient 1) was presented with fever, soft abdomen, severe pain, and redness and tenderness in the anal region for 3 days. He had been diagnosed previously with diabetes (type 2) with high blood pressure. His laboratory findings revealed high TLC, DLC, and ESR and raised CRP levels. His blood serology report revealed negative for both malarial and dengue antigen. He had perineal abscess (as increased risk in diabetic patient). He had a palpable inguinal lymph node and an inflamed perineal region with fever. He had neutrophilic leukocytosis with high CRP and ESR. His procalcitonin level revealed absence of bacterial infection (Table 9.3). He was treated surgically by incision and drainage of perineal abscess under general anesthesia, and intravenous followed by oral antibiotics, tight control of sugar by regular insulin, adequate analgesic, and antihypertensive were given.

A 64-year-old male (Patient 2) was presented with an intermittent severe-pain soft abdomen for the last 3 months. He had been diagnosed previously with diabetes (type 2). The glucometer revealed high sugar level. His laboratory findings revealed high TLC, DLC, and ESR and raised CRP levels. His blood serology report revealed negative for both malarial and dengue antigen. He had a big solitary stone and had no obstruction in the bile duct. He had cataract in right eye (as an increased risk in diabetic patients). He had no palpable inguinal lymph node. He had neutrophilic leukocytosis with high CRP and ESR (Table 9.3). He was treated surgically by incision and removal of stone from the gallbladder under general anesthesia, and intravenous followed by oral antibiotics, tight control of sugar by regular insulin, and adequate analgesic were given.

9.5 Discussions

The association between diabetes mellitus and vascular dysfunction is a hot topic of research. Several circulating biomarkers have been proposed as markers of endothelial dysfunction (Table 9.1). Important biomarkers of the vasculature system are associated with endothelial function, coagulation and inflammatory processes, and different degrees of oxidative stress. These biomarkers could analyze the different pathophysiological aspects in subjects with diabetes mellitus and, along with the newly investigated clinical biomarkers, could provide a useful strategy for the treatment of these suffering patients (Tousoulis et al. 2013).

It is widely known that diabetes mellitus impairs endothelial nitric oxide synthase activity. In diabetes mellitus the production of reactive oxygen species enhances, and consequent pro-atherogenic alterations and thus resulting in diminished nitric oxide bioavailability. Important biomarkers of the vasculature are mostly related to oxidative stress, endothelial dysfunction, and inflammatory and coagulation markers of diabetes. Thus, by inflammatory and oxidative processes (Fig. 9.2), the endothelial integrity was damaged in diabetes mellitus. Migration of varied leukocytes into the vessel wall and increased secretion of cytokines resulted in systemic inflammation. The elevated levels of circulating cytokines and other changes enhanced oxLDL production by scavenger macrophages and result in foam cell formation. Foam cell formation is a critical step in atherosclerotic especially in diabetes mellitus.

The use of insulin analogues, statins, hypoglycemic agents, and antihypertensive agents might put forth favorable outcomes on the vasculature of diabetic individuals as the prominent role of oxidative stress is known. But still several therapeutic strategies employing antioxidant therapy remains contentious. Thus, the use of definite biomarkers associated to vascular function could be a fruitful therapeutic strategy applied in such patients (Tousoulis et al. 2013; Mohanty et al. 2002; Mohanty et al. 2000).

CRP and Cancer

10

Abstract

C-reactive protein (CRP) as an acute-phase protein is a marker in inflammation, tissue injury and infection. The role of inflammation and CRP in different cancers is a recent matter of research. Increased CRP level is positively correlated with extent of disease and recurrence in advanced cancer. Its role as an interpreter of survival has been elucidated in melanoma, multiple myeloma, lymphoma, ovarian, renal, pancreatic and gastrointestinal tumors. The testing for CRP levels at the time of diagnosis could help clinicians determine severity of the disease with any adjacent inflammation, the tumor's state of aggressiveness, or other conditions in the body and help guide therapeutics to improve overall survival. The poor prognosis in patients with different types of solid cancers is correlated with elevated plasma CRP levels. The circulating CRP concentrations has stronger correlation with the severity, extent and progression of many different tumor pathologies and the prognostic significance of these associations are consistent with CRP level. Thus CRP is not just being a marker of disease or an inflammatory biomarker but also it contributes to the pathogenesis of cancer also.

Keywords

Breast cancer • Myeloma • Lymphoma • Sarcoma • Carcinoma • Metastasis

Chapter Highlights

1. Some organs of the body when chronically inflamed showed greater risk of having cancer.
2. Higher plasma CRP concentration in colon cancer as compared to healthy subjects helps in recommending the dose of anti-inflammatory drugs.
3. Increased CRP level is positively correlated with different types of multiple myeloma, melanoma, lymphoma, ovarian, renal, pancreatic and gastrointestinal tumors.
4. In the diagnosis of breast cancer, the severity of the disease, breast cancer outcomes, and the treatment protocol was guided by plasma CRP levels.

W. Ansar, S. Ghosh, *Biology of C Reactive Protein in Health and Disease*,
DOI 10.1007/978-81-322-2680-2_10

5. The CRP levels were higher among individuals who went on to develop colon cancer, but not rectal or prostate cancer, compared with controls.
6. The association between circulating levels of CRP cancer prognosis and cancer biomarkers was evaluated in a number of researches.

10.1 Introduction

Tumor or neoplasm is a malignant transformation of a normal tissue containing mass of abnormal cells. It is usually seen under pathological conditions induced due to mutational changes in the DNA base pairs or genetic units either through chemical or physical mutagens. *Oncogenic* or tumor-inducing viruses such as *Polyomavirus* and SV 40 can transform normal cells into cancerous cells. They display a tumor-specific antigen on the surface of transformed cells.

10.2 Classification of Tumors

A tumor is defined as a mass of cells that develop due to uncontrolled growth and division of cells but they are localized and do not spread into the surrounding. They are also termed as benign tumor. The tumor that grows rapidly by multiplying cells and invades into the tissues and spreads in the healthy areas is termed as malignant tumor, and the process of spreading is termed as malignancy. A tumor with features of malignancy is termed as cancer.

Based on the tissue embryogenesis and origin in the tissues, malignant tumors, also known as cancers, are classified as follows:

1. *Carcinoma* is (>80 %) the tumor arising from ectodermal and endodermal tissues or epithelial tissue layers, lining the internal organs or glands. Most of the cancers originating in the prostrate (males), breast (females), colon, and lungs are grouped under carcinomas.
2. *Sarcoma* (<1 %) is the tumor arising from the mesenchymal tissues and connective tissues comprising bones, fats, and cartilages.
3. *Lymphoma* (9 %) is the cancer of lymphoid tissue that proliferates as solid tumor.
4. *Metastatic tumors* are a small cluster of cancer cells that dislodge from their site of origin, enter the circulation via the blood or lymphatic vessels, and are carried through the circulatory system to other distantly located healthy tissues, where they continue to proliferate, leading to the spread of cancer.
5. *Leukemia* is a cancer arising in any class of hematopoietic cells that tend to proliferate as single cells within the lymph or blood.

10.3 Tumor Antigens

Tumor cells express completely a new set of antigens called tumor antigens. These sets of new antigens are tumor specific and are known as *tumor-specific transplantation antigens (TSTA)* that arise on the surface of tumor cells considered as new and foreign to the host and tumor-associated transplantation antigens (TATA) which are qualitatively and quantitatively different on tumor cells and normal cells.

10.4 Tumor-Associated Antigens

Expression of tumor-associated antigens (TAA) is observed on both normal cells and those of a tumor. During development, as early as in the fetus, although the immune system remains immature, TAA are synthesized and expressed. As they appear during fetal development, they are named as *oncofetal tumor antigens*. The human carcinoembryonic antigen (CEA) and human alpha-fetoprotein (AFP) are among the oncofetal antigens most studied so far. This expression of TAA however is mostly restricted in the adult. In the normal cells, a very low expression of TSA may be observed, which overshoots and is overexpressed on the tumor cells in high level.

Human *AFP* is associated with liver cancer, viral hepatitis and in minority of patients with cancer of the stomach and colon. It was first demonstrated in the serum of mice and later found to be present in humans. It is present in high concentration in fetal blood (2–3 mg/ml) and a very low level in adult blood. Elevated levels of AFP are found in most patients suffering from liver cancer. Detection of elevated levels of AFP during 18–28 weeks of gestation during pregnancy is valuable in prenatal diagnosis of congenital defects and in early detection of high-risk pregnancy.

CEAs are glycoproteins antigens found on the surface of tumor of gastrointestinal origin and embryonic tissue. *CEA* is mainly associated with gastrointestinal malignancies. Significantly elevated levels are seen in patients with colorectal cancer as well as in patients with noncancerous diseases such as chronic lung diseases, pancreatitis, and cirrhosis of the liver. Levels above normal are seen in individuals who are heavy smokers. *CEA* is not a reliable marker for the presence of tumor but useful in monitoring the response to the therapy of colon tumors.

10.5 Tumor-Specific Antigens

TSA or tumor-specific antigens have been found to be present on tumor cells that owe its origin on tumors that are induced by exposure to carcinogens of both chemical and physical nature and also those induced by oncogenic viral infections. TSA was initially demonstrated in mice immunized with methylcholanthrene-induced sarcoma. The tumor-specific antigens were not the same in inbred strains. These two distinct antigens were known as tumor-specific transplantation antigens. Tumors induced by chemicals contain a new type of cell surface transplantation antigens called as tumor-specific cell surface antigens *(TSCSA)*. Both TSTA and TSCSA in chemically induced tumors have identical determinants. Virally induced tumors express same tumor antigens commonly expressed on all tumors induced on the exposure of the same oncogenic virus irrespective of the species of the animal or that of the tissue in which it is expressed. A herpesvirus (Epstein–Barr) isolated from a human lymphoid tumor called Burkitt's lymphoma has shown to contain same antigens even with histologically distinct tumors. DNA viruses such as *Polyoma, Papilloma,* and *SV40* are not oncogenic in their natural host but can transform cells and produce tumors in other species. Similar to the chemical induced tumors, the tumors induced by DNA viruses bear TSTA and TSCSA and embryonic antigens. These antigens are expressed intracellularly as predominant nuclear proteins. These TSA antigens are termed as *T* antigen and *U* antigen. Oncogenic RNA viruses produce tumors in their natural hosts. They produce tumors in a variety of hosts including birds (Rous sarcoma virus and avian leukemia virus), rodents (murine leukemia virus), and cats (feline leukemia virus) and in higher animals, like monkeys and apes. Many of the RNA viruses are part of the host genome and are transmitted by germ cells. Tumor cells infected with murine leukemia virus exhibit both cell surface and internal antigens. The most important is viral envelope antigen *(VEA)* specified by a viral genome and are expressed as a part on the cell membrane. Both *VEA* and viral-related cell surface antigens are detected by cell cytotoxicity test, immunofluorescence, and rejection of transplant tumor cells.

10.6 Cancer

Cancer is known as a deadly disease with features of continuous uncontrolled cell growth and proliferation coupled with metastasis. Cancers are classified into over 100 different types, depending on the origin and cell type from where it starts and is affected. Cancer can spread and affect the whole system, thereby showing features of systemic cancer and affect the entire system of the body including reproductive, circulatory, endocrinal, digestive, and other systems of our body. The affected organs can malfunction like releasing hormones that in turn can lead to alter body function. Localized tumors restricted to their site of origin are considered to be benign, while those that have metastatic potential are termed as cancer.

10.6.1 Cancer Causes and Progression

Cancer is single cell in origin that has lost the capability of controlled cell division. Several sequential steps are involved in the gradual transformation of a cancer cell from a normal one.

A transformation leading to the progression from a precancerous state to malignant state is a cumulative effect involving an individual's genetic components, also termed as genetic predisposition and external factors, including physical, chemical, and biological agents including:

1. Carcinogens of physical origin including ultraviolet (UV) rays and ionizing radiation.
2. Carcinogens of chemical origin like asbestos, tobacco smoke, excess alcohol, unhealthy diet, aflatoxin, arsenic, and drugs.
3. Carcinogens of biological origin including parasites or infectious agents like oncogenic viruses, bacteria, etc. Hepatitis B virus (HBV), hepatitis C virus (HCV), and human papillomavirus (HPV) are examples of viruses that are capable of increasing the risk for liver and cervical cancer, respectively. HIV infection has been reported to be a major risk factor in patients suffering from cervical cancer.
4. Biological factors like mutations and aging are also a fundamental factor for cancer progression. Cancer rises have been recorded to escalate with increase in age. This is attributed to the fact of gradual increased accumulation of the associated risk factors for specific cancers with age. Aging combined with fewer tendencies for effective cellular repair mechanisms adds to the progression of cancer.

Malignancy is the chief concern of cancer including the features that (1) a cancer cell undergoes *uncontrolled growth;* (2) growth of new blood vessels in the adjoining vicinity of cancerous growth that provide nutrition to the cancer tissue, by a process called *angiogenesis*; and then finally (3) *metastasis*, wherein the cancerous cell evades the circulatory systems, enters the blood or lymph, and moves throughout the body, establishing a fresh growth in a distant tissue, thereby completely causing a breakdown of the healthy tissue by invasion.

10.6.2 Cancer: The World Scenario

As per reports from WHO World Cancer Report 2014, cancer is a deadly disease, leading to death globally. The WHO, from a report in 2012, revealed that cancer was the cause of the global death in as high as 8.2 million people. Lung cancer contributing to deaths of 1.59 million people, liver cancer contributing to deaths of 7.4 million people, followed by stomach cancer causing deaths of 7.2 million patients, colorectal cancer leading to 6.9 million deaths, breast cancer causing deaths of 5.2 million patients, and esophageal cancer contributing to deaths of 4 million patients globally were the major cancer killers.

10.7 Tumor Immunity

Tumor immunity is both T-cell and antibody mediated. The antitumor antibodies are effective in destruction of tumors in the presence of complement. T-cell response involved in the destruction of tumors is dependent on following mechanisms:

1. Cellular immunity mediated by cytotoxic T lymphocytes (CTLs)
2. Humoral Immunity mediated by antibody-dependent cell-mediated cytotoxicity (ADCC)
3. Immune functions mediated by natural killer (NK) cells and lymphokine-activated killer cells (LAK)
4. Activated macrophages and neutrophils

10.7.1 Destruction by Cytotoxic T Lymphocytes

Destruction of viral-induced tumors is primarily mediated by cytotoxic T lymphocytes, which are MHC class II and I restricted. Since MHC class II molecules expressed by T-helper cells (**T$_H$**) cells do not recognize MHC class I antigens on tumor cells directly, they are dependent on major APCs like macrophages to present antigens in association with class II MHC molecules. This recognition by class II MHC molecule-restricted **T$_H$** cells leads to its activation and subsequent release of interleukin-2 (lL-2). IL-2 in turn activates cytotoxic T lymphocytes.

NK cells or the natural killer cells, B cells, and macrophages contribute to the immune response. The activated cytotoxic cells kill tumor cells directly.

10.7.2 Antibody-Dependent Cell-Mediated Cytotoxicity

ADCC involves (1) binding of tumor-specific antibodies to the surface of tumor cell, (2) interaction of various cells such as macrophages and granulocytes which possess surface receptors for the Fe portion of an antibody bound to a tumor cell, and (3) destruction of tumor cells by the substances released from these cells.

10.7.3 Destruction by Natural Killer Cells and Lymphokine-Activated Killer Cells

Natural killer (NK) cells are activated by IL-2 and interferon and specifically kill tumor cells that do not express class I MHC molecules on their surface. They can kill different target cells they act on including virally infected cells, antibody-coated cells, and cells from a number of different tumors. NK cells secrete TNF-α, which induces hemorrhagic necrosis; however, the exact mechanism(s) by which the NK cells recognize and kill tumor cells is still not clear.

LAK cells are tumor-specific killer cells obtained from a patient. When grown in the presence of IL-2, they acquire more killing potential. They can kill a much broader range of tumor cells.

10.7.4 Destruction by Activated Macrophages and Neutrophils

Macrophages and neutrophils become activated by cytokines, notably INF-y produced by activated T lymphocytes. These activated T lymphocytes and macrophages attract to the area of antigen and also release other cytokines. INF-y also prevents migration of macrophages away from the antigen. The activated macrophages are cytotoxic to tumor cells, to self-cells, and to microorganisms in their vicinity. The killing effect is due to TNF-α release and lysosomal enzymes. More evidences are emerging in the destruction of tumors by activated macrophages based on the following observations:

- *Resistance to tumor can be abolished by specific depletion of macrophages.*
- *Increased resistance to tumor is associated with increase in the number of activated macrophages.*
- *Administration of TNF-α into tumor-bearing animals causes hemorrhage and tumor necrosis.*
- *Activated macrophages are found at the site of tumor regression.*

10.8 Immune Surveillance

The transformed or tumor cells display new antigenic determinants that evoke an immune response to cause their destruction. This is known as immunological surveillance. This immune surveillance serves to limit the growth of transformed cells. This concept was first reviewed by *Ehrlich*, and later confirmed by *L Thomas* (1950) and *Burnett* (1970). At times the transformed cell goes unchecked by the immune system. This escape from the immune system of the developed tumors may be attributed due to the following:

- Tumor-specific antigens are not immunogenic, and the degree of immunity developed is insufficient to reject rapidly growing tumors.
- Tumor-specific antigens are shed from immune complexes denying access of sensitized lymphocytes in their recognition and attachment.
- Masking of tumor antigenic determinants by antibodies causing immunological enhancement.
- Prostaglandins produced by macrophages of tumor-bearing host have immunosuppressive

effects like reduction in MHC II antigens on APC and NKC activity.

- Development of suppressor T-cell activity induces poor immunity.
- Reduction in MHC II antigens on APC by prostaglandin
- Resistance to tumor can be abolished by specific depletion of macrophages.
- Increased resistance to tumor is associated with increase in the number of activated macrophages.
- Administration of TNF-α into tumor-bearing animals causes hemorrhage and tumor necrosis.
- Activated macrophages are found at the site of tumor regression.

10.9 Cancer Biomarkers

Cancer biomarkers have been implicated for their role in risk assessment, screening, differential diagnosis, identification of the disease state, prognosis, and monitoring of treatment. Tumor markers are either expressed on the surface of the tumor cells or tissues or are released in body fluids. They include different types of molecules and comprise (1) endogenous proteins or metabolites, (2) factors for transcription, (3) cell surface receptors, and (4) proteins that are secreted.

Their levels of expression or modifications in the cancer are indicative of disease (tumor/cancer) state, progression characteristics/state of the disease, and response to therapies. Tumor markers find great importance in enabling early detection and diagnosis of the disease, thereby reducing the chances of mortality in cancer by facilitating early diagnosis of cancers thereby enabling effective treatments. The last decade records a considerable surge in the number of new and effective cancer biomarkers utilized in the effective diagnosis and prognosis of the disease. In-depth knowledge of the disease dynamics has been possible due to the discovery of a considerable potential number of tumor markers from different cancers. A list of biomarkers is included in the following table (Table 10.1).

10.10 Cancer and CRP

C-reactive protein has been observed to play a dominant role in diagnosis and prognosis in a diverse range of cancer from localized to systemic cancer. CRP immunoreactivity in inflammatory hepatocellular adenomas has been correlated with a higher risk of malignant transformation. An elevated serum level of CRP has been observed to denote poor prognosis with poor cancer-specific survival in hepatocellular carcinoma (HCC) patients (Shin et al. 2015). The serum CRP level has been reported to correlate with the clinical prognosis in patients with kidney, penile, and metastatic castration-resistant prostate cancer (PC) (Schnoeller et al. 2015). In melanoma patients, CRP finds importance as a prognostic marker (Fang et al. 2015) and finds potential applications as markers in monitoring prognosis with treatment in melanoma patients. A high CRP level on postoperative day (POD) may be used to predict or detect site-specific infection in postsurgery after intraoperative radiotherapy (IORT) for spinal metastasis (Sugita et al. 2015).

Overexpression of CRP and its inflated levels are reported to significantly correlate the worse outcome in prostate cancer patients and the elevated CRP level has direct correlation with risk of death in prostate cancer and finds applications in predicting mortality (Graff et al. 2015). In colorectal cancer, elevated preoperative serum CRP levels have been observed to correlate with poor survival in patients suffering from colorectal cancer(Shibutanietal.2014).Chemoradiotherapy-treated patients suffering from advanced-level nasopharyngeal carcinoma reveal that serum CRP may find importance in predicting poor prognosis (Zeng et al. 2015). Elevated levels of serum CRP were reported to correlate significantly with overall poor disease-free survival rate in esophageal cancer (EC) patients (Huang et al. 2015). Elevated levels of hs-CRP have been shown to be correlated well with an elevated risk of breast cancer in nondiabetic female individuals (Wang et al. 2014). The elevated CRP level is found to indicate poor prognosis after treatment with radiotherapy in patients suffering from prostate cancer (Thurner et al. 2015).

Table 10.1 Cancer biomarkers [Bhatt et al. 2010; Carcereny et al. 2007]

Diagnostic and prognostic cancer biomarkers	Tumor or cancer	Gland/organ
Prostate-specific antigen (PSA)	Prostate cancer	Prostate gland in males
α-Fetoprotein (AFP)	Hepatocellular carcinoma (HCC)	Malignancy of the hepatocytes in the liver
Cancer antigen 125 (CA-125)	Ovarian and fallopian tube cancer	Ovary and fallopian tube malignancy in females
Cancer antigen 15-3 (CA15-3) breast cancer 1, early onset (BRCA-1); breast cancer 2, early onset (BRCA-2)	Breast cancer	Breast in females
Cancer antigen 19-9 (CA 19-9)	Pancreatic and urinary bladder cancer	Pancreas and urinary bladder
Carcinoembryonic antigen (CEA)	Colorectal cancer	Colon and rectum of the large intestine
Human chorionic gonadotropin (hCG)	Germ cell tumors (ovarian and testicular)	Ovary and testicular germ cells
Thyroglobulin (Tg) papillary	Follicular thyroid cancer	Thyroid
Heat shock proteins (HSPs) Hsp27; Hsp70	Gastric, prostate carcinoma, bone cancer (osteosarcomas), cancer of the uterus, cervix, and carcinoma of the bladder	Stomach, prostate gland, bone, uterus, cervix, and urinary bladder
Cytokines like transforming growth factor β (TGF-β)	Malignant tumors	Different tissues
Interleukins	Oral cancer	Mouth
Glucose metabolism	All cancers	Different tissues
Sialic acids	Endocrinal cancer	Endocrines
9-0-Acetylated sialoglycoconjugates	Childhood acute lymphoblastic leukemia (ALL)	Blood
Philadelphia chromosome, Bcl2, and other gene translocation fusion products	Acute myeloid leukemia (AML), acute lymphoblastic leukemia (ALL), chronic myeloid leukemia (CLL), myelodysplastic syndrome (MDS), and Burkitt's lymphoma	Malignancy of the blood cells
Methylation of genome	Ovarian cancer	Ovary of females
Adenomatous polyposis coli (APC)	Adenocarcinoma, squamous cell carcinoma, stomach carcinoma, pancreatic, thyroid, and ovarian cancer	Squamous cell, stomach, pancreas, thyroid, ovary
microRNA	Pancreatic cancer, esophageal cancer, nephroblastoma, prostate cancer, hepatocellular carcinoma, lung cancer	Pancreas, esophagus, prostate, liver, kidney, lung
mRNA	neuroendocrine tumors	Nervous system
Circulating tumor cells (CTCs)	Metastatic breast cancer	Breast
Cancer stem cells (CSCs)	Acute myeloid leukemia (AML), melanoma, cancer of the brain, breast, and prostate	Malignancy of blood cells, brain, breast, and prostate

Elevated serum CRP levels were reported to indicate the subsequent incidence of liver cancer and could be accounted for as a factor for death from chronic liver disease (Chen et al. 2015). Elevated CRP levels on postoperative day 3 have been reported to help physicians predict the post-operative course in both comprehensive clinical searches and therapeutic approaches, against postoperative infectious complications (PICs) during postoperative cancer resection in gastric cancer (Shishido et al. 2015). Meta-analysis suggests that an elevated CRP level has a detrimental impact on the overall survival (OS) in metastatic prostate cancer (MPC) (Rocha et al. 2014). CRPs were reported as a marker in the serum known for its sensitivity and easily detectable nature and useful for better clinical stratification in colorectal cancer patients (Lumachi et al. 2014). Increased hs-CRP levels in serum have been reported to contribute as a major risk factor in breast cancer within the cohort of younger Chinese women (Wang et al. 2014).

CRP has been studied for its efficiency as a diagnosis and prognosis marker in non-small cell lung cancer (NSCLC, Jin et al. 2014). Meta-analysis studies have revealed the probable role of inflammatory markers including CRP in pancreatic cancer management (Stevens et al. 2015). Serum concentration of CRP levels has been shown to effectively predict disease-induced mortality in patients suffering from renal cell carcinoma (Hsiao et al. 2015). In a study reported from a Japanese pancreatic cancer cohort, neutrophil–lymphocyte ratio and CRP have been reported to be of prognostic value (Inoue et al. 2015). Preoperative CRP has been reported to correlate in patients suffering from breast cancer with node negativity (Sicking et al. 2014).

Elevated pretreatment concentration of CRP in the serum is indicative of its association with poorer prognosis in esophageal cancer patients (Zheng et al. 2014) and together with IL-6 in unresectable NSCLC (Liao et al. 2014) shows potential as a prognostic marker to monitor clinical outcome (Zheng et al. 2014). The elevated serum CRP level in predicting prognosis of patients with metastatic cancer of the prostate shows its potential as an independent predictor of worse overall survival in these patients (Xu et al. 2015). The Multinational Association of Supportive Care in Cancer (MASCC) risk score combined with the mean CRP value has been reported to successfully identifying patients suffering from febrile neutropenia and other hematological malignancies with higher death risk (Combariza et al. 2015). Elevated CRP levels in the preoperative serum CRP have been reported as a biomarker for prognosis in patient posttreatment with radical nephroureterectomy suffering from upper urinary tract urothelial carcinoma (Aziz et al. 2014; You et al. 2015) and are indicative of poor survival in chronic hemodialysis patients undergoing nephrectomy for renal cancer (Szturmowicz et al. 2014; Omae et al. 2015).

CRP in mice has been reported to downregulate the expression of tumor-associated M2 macrophages and thereby reduce the effects of intratumoral angiogenesis (Kuribayashi et al. 2014). Together with IL-6, CRP has potential in predicting risk of patients suffering from colorectal cancer (Zhou et al. 2014). In a study reported from the European Prospective Investigation into Cancer and Nutrition (EPIC), high blood concentrations of CRP have been reported to indicate elevated risk in colorectal cancer patients (Nimptsch et al. 2015).

In a recent study, CRP and Wnt signaling pathway interaction has been reported in colorectal cancer (Su et al. 2014).

CRP is known for its importance as a biomarker in indicating prognosis and has been reported from metastatic renal cell carcinoma undergoing molecular targeted therapy treatment with intermediate risk (Teishima et al. 2014).

Associations between pain and inflammatory processes in patients suffering from head and neck cancer (HNC) have been indicative of potential implications for future treatment strategies (Oliveira et al. 2014). An elevated CRP level and hypoalbuminemia have been reported to influence the survival in inoperable esophageal cancer patients under palliative treatment in advanced stages (Lindenmann et al. 2014).

An elevated CRP is a practical prognostic marker in patients with hepatocellular carcinoma (HCC, Kinoshita et al. 2014). Inflammatory biomarker CRP also finds importance in its association with radiation therapy-induced early adverse skin reactions (EASRs), together with obesity (Rodriguez-Gil et al. 2014). A contrasting study has also reported that elevated serum CRP, CEA (carcinoembryonic antigen), and mediastinal lymph node metastases also termed as N2 disease are poor indicators of prognosis in patients suffering from non-small cell lung cancer (Ni et al. 2014). In patients suffering from a disease termed as diffuse B-cell lymphoma, the CRP level finds significance as a biomarker of prognosis and a biomarker indicating disease-free survival and improves the ability of prediction (Troppan et al. 2014). A meta-analysis reveals that CRP has a prognostic significance in urological cancers (Dai et al. 2014).

CRP has been shown to have positive effects in identifying high-risk groups of patients with colorectal cancer screening and those who might benefit most from prophylactic anti-inflammatory therapy (Goyal et al. 2014).

CRP with slight genetic variation has been reported to find importance due to its predictive potential of enhanced risk prostate cancer (Markt et al. 2014). Preoperative serum CRP has also been reported of its implications in significant prognosis of stage IV colorectal cancer patients (Shibutani et al. 2014, 2015) and osteosarcoma (Yi et al. 2014).

10.11 Treatment

Types include (1) chemotherapy, (2) immunotherapy, (3) radiation therapy, (4) targeted therapies, (5) transplantation, and (6) other treatment methods like use of angiogenesis inhibitors. Cancer vaccines, cryosurgery, hyperthermia, laser treatment, and photodynamic therapy for cancer are other modes of therapy in cancer.

10.11.1 Immunotherapy

Immunotherapy, including both active and passive forms, includes interplay of the nonspecific and specific immune reactions (Table 10.2).

Immunopotentiating agents including biological response modifiers (BRMs) are capable of enhancing antitumor immunity. These include bacterial products, synthetic chemicals, and cytokines (Table 10.3). They are reported to function by activating macrophages and natural killer (NK) cells, leading to downstream cytokine production and enhancing T-cell functions.

In cytokine therapy, cytokines are being employed to potentiate the host immune system (Table.10.4).

Table 10.2 Immunotherapy [Schuster et al. 2006]

Active immunotherapy	Nonspecific	Bacillus Calmette–Guérin (BCG), levamisole, cytokine genes
	Specific	Tumor cell lysate, killed tumor cells, recombinant antigens, idiotype, stimulatory molecule genes
Passive immunotherapy	Nonspecific	Lymphokine-activated killer (LAK) cells, cytokines
	Specific	Individual antibodies or those coupled to drugs, prodrug toxins or bispecific antibodies, T cells
	Combined	LAK cells and bispecific antibody

Table 10.3 Active immunotherapy (nonspecific) mediated by biological response modifiers (BRMs) [Reang et al. 2006]

BRM categories	Types	Effects
Bacterial product	BCG, acnes, muramyl dipeptide, trehalose dimycolate	Activation of macrophages and NK cells mediated by cytokines
Synthetic molecules	Pyran, poly I:C, pyrimidines	Induction of interferon (IFN) production
Cytokines	Interferon-α, interferon-β, interferon-γ, IL-2	Activate macrophages and NK cells

Table 10.4 Cytokines and immunotherapy

Cytokine	Tumor/cancer	Antitumor effect
IFN-α, IFN-β	Hairy cell leukemia in remission stages	Overexpression of class I MHC molecules
IFN-γ	Peritoneal carcinoma in remission stages	Overexpression of MHC class I and II molecules; macrophages, activation of cytotoxic T cell and NK cell
IL-2	Renal carcinoma and melanoma in remission stages	Proliferation and activation of T cells, activation of NK cells
TNF-α	Can reduce malignant ascites	Macrophage and lymphocyte

Role of CRP in Malaria

11

Abstract

The worldwide situation of malaria is alarming. Increasing new cases of transmission of malarial infections in areas where it had not occurred before and the tendency to reinfect the areas from where it had been eradicated earlier remain the major concern with malaria. The mechanism of resistance to antimalarial drugs has increased; there are cases of mixed and single infections, emergence of malarial infections in non-endemic areas, and complication of infections in endemic areas. The vector resistance to insecticides has also become widespread. Nearly 220 million new infections annually are reported to occur globally. It is mostly endemic in continents of Africa, Asia, and Central and South America. Malaria turned out to be a major threat to public health today in more than 90 countries. C-reactive protein (CRP) is an acute-phase reactant and also a biomarker for inflammation. CRP can bind to different immune cells and also can activate various immune cells in various downstream signaling pathways. It plays various immunomodulatory roles and acts as a key molecule in the immune system. It can bind to a few kinds of Fcγ receptors present on the cell surface of various immune cells. The plasma concentration of CRP increases in different diseases during the inflammatory conditions. Its concentration also varies in acute versus chronic conditions. The role of CRP as an inflammatory biomarker and as an index/biomarker in a disease state,or its binding to different cells subsequently modulates the function of those cells. The various immunomodulatory roles of CRP in malaria are detailed in this chapter.

Keywords

Plasmodium • Cerebral malaria • Inflammation • Biomarker • Interleukins • Cytokines • Tumor necrosis factor • *Anopheles* • Parasite • Africa • Fatal • Prevalent • Chloroquine • Antimalarial • Drug resistance

W. Ansar, S. Ghosh, *Biology of C Reactive Protein in Health and Disease*,
DOI 10.1007/978-81-322-2680-2_11

Chapter Highlights

1. Malaria poses a major threat to life.
2. The infection is caused by parasites (*Plasmodium* sp.) that are transmitted by infected bites of *Anopheles* mosquitoes, also called as "malaria vectors." They bite mainly between dusk and dawn.
3. In 2013, malaria claimed around 584 000 lives from children in Africa. It poses risk to half of the global population but mortality rate is recorded highest in sub-Saharan Africa.
4. Malaria can be cured and prevented.
5. The malaria prevention steps and disease control measures can reduce malaria burden.
6. Non-immunized travelers from malaria-free areas are vulnerable to the disease or infection.
7. Children are mostly affected in Africa, with single death recorded every minute. However, since 2000, the pediatric mortality rates have reduced by nearly 58 %.
8. CRP recognizes host cells that are damaged which leads to their elimination by (1) complement activation (2) or opsonization after ligand binding. CRP promotes lysis of diseased RBC and functions as an innate immune molecule. CRP also plays a major beneficial role in complement-independent phagocytosis.
9. CRP is a sensitive indicator of the early phase of inflammation or process of destruction of tissue; CRP testing has therefore been advocated. For the general assessment of well-being in both human and veterinary screening studies, CRP levels form (1) a simple measure of extent of severity of the disease, (2) the efficiency of therapy, (3) the diagnostic and prognostic significance, and the (4) severity of complications.
10. Although CRP is a nonspecific inflammatory biomarker, a very high CRP concentration is measured during attacks of malaria. A strong association between CRP levels and malaria parasite densities was revealed in various studies. The high CRP levels associated with high levels of malaria parasitemia might provide a useful case definition for clinical malaria in areas where asymptomatic parasitemia is common.
11. This definition would be most appropriate in the assessment of morbidity in the community. High CRP levels could represent a more objective definition of morbidity, as an indicator of inflammation, and are more sensitive than body temperature measurements.
12. However, as an indicator of clinical malaria, its utility in a given setting depends both on the malaria-attributable CRP level being measurable against a background of increased CRP levels from other causes and upon the raised CRP concentration showing clear association with clinical malaria and not with asymptomatic parasitemia.

11.1 Introduction

Malaria is the one of the most common globally occurring diseases affecting 515–600 million people annually. Globally 40 % of the population is revealed to be at risk of malaria infection. Malaria is turning out to be a worldwide menace. The situation is deteriorating throughout the world. The causative agent of malaria is *Plasmodium* parasites transmitted through the bites of infected *Anopheles* mosquitoes (Malaria fact sheet N°94. www.who.int/mediacentre/factsheets/fs094/en/).

The five different types of malarial parasites including *P. falciparum, P. vivax, P. ovale, P. malariae*, and *P. knowlesi* are known to infect humans of which *P. falciparum* and *P. vivax* are the major vectors. *P. falciparum* infection is recorded as the most dangerous form, with highest incident rates of complications and mortality. The severest form of malaria infection is attributed to *Plasmodium falciparum*, but some severe form can also be caused by *P. vivax* and *P. knowlesi* (Malaria fact sheet N°94. www.who.int/mediacentre/factsheets/fs094/en/). The three severe complications of malaria are noted in Fig. 11.1.

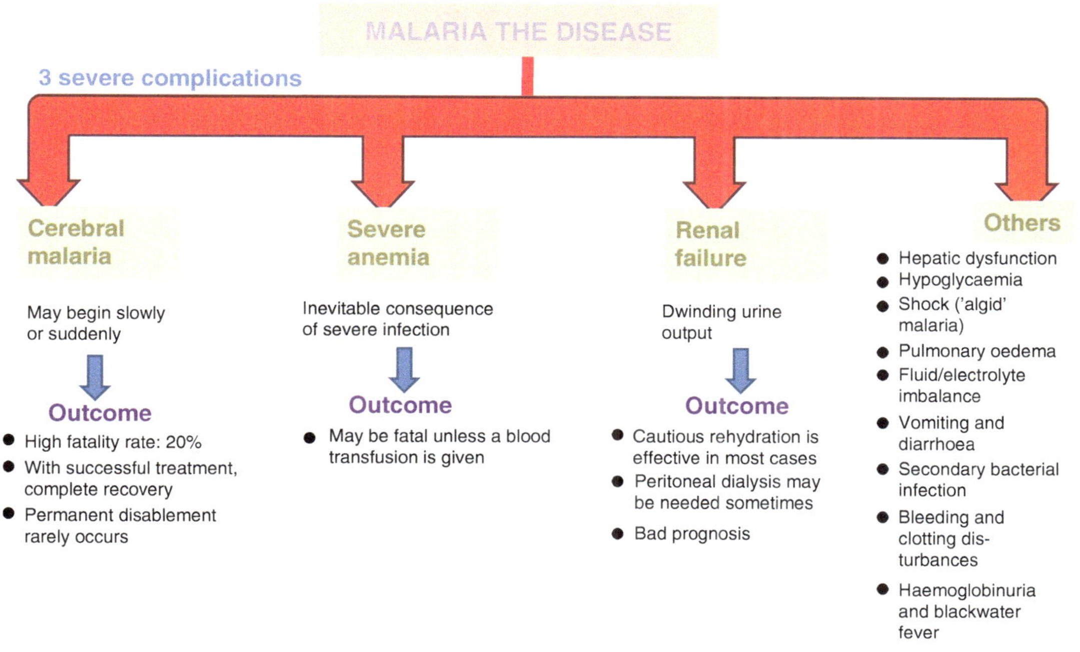

Fig. 11.1 *Complications of malaria*. Apart from the three severe complications of malaria, other complications were also noted with their outcome also

Also, *Plasmodium knowlesi*, mostly known to infect Southeast Asian long-tailed macaques, is also capable of infecting humans apart from the four known species of malaria. *Plasmodium* species of clinical importance in transmission are *P. falciparum* and *P. vivax*. In sub-Saharan Africa, *P. falciparum* claimed about one million lives annually, mainly children below 5 years. Eastern and central parts of Africa, South America, and Asia show prevalence of *P. vivax*. *Plasmodium cynomolgi*, also a simian malaria parasite, can transmit malaria to humans. *Plasmodium cynomolgi* is phenotypically and phylogenetically closer to *P. vivax*. It serves as a biological model system for the study of *P. vivax* (van Duivenvoorde et al. 2010; Cox-Singh et al. 2008; Schmidt et al. 2008).

The levels of transmission are increasing and there are some regions where it has returned even after its previous eradication. The vector is increasingly turning out to be resistant to insecticides, and with also increasing phenomenon of drug resistance, some multidrug-resistant (MDR) variety is coming out. There are increasing cases of mixed infections along with single infections. There are so many strategies, plans, and programs taken by different countries to restrict the geographical areas affected by malaria. But on the contrary, in recent years, due to changes in global climate, land use, easy international travel, and development of MDR strains of parasite, malaria transmission has become widespread. An estimated 219 million people are infected with malaria infection annually across the globe. Nearly, 660,000 African children died in 2010, with an uncertainty range of 490,000–836,000 cases (Gema Ruíz López del Prado et al. 2014).

Different drugs are being tested for malaria treatment, but not a single drug is able to overcome the burden. In endemic areas, one of the major concerns for malaria infection is its resistance to its oldest, cheapest treatment product, i.e., chloroquine (Talisuna et al. 2004). Along with chloroquine resistance in most affected areas, resistance to sulfadoxine pyrimidine and mefloquine is emerging. MDR varieties are mostly observed in Southeast Asia. The drug artemisinin (Afonso et al. 2006), although effective in treatment of malaria, produces resistant strains of *P. falciparum*. The pathophysiology of the disease mechanism is still not clearly understood, and therefore the factors contributing to generation of drug-resistant strains still remain to be addressed. The key players in the physio-pathogenesis of malaria have been implemented to present free radicals through oxidative stress (Wilmanski et al. 2005; Narsaria et al. 2012). The discovery of biocompatible polymeric nanoparticles has attracted considerable attention in pharmacology. In malaria treatment, the delivery of a nano-drug seems to be a promising and viable approach (Swai et al. 2008). Experimental nano-drug has been trialed in animal models since comparable life cycles are found in most mammalian *Plasmodium* species and show sensitivity to the similar groups of drugs. The use of drugs in experimental rodent models could be potentially trialed for human malaria or may provide additional information about their mode of action in humans. So a thorough in vivo drug screening system is the need of the hour to test the efficacy of different antimalarial drugs (Peters et al. 1975).

Malaria is a curable and preventable disease. The symptoms of malaria infection make their appearance almost after a week. But without treatment, the disease can lead to death. To reduce malaria transmission, the main way is the prevention of mosquito bites. The most susceptible group includes pregnant women, children, and immunosuppressed travelers also. Preventive measures and chemoprophylaxis recommendations for transmission of malaria infection include knowledge of travelers' age, destination, type of traveling, or his/her length of stay. The rates of malaria infection of travelers without immunization from malaria-free countries into malaria-infested areas are very high when they get infected. Areas showing high transmission rates of malaria are a major threat to travelers, particularly Papua New Guinea, sub-Saharan Africa, and islands of South Pacific areas (Lopez et al. 2014).

CRP shows promise as an early defense system against (1) malarial infection and (2) acute inflammation. CRP concentration escalates about 50,000 times as compared to that of normal individuals within 6 h of infection and reaches the peak at 48 h in infected individuals. As an accurate inflammation indicator, the production of CRP is only known to be affected in cases of liver failure. CRP is well characterized as an acute-phase reactant in the serum and plays a role in multiple immunoregulatory functions like (1) classical complement cascade activation, (2) bacterial opsonization and eventual phagocytosis, and (3) phagocytic cell stimulation (Dong and Wright 1996).

Although predominantly produced and secreted by liver hepatocytes, CRP is also produced by (1) lymphocyte subsets, (2) Kupffer cells, and (3) blood monocytes (Dong and Wright 1996). It was first identified for its role in defense against human infection and property to bind to phosphorylcholine on microorganisms' membranes (Gotschlich and Edelman 1967).

Pro-inflammatory cytokines released by host mononuclear cells induce CRP secretion in malaria (Harpaz et al. 1992) and its elevated level shows strong correlations with parasitemias.

CRP, regulated at the transcriptional level and induced by cytokines (Kremsner et al. 1996; Ablij and Meinders 2002), is a marker of inflammatory reactions, and cytokine activation (Jakobsen et al. 1998) is produced as a consequence to infection (McCarty 2004). It also leads to the expression of different adhesion molecules at inflammation sites. Elevated CRP levels, as a response to acute events, bear a strong positive correlation between the duration and the intensity of the stimulus and the number of hepatocytes synthesizing CRP (Ablij and Meinders 2002).

This chapter will encompass issues related to epidemiology of malaria, immune responses in the infection, special immune responses in the presence of CRP, and the detailed role of CRP in malaria infection.

11.2 Epidemiology of Malaria

Worldwide, the cases of new infections every year have been estimated to be nearly 220 million. In Asia, Africa, and Central and South America, the disease is endemic. Nearly 40 % of the population globally suffers from malaria. In today's perspective, it has turned out to be a major public health issue in nearly 90 countries, with a number of infections of about 240 million people worldwide. Each year, from around 300 to 500 million cases reported of infection, 1.5 to 2.7 million people undergo death. Children are the worst sufferers. In Africa, children under the age of 5 died quite often from malaria infection. In our world, in every 30 s, malaria kills one child. The endemicity of malaria infection is mostly seen in the tropics. According to the World Health Organization (WHO) report, 2012, nearly 3.4 billion people globally are annually exposed to malaria infection. Of the 3.4 billion people, approximately 1.2 billion individuals are at high risk. Nearly 34 malaria-endemic countries were estimated by WHO in 2012. It was also stated in the 2012 report that more than 207 million people, in 2012, developed symptomatic malaria. The curve of malaria infection worldwide is very fluctuating. Between 2000 and 2010, the number of reported annual malaria cases decreased by 85 % in 2004 with a height of 1.82 million people which lowered in 2010 to 1.24 million people. During this period, most of the deaths, i.e., nearly 80 % of the deaths, occur in sub-Saharan Africa. Nearly 627,000 people died of malaria infection in 2012, with an uncertainty range of also 473,000 to 789,000. Out of the 219 million cases of illness, the estimated deaths each year due to malaria infection are 660,000 individuals (Lopez Del Prado et al. 2014). Malaria continues to be the major killer globally, with nearly 250 million clinical cases reported per annum (Global status report on noncommunicable diseases 2014 (2012)).

In low-malaria-incidence areas of Panama, parasites were found to be clustered into three clonal subpopulations whose chloroquine resistance genotype matches the haplotype of Colombian origin. Thus resurgent parasite populations are highly clonal and are likely resulted from imported or vestigial cases of epidemic expansion. Genetic tools can be used to guide strategies to prevent further resurgence in areas of malaria elimination (Obaldia et al. 2014).

In India, malaria is one of the prevalent diseases and it is endemic in few areas also. It is endemic in the state of Arunachal Pradesh, India. From a retrospective surveillance study in 1995 to 2012, it was seen that *P. falciparum* caused 17.7 % and *P. vivax* caused 80.8 % of total malaria cases. But the prevalence rates of infection by *P. falciparum* remained constant while infection by *P. vivax* declined significantly during the study period. The decline of prevalence rates may be implemented to constant vector and disease management programs. The seasonal prevalence of malaria transmission was also noted in this malaria epidemiological study (Mutheneni et al. 2014).

In the Adana province of Turkey, a malaria-endemic area, between 2002 and 2012, 252 patients were diagnosed with malaria, where 229 patients had *P. vivax* infections, and *P. falciparum* was detected in about 23 (8.1 %) patients. This epidemiological study was aimed in surveillance studies without any interruption in malaria treatment to prevent reemerging of malaria in this region (Kuşcu et al. 2014).

People of the Fulani ethnic group are less susceptible to malaria infection than other African populations due to their difference in genotype from other Africans. The differential plasma concentration of CRP could be attributed to its polymorphisms in the promoter region of the *CRP* gene. The low susceptibility of malaria in the Fulani ethnic group is contributed toward the polymorphisms in the CRP gene. This study indicates the CRP's major role in the malaria immune responses by the host (Israelsson et al. 2009).

The highly polymorphic *CRP* gene has SNPs at different loci. The triallelic −286 (C>T>A) SNP (rs3091244) is strongly correlated with the CRP concentration in the plasma. The higher basal CRP concentration was found to be associated with allele-A. The distribution genotypes and alleles of *CRP* are ethnically and geographically different. Earlier researches reveal that allele-A in Africa is more common than in white populations, and it is hypothesized that allele-A can confer a survival advantage to the individuals of malaria-endemic areas. Three SNPs, −286 C>T>A, −717 T>C, and +1444C>T, were identified from a village in eastern Sudan. The −286 SNP differentially varies in different ethnic groups and is strongly correlated with varying susceptibility to malaria (Giha et al. 2010a, b, c, 2011; Morita et al. 2007; Israelsson et al. 2009; Kovacs et al. 2005; Kathiresan et al. 2006).

The role of nonspecific infection and inflammation markers linked with malaria infections is a major topic of research. The basic hematologic changes of these nonspecific markers and their contributions during the disease and after treatment remain unclear. In Brazilian endemic areas, *Plasmodium vivax* infection alone contributes to more than 80 % of malaria cases; the role of these nonspecific markers and their diagnostic values as hematological parameters has not been well established. The changes in neutrophils, platelets, CRP, and NO in adults with uncomplicated *P. vivax* and *P. falciparum* malaria from the Brazilian Amazon were well researched during the acute and convalescent phases of infection (Oliveira-Ferreira et al. 2010).

Malaria poses a disease burden in different communities, mostly in sub-Saharan Africa and other equatorial regions. Healthy African children with higher plasma levels of CRP as compared to control group of children from Europe and Papua New Guinea might indicate that the continuous cyclical malaria infection retains always a clinical level of CRP in African populations (Giha et al. 2010a, b, c; Imrie et al. 2007; Hurt et al. 2006).

Plasmodium falciparum infection contributes to 10–14 % mortality rate and is known as cerebral malaria (CM). Nearly 1–2 million deaths per annum in young children predominantly in sub-Saharan Africa and Southeast Asia were reported due to CM (Imrie et al. 2007).

Proper documentation of the epidemiology and burden of malaria is far from complete. Measurement of malaria-related mortality suffers difficulty as the disease symptoms are highly nonspecific. Drugs ineffective against potentially lethal malaria lead to increased mortality. During the year 2000, nearly 2 million cases were reported in India, of which 1.04 million (50 %) were due to *P. falciparum* infection (Imrie et al. 2007).

The strong association between CRP levels and parasite load in malaria in pediatric patients was investigated in the Kilombero district of Tanzania, where perennial *P. falciparum* is transmitted intensely. The relationships between CRP levels with clinical morbidity or with *P. falciparum* parasitemia only were analyzed to understand the effectiveness of CRP as a marker in field studies on malaria morbidity in areas where malaria is endemic. The study was conducted taking a few communities of Western Europe as the control group (Hurt et al. 1994a, b).

The estimated number of deaths from malaria, the most important parasitic disease in 2010, was 1.2 million, especially in children below 5 years of age. Nearly annual visit of 125 million of travelers to malaria-endemic regions is recorded and interestingly more than 10,000 return with malaria. For nonimmune travelers, *P. falciparum* malarial infection is a medical emergency and early diagnosis is important, in order to prevent complications during rapid initiation of treatment with antimalarial agents (Stauga et al. 2013; Murray et al. 2012; WHO 2011a, b).

P. vivax, the most widespread *Plasmodium* species, lead to almost 400 million infections per annum and cause the majority of malaria cases within the Brazilian Amazon,where there are high rates of asymptomatic infection. *P. vivax* malaria infections with severe complications have been reported in the Amazon region. Recent studies in Melanesian populations show positive association of *P. vivax* malaria with severe complications and mortality rate. Along with *P. falciparum* malaria, with records of drug resistance worldwide, the complications of *P. vivax* infection are turning out to be a global health menace (Andrade et al. 2010; Price et al. 2007; da Silva Jr 2006; Alves et al. 2002; Ladeia-Andrade et al. 2009; Barcus et al. 2007; Genton et al. 2008).

The co-association pattern of CRP and NO is studied in malaria patients and in control individuals from NW and SE malaria-endemic regions of Iran (Nahrevanian et al. 2008).

CRP genotypes and malariometric indices were studied from separate groups of sympatric ethnic populations from countries like Mali and Sudan, where malaria endemicity is markedly different (Israelsson et al. 2009).

From 45 countries, more than 2 billion people are inhabitants of malaria-endemic areas. Of them, around 50 %, i.e., 1 billion people, are reported to carry parasites at some or another point in time. Per annum malaria mortality rate ranges from 0·5 to 3·0 million people (Marsh 1998).

A total of 2547 women in rural northern Tanzania were screened for blood malaria parasites and hemoglobin (Hb) and urine infections. The cross-sectional study was carried out to find out the relation between anemia during pregnancy and (1) nutritional characteristics like (2) iron content, (3) level of ferritin, (4) total iron-binding capacity (TIBC), (5) cobalamin, (6) folate, (6) vitamin A, and (7) infections (raised concentrations of CRP) (Hinderaker et al. 2002).

In western Nigeria, endemic malaria is reported from 12 to 36-month-old children. In 1999, around Ile-Ife in Nigeria, 207 children (aged 0–60 months) presented with fever. Their medical and anthropometric data, body temperature, plasma CRP, retinol, carotenoids, tocopherols, and parasitemia were measured (Cooper et al. 2002a, b).

11.3 Immune Responses in Malaria: Brief Pathophysiology

Immunological responses in malaria infection involve a complex intertwined circuit of cells, pro- and anti-inflammatory cytokines, and also some regulatory mechanisms (Fig. 11.2). The immune system is always fighting to bar the

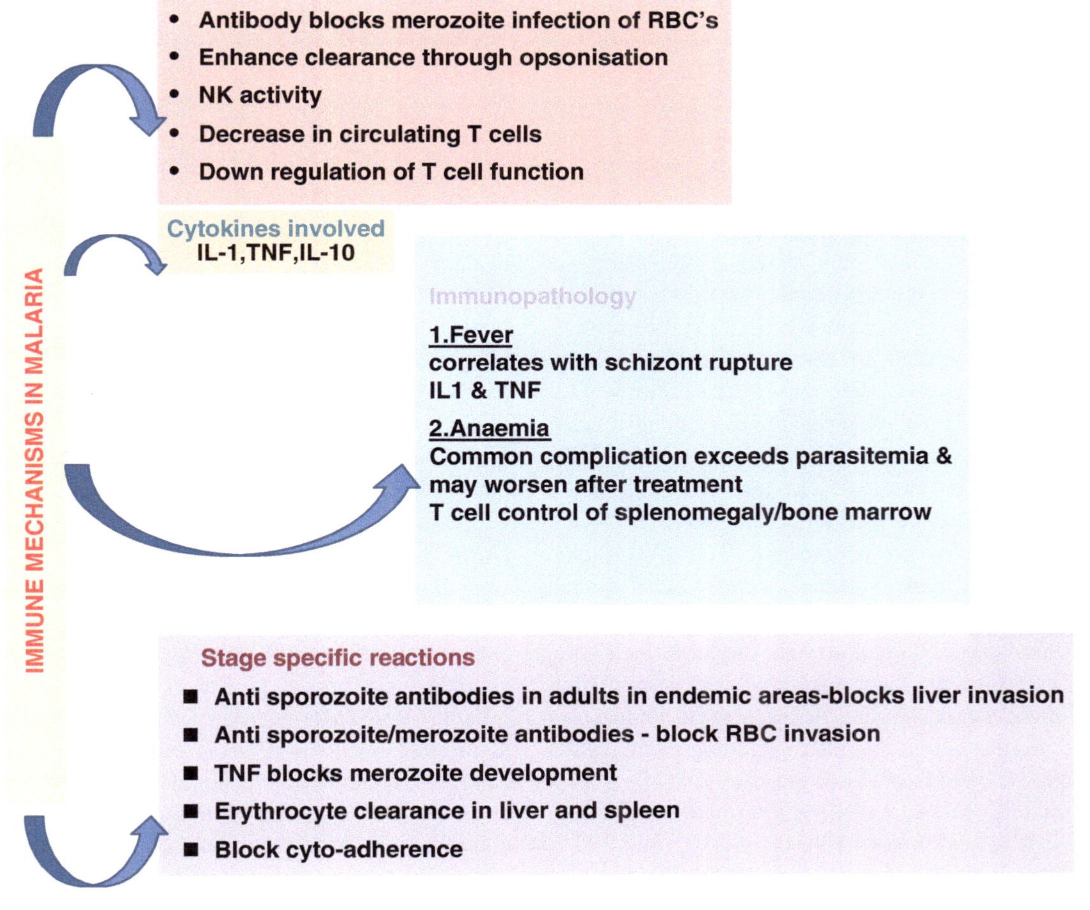

Fig. 11.2 *Immune mechanisms in malaria.* It involves a network of cytokines, immune cells, and different stage-specific reactions

constant inflow of pathogens in its own territory. To avoid excessive collateral, for efficient control of pathogens, a potent arsenal along with a tight regulatory mechanism is required. Thus, to maintain the homeostasis and equilibrium forms the major challenge of the immune system. Pathogens and their interaction with antigen-presenting cells (APCs) and lymphocytes maintain the equilibrium. Immune protection is conferred by anti-inflammatory cytokines secreted by T cells and APCs (Bachmann and Kopf 2002). A low plasma ratio of IL-10 which is a Th2 cytokine and TNF which is a Th1 cytokine TNF is observed in the pathophysiology of severe malarial anemia (Wilmanski et al. 2005).

The role of pro-inflammatory cytokines can be triggered by the pathogenic mechanisms of the parasite (Mitchell et al. 2005) or may be contributed to the production of free radicals in host cells in response to infection (Keller et al. 2004). The primary infection of malaria can be controlled by the rapid pro-inflammatory responses in innate immunity. But the rapid and potent development of cytokine secretion can increase the severity in malaria or may amplify the adaptive immune response (Riley 1999). Acquired immunity to malaria enables the control of circulating pro-inflammatory cytokine concentrations thereby leading to parasite clearance without triggering any deteriorating pathology. The mechanism by which acquired immunity maintains its constrained behavior is not known. The downstream effect of Th1 and Th2 cytokines in humans is controlled and regulated by IL-10 and transforming growth factor-β (TGF-β). In innate immunity, both cytokines are produced by innate immune cells like macrophages, and in acquired immunity, the cytokines are produced by T cells. Thus, the regulatory immune responses of both these cytokines are maintained in innate and adaptive immunity also (Schmieg et al. 2003). IL-10 cytokines are produced by CD4+ T cells and all leukocytes (Niikura et al. 2011).

The cytokines TNF-α, IFN-γ, and IL-12 secreted by T_H lymphocytes help in controlling the infection, while Th2 cytokines like 1 L-10 and TGF-β aggravate the disease. However, in different immune responses against parasite elimination, the consensus of these cytokine profiles varies (Percário et al. 2012).

In the first stage of parasite elimination, at the initial stages of infection, the host mounts an adaptive immune response against the parasite. However, a chronic or exaggerated condition may favor the severe anemic conditions. At the second stage, the host promotes clearance of the initial immune responses (Th1) and replaces them by adaptive immune response. With obvious consequences and improper activation of these responses, the parasite tends to resist the host defense (Percário et al. 2012). Recent studies revealed relatively low levels of pro-inflammatory cytokines (TNF-α, IFN-γ, and IL-12) and high-level anti-inflammatory cytokines (IL-10 and TGF-β) after malaria treatment (Percário et al. 2012; Imrie et al. 2007).

Malaria infection has been known to be associated with hematologic changes, like anemia, thrombocytopenia, and leukocytosis or leukopenia. These clinical factors alter with patient genetic background, the level of malaria endemicity, demographic factors, nutritional status, and immunity in malaria (Price et al. 2001; Erhart et al. 2004).

Altered numbers of white blood cells and platelet count are well documented in both *P. vivax* and P. *falciparum* malaria cases in immune and naïve patients of different ages (Martelo et al. 1969; Winters and Murray 1992; Ladhani et al. 2002; Taylor et al. 2008). Mild or moderate thrombocytopenia in patients from children to adults from malaria-endemic areas is also noted (Josué da Costa Lima-Junior et al. 2012).

A highly coordinated differentiation process generates neutrophils from myeloid precursor cells in the human bone marrow. Neutrophils act as receptors for components of mediators of inflammation (Mora-Jensen et al. 2011). The neutrophil precursor count is an important indicator of inflammation (Zhu et al. 2010; Corti et al. 2012). Variations in neutrophil precursor count are routinely quantitated in the laboratory in malaria-endemic areas as a sensitive but nonspecific indicator of malaria infection (Maina et al. 2010; Habeeb et al. 2012).

Cytokines and chemokines act as biomarkers of infection in malaria. Different nonspecific inflammatory mediators are also biomarkers in malaria infection. Several soluble mediators are released during malaria infection. Soluble mediators like CRP and nitric oxide (NO) find importance as important inflammatory biomarkers in malaria. The NO and CRP levels are important in assessing the severe malaria and as a prognostic tool in the follow-up of malaria infection (McGuire et al. 1996; Armah et al. 2007; Conroy et al. 2011).

In malaria infection, when schizonts rupture, host monocytes and macrophages secrete pro-inflammatory cytokines to stimulate CRP production from the liver (Imrie et al. 2007; Karunaweera et al. 1992; Kwiatkowski et al. 1989). In febrile individuals, a strong correlation between CRP and malaria parasitemia has been seen. Pediatric populations in malaria-endemic areas of Africa show a chronic inflammatory acute-phase response after continued and prolonged exposure to *Plasmodium* infections (Imrie et al. 2007; Naik and Voller 1984; Ree 1971; Chagnon et al. 1992; Verhoef et al. 2001).

Different schools of thought exist regarding the CRP's role in malaria. Studies showed that increased CRP levels are associated with malaria mortality and morbidity (Bruce et al., 2000). Other studies reported that reduced CRP levels may be an important pathological mechanism in severe malaria (Cox et al. 1994). Differences in ethnic background also pose different gradations in inflammation (Imrie et al. 2006). The level of CRP is reported to be high in blacks as well as African descendants compared to whites (Cattani et al. 1986; Burkot et al. 1987).

The protection against preerythrocytic stages of malaria parasites depends on both CMI (cell-mediated immunity) and humoral immune responses of the host (O'Donnell et al. 1998; Imrie et al. 2006). The possible role of cytokines in protective mechanisms at the preerythrocytic stage of the infection has also been studied. IL-1 and IFN-γ strongly inhibit *P. falciparum* sporozoite development in primary cultures of functional hepatocytes without directly affecting free sporozoites. IL-1 exerts its effect at the very early phase of malaria infection and IFN-γ prevents post-penetration (intra-hepatocyte) cellular mechanism of the parasite. Pro-inflammatory IL-1-stimulated hepatocytes secrete CRP. CRP concentration increases in different infections including infectious disorders (Burkot et al. 1987). CRP poses deleterious effects on sporozoites and on hepatic stages of *Plasmodium spp*. Even with nonspecific induction, CRP protects the host from sporozoitic stages of malaria (Imrie et al. 2006).

Antibodies, cytotoxic T cells, cytokines, and a variety of other soluble mediators are the key molecules involved in the potential mechanisms of immunity against malaria. In antimicrobial immune responses, macrophages generate toxic molecules, reactive oxygen, nitrogen intermediate (RNI) species, H_2O_2, and nitric oxide (NO) (Bogdan et al. 2000).

As a host nonspecific innate defense mechanism in malaria infection, CRP concentration also increases, in addition to NO (Nahrevanian et al. 2008). During malaria infection, for a protective immunity, the importance of the quantity of cytokine released and the rate, time, and site of production of cytokines should be balanced and maintained in the chemokine network (Nahrevanian et al. 2008).

In the malaria disease's clinical manifestation and severity, the host's immune system and the parasite's virulence factors play an important role. *P. falciparum* malaria affects every organ of the body and thus is a multisystem disorder. In *P. falciparum* infection, circulatory dysfunction, coagulopathy, specific pathophysiology, systemic inflammation, and endothelial activation often occur (Stauga et al. 2013).

The pathogenesis of cerebral malaria involves the secretion of cytokines and chemokines; mechanical blockage caused by sequestration of parasitized red blood cells (pRBCs), leukocytes, and platelets; angiogenic failure; and genetic and/or immune status of the host along with a network of parasite factors concerned The elevated levels of pro-inflammatory cytokines, TNF-α, IFN-γ, and lymphotoxin were also seen (Fig. 11.3). The sequestration of pRBCs, leukocytes, and platelets within brain vessels leads to

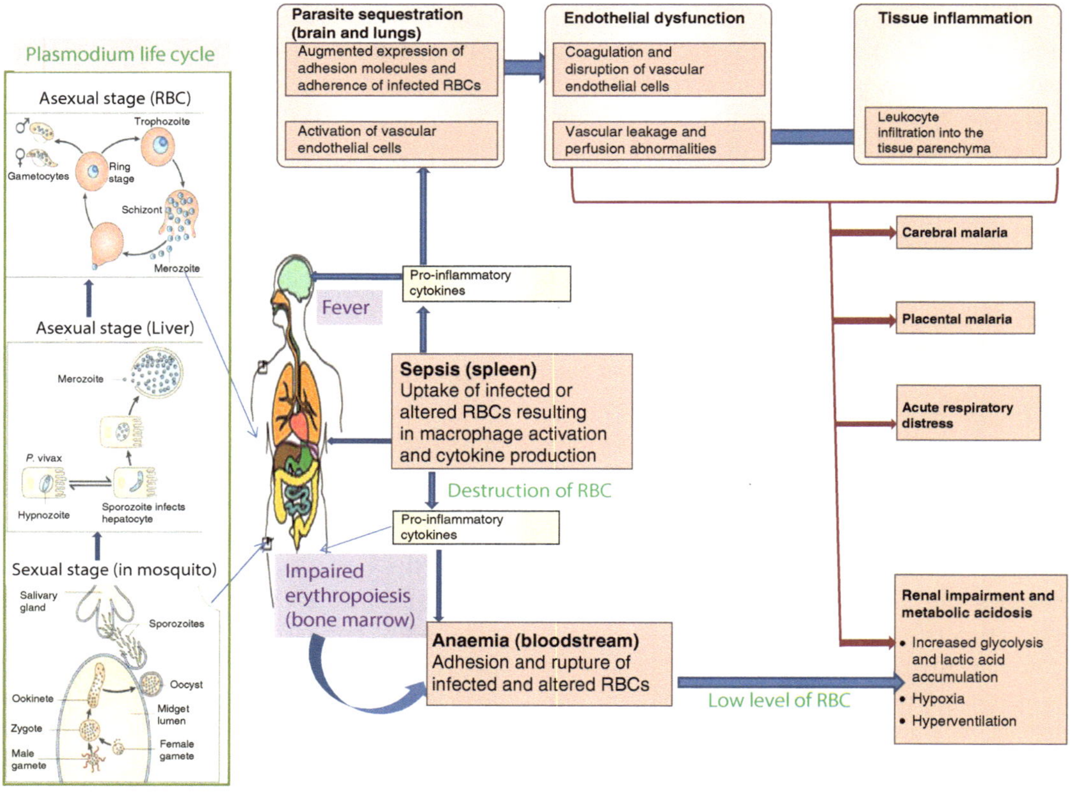

Fig. 11.3 *Pathogenesis of malaria*. Detailed pathogenesis of malaria and its relation with different stages of life cycle are shown

metabolic insufficiencies, upregulated adhesion molecules, and immunological responses with the local involvement of monocytes and T cells activated by *Plasmodium* antigens were the major mechanisms to exacerbate CM. The complex network of cytokines and chemokines appears to act together to control the immune responses in CM (Armah et al. 2007).

It is reported that *Plasmodium malariae* infection showed greater frequency in women with alpha(+)-thalassemia (Mockenhaupt et al. 2001).

11.4 Relation between CRP and Malaria: Role of CRP in the Pathology of Different Malarias

CRP is a classic marker for inflammation as compared to other acute-phase proteins whose concentration rises by two- to fourfold while CRP concentration rises by 100–1000 fold during an acute-phase reaction.

CRP competes with IgG receptor binding and thus plays a role in antigen-specific immune response. The differential plasma concentration of CRP could be attributed to its genetic polymorphic promoter region. The low susceptibility of malaria in the Fulani ethnic group is contributed toward the polymorphisms in the CRP gene. CRP plays an important role in the immune responses to malaria and is confirmed from experimental studies (Ansar et al. 2009a; Israelsson et al. 2009).

CRP bears a sequence homology to FcγR-binding regions of IgG (Bang et al. 2005). The high affinity of CRP to the R variant of FcγRIIa-H/R131 is indicative of the fact that CRP competes with antimalarial IgG antibodies for binding to phagocytic cells expressing FcγR receptors (Stein et al. 2000). The differential blood concentration of CRP has been attributed to a triallelic upstream mutation (_286 C>T>A) of the CRP (Israelsson et al. 2009). The strong association between the CRP-286A allele and increased susceptibility to malaria has been reported (Giha et al. 2010c, 2011).

In malaria infection, CRP secretion is induced by pro-inflammatory cytokines secreted by host mononuclear cells (Harpaz et al. 1992). CRP levels and parasitemias bear strong association. NO also is highly toxic to intraerythrocytic malaria parasites and has strong association with the severity of *Plasmodium falciparum* malaria. The high level of NO production by neutrophils has been correlated with fast parasite clearance of *P. falciparum* infection. Thus nonspecific inflammatory markers find importance in conferring immune protection from malaria (Greve et al. 1999). Thus CRP and NO act as inflammatory mediators in malaria. Their concentration is also strongly correlated with severity and parasitemia in *P. falciparum* malaria.

Several soluble mediators are released during malaria infection. In malaria, soluble mediators like CRP and NO find role as important inflammatory biomarkers (McGuire et al. 1996; Armah et al. 2007; Conroy et al. 2011). The NO and CRP levels find application in assessing the severity of malaria and as prognostic factors during follow-up of malaria infection.

In malaria infection, host monocytes and macrophages release pro-inflammatory cytokines to stimulate CRP synthesis from the liver when the schizonts rupture (Karunaweera et al. 1992; Kwiatkowski et al. 1989). The strong correlations between CRP and malaria parasitemia have been seen in febrile individuals. After continued and prolonged exposure to *Plasmodium* infections, pediatric populations in malaria-endemic areas of Africa have a chronic inflammatory acute-phase response (Imrie et al. 2007).

11.5 Association of CRP with Malaria Infection

CRP is an important component of innate immunity and is believed to play an important role as an early defense mechanism against infection in the host. During acute inflammation, CRP levels escalate to as more than 1000-fold beyond normal, typically within 6 h, and reach their peak at 48 h of inflammation. The serum CRP level is an accurate indicator of inflammation. During liver failure, only the production of CRP is interfered. CRP is well characterized as a serum acute-phase protein and a well-known clinical biomarker (Ansar and Ghosh 2013). CRP has been shown to be involved in multiple immunoregulatory functions. CRP activates the classical complement pathway, acts as a lectin and activates the lectin complement pathway (Ansar et al. 2009a; Ansar et al. 2009b), opsonizes microbes for phagocytosis, and stimulates phagocytic cells (Dong and Wright 1996). Hepatocytes are cells that produce CRP predominantly. Other cells also capable of synthesizing this protein include (1) lymphocyte subsets, (2) Kupffer cells, and (3) monocytes (Dong and Wright 1996). The local secretion of CRP in inflamed condition is also noted (Gould and Weiser 2001). It was first identified in defense against infection in humans by its property to bind to phosphorylcholine expressed on membranes of microorganisms (Gotschlich and Edelman 1967).

CRP and NO act as inflammatory mediators in malaria. Their concentration is also strongly correlated with severity and parasitemia in *P. falciparum* malaria. NO shows high toxicity toward intraerythrocytic malaria parasites and has strong association with the severity of *Plasmodium falciparum* infection. The changes in blood neutrophils, platelets, CRP, and NO in adults from the Brazilian Amazon with uncomplicated *P. vivax* and *P. falciparum* malaria were well studied during the acute and convalescent phases of infection (Oliveira-Ferreira et al. 2010).

With the evolution and gradual changes of malaria infections, the human genome reveals genetic SNPs. SNPs were found to confer protection against malaria or are strongly correlated with susceptibility to malaria. Both *CRP* and *FcγR* genes are located on chromosome 1. CRP bears sequence homology to the FcγR-binding regions of IgG and shows high affinity to the R131-allele product of the *FcγRIIa* gene. *FcγRIIa* gene is studied to be associated with protection against malaria. The antimalarial IgG binding to phagocytic and other FcγR-containing cells competes with CRP to bind with these immune cells (Giha et al. 2010a, b, c; Braga et al. 2005).

CRP protects against infection due to its property of binding to phosphorylcholine (PC) present in the membranes of microorganisms. In malarial infection, the secretion of CRP is mostly induced by pro-inflammatory cytokines (like IL-6 and IL-1) secreted by host monocytes/macrophages. The plasma level of CRP has strong correlations with malarial parasitemias. In African sub-Saharan regions and in the equatorial zones, malaria infection imposes an unlimited burden on the health of these people. Apparently, healthy children from Africa revealed higher plasma CRP levels as compared with control children in Europe and Papua New Guinea, which might indicate that the continuous cyclical malaria infection retains always a clinical level of CRP in African populations (Giha et al. 2010a, 2010b, 2010c; Gotschlich and Edelman 1967; Volanakis and Kaplan 1971; Hurt et al. 1994a, b; Imrie et al. 2007).

The relationship of CRP in acute-phase response (APR) was examined with demography, malariometric indices, and genetic polymorphisms in a cohort of children from a selected area of Papua New Guinea (Imrie et al. 2007; Cox et al. 1994). From this study, in children of the north coastal region of Papua New Guinea, a chronic, asymptomatic malaria infection with clinical symptoms of fever, high parasitemia, and splenomegaly was revealed.

11.6 Role of CRP in Malaria

Though not many reports were available on the role of CRP in malaria, but still nowadays it had already shown its importance in different types of malaria. The earliest report of the relation of CRP and malaria dates back to the 1980s. In 1984, Naik and Voller developed a microplate ELISA to correlate the relationship between CRP and malaria. This microplate ELISA was developed to quantitate CRP. It was shown that African patients with high *Plasmodium falciparum* parasitemia had the highest serum levels of CRP. Strikingly, even African children with lower parasitemia had higher serum CRP levels as compared to other children without parasitemia. The control group of people from the UK had lower CRP levels than those of all African groups (Naik and Voller 1984).

In 1988, the level of CRP was measured in severe malarial infection in 19 patients along with other 35 patients suffering from bacterial meningitis, tuberculosis (TB), meningitis, and convulsions without CNS (central nervous system) infection. The level of CRP was studied in the cerebrospinal fluid (CSF) of all together 54 patients. Nearly five out of 19 patients suffering from severe malarial infection had CSF CRP levels above 1 mg/l. In contrast, CSF CRP was above 1 mg/l in all patients suffering from TB meningitis. It is known that malaria and TB are closely related to each other with some symptoms resembling to each other. Malaria and TB coinfection are also common (Ansar et al. 2009a). Thus CSF CRP was used as a useful differentiating marker in areas where malaria and TB meningitis are common (Cuevas et al. 1988).

The level of CRP was measured in 258 patients with acute malaria in a malaria-endemic area as compared to a control group of 120 patients with other febrile illness. The malaria patients had lower WBC counts and marginally lower hemoglobin level as compared to control patients. In 80 % of the malaria patients, thrombocytopenia was common, whereas it is only 13 % in control groups (p less than 0.01). Interestingly, no major differences were found in erythrocyte sedimentation rate (ESR) or CRP values in malaria patients as well as in patients with different ethnic background in the same endemic area. Thus, in these patients and in control groups, CRP was not noted as a clinical marker, whereas thrombocytopenia was estimated as a probable risk factor in malaria (Eriksson et al. 1989).

In the same year of 1989, in the journal *Infection and Immunity*, another paper cited the protective role of CRP in malaria. Pied et al. (1989) showed that interleukin-1 (IL-1)-stimulated human hepatocytes release an increased amount of major acute-phase reactant, the CRP. The authors found that CRP bound to the sporozoite surface membranes of *Plasmodium falciparum* and *P. yoelii*, through a PC (phosphorylcholine)-binding site. This report was strongly correlated with other findings where CRP binds with damaged RBC in malaria (Ansar et al. 2006; Ansar et al. 2009b) through the PC-binding site. In 1989, it was also confirmed in in vivo and in vitro animal models of rats that CRP could inhibit at the very early phase of infection (preerythrocytic stages).

The relationship of different cytokines (IL-1α, IL-1β, IL-2, IFN-γ, TNF-α, IL-4, IL-6, granulocyte macrophage-colony-stimulating factor, and soluble CD4, CD8, and IL-2 receptor (sCD4, sCD8, and sIL-2R, respectively) profile with CRP was correlated in immunized, non-parasitemic individuals and in unvaccinated, parasitemic individuals after challenge with infectious *P. falciparum* sporozoites. In a period of 28 days, the level of CRP and IFN-γ was higher in parasitemic individuals, whereas levels of other cytokines or soluble receptors were either very low or at baseline values. The increase of CRP and IFN-γ was abrupt. In contrast, in protected individuals, only the level of IL-6 was high. Thus high levels of CRP in unprotected parasitized individuals were systematically correlated with the individual's cytokine profiles (Harpaz et al. 1992).

In *P. falciparum* infection, the correlation of CRP and cytokine level was noted, and thus, in the same year, the relation of serum proteins and cytokine profile was also established. In uncomplicated *P. falciparum* infection, the level of serum protein like CRP, ceruloplasmin (COE), β2-microglobulin (B2M), α1-acid glycoprotein (AAG), α1-antitrypsin (AAT), haptoglobin (HPT), prealbumin (PRE), retinol-binding protein (RBP), albumin (ALB), and transferrin (TRF) were measured in an endemic area of the Amazonian rain forest in patients with cytokine-mediated *P. falciparum* infection. Before treatment, the level of CRP was high in both semi-immune (SI) and nonimmune (NI) patients. But the concentrations of other serum proteins like HPT, PRE, RBP, ALB, and TRF were low in both patient groups during the acute phase of the disease. After follow-up, the concentrations of HPT, AAT, and B2M were significantly altered in NI patients. Thus, along with CRP, the concentrations of other serum proteins were altered in *P. falciparum* infection (Graninger et al. 1992).

In highly malaria-endemic area of Tanzania, 629 children under the age of 6 years were selected to estimate the acute-phase reaction of CRP in malaria of children of different ages and to correlate its serum concentrations as a useful quantitative marker of morbidity in such children. The median CRP level was much higher in the blood of children than the reported CRP concentration from normal children in a non-endemic zone. Once more, they strongly correlated the high CRP levels with *P. falciparum* parasitemia in children under 1 year of age. In areas with very high levels of *P. falciparum* transmission, the concentration of CRP alters with malaria infection. This affects the parasitological and clinical scenario in the patient. Thus, determination of CRP levels would enable quicker assessment of the overall burden of morbidity, especially in children. Increased parasite densities increase the concentration of CRP and thus may be associated in a measure of malaria-specific morbidity and its estimated malaria morbidity risks in serological surveys (Hurt et al. 1994a, b).

For the assessment of malaria intervention program in large numbers of people, CRP and haptoglobin are used as they are simple and inexpensive. McGuire in 1996 showed that elevated CRP (>8 mg/L) was seen when parasitemia is high. Thus CRP was again correlated as marker of malarial morbidity along with haptoglobin in 1505 Gambian children The acute-phase response was again correlated with increased parasitic burden and with malarial morbidity (McGuire et al. 1996).

In a study conducted on Gabonese patients infected with *P. falciparum* parasite, children with uncomplicated malaria and severe malaria were taken along with adults with uncomplicated malaria before and after malaria therapy. The plasma concentration of neopterin, nitrogen oxide, and CRP was significantly more in severe malaria as compared to uncomplicated malaria. Due to the acute-phase response of CRP in malaria, the patients had slow parasitological and clinical cure after therapy (Kremsner et al. 1996). Thus CRP is playing a detrimental effect during the acute phase of malaria infection. This report is highly correlated with the report of Hurt et al. (1994a, b). In a rural Zambian malaria-prevalent population, the same kind of report was seen where, along with CRP, AGP was also taken in 210 children under 2 years of age as one of the serum protein markers. As noted before by other researchers, CRP and AGP concentration was high in patients with high parasite burden. But the serum CRP and AGP levels were negatively correlated with vitamin A uptake. In high parasitic burden, the acute-phase response of CRP and AGP was high, whereas the serum retinol level was low (Hautvast et al. 1998).

In Danish malaria patients, elevated concentration of serum CRP and hemoglobin (Hb) and reduced concentration of haptoglobin were recorded during the acute phase of the reaction. After initiation of treatment, CRP levels were significantly high in patients for nearly 3 days which gradually lower in due course of treatment. In *P. falciparum* malaria, soluble Hb levels showed correlation with malariometric parameters, whereas CRP was not taken as a marker in epidemiological studies of malaria (Jakobsen et al. 1997).

Malaria occurrence is relatively rare in Japan. Patients in the laboratory for malarial screening were screened for malarial parasites using conventional and modernized technologies. In the past 4 years, malarial parasites were detected in only eight patients. Conventional screening techniques include identification of parasites in blood smear by microscopy. The common laboratory and clinical findings include the detection of hepatic dysfunction, elevated LDH activity, thrombocytopenia, and increased CRP levels in serum of malaria patients (Okayama et al. 1999).

The effects of submicroscopic *P. falciparum* infections in pregnancy were detected by microscopy methods and PCR assays among 530 pregnant women in Ghana with plasmodial infections. The prevalence of anemia, fever, and signs of inflammation with CRP levels > 0.6 mg/dl was observed to be more frequent in pregnant women. Antimalarial drugs (chloroquine, pyrimethamine) reduce the prevalence of *P. falciparum* infections and at the same time raise the proportion of submicroscopic parasitemia. These patients had slightly reduced Hb levels and significantly high serum CRP levels. The marker of inflammation, CRP, was frequently used along with parasite-specific PCR to detect submicroscopic *P. falciparum* infections. Elevated CRP levels also contribute toward mild anemia in aparasitemic pregnant women (Mockenhaupt et al. 2000a, b). The relation of anemia and CRP was also validated in 2009 (Ansar et al. 2009a; Ansar et al. 2009b).

CRP was used as a surrogate marker of malaria morbidity in malaria-holoendemic Magoda, Tanzania, where the prevalence of *Plasmodium* infections was 96.0 %. The reactivity of the IgG antibody against two recombinant protein fragments from *P. falciparum* rhoptry-associated protein-1 (rRAP-1) was detected in children (1–4 years) with asymptomatic malaria infections and also with acute clinical disease. The reactivity of IgG to rRAP-1 was higher in asymptomatic malaria infections and lower in mixed *Plasmodium* infections or in a single infection. IgG bears a similar structural and functional entity to CRP (Ansar et al. 2009a) and showed reactivity toward *P. falciparum* proteins (Alifrangis et al. 1999).

In children below 2 years of age in rural Zambia, the relationships of infections with subsequent 3-month-long increment were shown through anthropometric measurements. Other factors included in the study were CRP, α1-acid glycoprotein, retinol, and malaria parasitemia. Nearly 50–71 % of the children showed elevated acute-phase reactant concentration of CRP; 79–83 % had malaria parasitemia, whereas 55–64 % individuals showed low retinol concentrations. Thus, due to multifactorial effects, malaria infections in these children contribute to their short-term retardation of linear growth (Hautvast et al. 2000).

From the previous study, in Zambia, the relations of infant length with other factors were correlated in malaria infection. In this piece of work, the author pointed the degree of the association between APPs with malaria-induced hyporetinemia with a serum retinol level <0.70 micromol/L. Nearly 90 children undergoing a clinical trial of vitamin A supplementation were assessed, and the serum concentrations of CRP, retinol-binding protein (RBP), haptoglobin, α(1)-acid glycoprotein (AGP), α(1)-antichymotrypsin, transthyretin, and albumin were determined. These children showed symptoms of splenomegaly and were positive for *Plasmodium falciparum* infection. The concentration of CRP and AGP showed inverse correlation with retinol but positively with RBP, transthyretin, and albumin. All APPs, except α(1)-antichymotrypsin, were significantly correlated with symptoms of splenomegaly. Of the positive APPs, AGP showed strong correlation with CRP indicating chronic inflammation. Thus, RBP or transthyretin increases retinol, whereas an equivalent increase in AGP decreases retinol concentration. They showed that, apart from AGP, positive APPs are promising markers of type and severity of inflammation and can be used as a mode to correct for malaria-induced hyporetinemia (Rosales et al. 2000).

During malaria infection, due to acute-phase responses to infection, a major metabolic change occurs during clinical malaria. Mostly biochemical measurement of iron status becomes difficult. European patients developing uncomplicated *P. falciparum* malaria after their visits to endemic areas were included in a study to measure plasma concentrations of CRP, soluble transferrin receptors (TfR), ferritin, AGP (α 1-acid glycoprotein), and ACT (α 1-antichymotrypsin) along with degree of parasitemia. The plasma concentrations of CRP, AGP, and ACT correlated strongly with each other. The concentration of CRP and AGP negatively correlated with TfR. Thus this study concludes that, when both clinical episodes of malaria and iron deficiency occur simultaneously, the plasma concentration of TfR values must be interpreted carefully as indicators of iron status (Beesley et al. 2000).

The relation of anemia in pregnancy in sub-Saharan Africa was correlated with iron deficiency, hemoglobinopathies, and malaria. Anemia in pregnancy is a multifactorial process. The levels of ferritin were influenced both by malaria and inflammatory processes. The independent risk factors for anemia in multivariate analysis include raised CRP levels, low gravidity, malaria with high parasitic densities, second trimester of pregnancy, and homozygous alpha(+)-thalassemia (Mockenhaupt et al. 2000a, b).

Interestingly, the level of CRP is low in parasitemic individuals with infections from both *P. falciparum and P. malariae. P. malariae* was thought to influence the pathophysiology of *P. falciparum* malaria and lower the prevalence of inflammation and febrile responses in women with alpha(+)-thalassemia (Mockenhaupt et al. 2001).

Pregnancy related anemia (apart from nutritional deficiencies) increases plasma concentrations of CRP in most of the women studied in rural northern Tanzania (Hinderaker et al. 2002).

In Nigeria, 207 children of ages ranging from 0 to 60 months who presented high fever were studied where parasite density and CRP concentration in plasma were taken as markers of disease severity (Cooper et al. 2002a, b).

CRP competes with IgG antibody binding to its receptors, thus interfering with the antigen-specific immune response. The differential plasma concentration of CRP could be attributed to its polymorphisms in the CRP promoter region. The low susceptibility of malaria in the Fulani ethnic group is contributed toward the polymorphisms in the CRP gene. This study points to the inference that CRP may play a pivotal role in the immune responses to malaria (Israelsson et al. 2009).

In malaria treatment, the scenario is continuously worsened by resistance to drug and insecticide. So vaccine preparation focused on new immune system targets was urgently needed. Different soluble mediators like cytotoxic T cells, antibodies, and cytokines are the key molecules involved in the potential mechanisms of immunity against malaria. Considerable amounts of molecules that are toxic, like reactive oxygen and nitrogen intermediates (RNI), H_2O_2, and nitric oxide (NO), are released by macrophages during antimicrobial immune responses (Bogdan et al. 2000).

Being a nonspecific innate defense component of the host in malaria infection, CRP concentration also increases, in addition to NO (Nahrevanian et al. 2008). During malaria infection, the CRP levels are elevated (Kremsner et al. 1996) and so are associated with clinical forms of malaria (Jakobsen et al. 1998). Thus CRP plays a key element in malaria infection. However, in the same infection, in humans, such co-association between NO and other immune factors remains controversial (Balmer et al. 2000; Nahrevanian et al. 2006; Cramer et al. 2004).

RNI, NO, and NO synthase (NOS) and their role in malaria of the residents were investigated (Nahrevanian and Dascombe 2001, 2002; Dascombe and Nahrevanian 2003). Co-association of CRP and NO was studied in Iranian malaria patients and in control individuals from NW and SE malaria-endemic regions of Iran (Nahrevanian et al. 2006; Nahrevanian et al. 2008).

CRP is a sensitive marker for early phase of inflammatory or tissue destruction processes in the host. The strong association between CRP concentrations and malaria parasite densities in pediatric patients was investigated in the Kilombero district of Tanzania, an area of intense perennial *P. falciparum* transmission. In areas where asymptomatic parasitemia is common, the high correlation between CRP levels with high levels of malaria parasitemia could serve as a marker for clinical malaria. Such a marker could be more sensitive than other signs of inflammation and thus can be used in assessing morbidity or mortality of a community. Raised CRP levels should clearly be associated with clinical malaria and not with asymptomatic parasitemia. The relationships between CRP levels with clinical morbidity or with *P. falciparum* parasitemia only were analyzed to determine CRP and its role as a marker in field studies on malaria morbidity in areas where malaria is endemic. The study was conducted taking a few communities of Western Europe as the control group (Hurt et al. 1994a, b).

Differences in ethnic background also pose different gradations in inflammation [15]. The level of CRP is reported to be high in blacks and descendants from Africa as compared to the whites [16, 17] [20]. A probable hypothesis proposed is that they have a more pro-inflammatory disposition in African descendants due to different evolutionary histories. Currently our understanding of "natural" CRP levels in traditional African populations simulating the recent evolutionary past is highly limiting. CRP is a biomarker of different pro-inflammatory conditions. Circulating CRP level was reported to be elevated in a contemporary population as compared to traditional people living under the same adverse environmental conditions in a malaria-endemic area of Ghana (Eriksson et al. 1989; Khera et al. 2005).

The protection against preerythrocytic stages of malaria parasites is conferred by both cell-mediated and humoral immune responses of the host. The possible role of cytokines in protective mechanisms at the preerythrocytic stage of the infection has also been studied. Interleukin-1 (IL-1) and interferon gamma (IFN-γ) strongly inhibit *P. falciparum* sporozoite development in primary cultures of functional hepatocytes without directly affecting free sporozoites. IL-1 exerts its effect at the very early phase of malaria infection and IFN-γ prevents post-penetration (intra-hepatocyte) cellular mechanism of the parasite. Pro-inflammatory IL-1-stimulated hepatocytes secrete CRP. CRP concentration increases in different infections including infectious disorders. CRP poses deleterious effects on sporozoites and on hepatic stages of *Plasmodium spp*. Even with nonspecific induction, CRP protects the host from sporozoitic malarial stages (Pied et al. 1989; Ferreira et al. 1986; Mellouk et al. 1987; Pepys and Baltz 1983).

11.7 Clinical History of Patients

The early symptoms like (1) ache in the head and body, (2) sickness, (3) fatigue, (4) abdominal problems, and (5) fever are very nonspecific. Sometimes (1) chest pain and abdominal pain, (2) diarrhea, (3) arthralgia, (4) urticaria, and (5) myalgia may also be present. Although symptoms are variable at clinical presentation, most patients with uncomplicated malaria may present symptoms like fever, malaise, and mild anemia. In young children, clinical presentation includes irritability, food refusal, and vomiting.

Sometimes physical examination of the patient shows only the sign of fever. Mostly, in a large number of patients, tertian fever, i.e., spikes of fever, on alternate days along with chill and rigor is observed. The fever may even be absent in some cases and is usually persistent to start with rather than the tertian one. Sometimes the fever may be accompanied by rigors. In some patients, palpable liver and spleen are observed. Features to diagnose malaria clinically are highly nonspecific and common to other related infectious diseases of the tropical countries.

In malaria-endemic areas, due to nonspecific nature of the disease, clinical presentation may cause overtreatment of malaria while the diagnosis of actual malaria in low-transmission areas is missed. Eventually, in malaria-endemic areas, this may lead to misdiagnosis and nontreatment of other infectious diseases. So, clinical assessment is important to diagnose malaria.

Early diagnosis of malaria (1) reduces the sufferings of subjects together with reduction in the chances of parasite transmission in the society. The diagnosis of malaria is made from clinical symptoms of the patients. Light microcopy and other microscopic techniques are the most common methods used to detect malaria parasites in the peripheral blood of the patient. Generally, to diagnose malaria, a drop of blood from the infected person is collected by small aseptic finger prick or phlebotomy. The thick and thin blood films and smears are prepared.

The history, clinical presentation, and treatment summary of two patients were noted in Table 11.1. Both the patients were admitted to the MB Nursing Home, Park Circus in Kolkata for their diagnosis.

11.8 Discussion

Since 2000, due to strategic prevention and control measures, mortality rates in malaria have reduced by 25 % globally. Malaria is preventable and curable. The approach to eliminate or control malaria includes some basic health and hygienic measures, mosquito control, and improved personal and community protection. But the future development in malaria eradication and its control remains in the development of (1) diagnostic tools that are sensitive, (2) better and efficient antimalarial drugs, (3) improved researches to develop malaria specific vaccines,(4) improved monitoring of human infection, and (5) parasite surveillance, together with effective delivery of drugs. Due to long years of exposure, partial immunity is developed in individuals while it never provides complete protection. At present, there are no licensed vaccines against malaria. It is expected that, by 2015, results about a research vaccine candidate known as RTS,S/AS01 will pave new dimensions in malaria treatment (Lopez et al. 2014).

Presently used drugs, chloroquine (Talisuna et al. 2004) and artemisinin (Afonso et al. 2006), have responded negative to drug-resistant strains of *Plasmodium* sp. Better understanding of disease physio-pathogenic mechanisms will lead to understanding of developing a mechanism of such resistance. The key players in the physio-pathogenesis of malaria have been implemented to present free radicals through oxidative stress (Wilmanski et al. 2005; Narsaria et al. 2012). To reduce drug-induced side effects and maximize the effectiveness of pharmaceutical delivery to targeted locations, nanotechnology is used to generate nano-drugs as potential carriers for the controlled and site-specific delivery of drugs (Nakache et al. 2000). The discovery of polymeric nanoparticles that are biocompatible has attracted considerable attention in pharmacology. In malaria treatment, the nano-drug delivery system shows great promise and is a viable approach (Swai et al. 2008). An experimental nano-drug has been trialed in animal models as all *Plasmodium* species that infect mammals with similar life cycles and are sensitive to the similar group of drugs. The drugs used in experimental rodent models could be potentially trialed for human malaria or may provide additional information about the mechanism of action of a compound in human. So a thorough in vivo screening system finds importance in efficacy

Table 11.1 Clinical history and laboratory parameters

Parameters	Patient 1	Patient 2
A. Physical parameters		
Age (years)/sex (M/F)/weight (kg)	32/male/56	56/female/84 (obese)
Chief complaint	1. Fever for 10 days 2. Chills and rigor 3. Vomiting and weakness for 3 days	1. Fever for 8 days 2. Body ache for 2 days 3. Headache for 3 days 4. Abdominal discomfort
Past medical history	1. Nothing significant	1. Diabetes mellitus type 2 for 2 years
Past surgical history	1. Nothing significant	1. Hysterectomy and bilateral salpingo-oophorectomy—8 years back
Birth history and vaccination	NA	NA
Drug allergy	Name not known	Not known till now
On examination		
	Patient conscious, alert, and cooperative	Patient conscious, alert, and cooperative
GCS	15/15	15/15
Pulse/min	132	128
Temperature (°F)	102.4	103.1
Pallor	Mild	Positive
BP (mm of Hg)	90/60	99/72
JVP	Normal	Normal
Palpable lymph node	No palpable lymph node	Not palpable
Hydration status	Severe dehydration	Severe dehydration
Clubbing	Not present	Not present
B. Systemic examination		
Chest (Air entry)	Bilateral adequate	Adequate
CVS (S1 and S2)	Audible	Audible
Abdomen	1. Soft 2. Liver palpable 2 cm below costal margin 3. Spleen palpable by hooking method	Soft
Intestinal peristaltic sound	Present	Present
C. Local examination		
1. Body red flushed 2. No secretion from ears 3. No sore throat 4. Conjunctiva yellowish		1. Body red flushed 2. No secretion from ears 3. No sore throat
D. Laboratory investigations		
Hb (gm/dl)	9.2	8.4
D-1. Inflammatory and infective markers		
TLC (cells/cu.mm)	14.700	7900
DLC (%)	N (86), L (10), M (3), E (1)	N (62), L (34), M (1), E (3)
ESR (mm/first hour)	92	96
CRP (mg/L)	78	126
Procalcitonin (ng/ml)	12	14

(continued)

Table 11.1 (continued)

Parameters	Patient 1	Patient 2
Blood culture (aerobic)	No bacterial growth till 48 h	ND
Widal test	Negative	Negative
Blood culture (anaerobic)	No bacterial growth till 48 h	ND
Pro-BNP (pg/mL)	ND	ND
Malarial antigen	*P. vivax* positive	*P. vivax* positive
Dengue serology	Negative	Negative
D-2. Sugar level tests		
Blood sugar (mg/dl)	RBS-112	RBS-236
Glycosyated hemoglobin test (Hb A1c) (%)	8.6	7.6
D-3. Renal function tests		
Urea (mg/dl)	36	30
Creatinine (mg/dl)	1.2	1.3
D-4. Electrolytes		
Sodium (Na+) (mEq/L)	134	139
Potassium (K+) (mEq/L)	4.2	3.8
Calcium (serum) (Ca++) (mEq/dl)	ND	9.3
D-5. Autoimmune markers		
Antinuclear antibody	ND	ND
RA factor (above 20 IU/mL)	ND	ND
HLA (B-27)	ND	ND
D-6. Coagulation markers		
PTT (sec) (control: 12 s)	13	13.7
APTT (sec) (control: 26 s)	28	32
Liver function test **1. Bilirubin (mg/dl)** **2. Albumin (gm/dl)** **3. Other liver enzymes**	1. Bilirubin—2.2 2. Albumin—3.4 3. Liver enzymes are marginally increased	Mildly deranged
ABGA	ND	
D-7. Chest X-ray		
	Normal profile seen	Normal profile seen
D-8. USG of abdomen		
	1. **Liver**—Mildly enlarged **2**. Spleen—marginally enlarged	Hepatomegaly
D-9. Cardiac function tests		
ECG	Sinus tachycardia	Sinus tachycardia
Echocardiography	ND	ND
EF (%)	ND	ND
D-10. Bronchoscopy and BAL/ET suction		
ND		ND
D-11. Urine routine examination and culture sensitivity		
1. Pus cells **1–2/HPF** 2. No bacterial growth within 72 h		Pus cells 2–3/HPF

(continued)

Table 11.1 (continued)

Parameters	Patient 1	2
E. Treatment summary		
1. Adequately hydrated by i.v. fluid and oral fluid 2. Oral chloroquine as per recommended dose 3. Blood for G6PD was checked and it was normal. Primaquine was started 4. Cold sponging and antipyretic and antiemetic drugs were given 5. Patient became afebrile on day 2 after hospitalization 6. Repeat parasitic load was checked and it was negative 7. Repeat LFT became normal		1. On sliding dose of human insulin (human Actrapid) subcutaneously
F. Output		
Successfully discharged with clinical advices		Successfully discharged

GCS Glasgow coma scale, *min* minutes, °*F* degree Fahrenheit, *BP* blood pressure, *JVP* jugular venous pressure, *CVS* cardiovascular system, S_1 first heart sound, S_2 second heart sound, *TLC* total leukocyte count, *DLC* differential leukocyte count, *ESR* erythrocyte sedimentation rate, *CRP* C-reactive protein, *N* neutrophil, *M* monocyte, *L* lymphocyte, *E* eosinophil, *ND* not determined, *FBS* fasting blood sugar, *RBS* random blood sugar, *RA* rheumatoid arthritis, *HLA* human leukocyte antigen, *PTT* prothrombin time test, *APTT* activated partial thrombin time, *ECG* electrocardiogram, *RBBB* right bundle branch block, *WNL* within normal limit, *PAP* pulmonary artery pressure, *EF* ejection fraction, *BNP* brain natriuretic peptide, *μg* microgram, *mg* milligram, *pg* picogram, *i.v.* intravenous, *i.m.* intramuscular, *SOS* as and when required, *GTN* glyceryl trinitrate, *CBG* capillary blood glucose, *NA* not applicable, *MDR* multidrug resistance, *HPF* high-power field, *AMBU* artificial manual breathing unit, *ET* endotracheal, *CBG* capillary blood glucose, *mEq* milliequivalents, *dl* deciliter, PaO_2 partial pressure of arterial oxygen, *USG* ultrasonography, *ST* segment that represents the period when the ventricles are depolarized in ECG, *BAL* bronchoalveolar lavage

testing of different antimalarial drugs (Peters et al. 1975).

Treatment failure (TF) occurs when there is resurgence of malarial parasitemia post-antimalarial treatment. Disease-resistant parasites along with host factors contribute toward TF. Host factors and immunity varies from one endemic setting to other. The predominant parasite factors include mutations of genes specific for drug metabolism (WHO 1996; Wongsrichanalai et al. 2002; Hayton and Su 2004). Chloroquine (CQ) resistance shows strong correlation with mutation of an amino acid at position 76 of the *P. falciparum* chloroquine resistance transporter (Pfcrt) gene (Fidock et al. 2000). Detection of the parasite mutations is a promising tool for global mapping of antimalarial drug resistance in the future (Plowe et al. 2007). Host immune response modulates its activity according to the duration and intensity of exposure to malaria. Host immunity is also modulated by host genetic factors like Fc gamma-receptor-II (FcγRII) and genetic mutations of CRP. CRP shares functional similarity with IgG and like IgG competes with the binding of Fcγ receptors. The binding ability of certain IgG subclasses was altered by the polymorphism in the FcgRIIA gene, (R) or (H), at position 131 (Braga et al. 2005). Interestingly, the R131 allele is known to associate with malaria protection of the host (Braga et al. 2005), and CRP has high affinity to the product of the R131 allele (Stein et al. 2000). Inflammation and polymorphism of CRP in malaria infection (Naik and Voller 1984) were reported through the association between single-nucleotide polymorphism (SNP) of CRP and susceptibility to malarial infection (Giha et al. 2009a, b).

ERRATUM

Biology of C Reactive Protein in Health and Disease

Waliza Ansar and Shyamasree Ghosh

W. Ansar, S. Ghosh, *Biology of C Reactive Protein in Health and Disease*,
DOI 10.1007/978-81-322-2680-2

DOI 10.1007/978-81-322-2680-2_12

In the original version, the affiliation of Dr. Ghosh which was published on page iv was incomplete. Here's the complete affiliation of Dr. Ghosh.

Shyamasree Ghosh
School of Biological Sciences
National Institute of Science, Education and Research (NISER)
Bhubaneswar, Odisha, India

The present version is updated with complete affiliation of Dr. Ghosh.

The online version of the updated book can be found at
http://dx.doi.org/10.1007/978-81-322-2680-2

Glossary

α_1-Acid glycoprotein A plasma protein of molecular weight 39.5 kD. As it is an acute-phase reactant, its concentration increases during acute inflammation. It binds nonspecifically to vitamin B12 and progesterone.

2D PAGE (Two-dimensional polyacrylamide gel electrophoresis) 2D PAGE is the most common technique employed for protein separation. First, proteins are separated in one dimension according to their molecular mass/size and then separated according to their isoelectric point (or their charge) in the second dimension. After separation, to visualize the protein spots, the gel is stained also. It is used in protein chemistry and in proteomics.

A1C A person's average blood glucose level over a period of 2–3 months is measured by this test. It is also known as hemoglobin A1C or glycosylated hemoglobin, where the amount of glucose that sticks to the red blood cell is tested. The stuck glucose is proportional to the amount of glucose in the blood.

ABER (annual blood examination rate) It is calculated as number of slides examined/population ×100. WHO recommends this for malaria-infected areas where the number of slides examined per month should equal at least to 1 % of the population.

Ablation In medicine, the removal or destruction of a body part or tissue or its function. Ablation may be performed by surgery, hormones, drugs, radiofrequency, heat, or other methods.

Abnormal Not normal. It describes a state, behavior, or condition that is unusual or quite different from what is considered normal. An abnormal growth or lesion in or on the body may be benign (not cancerous), premalignant or precancerous (likely to become cancerous), or malignant (causing cancer).

ABO blood group system Proposed by Landsteiner, it is a system used to group human blood into four different types, based on the presence or absence of some markers on the surface of red blood cells. The four main blood types are A, B, O, and AB. During blood transfusion, grafting, and other surgical operations, the ABO blood group system is used to match the blood type of the donor and the person receiving the transfusion or grafting. Universal donors are people with blood type O who can donate blood to anyone. Universal recipients are people with blood type AB who can accept blood from all donors. People with type A or B can receive type O blood or matching blood.

Abscess Accumulation of products of inflammation in a localized part in tissues, organs, or confined spaces in the body with an enclosed collection of pus. It is usually swollen and inflamed. An abscess is a sign of infection.

Acanthosis nigricans It is a skin condition characterized by darkened skin patches. It is common in people who have insulin resistance. This condition is also seen in prediabetes or type 2 diabetes people.

Acarbose (brand name, Precose) An oral medicine which blocks the enzymes that digest starches in food and is alpha-glucosidase inhibitors. It is used to treat type 2 diabetes. The result is a slower and lower rise in blood glucose level throughout the day, specially right after meals.

W. Ansar, S. Ghosh, *Biology of C Reactive Protein in Health and Disease*,
DOI 10.1007/978-81-322-2680-2

Accelerated-fraction radiation therapy Radiation treatments are given more than once a day where the total dose of radiation is divided into small doses. The total dose of radiation is also given for fewer days compared to standard radiation therapy.

ACE inhibitor An oral medicine which is angiotensin converting enzyme that actually lowers blood pressure; ACE stands for an-gee-oh-TEN-sin. Also used in slowing down kidney damage in diabetic people with albumin in their urine.

Acellular vaccine A vaccine containing partial cellular material as opposed to complete cells.

Acesulfame potassium (brand name, Sunett) A dietary sweetener with no nutritional value and no calories. Also known as acesulfame K.

Acetohexamide (brand name, Dymelor) An oral medicine belonging to the class sulfonylureas, used to treat type 2 diabetes. It lowers blood glucose by helping the pancreas secrete more insulin and helps the body to use the insulin better.

Acetylcholine A type of neurotransmitter chemical made by some types of nerve cells to send messages/signals to other cells on the other side of a synapse like other nerve cells, muscle cells, and gland cells. It is released from the nerve ending and helps control memory and the action of certain muscles.

Acne A disorder of the skin in which oil glands and hair glands become inflamed.

Acute-phase protein Their concentrations can decrease (i.e., serum albumin, transferrin, RBP4) as well as increase (i.e., C-reactive protein, alpha-2-macroglobulin, serum amyloid) following exposure to inflammatory conditions.

Acquired (adaptive) immune system A part of the immune system that acquires and matures and comprises of particularly B and T lymphocytes, characterized by a slower response to the first encounter with an infection and then a rapid recall memory for any future encounters with the same infection. Memory is sustained for long periods throughout its life.

Acquired immune deficiency syndrome (AIDS) A life-threatening disease caused by a human immunodeficiency virus (HIV) and characterized by breakdown of the body's immune defenses, and the body cannot defend itself against different infections (like pneumonia). HIV is spread through direct contact with the blood and body fluids of an infected individual. High-risk activities include unprotected sexual intercourse and intravenous drug use (sharing needles). There is no cure for AIDS; however, research efforts are ongoing to develop a vaccine. People with AIDS are at an increased risk for developing certain cancers and for infections that usually occur only in immunocompromised individuals.

ACTH (or adrenocorticotropic hormone and corticotrophin) A hormone secreted from pituitary gland which acts on the outer part of the adrenal gland to control the release of corticosteroid hormones. It is mostly secreted during times of stress.

Active immunity Long-lasting immunity that is acquired through the production of antibodies and memory T cells in response to stimulation by a disease-causing organism, antigens, or a vaccine. Active immunity is usually permanent in an individual throughout the duration of his lives.

Acute bacterial prostatitis Inflammation of the prostate gland caused by bacteria that begins suddenly and gets worse quickly. Symptoms include fever and chills, body aches, pain in the lower back and genital area, a burning feeling during urination, and problems with emptying the bladder all the way.

Acute leukemia A rapidly progressing cancer that starts in blood-forming tissue like the bone marrow. The disease causes large numbers of white blood cells to be produced and enter the blood stream.

Acute lymphoblastic leukemia (acute lymphocytic leukemia or ALL) A fast-growing (aggressive) type of leukemia (blood cancer) where too many lymphoblasts (or immature white blood cells) are found in the bone marrow and blood.

Acute myelogenous leukemia (acute myeloblastic leukemia or acute myeloid leukemia or acute nonlymphocytic leukemia or AML or ANLL) A fast-growing (aggressive) disease in which too many myeloblasts (immature white blood cells excluding the lymphoblasts) are found in the bone marrow and blood.

Acute pain Pain that comes on quickly, lasts for a relatively short time, and can be severe also.

Acute-phase proteins (APPs) APPs are a large group of biochemically and functionally unrelated proteins whose plasma concentrations increase (positive acute-phase proteins) or decrease (negative acute-phase proteins) severalfold in response to acute infections, tissue injury, burns, or inflammation, i.e., after the onset of a systematic inflammatory reaction. The examples of APPs are C-reactive protein (CRP), serum amyloid A (SAA), fibrinogen, mannose-binding proteins, complement components, alpha 1-acid glycoprotein (AGP), etc.

Acute-phase response When the body is invaded by a pathogen, macrophages release the protein signals interleukin-1 (IL-1) and IL-6 to fight against the infection. Interleukins raise the temperature of the body, causing the fever as a general innate acute-phase response that often accompanies infection. Fever helps to eliminate infections because most bacteria grow optimally at temperatures lower than normal body temperature. Interleukins stimulate liver cells to secrete increased amounts of several plasma proteins like acute-phase proteins (APPs) into the bloodstream. APPs bind to bacteria and by activating the complement proteins lyse the pathogen. Interleukins increases the number of circulating neutrophils and eosinophils, which help to combat infection.

Acute promyelocytic leukemia (promyelocytic leukemia or APL) An acute myeloid-type fast-growing (aggressive) leukemia where too many immature blood-forming cells are found in the blood and bone marrow. It is usually marked by an exchange of parts of chromosomes 15 and 17.

Acute radiation sickness (acute radiation syndrome or radiation poisoning or radiation sickness or radiation sickness syndrome) Serious illness caused by being exposed over a short period of time to high doses of certain types of radiation. The symptoms usually occur right after exposure and continue for time with some relapses also. Symptoms include nausea and vomiting, diarrhea, headache, dizziness, weakness, fatigue, bleeding, hair loss, swelling, itching and redness of the skin, and other skin problems. Very large doses of radiation may cause death.

Acute respiratory distress syndrome (ARDS) A life-threatening condition that develops due to inflammation and injury to the lungs causing fluid buildup in the airways.

Acute inflammation Sudden onset of inflammation, usually lasts for few days to weeks, marked by the classical signs of inflammation and marked by vascular and exudative processes.

Acute A short-term, intense health effect. Describes something that happens suddenly and for a short time. Opposite of chronic.

Addiction An uncontrollable craving, seeking, and use of a substance, like a drug or alcohol.

Addison disease (adrenal insufficiency) A rare disorder in which the adrenal glands do not make enough of certain hormones due to immune system problems or infection, cancer, or other diseases. Symptoms include weight loss, loss of appetite, nausea and vomiting, diarrhea, muscle weakness, fatigue, low blood sugar, low blood pressure, and patchy or dark skin.

Adenocarcinoma in situ (or AIS) A condition in which abnormal cells are found in the internal lining of glandular tissue like uterus, cervix, lung, pancreas, and colon. Adenocarcinoma in situ occurs most often in the cervix and may become cancerous and spread to nearby normal tissue.

Adenocarcinoma A cancerous tumor that begins in glandular (secretory) cells and arises in or resembles glandular tissue. Most cancers of the breast, pancreas, lung, prostate, and colon are adenocarcinomas.

Adenoid cystic carcinoma A rare type of cancer that usually begins in the salivary glands.

Adenoma A tumor, not a cancer, that starts in gland-like cells of the epithelial tissue.

Adenosarcoma A tumor that is a mixture of an adenoma (a tumor that starts in the gland-like cells of epithelial tissue) and a sarcoma (a tumor that starts in bone, cartilage, fat, muscle, blood vessels, or other connective or supportive tissue). An example of an adenosarcoma is Wilms tumor.

Adenosine deaminase (ADA) An enzyme found in mammalian tissues that catalyzes the breakdown of adenosine into inosine and ammonia. ADA deficiency can lead to one form of severe combined immunodeficiency disease. ADA deficiency leads to the destruction of B and T cells which can also cause metabolic problems in cells.

Adenosquamous carcinoma A type of cancer that contains a mixture of two types of cells, the squamous cells and gland-like cells.

Adhesive capsulitis The loss of the ability of the shoulder to move in all directions causing pain. It is a condition of the shoulder associated with diabetes.

Adjunct agent In addition to the primary therapy, a drug or chemical substance used in cancer therapy.

Adjuvant chemotherapy It is a term used to describe the role of chemotherapy in relation to other cancer treatments. It is typically given alone or with radiation after surgical resection.

Adjuvant radiation therapy After cancer treatment, radiation is used in order to prevent a cancer from recurring.

Adjuvant A substance (e.g., aluminum salt) that is added along with other substances during production to increase the body's immune response to a vaccine.

Adrenal cancer Cancer that is formed in the tissues of the adrenal glands. Adrenal cancer that starts in the outside layer of the adrenal gland is called adrenocortical carcinoma. Adrenal cancer that starts in the center of the adrenal gland is called malignant pheochromocytoma.

Adrenal gland A gland located on each kidney that secretes hormones regulating metabolism and controls heart rate and blood pressure, sexual function, water balance, and stress.

Adrenocortical cancer (adrenocortical carcinoma or cancer of the adrenal cortex) A rare cancer that forms in the outer layer of tissue of the adrenal gland.

Adult progeria (Werner syndrome or WS) An autosomal recessive gene disease, an inherited disorder, caused by a mutation in a gene involved in cell division and marked by rapid aging that begins in early adolescence. Patients may be shorter than average and have health problems such as loss and graying of hair, hardening of the arteries, thinning of the bones, diabetes, and thin, hardened skin. They also have an increased risk of cancer, especially osteosarcoma.

Adult T-cell leukemia/lymphoma (ATLL) A fast-growing (aggressive) type of T-cell non-Hodgkin lymphoma caused by the human T-cell leukemia virus type 1 (HTLV-1). It is marked by bone and skin lesions, high calcium levels, and enlarged lymph nodes, spleen, and liver.

Adult-onset diabetes Former term for type 2 diabetes.

Adverse events Undesirable experiences occurring after immunization that may or may not be related to the vaccine.

Advisory Committee on Immunization Practices (ACIP) A panel of ten experts who make recommendations on the use of vaccines in the United States. The panel is advised on current issues by representatives from the Centers for Disease Control and Prevention, Food and Drug Administration, National Institutes of Health, American Academy of Pediatrics, American Academy of Family Physicians, American Medical Association, and others. The recommendations of the ACIP guide immunization practice at the federal, state, and local level.

Affinity chromatography A method for separating proteins/molecules by using their ability to bind specifically to other molecules. A biological/affinity molecule can be immobilized and a protein will bind specifically to it. If an antibody is used as the immobilized molecule to "capture" its antigen, then it is called immunoaffinity chromatography.

Agammaglobulinemia An almost total lack of immunoglobulins or antibodies.

Age-related macular degeneration or AMD or ARMD or macular degeneration A condition most often seen in people who are over the age of 50, in which there is a slow breakdown of cells in the center of the retina. This blocks vision in the center of the eye and can cause difficulties in reading and driving.

AGEs Stands for advanced glycosylation endproducts. AGEs are produced in the body when glucose links with protein. They play a role in damaging blood vessels, which can lead to diabetes complications.

Aggravating factor Something that makes a condition worse. For example, tobacco smoke is an aggravating factor for asthma.

Aggressive lymphoma (high-grade lymphoma or intermediate-grade lymphoma) A type of lymphoma that grows and spreads quickly and has severe symptoms.

Aggressive In medicine, describes a tumor or disease that forms, grows, or spreads quickly. It may also describe treatment that is more severe or intense than usual.

Agnogenic myeloid metaplasia (chronic idiopathic myelofibrosis or idiopathic myelofibrosis or myelosclerosis with myeloid metaplasia or primary myelofibrosis) A progressive, chronic disease in which the bone marrow is replaced by fibrous tissue and blood is made in organs like liver and the spleen, instead of in the bone marrow. This disease is marked by an enlarged spleen and progressive anemia.

Agonist A drug or substance that binds to a receptor inside a cell or on its surface and causes the same action as the substance that normally binds to the receptor.

AIDS-related cancer Types of cancer that is more likely to occur in people who are infected with the human immunodeficiency virus (HIV) and thus considered to have AIDS. The most common AIDS-related cancers are non-Hodgkin lymphoma, Kaposi sarcoma, and cervical cancer along with less common types of cancers of the mouth, throat, liver, lung, colon, rectum, anus, testes, and skin. People infected with HIV who develop any one of these cancers are.

Albuminuria A condition in which the urine has more than normal amounts of albumin. Albuminuria may be a sign of nephropathy, a kidney disease.

Allergen Any substance that causes an allergy.

Allergic reaction Secondary immune response to an environmental allergen.

Allergy (or hypersensitivity) An inappropriate, misguided, and harmful response of the immune system to normally harmless substances where the body has an exaggerated response to a substance (e.g., food or drug).

Allogeneic stem cell transplantation A procedure in which a person receives blood-forming stem cells from a genetically similar but nonidentical donor like a sister or brother.

Alpha cell A type of cell in the pancreas, which releases glucagon hormone called. When blood glucose falls too low, the body sends a signal to the alpha cells to make glucagon. Then glucagon reaches the liver where it tells it to release glucose into the blood for energy.

Alpha-1 antitrypsin It is produced mostly in the liver to protect the lungs from neutrophil elastase. It destroys neutrophil elastase before it has a chance to begin damaging the delicate lung tissue. However, if an individual does not have enough alpha-1 antitrypsin, the enzyme goes unchecked, and it may attack the lung.

Alpha-glucosidase inhibitor (generic names, acarbose and miglitol) A class of oral medicine for type 2 diabetes that blocks slowly the enzymes that digest starches in food. The result is a lower rise in blood glucose throughout the day, especially right after meals.

Alternative medicine Treatments that are used instead of standard treatments. Less research has been done for most types of alternative medicine and is less accepted and widely used. Alternative medicine may include megadose vitamins, special diets, herbal preparations, special teas, and magnet therapies also for treatment of different diseases or conditions. For example, a special diet may be used instead of anticancer drugs in cancer treatment.

Alveolar rhabdomyosarcoma (or ARMS) A soft tissue tumor that is most common in older children and teenagers. It begins in embryonic muscle cells and can occur at many places in the body but usually occurs in the trunk, arms, or legs.

Amastigote A form of Trypanosomatidae with a short flagellum.

Amelanotic melanoma A type of skin cancer in which the cells do not make the pigment melanin. Skin lesions are often irregular and may be pink or red or have light brown, tan, or gray at the edges.

Amoeboid Irregular and changeable in shape; movement by extension of pseudopodia.

Ampulla of Vater cancer (or ampullary cancer) Cancer that forms in the ampulla of Vater with symptoms of jaundice, abdominal pain, nausea, vomiting, and weight loss.

Ampulla of Vater An enlargement of the ducts from the liver and pancreas at the point where they enter the small intestine.

Amylin A hormone formed by beta cells in the pancreas. Amylin regulates the timing of glucose release into the bloodstream after eating by slowing the emptying of the stomach.

Amyotrophy A type of neuropathy resulting in pain, weakness, and/or wasting in the muscles.

Anemia A condition in which the number of red blood cells is less than normal, resulting in less oxygen being carried to the body's cells. A reduction in the number of circulating red blood cells or in the quantity of hemoglobin. In many diseases like malaria, anemia is quite common. Pallor may be visible in the patient.

Anal cancer Cancer that forms in tissues of the anus.

Anal Pap smear (or anal Pap test) A procedure in which cells are scraped from the lining of the anus and observed under microscope to find cancer and changes in cells that may lead to cancer. An anal Pap smear can also show conditions that are not cancerous, such as infection or inflammation.

Anaphylactic shock A severe and sometimes life-threatening allergic immune reaction to an antigen that a person has been previously exposed to. It is characterized by a swelling of body tissues including the throat, difficulty in breathing, and a sudden fall in blood pressure due to collapsed blood vessels, itchy skin, edema, fainting, and death.

Anaphylaxis An immediate and severe allergic reaction to a substance (e.g., food or drugs) which can be fatal and requires immediate medical attention. Symptoms of anaphylaxis include breathing difficulties, loss of consciousness, and a drop in blood pressure.

Anaplastic large cell lymphoma (or ALCL) An aggressive type of non-Hodgkin lymphoma that is usually of the T-cell type and may appear in the lymph nodes, skin, bones, soft tissues, lungs, or liver. The cancer cells express a marker called CD30 or Ki-1 on the surface of lymphoma cells.

Anaplastic thyroid cancer A rare, aggressive type of thyroid cancer in which the malignant cells look very different from normal thyroid cells.

Anaplastic A term used to describe cancer cells that divide rapidly and have little or no resemblance to normal cells.

Anastomosis The joining together of two ends of healthy bowel after diseased bowel has been cut out by the surgeon. This may be contrasted to a colostomy, when the bowel ends may be permanently diverted or anastamosed at a later surgery.

Androblastoma (or arrhenoblastoma or Sertoli-Leydig cell tumor of the ovary) A rare type of ovarian tumor in which the tumor cells secrete a male sex hormone. This may cause virilization (the appearance of male physical characteristics in females).

Anergy A state of unresponsiveness, induced when the T cell's antigen receptor is stimulated that effectively freezes T-cell responses pending a "second signal" from the antigen-presenting cell (co-stimulation).

Angiogenesis Blood vessel formation. Tumor angiogenesis is the growth of new blood vessels that tumors need to grow. This process is caused by the release of chemicals by the tumor and by host cells near the tumor.

Angioimmunoblastic T-cell lymphoma An aggressive type of T-cell non-Hodgkin lymphoma marked by enlarged lymph nodes and hypergammaglobulinemia (increased antibodies in the blood). Other symptoms may include a skin rash, fever, weight loss, or night sweats.

Angiopathy Any disease of the blood vessels (veins, arteries, capillaries) or lymphatic vessels.

Angioplasty A nonsurgical procedure for treating diseased arteries.

Angiosarcoma A type of cancer that begins in the cells that line blood vessels or lymph vessels. Cancer that begins in blood vessels is called hemangiosarcoma. Cancer that begins in lymph vessels is called lymphangiosarcoma.

Animal trap A cage, generally made of cloth, that is baited with an animal such as a cow, goat, etc. Collections of mosquitoes are made on the walls of this trap to assess and compare populations biting domestic animals with populations in dwellings.

Anopheles A genus of mosquito, some species of which can transmit human malaria.

Anorexia Lack of appetite, lack of desire or interest in food.

Anterior urethral cancer A disease in which malignant cells are found in the part of the urethra that is closest to the outside of the body.

Anthrax An acute infectious disease caused by the spore-forming bacterium *Bacillus anthracis*. Anthrax most commonly occurs in hoofed mammals and can also infect humans.

Anthropophagy The process of feeding on people as opposed to animals. Similar to anthropophilic.

Anthropophilic Describes mosquitoes that prefer to take blood meals from humans.

Antibiotic A drug that kills or slows the growth of bacteria, for example, penicillin.

Antibodies (Also known as immunoglobulins.) Antibodies are specialized Y-shaped circulating proteins that are part of the immune system to protect itself from "foreign" substances such as bacteria or viruses. They are created when an antigen (such as a virus or bacteria) is detected in the body. The antibodies bond with the specific antigen that triggered their production and that action neutralizes the antigen, which is a threat to the body. Antibodies are created to fight off whatever has invaded the body. Proteins that are found in blood or other bodily fluids of vertebrates. Antibodies are produced by a kind of white blood cell called a B cell.

Antibody-dependent cell-mediated cytotoxicity (ADCC) An immune response in which antibody, by coating target cells, makes them vulnerable to attack by immune cells.

Antibody-mediated immunity (or humoral immunity) Immunity that results from the activity of antibodies in blood and lymphoid tissue.

Antigen A foreign substance (usually a protein or carbohydrate) capable of triggering an immune response in an organism. When the body detects it, it produces specific antibodies to fight off the antigen. Antigens may be soluble substances, such as toxins and foreign proteins, or particulate, such as bacteria and tissue cells; however, only the portion of the protein of polysaccharide molecule known as the antigenic determinant combines with antibody or a specific receptor on a lymphocyte. Antigens are usually proteins or polysaccharides and include the coats, capsules, cell walls, flagella, fimbriae, and toxins of parasites, bacteria, viruses, and other microorganisms.

Antigen-presenting cells (APCs) Cells (B cells, dendritic cells, and macrophages) that process and present exogenous antigens to T lymphocytes.

Antihypertensive A medication or other therapy that lowers blood pressure.

Anti-infection vaccines Vaccines that prevent people from becoming infected after being bitten by infected mosquitoes.

Antimicrobial agents The drugs, chemicals, or other substances that kill or slow the growth of microbes. They include antibacterial drugs (which kill bacteria), antiviral agents (which kill viruses), antifungal agents (which kill fungi), and antiparasitic drugs (which kill parasites).

Antimicrobial resistance Antimicrobial resistance is the result of microbes changing in ways that reduce or eliminate the effectiveness of drugs, chemicals, or other agents to cure or prevent infections.

Antinuclear antibody (ANA) An autoantibody directed against a substance in the cell's nucleus.

Antiretroviral drugs Drugs that act against retroviruses (such as HIV).

Antiserum Serum that contains antibodies against a particular microbe.

Antitoxin Antibodies capable of destroying and inactivate toxins generated by microorganisms including viruses and bacteria.

Antiviral Literally "against virus" – any medicine capable of destroying or weakening a virus.

API Annual parasite incidence. API = confirmed cases during 1 year/population under surveillance ×1000.

Apical complex A structure in merozoites and sporozites of Apicomplexa, comprising a polar ring, conoid, and rhoptries, that aids penetration of host cells.

Appendix Lymphoid organ in the intestine.

Aralen A brand name for chloroquine phosphate.

ARB ARB stands for angiotensin receptor blocker. An oral medicine that lowers blood pressure.

Archived tumor sample A tumor sample that has been routinely preserved and stored Tumor tissue is commonly preserved for storage by treatment with a preservative called formalin and then embedded in paraffin wax.

Arm A group of participants in a clinical trial, all of whom receive the same treatment, intervention, or placebo. The other arm(s) receives a different treatment.

Aromatase Inhibitor A medication that reduces the amount of estrogen in the body. It may be used to treat women with estrogen-receptor-positive breast cancer.

Artemisinin-based combination therapies (ACTs) A group of malaria medications that produces a very fast response in people with malaria and are active against multidrug-resistant *Plasmodium falciparum* malaria. ACTs are well tolerated by people who have malaria and have the potential to reduce malaria transmission by decreasing the presence of the parasite in the bloodstream. It is used for the treatment (not prevention) of malaria usually as a part of a combination therapy, derived from the sweet wormwood or Qinghao plant (*Artemisia annua*).

Arteriosclerosis Hardening of the arteries.

Artery A large blood vessel that carries blood with oxygen from the heart to all parts of the body.

Arthralgia Joint pain

Arthritis The inflammation of a joint which results in pain and difficulty in moving. The inflamed joints developed stiffness, warmth, swelling, redness, and pain. There are over 100 types of arthritis like rheumatoid arthritis, osteoarthritis, psoriatic arthritis, ankylosing spondylitis, lupus, gout, and pseudogout.

Asexual A form of reproduction not involving the fusion of gametes.

Aspart Insulin A rapid-acting insulin. On average, insulin aspart starts to lower blood glucose within 10–20 min after injection. It has its strongest effect 1–3 h after injection but keeps working for 3–5 h after injection.

Aspartame (brand names, Equal, NutraSweet) A dietary sweetener with almost no calories and no nutritional value.

Assay Analytic procedure in laboratory medicine, pharmacology, environmental biology, continuous delivery, and molecular biology for qualitatively assessing or quantitatively measuring the presence, amount, or functional activity of a target entity (the analyte).

Association The degree to which the occurrence of two variables or events is linked. Association describes a situation where the likelihood of one event occurring depends on the presence of another event or variable. However, an association between two variables does not necessarily imply a cause and effect relationship. The term association and relationship are often used interchangeably.

Asthma Chronic inflammatory condition where the bronchial tubes (in the lungs) become easily irritated, causing difficulty in breathing. This leads to constriction of the airways resulting in wheezing, coughing, difficulty in breathing, and production of thick mucus. Inflammation is caused by over-reactive airways. The cause of asthma is not yet known, but environmental triggers, drugs, food allergies, exercise, infection, and stress have all been implicated.

Asymptomatic infection (inapparent or subclinical infection) The presence of an infection without symptoms.

Atherosclerosis A collection of a number of diseases where clogging, narrowing, and hardening of the body's large arteries and medium-sized blood vessels occur. Atherosclerosis can lead to stroke, heart attack, eye problems, and kidney problems. Fat builds up in the large- and medium-sized arteries, preventing the blood flow in the vessels.

Atovaquone A drug used against malaria. It is found in the combination atovaquone–proguanil which can be used for both prevention and treatment.

Attenuated vaccine (or live vaccine) A vaccine in which live bacteria or viruses are weakened through chemical or physical processes in order to produce an immune response without causing the severe effects of the disease. Attenuated vaccines currently licensed in the United States include measles, mumps, rubella, polio, typhoid, yellow fever, and varicella.

Autism A chronic developmental disorder usually diagnosed between 18 and 30 months of age. Symptoms include problems with social interaction and communication as well as repetitive interests and activities. At this time, the cause of autism is not known although many experts believe it to be a genetically based disorder that occurs before birth.

Autoantibodies Autoantibodies are a group of antibodies that "go bad" and mistakenly attack and damage the body's tissues and organs.

Autochthonous Regarding malaria, it refers to local transmission. This can either be indigenous or introduced (in a geographic area where malaria does not occur regularly). Differentiated from imported, congenital, or blood-borne malaria.

Autoimmune disease (or autoimmune disorder) Disorder of the body's immune system in which the immune system mistakenly attacks and destroys body tissue that it believes to be foreign. Patients with different autoimmune diseases frequently have unusual antibodies circulating in their blood. These antibodies target their own body tissues. In women the frequency of autoimmune diseases is more than in men. The estrogen of females may influence the immune system of some women to develop autoimmune diseases. The presence of one autoimmune disease furthermore increases the chance for developing another consequent autoimmune disease. Examples of autoimmune diseases include systemic lupus erythematosus, Addison disease, scleroderma, Hashimoto thyroiditis, Sjogren syndrome, rheumatoid arthritis, juvenile (type 1) diabetes, polymyositis, vitiligo, pernicious anemia, glomerulonephritis, and pulmonary fibrosis.

Autoinfection For parasites that normally transmit from host to host by a free-living stage, the rapid development and establishment of a parasite in the same host individual from which it originated.

Autonomic neuropathy A type of neuropathy affecting the lungs, heart, stomach, intestines, bladder, or genitals.

Autoreactive Describes immune cells that mount a response against the body's own cells or tissues.

Avidity A term used to describe the combined strength of multiple bond interactions. Avidity is a term that can be used to describe antibody-antigen interactions, where multiple, weak bonds form between antigen and antibody. Individually, each bond may be weak; however, when multiple individual bonds are present at the same time, the overall effect is strong binding of antigen to antibody.

B cell or B lymphocyte One type of lymphocyte generally considered part of the adaptive immune system. B cells come from bone marrow and develop into blood cells called plasma cells, which are the source of antibodies and mediate humoral immunity.

Background retinopathy (or simple or non-proliferative retinopathy) It is an early stage of diabetic retinopathy. A type of damage to the retina of the eye marked by bleeding, fluid accumulation, and abnormal dilation of the blood vessels.

Bacteria Tiny one-celled organisms present throughout the environment that require a microscope to be seen. While not all bacteria are harmful, some cause disease. Examples of bacterial disease include diphtheria, pertussis, tetanus, *Haemophilus influenzae*, and pneumococcal.

Barium enema (or lower GI series) The barium enema procedure consists of the insertion of barium (a radiolucent solution like a series of X-rays) to coat the lower gastrointestinal tract with barium. On X-ray, areas in which the barium "lights up" may indicate abnormal cell proliferation.

Basal rate A steady trickle of low levels of longer-acting insulin, such as that used in insulin pumps.

Basal secretion (basal insulin) We all should have a small amount of insulin that is constantly present in the blood; that is the basal secretion. People with type 1 diabetes must take a form of insulin that replicates the basal secretion throughout the day; that is basal insulin.

Basophil Basophil is a type of white blood cell. It contains the inflammatory mediators like histamine. Along with mast cells, basophils are responsible for the symptoms of allergy.

Bed nets Bed nets are used to prevent malaria transmission by forming a protective barrier around persons using them and therefore limiting their exposure to mosquito bites. Bed nets have repeatedly been shown to reduce severe disease and mortality due to malaria in endemic regions.

Beta-blocker An antihypertensive medication that limits the activity of hormone epinephrine.

Beta cells Beta cells are located in the islets of Langerhans in the pancreas. They are responsible for making insulin.

Beta-glucan Polymer of glucose molecules attached in a specific conformation. Beta-glucans can come from different sources such as cereals (oats and barley) and fungi such as mushroom and yeast.

Complement A system of proteins that are present in blood that act together to defend the body against pathogens.

Bias Flaws in the collection, analysis, or interpretation of research data that lead to incorrect conclusions.

Biguanide (generic name, metformin) A class of oral medicine used to treat type 2 diabetes that lowers blood glucose by reducing the amount of glucose produced by the liver and by helping the body respond better to insulin.

Biochemicals Chemicals produced within living organisms. Many coordinate to fight off invasion in an immune response.

Bioinformatics The use of data/information from experiments of genomics for research generation/creation, collection, and storage (in databases) of different collective information to efficiently accomplish an objective to discover a new pharmaceutical, a new medicine, or a new herbicide or in therapeutics.

Biological barriers The body's first layer of protection against harmful microbes.

Biological plausibility A causal association is consistent with existing medical knowledge.

Biological response modifiers (BRM) Substances, either natural or synthesized, that boost, direct, or restore normal immune defenses. BRMs include interferons, interleukins, thymus hormones, and monoclonal antibodies.

Biopsy A procedure where tumor tissue is removed from the body for laboratory examination to determine whether or not cancer is present. A biopsy can be performed using a needle to extract a small piece of tissue or as a surgical procedure to remove a larger piece of tissue.

Biotechnology The use of living organisms or their products to make or modify a substance. Biotechnology includes recombinant DNA technology and hybridoma technology.

Blood glucose level The amount of glucose in a given amount of blood. It is noted in milligrams in a deciliter or mg/dL. This level is very important for people with diabetes, and they must monitor their blood glucose level throughout the day. If the blood glucose level is too high (hyperglycemia), that means that there is not enough insulin in the blood. If it is too low (hypoglycemia), that means that there is too much insulin.

Blood glucose meter A small, portable electronic machine (biosensors) used by people with diabetes to check their blood glucose levels. After pricking the skin with a lancet, one places a drop of blood on a test strip in the machine. The monitor soon displays the blood glucose level as a number on the meter's digital display.

Blood glucose monitoring The simple blood test used to check the amount of glucose in the blood; a tiny drop of blood, taken by finger pricking, is placed on a test strip and inserted in the meter for reading.

Blood glucose Also known as blood sugar, glucose comes from food and is then carried through the blood to deliver energy to cells. The main sugar found in the blood and the body's main source of energy.

Blood meal Blood taken from a human or other host by a mosquito.

Blood pressure The amount of force exerted by the blood against the walls of the arteries. The units are expressed in millimeters of mercury. The systolic pressure is the top number and is the pressure in the arteries where the heart is forcing blood through them. The diastolic pressure is the bottom number and is the pressure in the arteries when the heart relaxes.

Blood urea nitrogen (BUN) Protein breakdown waste product in blood from which the urea is removed by the kidneys. Rise of BUN level is indicative of kidney malfunction.

Blood vessel A vessel that carries blood. Component of the circulatory system. Examples are artery, vein, or capillary.

Blood-brain barrier Layer of tissue separating the brain from the body. This blocks the entry of immune cells inside the brain.

Blood-forming stem cells Progenitor, multipotent cells found in the bone marrow that multiply profusely and differentiate into red blood cells (RBC), white blood cells (WBC), and platelets.

Body mass index (BMI) Used to evaluate body weight relative to a person's height. BMI indicates the physical status of an individual as underweight, normal weight, overweight, or obese.

Bolus secretion (bolus insulin) After food intake, the pancreas releases insulin to essential for carbohydrate digestion, also termed as the bolus secretion. Type 1 diabetes individuals however require supplementary insulin that replicates the bolus secretion; this is known as bolus insulin.

Bolus An extra insulin dose required to combat an expected blood glucose level rise after a meal or snack.

Bone marrow Includes soft tissue located in the bone cavities. It is the place of origin of all different lineages of blood cells.

Booster shots Doses of a vaccine administered to "boost" the immune system of host.

Borderline diabetes Also termed as type 2 diabetes or impaired glucose tolerance.

Brachial neuritis Nerve inflammation in the arm causing muscle weakness and pain.

Bradycardia Abnormally slow heartbeat.

BRCA1 and BRCA2 Genes that control cell growth. Mutations in the BRCA1 and/or BRCA2 gene(s) lead to higher risk in developing breast and ovarian cancer.

Breakthrough infection Development of an infectious disease by a vaccinated individual.

Brittle diabetes When the blood glucose level rises often from low to high and from high to low.

Bunion A bulge on the first joint of the big toe, due to swelling of a fluid sac under the skin which may become red, sore, and infected.

Cachexia Disease-related physical wasting with loss of weight and body musculature.

Callus Thick or hardened skin of the foot, developed due to rubbing or pressure.

Calorie Represents food energy from carbohydrate, protein, fat, and alcohol.

Cancer staging Includes numbers I to IV of how much a cancer has spread in the body. Criteria for staging include tumor size, portion of tissue penetrated, its presence in adjacent organs, and lymph affected.

Cancer A disease condition manifested by uncontrolled cell division, loss of cell death, and loss of normal cell's life cycle. Cancer cells can also invade nearby tissues and can spread through the bloodstream and lymphatic system to other parts of the body through a process called metastasis.

Capillary Smallest of the body's blood vessels. They transport oxygen and glucose to the cells. Waste products such as carbon dioxide are transported back from the cells into the blood through capillaries.

Capsaicin Constituent of hot peppers. Finds use on the skin to relieve pain from diabetic neuropathy.

Carbamate Used as an insecticide.

Carbohydrate counting A method meal planning for diabetic individuals based on counting the number of grams of carbohydrate in food.

Carbohydrate One of the components of food that provides us energy. Source of carbohydrates include starches, vegetables, fruits, dairy products, and sugars. Examples are bread, rice and potatoes, starch, etc.

Carcinoma in situ Tumor of the epithelial cells restricted to the tissue of origin, without invasion the basement membrane.

Carcinoma Malignant cancer originating in the epithelial cells that is capable of invading the surrounding tissue and metastasizing to the different tissues.

Cardiac arrest When the heartbeat stops.

Cardiac catheterization Used for diagnosis. In this procedure a catheter which is a tiny hollow tube is allowed to enter a vessel in the groin through the aorta into the heart so as to do imaging studies or diagnosis of problems on the heart and blood vessels.

Cardiac output Blood that is pumped through the circulatory system in 1 min.

Cardiologist A doctor who treats cardiac disorders.

Cardiology Study of the heart and cardiac treatment.

Cardiometabolic risk factors Set of factors that poses risk for development of diabetes or heart disease.

Cardiomyopathy A disease of the heart muscle thereby leading to the loss of its pumping strength.

Cardiovascular (CV) Involves the components of the circulatory system, heart and blood vessel.

Cardiovascular disease (CVD) Diseases of the heart and circulatory system.

Cataract Clouding of the lens of the eye.

Catarrhal inflammation Inflammation affecting mucous surface.

Causal association One variable (e.g., smoking) may sometimes affect another variable (e.g., cancer).

CD (Cluster of differentiation) Molecules expressed on cell surface of immune cells.

CD4+ helper T cells T cells expressing CD4 receptors. Helper T cells are infected and killed by HIV.

CD8+ T lymphocyte T cells expressing CD8 receptors. They may be cytotoxic T lymphocytes (CTLs) or suppressor T cells.

Cell-mediated immunity (cellular immunity) Immune response mediated by helper T cells and CTLs. Its enables removal of cells infected with microorganisms such as viruses, fungi, and certain bacteria.

Cell The smallest and basic unit of life.

Central nervous system Composed of the brain and spinal cord. The central nervous system enables and controls the activity of the entire nervous system and interacts with the immune system.

Cerebral hemorrhage Bleeding within the brain.

Cerebral malaria Malaria caused at times with *P. falciparum* infection involves infection of the very small capillaries that flow through the tissues of the brain. It is highly fatal.

Cerebral thrombosis Blood clot in an artery that supplies blood to the brain thereby causing a block.

Cerebrovascular accident Apoplexy or stroke caused by an impeded blood supply to the brain.

Cerebrovascular disease Damage to blood vessels in the brain. Vessels can burst and bleed or become clogged with fatty deposits. When blood flow is interrupted, brain cells die or are damaged, resulting in a stroke.

Cerebrovascular Involving blood vessels in the brain.

Certified diabetes educator (CDE) A healthcare professional with expertise in diabetes education.

Challenge The deliberate exposure of an immunized animal or person to the infectious agent in vaccination protocols.

Charcot's foot In this condition the foot joints and soft tissue are destroyed. Results from nerve damage.

Chemokine Molecule released by immune cells that causes WBC neutrophils and monocytes to move through forming a network between immune cells.

Chemoprophylaxis The use of antimalarial drugs to prevent malaria.

Chemotherapy Cytotoxic drug treatment in cancer.

Chloroquine A drug used against malaria for both prevention and treatment. The mainstay of malaria treatment since 1945, but a large number of resistant strains of *P. falciparum* also are known.

Chlorpropamide Sulfonylurea's class of medicine used in type 2 diabetes treatment that acts by lowering blood glucose levels.

Cholecystitis Inflammation of the gallbladder.

Cholesterol Constituent of food and is also produced by the liver and found in the blood. There are two types of cholesterol, HDL and LDL. High levels are a risk factor for heart disease and stroke.

Chromosome Carriers of genetic information.

Chronic health condition A health-related state lasting for a considerable period.

Chronic inflammation It is a prolonged and persistent inflammation.

Chronic Has occurred for a considerable period of time.

Cinchonism Quinine or quinidine side effects. Includes tinnitus, headache, nausea, diarrhea, altered auditory acuity, and blurred vision.

Circulation Blood flow through the heart and blood vessels.

Cirrhosis of the liver Chronic liver disease where normal liver parenchyma replaces the fibrous tissue.

Clindamycin An antibiotic used for malaria treatment in combination with another drug, like quinineÂ or quinidine.

Clinical cure Elimination of malaria symptoms, sometimes without eliminating all parasites.

Clinical laboratory services Different examination of human body materials for the purpose of providing information for the diagnosis, prevention, or treatment.

Clinical trial A research study to test drugs, procedures, or testing technologies to determine whether these are effective and/or safe in human use.

Clinically validated Determination that a test is accurate including determination of sensitivity, specificity, and positive and negative predictive values.

Clonal anergy Switching off the ability of potentially harmful T or B cells to participate in immune responses. Essential for generating T and B cells tolerance to the body's "self" tissue antigens.

Clonal deletion The genetically controlled process of eliminating immune cells that could destroy the body's own cells and tissues.

Clonal selection Lymphocyte proliferation process producing clone of antibodies after recognition of a specific antigen receptor on a lymphocyte that binds to a given antigen.

Clone A group of genetically identical cells with a single common ancestor.

Colectomy Surgical resection of all or part of the colon.

Colitis Inflammation of the colon.

Colon polyp A fleshy growth on the inner lining of the colon.

Colonoscopy Inspection through a fiber-optic scope of the inside of the colon.

Colorectal Related to the colon and/or rectum.

Coma A sleep-like state in which a person is not conscious.

Combination oral medicines Includes two or more different medicines.

Combination therapy Different medicines like oral hypoglycemic agents or an oral hypoglycemic agent together with an insulin used to control high blood glucose levels of type 2 diabetic individuals.

Combination vaccine Two or more vaccines administered in a single dose in order to reduce the number of shots given. For example, the MMR (measles, mumps, rubella) vaccine.

Communicable Transmitted from one person or animal to another.

Community immunity A sufficient proportion of a population is immune to an infectious disease through vaccination to make its spread from person to person unlikely. Also known as herd immunity.

Complement cascade Sequence of steps triggered by an antigen-antibody complex, in which each component of the complement system is activated in turn.

Complement receptor A cell surface protein that binds to activated complement proteins.

Complications Harmful effects of disease.

Congenital defects Problems or conditions that are present at birth.

Congenital malaria Malaria in a newborn or infant, transmitted from the mother.

Congestive heart failure A condition in which the heart cannot pump out all of the blood that enters it, which leads to an accumulation of blood in the vessels and fluid in the body tissues.

Conjugate vaccine Composed of two compounds, a protein and polysaccharide, to increase a vaccine's effectiveness.

Conjunctivitis Inflammation of the mucous membranes surrounding the eye caused by viruses, bacteria, or allergies causing the area to become red and irritated.

Constant region That part of an antibody's structure that is characteristic for each antibody class.

Contraindication A condition in a recipient which is likely to result in a life-threatening problem if a vaccine were given.

Conventional therapy Conventional therapy includes use of medication, meal planning, and exercise, along with regular visits to healthcare providers.

Core biopsy A needle is used to remove a small, intact sample of tissue from an identified breast mass in order to examine it and obtain a preliminary diagnosis.

Coronary artery disease Is a type of heart disease caused by the generation of plaque in the arteries that supply blood to the heart.

Coronary heart disease Heart disease caused by narrowing of the arteries that supply blood to the heart.

Coronary thrombosis The formation of a clot in one of the arteries that carry blood to the heart muscle.

Co-stimulation The delivery of a second signal from an antigen-presenting cell to a T cell. The second signal rescues the activated T cell from anergy, allowing it to produce the lymphokines necessary for the growth of additional T cells.

C-peptide "Connecting peptide," a substance the pancreas releases into the bloodstream in equal amounts to insulin.

CR3 receptors Protein on the immune cells surface including neutrophils which when bound to beta-glucan helps to prime the neutrophils to respond to infectious challenges.

Creatinine A protein waste product in the diet and from the muscles of the body. Creatinine is removed from the body by the kidneys; as kidney disease progresses, the level of creatinine in the blood increases.

Crib or cot death See sudden infant death syndrome (SIDS).

Crohn's disease A chronic medical condition characterized by inflammation of the bowel. Symptoms include abdominal pain, diarrhea, fever, loss of appetite, and weight loss.

Cryptic malaria A case of malaria where epidemiologic investigations fail to identify an apparent mode of acquisition.

Cutaneous leishmaniasis A form of leishmaniasis involving the skin at the site of sand fly bite.

Cytokine Secreted by immune cells. They are the key component of inflammation. Cytokines include lymphokines produced by lymphocytes and monokines produced by monocytes and macrophages.

Cytoplasm The living matter within a cell.

Cytotoxicity Degree to which a substance is poisonous to cells.

Dawn phenomenon The early-morning (4–8 am) rise in blood glucose level.

DDT DDT (dichlorodiphenyltrichloroethane) was the main insecticide to eradicate mosquitoes used during the 1950s and 1960s.

Death The cessation of life.

DEET N,N-Diethyl-meta-toluamide, an ingredient of insect repellents.

Defervescence The reduction of a patient's abnormally elevated temperature into the normal range.

Dehydration The loss of too much body fluid through frequent urinating, sweating, diarrhea, or vomiting.

Deltamethrin An insecticide.

Deltoid A muscle in the upper arm where shots are usually given.

Demyelinating disorders A medical condition where the myelin sheath is damaged. Examples of demyelinating disorders include multiple sclerosis (MS), optic neuritis, transverse neuritis, and Guillain-Barré syndrome (GBS).

Dendritic cell White blood cells found in the spleen and other lymphoid organs. They initiate the immune response by activating lymphocytes and stimulating the secretion of cytokines and also prevent autoimmune reactions.

Dermopathy Disease of the skin.

Desensitization Reduce or stop a response such as an allergic reaction to something.

Dextrose Simple sugar found in blood that serves as the body's main source of energy.

Diabetes Control and Complications Trial (DCCT) A study by the National Institute of Diabetes and Digestive and Kidney Diseases conducted from 1983 to 1993 in people with type 1 diabetes.

Diabetes educator A healthcare professional who teaches people who have diabetes how to manage their diabetes.

Diabetes insipidus A condition characterized by frequent and heavy urination, excessive thirst, and an overall feeling of weakness, with normal blood glucose levels.

Diabetes mellitus A condition characterized by hyperglycemia resulting from the body's inability to use blood glucose for energy. In type 1 diabetes, the pancreas no longer makes insulin, and therefore, blood glucose cannot enter the cells to be used for energy. In type 2 diabetes, either the pancreas does not make enough insulin or the body is unable to use insulin correctly.

Diabetes Prevention Program (DPP) A study by the National Institute of Diabetes and Digestive and Kidney Diseases conducted from 1998 to 2001 in people at high risk for type 2 diabetes.

Diabetes specialist nurses (DSNs) Nurses trained in diabetic care.

Diabetes A chronic health condition where the body is unable to produce insulin and properly breakdown sugar (glucose) in the blood. Symptoms include hunger, thirst, excessive urination, dehydration, and weight loss. The treatment of diabetes requires daily insulin injections, proper nutrition, and regular exercise. Complications can include heart disease, stroke, neuropathy, poor circulation leading to loss of limbs, hearing impairment, vision problems, and death.

Diabetic diarrhea Loose stools, fecal incontinence, or both that result from an overgrowth of bacteria in the small intestine and diabetic neuropathy in the intestines. This nerve damage can also result in constipation.

Diabetic ketoacidosis (DKA) An emergency condition in which extremely high blood glucose levels, along with a severe lack of insulin, result in the breakdown of body fat for energy and an accumulation of ketones in the blood and urine. Signs of DKA are nausea and vomiting, stomach pain, fruity breath odor, and rapid breathing. Untreated DKA can lead to coma and death.

Diabetic mastopathy A rare fibrous breast condition occurring in women and sometimes men, with long-standing diabetes. The lumps are not malignant and can be surgically removed, although they often recur.

Diabetic retinopathy Diabetic eye disease; damage to the small blood vessels in the retina. Loss of vision may result. The eye disease that occurs in someone with diabetes when the small blood vessels of the retina become swollen and leak liquid into the retina, causing blurring vision; it can sometimes lead to blindness.

Diabetogenic Causing diabetes. For example, some drugs cause blood glucose levels to rise, resulting in diabetes.

Diabetologist A doctor who specializes in treating people with diabetes.

Diagnosis The determination of a disease from its signs and symptoms, tests, and reports.

Dialysis The process of cleaning wastes from the blood artificially, done by the kidneys.

Dietitian A healthcare professional who advises people about meal planning, weight control, and diabetes management.

Dilated eye exam A test done by an eye care specialist in which the pupil (the black center) of the eye is temporarily enlarged with eyedrops to allow the specialist to see the inside of the eye more easily.

Diphtheria A bacterial disease marked by the formation of a false membrane, especially in the throat, which can cause death.

Disease Sickness, illness, or loss of health.

Disseminated infection An infection where the germ enters the body through a single entry point and then disperses throughout the body.

Distant recurrence The spread of cancer to parts of the body other than the place where the cancer first occurred. In breast cancer, the cancer can spread to the lungs, liver, brain, or bones.

Diurnal During the daytime.

DNA (deoxyribonucleic acid) The double-stranded, helical molecular chain found within the nucleus of each cell. DNA carries the genetic information that encodes proteins and enables cells to reproduce and perform their functions.

DNA vaccine (nucleic acid vaccine) Direct injection of gene(s) coding for specific antigenic protein(s).

Doxycycline An antibiotic drug by itself for prevention or in combination with either quinine or quinidine used in malaria treatment

D-phenylalanine derivative A class of oral medicine for type 2 diabetes that lowers blood glucose levels by helping the pancreas make more insulin right after meals. (Generic name, nateglinide)

Drug resistance Drug resistance is the result of microbes changing in ways that reduce or eliminate the effectiveness of drugs, chemicals, or other agents to cure or prevent infections.

Dukes staging system A system of staging rectal cancers developed by Cuthbert Duke in 1932. The original system had three stages but has been modified over time to include four stages with variations on two of the four stages.

Dupuytren's contracture A condition associated with diabetes in which the fingers and the palm of the hand thicken and shorten, causing the fingers to curve inward.

Dysplasia Abnormal cells that progress to cancer.

Dyspnea Shortness of breath.

Dysrhythmia An abnormal heart rhythm.

E. coli A gram-negative, facultative anaerobic bacterium.

Early-stage invasive breast cancer Include stage I and II and some stage III breast cancer.

Ectoparasite Ectoparasites are usually arthropods which parasitize the skin.

Edema Swelling caused by excess fluid in the body caused by the excessive accumulation of fluid in body tissues.

Effector arm The part of the immune system that recognizes and responds to infection.

Efficacy rate A measure used to describe how good a vaccine is at preventing disease.

Efficacy In vaccine research, the ability of a vaccine to produce a desired clinical effect, such as protection against a specific infection or disease, at the optimal dosage and schedule in a given population. A vaccine may be tested for efficacy in phase 3 trials if it appears to be safe and shows some promise in smaller phase 1 and 2 trials.

EIR Entomological inoculation rate = mas, where ma = number of mosquito bites per night and s = proportion of those bites positive for sporozoites.

Electromyography (EMG) A test used to detect nerve function, measuring the electrical activity generated by muscles.

Electrophoresis DNA or proteins are separated by a technique using an electric current.

Elephantiasis Syndrome developed due to long-term obstruction of lymphatic vessels that leads to engorgement and thickened skin. It causes disfigurement, often of the leg.

Elimination The process of removing something on a temporary or semipermanent basis.

ELISA (enzyme-linked immunosorbent assay) A biochemical test that detects antibodies to a specific antigen (foreign substance in the body) based on a reaction that leads to a detectable color change in the test tube.

Encephalitis Inflammation of the brain caused by a virus. Encephalitis can result in permanent brain damage or death.

Encephalopathy A general term describing brain dysfunction. Examples include encephalitis, meningitis, seizures, and head trauma.

Endemic The constant presence of a disease or infectious agent within a given geographic area or population group; can also refer to the usual prevalence of a given disease within such area or group.

Endocrine gland A group of specialized cells that release hormones into the blood. For example, the islets in the pancreas, which secrete insulin, are endocrine glands.

Endocrinologist A doctor who treats people who have endocrine gland problems such as diabetes.

Endogenous Anything arising or originating within the body or derived from the body.

Endophagic Feeds indoors. An endophagic mosquito is a mosquito that feeds indoors.

Endophilic Tends to inhabit/rest in indoor areas.

Endoscope A long slender medical instrument for examining the interior of a bodily organ or performing minor surgery.

Endoscopy Visual examination of a bodily orifice, canal, or organ using an endoscope.

Endotoxin Poison in bacterial outer membranes like lipopolysaccharide that is harmful to the body.

Enzyme Biochemical catalyst that speeds up a biochemical reaction.

Eosinophil A granulocytic white blood cell.

Epidemic The occurrence of disease within a specific geographical area or population that is in excess of the normal level.

Epidemiology The study of the distribution and determinants of health-related states or events in specified populations and the application of this study to the control of health problems.

Epithelial cells Cells that make up the epithelium, the covering for internal and external body surfaces.

Epitope A specific site on an antigen that stimulates specific immune responses, such as the production of antibodies or activation of immune cells.

ER (estrogen receptor) Protein receptors for estrogen in the cell.

Eradication The process of removing something permanently.

Erectile dysfunction (or impotence) The inability to get and/or keep an erection for sexual intercourse, in men with diabetes. This may be due to poor blood supply to the penis and/or nerve damage caused by diabetes complications.

Erythema multiforme A medical condition characterized by inflammation of the skin or mucous membranes (including the mouth, throat, and eyes). Erythema multiforme has been reported following infection. Symptoms persist anywhere from 2 days to 4 weeks and include skin lesions, blisters, itching, fatigue, joint pain, and fever.

Erythrocyte A red blood cell (RBC).

Erythrocytic schizogony The process of asexual reproduction of malaria parasites within red blood cells.

Erythrocytic stage A stage in the life cycle of the malaria parasite found in the RBC. Erythrocytic stage parasites cause the symptoms of malaria.

Estrogen A steroid hormone produced by the ovaries, responsible for promoting development and maintenance of female secondary sex characteristics. Estrogen may play a role in certain immune system diseases.

Etiology The cause or origin of a disease or disorder; the study of the factors that cause disease and of the method of their introduction into the host.

Euglycemia A normal level of glucose in the blood.

Exchange lists One of the several approaches for diabetes meal planning. Foods are categorized into three groups based on their nutritional content. Lists provide the serving sizes for carbohydrates, meat and meat alternatives, and fats. These lists allow for substitution for different groups to keep the nutritional content fixed.

Exoerythrocytic schizogony The process of asexual reproduction of malaria parasites outside of red blood cells, usually in the liver. This process is asymptomatic.

Exit trap A trap constructed to capture mosquitoes that are exiting a house or structure. Often used in studies that compare the tendency of mosquitoes to rest indoors after feeding versus to fly outside after feeding.

Exoerythrocytic stage A stage in the life cycle of the malaria parasite found in liver cells (hepatocytes). Exoerythrocytic stage parasites do not cause symptoms.

Exogenous Anything arising or originating outside the body.

Exophagic Feeds outdoors.

Exophilic Tends to inhabit/rest in outdoor areas. After biting, an exophilic mosquito flies outside and rests in the woods, grass, or other outside areas. Exophilism makes use of residual insecticides in buildings less effective.

Exposure Contact with infectious agents (bacteria or viruses) in a manner that promotes transmission and increases the likelihood of disease.

Expression system In genetic engineering, the cells into which a gene has been inserted to manufacture desired proteins.

Exudative inflammation One type of inflammation in which the prominent feature is an exudate.

Fansidar Brand name of sulfadoxine–pyrimethamine, a drug used against malaria. Its value has been compromised by the emergence of drug-resistant malaria parasites.

Fasting blood glucose A blood glucose test taken before eating, usually in the morning, and is used when diagnosing diabetes.

Fat Fat is an energy source for your body (the other two energy sources are carbohydrates and protein).

Febrile Relating to fever; feverish.

Fibrillation Rapid contractions of the heart muscles.

Fibrinous inflammation Inflammation marked by an exudate of coagulated fibrin.

Functional antibody An antibody that binds to an antigen and has an effect that can be demonstrated in laboratory tests.

Fungal infections Any infection caused by fungus.

Fungus Member of a class of relatively primitive vegetable organism. Fungi include mushrooms, yeasts, rusts, molds, and smuts.

G6PD deficiency An inherited abnormality that causes loss of a red blood cell enzyme. It may give a person some protection against malaria, but it also means that person cannot take the antimalarial drug primaquine.

Gametocyte count Number of gametocytes per mm^3 of blood. The lower the gametocyte count, the lower the infectivity of the human to the mosquito.

Gametocyte rate Percentage of persons in an area who carry gametocytes. Expressed as a percentage.

Gametocyte The sexual reproductive stage of the malaria parasite.

Hemoglobin A part of a red blood cell or erythrocyte that carries oxygen to the cells. Sometimes it joins with the glucose in the blood flow.

Infection The invasion of an organism by a pathogen such as bacteria, viruses, or parasites, some of which lead to disease.

Infectious agents Organisms (e.g., bacteria or viruses) capable of spreading disease.

Infectious (or communicable) Capable of spreading disease.

Inflammation It is an accumulation of fluid and cells that occurs as the immune system fights a hostile invader. Symptoms include redness, swelling, heat, and pain which are induced by chemicals released by macrophages, thus resulting from injury to tissue.

Inflammatory bowel disease (IBD) A general term for any disease characterized by inflammation of the bowel. Symptoms include abdominal pain, diarrhea, fever, loss of appetite, and weight loss. Crohn's disease and colitis are the types of IBD.

Inflammatory response Infection often results in tissue damage, which may trigger an inflammatory response. This response promotes blood flow to the area, increases the permeability of capillaries, redness, and induces coagulation. The increased blood flow is responsible for the leakiness of the capillaries which allows cells and fluids to enter tissues, causing pain and swelling. These effects bring more phagocytic cells to the area to help eliminate the pathogens. The first cells to arrive, usually within an hour, are neutrophils and eosinophils, followed a few hours later by macrophages. Macrophages not only engulf pathogens but also help the healing process by disposing of cellular debris which accumulates from destroyed tissue cells and neutrophils that self-destruct after ingesting microorganisms. If infection persists, components of specific immunity – antibodies and T cells – arrive at the site to fight the infection.

Influenza A highly contagious viral infection characterized by sudden onset of fever, pains and severe aches, and inflammation of the mucous membrane.

Informed consent An agreement signed by prospective volunteers for a clinical research trial that indicates their understanding of (1) why the research is being done; (2) what researchers want to accomplish; (3) what will be done during the trial and for how long; (4) what risks are involved; (5) what, if any, benefits can be expected from the trial; (6) what other interventions are available; and (7) what are the participant's right to leave the trial at any time.

Innate immune system Component of the immune system that consists of a set of genetically encoded responses to pathogens and does not change or adapt during the lifetime of the organism. Innate immunity involves quickly mobilized defenses triggered by receptors that recognize a broad spectrum of microbes; in contrast to adaptive immunity, it does not acquire memory for an improved response during a second exposure to infection.

Insulin resistance It is an early sign of type 2 diabetes. Insulin resistance is when the body does not respond, as it should, to insulin. This affects adipose tissue particularly around the abdomen, hips, and thighs.

Insulin A hormone produced in the beta cells of the pancreas naturally in humans and animals. Insulin helps glucose in the blood enter your body's cells where it is used as fuel by your body. People with type 1 diabetes do not have this hormone; people with type 2 diabetes either do not have it or their bodies are not able to use it.

Interferon Protein molecules produced by virally infected cells. It helps the body fight off viral infections.

Interleukin A major group of lymphokines and monokines that act as inflammatory mediators.

Interstitial inflammation One type of inflammation affecting mainly the stroma of an organ.

Introduced malaria Mosquito-borne transmission of malaria, from an imported case in a geographic area where malaria does not occur regularly.

Intussusception A type of bowel blockage that happens when one portion of the bowel slides into the next, looking like the pieces of a telescope. It is treated in a hospital and may require surgery.

Invasive breast cancer (or infiltrating breast cancer) Cancer that has spread from breast into surrounding healthy tissue. Most invasive breast cancers start in the ducts that carry milk from the lobules to the nipple. Invasive breast cancer can spread to other parts of the body through the blood and lymphatic systems.

Investigational vaccine A vaccine that has been approved by the Food and Drug Administration (FDA) for testing and evaluation phase in clinical trials on humans and is not licensed for use in the general public.

IRB (institutional review board) A committee of physicians, statisticians, community advocates, and others that reviews clinical trial protocols before they can be initiated and is responsible for monitoring the safety of ethical rules during clinical trials at that institution and ensures the rights of participants.

Ischemia Decreased flow of oxygenated blood to an organ due to obstruction in an artery.

Ischemic heart disease Coronary artery disease or coronary heart disease caused by narrowing of the coronary arteries and decreased blood flow to the heart.

Jaundice Yellowing of the skin and eyes. This condition is often a symptom of hepatitis infection.

Juvenile diabetes Diabetes in childhood and adolescence.

Ketoacidosis A condition where ketones are released into the urine and blood, leading to diabetic ketoacidosis. It is often caused by an infection or other illness (like dehydration or from taking too little insulin) where the body begins to break down muscle and fat to regain energy.

Ketone test Strip test for the presence of ketones.

Ketones Ketones are a by-product when the body starts to break down fat in order to get energy. Accumulation of ketones in the blood makes the blood acidic and can lead to diabetic ketoacidosis.

Killer T cell A type of lymphocyte that directly attacks and kills infected cells or other targets (like tumor cells and even one's own tissues), generated by the coordinated action of dendritic cells and CD4+ helper T cells.

Knockout Knockouts are used by immunologists in genetic engineering to determine the functions of specific genes that encode immune proteins. A specific gene is deliberately removed in order to create a model organism which is unable to carry out the functions the gene codes.

Kupffer cells Specialized macrophages in the liver.

LAK cells Lymphocytes transformed in the laboratory into lymphokine-activated killer cells, which attack tumor cells.

Langerhans cells Dendritic cells in the skin that pick up antigen and transport it to lymph nodes.

Lariam (brand name, mefloquine) A drug used against malaria for both prevention and treatment.

Larvae Immature wingless forms of insects such as mosquitoes.

Latency The state or period in which a virus has invaded a host but is not actively multiplying, and during which symptoms of the infection are not seen. "Microbial latency" means the microbe is not multiplying, as occurs in some cells in HIV infection, while "clinical latency" means that the patient does not have symptoms of disease even though the virus is multiplying and damaging the immune system. In HIV, clinical latency precedes the AIDS stage.

Lectins Lectins are proteins that share the common characteristic of binding to carbohydrates. Plant lectins or hemagglutinins are primarily used for identified by their hemagglutinating activity. A variety of lectins occur in animal species also and they have differential carbohydrate recognition.

Leishman test Intradermal injection of leishmanial antigen causes a delayed tuberculin type of reaction.

Leishmaniasis A rare infectious disease caused by any of a number of parasitic *Leishmania* species. Infection can cause any of the three different manifestations, cutaneous leishmaniasis, mucosal leishmaniasis, and visceral leishmaniasis.

Leprosy A chronic, progressive infectious disease caused by *Mycobacterium leprae* which causes skin sores and also affects the eyes, mucous membranes, and peripheral nerves.

Lesion An abnormal change in the structure of an organ, due to injury or disease.

Leukemia Cancer of the blood cells, usually white blood cells.

Leukocyte A white blood cell that destroys invading cells and pathogens and removing cellular debris from the body.

Leukocytosis Increase in total white blood cell count.

Leukopenia Decrease in total white blood cell count.

Leukotriene Inflammatory molecule that helps to mediate different allergic responses. It causes muscle contraction and lung constriction in asthma.

Lipids Fatty substances in the blood (e.g., cholesterol and triglycerides).

Lipohypertrophy Lipohypertrophy occurs when an injection site is overused. The skin swells and becomes tough and a node can develop.

Lipoproteins Transporters of fatty substances in the blood.

Live vaccine (or attenuated vaccine) A vaccine in which live virus is attenuated through chemical or physical processes in order to produce an immune response without causing the severe effects of the disease. Attenuated vaccines currently licensed in the United States include measles, mumps, rubella, shingles (herpes zoster), varicella, and yellow fever.

Live vector vaccine A vaccine that uses a non-disease-causing virus or bacterium to transport foreign genes into the body, thereby stimulating an effective immune response to the foreign products.

Local recurrence The reappearance of cancer in the part of the body where it first occurred.

Long-acting insulin Insulin with a very long action time, up to 24 h.

Longevity The longevity or length of lifespan of the mosquito is of considerable importance in malaria control. There are two reasons for this. The first is that the reproductive cycle of malaria in the mosquito takes 10–11 days, and the second is that if the mosquito lives a long time, it will be able to take several blood meals and will have a higher chance of biting a human who has malaria parasites.

Low-density lipoprotein (LDL) The primary cholesterol-carrying blood substance.

Lumpectomy A surgical procedure that removes a localized mass of tissue, including the breast cancer tumor and a small amount of normal, noncancerous tissue surrounding the tumor.

Lupus A disease characterized by chronic inflammation of the connective tissue. The swelling of the connective tissue causes damage to the skin, joints, kidneys, nervous system and mucous membranes. The disease begins with fever, joint pain, and fatigue, with additional symptoms of nausea, fatigue, weight loss, arthritis, headaches, and epilepsy to continue over the years. This condition is diagnosed most frequently in young women but also occurs in children. Problems with heart, lung, and kidney function may also result.

Lyme disease A bacterial disease transmitted by infected ticks. Human beings may come into contact with infected ticks during outdoor activities. Symptoms include fatigue, chills, fever, headache, joint and muscle pain, swollen lymph nodes, and a circular pattern skin rash. Long-term problems include arthritis, nervous system abnormalities, irregular heart rhythm, and meningitis. Lyme disease can be treated with antibiotics. A vaccine was available from 1998 to 2002.

Lymph node A small bean-shaped organ of the immune system distributed widely throughout the body and linked by lymphatic vessels. Lymph nodes are garrisons of disease-fighting B and T cells, dendritic cells, macrophages, and other kinds of immune cells and filter out harmful microbes and toxins. Lymph nodes may become enlarged when they are actively fighting infection.

Lymph A transparent, slightly yellow fluid that carries lymphocytes, bathes the body tissues, and drains into the lymphatic vessels.

Lymphatic vessels A body wise network of channels, similar to the blood vessels, which transport lymph to the immune organs and into the bloodstream.

Lymphedema Swelling from fluid buildup caused by improper functioning of the lymph system.

Lymphocyte cells Cellular components of the adaptive immune system including B and T cells which are highly specific for antigens associated with microbes, tumor cells, transplants, allergies, and tissues attacked in autoimmune diseases. The clones of lymphocytes, each with a single specificity, and exposure to antigens lead to clonal expansion, the acquisition of helper and killer functions, and formation of immune memory.

Lymphoid organs The organs of the immune system, where lymphocytes develop and congregate. They include the bone marrow, thymus, lymph nodes, spleen, and various other clusters of lymphoid tissue along with blood and lymphatic vessels.

Lymphokines Powerful soluble chemical substances secreted by lymphocytes which help direct and regulate the immune responses.

Lymphoma Cancer involving lymph nodes and the immune system.

Lysis Bursting and death of a cell.

Lysozyme An enzyme in saliva and tears that destroys bacteria.

Macrogametocyte The female form of the gametocyte.

Macrophage Cells important in both innate and adaptive immunities that engulf potentially infectious agents and trigger innate and adaptive immune responses.

Macrovascular complications Over time, poor blood glucose control can lead to serious complications, including damage to major blood vessels – to the macrovascular system. Macrovascular complications cause plaque to build up in the arteries, which can lead to a heart attack or stroke.

Macular Skin lesions, normally red colored.

Major histocompatibility complex (MHC) Cell surface molecules that present antigen to T cells. MHC class I molecules present endogenous antigen to cytotoxic T cells; MHC class II molecules present exogenous antigen to helper T cells.

Malaise ("feel achy all over" or "flu-like symptoms") Subjective generalized feeling of being sick, ill, or not healthy varying from mild to severe in intensity. It may be the lone clinical manifestation of malaria or may accompany other signs and symptoms like fever, headache, or nausea.

Malaria A disease caused by parasites that are transmitted to humans via mosquito bites. Symptoms of infection may include fever, chills, headache, muscle pain, fatigue, nausea, and vomiting. These symptoms usually appear between 9 and 14 days after a person is bitten by an infected mosquito. In severe cases, the disease can be life-threatening.

Malarone Brand name of atovaquone–proguanil, a drug used against malaria for both prevention and treatment.

Malignant A malignant tumor tending to be severe, which invades and becomes progressively worse and has the ability to destroy nearby tissue and/or metastasize to other parts of the body.

Mast cell A type of leukocyte found mainly in the mast cells, basophils of connective tissues, which produces histamine and other inflammatory molecules during allergy.

Mastectomy A surgical procedure to remove all or a large part of the breast.

Metastasize A term that is used to refer to cancer spreading from its site of origin to other sites in the body.

Maturity-onset diabetes of the young (MODM) A rare type of diabetes that develops before the age of 25, runs in families, and can often be controlled by diet and physical activity alone or by tablets.

M-cells Cells located in the immune tissue of the intestines to continuously sample the contents of the intestines to protect the body from potential pathogens.

Measles A contagious viral disease marked by the eruption of red circular spots on the skin.

Mefloquine A drug used against malaria for both prevention and treatment.

Memory B and T cells B and T cells that remain in the body after the completion of an immune response; memory is imparted to improve function of individual cells within the clone.

Memory lymphocytes Long-lived activated lymphocytes that respond quickly when specific antigen is encountered again.

Meningitis Inflammation of the brain and spinal cord that can result in permanent brain damage and death.

Meningoencephalitis Inflammation of the brain and meninges that involves the encephalon and spinal column.

Merozoite The form of the malaria parasite that invades human red blood cells; one of the organisms formed by multiple fission of a sporozoite within the body of the host during the asexual phase of reproduction of a malarial plasmodia and other Sporozoa. Liver-stage and blood-stage malaria parasites develop into schizonts which contain many merozoites. When the schizonts are mature, they (and their host cells!) rupture; the merozoites are released and infect red blood cells.

Metastatic breast cancer Advanced breast cancer that has spread beyond the breast and local lymph nodes to other parts of the body like the lungs, liver, brain or bones, or other tissues.

Microalbuminuria The presence of small amounts of protein in urine during the first stages of kidney disease.

Microbe (or microorganism) A microscopic living organism. Examples include bacteria, protozoa, viruses, and some fungi and parasites.

Microencapsulated It is a method of protecting a drug or vaccine antigen, enhances an antigen's absorption, protects from rapid breakdown, and improves the immune response to that antigen. It involves a thin layer of biodegradable substance over the antigen.

Microgametocyte The male form of the gametocyte.

Microglia Specialized immune cells, related to macrophages that protect the central nervous system.

Microheterogeneity Denotes the variation in the chemical structure of a substance that does produce some minor change in its properties in physiological or in disease state.

Microvascular complications Over time, poor blood glucose control can lead to serious complications, including damage to tiny blood vessels – to the microvascular system. These microvascular complications of diabetes can lead to problems with the retinopathy or cataracts of eyes, nephropathy, and neuropathy.

Molecular diagnostics The measurement of DNA, RNA, proteins, or metabolites to detect genotypes, mutations, or biochemical changes.

Molecular methods Laboratory techniques that are based on identification and characterization of certain molecules and gene sequences of a pathogen's genetic makeup.

Molecular mimicry In many autoimmune disorders, a microbe carries antigens that resemble/mimic those on a particular organ, causing the immune system to attack the body.

Molecule The smallest amount of a specific chemical substance that can exist alone.

Monoclonal antibodies Antibodies derived from a single cell and used against a specific antigen such as a cancer cell. As a tool for binding to specific protein molecules, monoclonal antibodies are invaluable in research, medicine, and industry. Rituxan and Herceptin are monoclonal antibodies used in the treatment of lymphoma and breast cancer, respectively. These antibodies are used during controlled malaria challenge trials to validate target antigens that hold promise for preventing infection and transmission.

Monocyte Leukocyte with a large, usually kidney-shaped nucleus. A large white blood cell in the blood that ingests microbes or other cells and foreign particles. When a monocyte passes out of the bloodstream and enters tissues, it develops into a macrophage.

Monokines Powerful soluble chemical molecules secreted by monocytes and macrophages to immune responses.

Mononucleosis Common infectious virus.

Monounsaturated fats Dietary fats, such as olive oil or canola oil, that do not seem to have any effect on blood cholesterol.

Monovalent vaccine A vaccine that contains only one antigen.

Morphology The study of the form, structure, and configuration of an organism, including parasites. This includes shape, structure, color, and pattern as well as the form and structure of internal and external parts.

Mucocutaneous leishmaniasis A form of leishmaniasis involving mucous membrane of the mouth and nose that spread from a nearby cutaneous lesion.

Mucosal immunity Resistance to infection across the mucous membranes present in the linings of respiratory tract, reproductive tract, gastrointestinal tract, and other moist surfaces of the body exposed to the outside environment. Mucosal immunity depends on immune cells and antibodies.

Mucosal membranes The soft, wet tissue that lines body openings specifically the mouth, nose, rectum, and vagina.

Multiple injection treatment A treatment with injections where short (or fast)-acting insulin is used before meals and intermediate (or long)-acting insulin usually is used before bedtime.

Multiple sclerosis (MS) MS is a progressive and usually fluctuating disease of the central nervous system characterized by the destruction of the myelin sheath surrounding neurons, resulting in plaques. MS has exacerbations and remissions which may not reach baseline levels; thus, permanent disability and sometimes death may occur. The cause of MS is unknown.

Mumps Acute contagious viral illness of the parotid glands marked by swelling.

Murmur A rasping or blowing sound heard while listening to the heart that may or may not indicate problems within the heart or circulatory system.

Mutation A process in which a microbe or organism undergoes a permanent change in hereditary material.

Myelin A white, fatty material that sheathes nerves and enhances the transmission of signals between the brain and the body.

Myocardial infarction (or heart attack) Occurs when one or more regions of the heart muscle experience a severe or prolonged decrease in oxygen supply caused by a blocked blood flow to the heart muscle.

Myocardial ischemia Insufficient blood flow to part of the heart.

Myocardium The muscle wall of the heart.

Nasopharyngeal carcinoma A malignant cancer that occurs in the nasopharynx area with apparent no symptoms until the cancer has metastasized to other parts of the body.

Natural killer (NK) cells They are known as "natural" killers because they attack without first having to recognize specific antigens. A type of lymphocyte which is able to directly kill virally infected cells and/or cancer cells, generally not thought to have a memory function and therefore considered part of the innate immune system.

Necrosis Pertaining to the death of tissue.

Nephropathy Nephropathy is damage to the kidneys, due to long-term complication of diabetes.

Neuritis Inflammation of the nerves.

Neuropathy Damage to the nerves caused by many years of high blood glucose levels leading to permanent disability. Symptoms include pain, muscle weakness, numbness, loss of coordination, and paralysis.

Neutropenia Reduced number of granulocytes in the blood.

Neutrophil A type of major effector immune cell that kills microorganisms by engulfing them and digesting them with enzymes and chemicals.

Newly diagnosed A term used to describe breast cancer that has recently been identified.

Nitroglycerin A medication used to relax or dilate arteries and veins.

NK T cell A T cell with some characteristics of NK cells producing large amounts of cytokines when stimulated and is activated by lipids bound to non-MHC molecules called CD1d.

Node-positive breast cancer Breast cancer that has spread to the lymph nodes.

NOD-LRR family Pattern recognition receptors composed of C-terminal leucine-rich repeats, a central nucleotide-binding oligomerization domain, and N-terminal protein–protein interaction motifs, such as caspase recruitment domains, pyrin domains, or a TIR domain.

Noninvasive procedures A diagnostic treatment that does not require entering the body or puncturing the skin.

Obesity An excessive accumulation of fat in the body. A person with a body mass index (BMI) greater than 30 is considered obese.

OKT3 A monoclonal antibody that targets mature T cells.

Oncologist A physician who specializes in the study and treatment of tumors.

Oncology The study and treatment of cancer.

Oncotype DX® test The Oncotype DX breast, colon, and prostate cancer diagnostic tests that help patients and their physicians make informed, individualized treatment decisions.

Oocyst Oocysts are *Plasmodium* cysts located in the outer stomach wall of mosquitoes, where sporozoite development takes place. When mature, the oocysts rupture and release sporozoites. Sporozoites subsequently migrate to the mosquito's salivary gland and are injected into the host when the mosquito feeds.

Ophthalmologists Doctors with specialist training in the diagnosis and treatment of diseases affecting the eyes.

Opportunistic infection An infection that occurs more frequently or is more severe in people with weakened immune systems than in people with healthy immune systems.

Opsonize To coat an organism with antibodies or a complement protein so as to make it palatable to phagocytes.

Optic neuritis A medical condition where one or both eyes may be affected and vision deteriorates rapidly over hours or days due to demyelination of optic nerves.

Oral glucose tolerance test The oral glucose tolerance test measures the blood glucose level five times over a period of 3 h after you drink a high glucose mixture during diabetes.

Orchitis A complication of mumps infection occurring in young males with symptoms of inflammation of the testicles, headache, nausea, vomiting, pain, and fever. It begins 7–10 days after onset of mumps and in rare cases sterility occurs.

Orthostatic hypotension Large decrease in blood pressure occurring when an individual arises from a seated or lying position accompanied by clinical manifestations like faintness, light-headedness, dizziness, or increased pulse. Found mostly in patients with malaria infections.

Otitis media A viral or bacterial infection that leads to inflammation of the middle ear along with an upper respiratory infection with symptoms of earache, high fever, nausea, vomiting and diarrhea, hearing loss, facial paralysis, and meningitis also.

Outbreak Sudden appearance of a disease in a specific geographic area or in a population.

Oxidative burst Method through which some cells of the innate immune system, like neutrophils, produce antimicrobial chemicals.

Pacemaker An electronic device that is surgically implanted into the patient's heart and chest to regulate heartbeat.

Palpitation Irregular heartbeat that can be felt by a person.

PAMPs Pathogen-specific molecular patterns recognized by host pattern recognition receptors.

Pancreas The pancreas is an organ of the endocrine system that produces the hormone insulin.

Pandemic An outbreak of disease occurring over a wide geographical area and affecting an exceptionally high proportion of the population.

Papular Marked by small red-colored elevation of the skin.

Parasitemia The status of having parasites or the quantity of parasites in the blood. It is referred to as "asymptomatic parasitemia," if no fever or other symptoms are present.

Parasite A plant or animal that lives, grows, and feeds on or within another living organism.

Parasitic conditions A condition that is characterized by another organism living on another organism.

Paratenic host A host in which a parasite survives and can be transmitted to another host but does not develop.

Parenchymatous inflammation One type of inflammation affecting mainly the essential tissue elements of an organ.

Parenteral Medications or vaccines administered intravenously or by injection into the subcutaneous, intramuscular region, or intravenous.

Paroxysm A sudden attack or increase in intensity of a symptom, at intervals. Malaria fever has paroxysms of sudden severe temperature elevations accompanied by profuse sweating.

Passive immunity Effective but limited protection against disease through antibodies produced by another human being or animal which diminishes over time.

Pathogen An organism (e.g., bacteria, viruses, parasites, and fungi) that causes disease in human beings.

Pathogenesis The origin and development of a disease. More specifically, it is the way a microbe causes disease in its host.

Pathologist Physician who identifies diseases by studying cells and tissues under a microscope.

Pathology report A report ordered by authorized healthcare professionals that describes what was found in tissue removed from the patient's body. It usually includes information on the tumor's grade and stage.

Penicillin A mixture of nontoxic antibiotics produced by mold and used regularly to treat harmful bacteria.

Percutaneous coronary intervention (PCI) Procedures performed in the cath lab, including all interventional procedures and stent placements.

Percutaneous transluminal coronary angioplasty (PTCA) A technique to treat heart disease and chest pain by using angioplasty in the coronary arteries to permit more blood flow into the heart.

Perianal abscess A local accumulation of pus that forms next to the anus causing tender swelling and pain on defecation.

Pericarditis Inflammation of the membrane that surrounds the heart.

Permethrin A pyrethroid insecticide.

Pertussis (whooping cough) Bacterial infectious disease marked by a convulsive spasmodic cough, sometimes followed by a crowing intake of breath.

Petechiae A tiny reddish or purplish spot on the skin or mucous membrane, seen during infectious diseases.

Peyer's patches Areas in the gut which contain different types of immune cells needed for initiation of an immune responses.

Phagocyte A white blood cell that engulfs and consumes foreign material.

Phagocytosis Process by which one cell engulfs another cell or large particle.

Pharmacokinetics The processes of absorption, distribution, metabolism, and excretion of a drug or vaccine.

Phase 1 vaccine trial A closely monitored clinical trial of a vaccine conducted in a small number of healthy volunteers to determine the vaccine's safety and immunogenicity in humans, its metabolism and pharmacologic actions, and side effects associated with increasing doses.

Phase 2 vaccine trial Controlled clinical study of a vaccine to identify common short-term side effects and risks associated with the vaccine, to collect additional information on its immunogenicity and efficacy and to collect initial information through live agent challenge of vaccinated volunteers.

Phase 3 vaccine trial Large controlled study to determine the ability of a vaccine to produce a desired clinical effect on the risk of a given infection, disease, or other clinical condition at an optimally selected dose and schedule. These trials also gather additional information about safety needed to evaluate the overall benefit-risk relationship of the vaccine and to provide adequate basis for labeling. Phase 3 trials usually include several hundred to several thousand volunteers.

Placebo An inactive substance administered to some study participants while others receive the agent under evaluation, to provide a basis for comparison of effects.

Plaque Deposits of fat or other substances attached to the artery wall.

Plasma cell An antibody-producing lymphocyte derived from a B cell upon reaction with a specific antigen.

Plasmodium The genus of the parasite that causes malaria. The genus includes four species that infect humans, *Plasmodium falciparum*, *Plasmodium vivax*, *Plasmodium malariae*, and *Plasmodium ovale*. *Plasmodium knowlesi* is a zoonotic species that naturally infects macaques in Southeast Asia that can also infect humans.

Platelets Granule-containing cellular fragments critical for blood clotting and sealing off of wounds.

Pneumonia Inflammation of the lungs characterized by fever, chills, muscle stiffness, chest pain, cough, shortness of breath, rapid heart rate, and difficulty in breathing.

Pneumonitis Inflammation of the lungs.

Podiatrists (also called chiropodists) Healthcare professionals with expert knowledge in feet and foot care.

Poliomyelitis (polio) An acute infectious viral disease characterized by fever, paralysis, and atrophy of skeletal muscles.

Polyethylene glycol An electrolyte-based laxative solution used to clean the bowel before a gastrointestinal exam.

Polymerase chain reaction (PCR) A technique in molecular biology, in which a single or few copies of a piece of DNA are amplified to yield many more copies of the DNA sequence of interest. This technique can make it easier to diagnose certain infections/diseases.

Polymorphic Literally meaning having more than one form. In terms of genes, it means that there are several alleles of a particular gene that occur simultaneously in a population.

Polymorphs Short for polymorphonuclear leukocytes or granulocytes.

Polyp A usually nonmalignant growth or tumor protruding from the mucous lining of an organ, such as the colon. Colon polyps are fleshy growths that occur on the inner lining of the large intestine.

Polysaccharide vaccines Vaccines that are composed of long chains of sugar molecules that resemble the surface of certain types of bacteria. Polysaccharide vaccines are available for pneumococcal disease, meningococcal disease, and *Haemophilus Influenzae* type b.

Polyunsaturated fat A type of fat found in vegetable oils and margarines that does not appear to raise blood cholesterol levels.

Positive acute-phase proteins The increased levels of major acute-phase proteins can change in excess of a thousandfold higher than normal values.

Positron emission tomography (PET) A nuclear scanning device that gives a three-dimensional picture of the heart to provide information about the flow of blood through the coronary arteries to the heart muscle.

Potency A measure of strength.

PR (progesterone receptor) A protein that may be present on certain cells to which progesterone molecules can attach. The term "PR positive" refers to tumor cells that contain the progesterone receptor protein. These cells are generally sensitive to hormone therapy.

Precaution A condition in a recipient which may result in a life-threatening problem if the vaccine is given or a condition which could compromise the ability of the vaccine to produce immunity.

Prediabetes (or glucose intolerance) Prediabetes is an early sign of type 2 diabetes, when a person has high blood glucose levels but they are not high enough yet to be diagnosed as diabetes.

Predispose To make more likely or render susceptible.

Preerythrocytic Prior to entering red blood cells.

Presumptive treatment Treatment of clinically suspected cases without, or prior to, results from confirmatory laboratory tests.

Prevalence The number of people in a given population affected with a particular disease or condition at a given time. Prevalence can be thought of as a snapshot of all existing cases at a specified time.

Primaquine A drug used against malaria for the prevention of *P. vivax* or for the eradication of the hypnozoites of *P. vivax* and *P. ovale*.

Prime To make ready for action without complete activation of an immune response.

Priming Giving one vaccine dose(s) first to induce certain immune responses, followed by or together with a second type of vaccine.

Prodromal An early symptom indicating the onset of an attack or a disease.

Proguanil A drug used for both prevention and treatment against malaria in the combination of atovaquone–proguanil.

Promastigote A form of Trypanosomatidae with the flagellum arising anterior to the nucleus and lacking an undulating membrane.

Prophylaxis Prevention of disease.

Proportional case rate The number of cases diagnosed in India as clinical malaria for every 100 patients attending hospitals and dispensaries.

Protease An enzyme that catalyzes the splitting of proteins into smaller molecules.

Proteinuria Large amounts of protein in the urine due to kidney damage in more advanced kidney disease from having high blood glucose levels.

Proteome All the proteins of the body made by a cell, organ, or organism at a particular time and under specific conditions.

Proteomics The study of the protein, the protein–protein interactions, DNA protein interactions, protein in complexes, and protein subcellular localization. It includes techniques like protein sequencing, mass spectrometers, 2D protein databases, and many others.

Protocol The detailed plan for a clinical trial that states the trial's rationale, purpose, vaccine dosages, routes of administration, length of study, eligibility criteria, and other aspects of trial design.

Protozoa Single-celled, free-living or parasitic, microscopic organisms of the kingdom Protista where the single cell performs all necessary functions of metabolism and reproduction.

PSA (prostate-specific antigen) A protein exclusively produced by the prostate in increased levels in the blood of men who have prostate cancer or other prostate diseases like benign prostatic hyperplasia or inflammation of the prostate.

Pseudomembranous inflammation An acute inflammatory response toward a powerful necrotizing toxin on a mucosal surface, composed of a false membrane of precipitated fibrin, necrotic epithelium, and inflammatory white cells.

Pyrethroid A class of insecticides derived from the natural pyrethrins.

Quarantine The isolation of a person or animal who has a disease (or is suspected of having a disease) in order to prevent further spread of the disease.

Quinine A drug, originally extracted from tree bark of the cinchona tree, which was the only available antimalarial treatment for nearly 300 years.

Radiation therapy The use of radiation to destroy cancer cells locally before or after surgery or in combination with chemotherapy.

Radical cure Primaquine treatment intended to achieve cure of *P. vivax* or *P. malariae* malaria by destroying latent exoerythrocytic stage parasites (hypnozoites).

Radioisotope A radioactive material injected into the body so that a nuclear scanner can make pictures.

Radionuclide ventriculography A diagnostic procedure used to determine the shape and size of the heart's chambers.

Randomized trial A study in which participants are assigned by chance to one of the two or more intervention arms or regimens. Randomization minimizes the differences among groups by equally distributing people with particular characteristics among all the trial arms.

Reactogenicity The capacity of a vaccine to produce adverse reactions.

Reagent Any chemical used in a laboratory test or experiment.

Reassortment The constant state of flux and rearrangement seen in the genes of viruses.

Receptor A molecule on the surface of a cell that serves as a recognition or binding site for antigens, antibodies, or other cellular or immunology components.

Recombinant DNA technology Technologies where a recombinant DNA molecule is introduced into a cell and it is replicated either autonomously or as an integral part of a cellular chromosome.

Recombinant Resulting from new combinations of genetic material or cells; the genetic material produced when segments of DNA from different sources are joined to produce recombinant DNA.

Recrudescence A repeated short-term attack/relapse of malaria due to the survival of malaria parasites in red blood cells. Characteristic of *P. malariae* infections.

Recurrence (or long-term relapse) A repeated attack for weeks, months, or occasionally years, after initial malaria infection due to reinfection of red blood cells from hypnozoites that persisted in hepatocytes.

Red blood cells Any of the hemoglobin-containing cells that carry oxygen to the tissues and are responsible for the red color of vertebrate blood.

Refractory malaria Malaria that is not responsive to residual treatment due to lack of response (other than physiological insecticide resistance) to residual treatment. Examples of causes of refractory malaria are vector exophily and zoophily with failure to enter houses.

Regulatory T cells (Treg cells) Special T cells that regulate or suppress immune responses, preventing autoimmunity.

Relapse Recurrence of disease after it has been apparently cured.

Relapsing malaria Renewed manifestation of clinical symptoms and/or parasitemia of malaria infection that is separated from previous manifestations of the same infection by an interval greater than any interval resulting from the normal periodicity of the paroxysms.

Replication Process by which an organism produces a copy of itself – for example, the way microbes reproduce.

Resection Surgery to remove a cancer and some surrounding tissue.

Reservoir host of leishmaniasis Dog in the Mediterranean countries, man in the Middle East, and wild rodents in Asia and Africa.

Residual treatment Treatment of houses, animal sheds, and other buildings where people or animals spend nighttime hours with insecticide that has residual efficacy. The goal of residual treatment is to block transmission by stopping human–vector contact.

Resistance The ability of a pathogen to reproduce despite the presence of drugs designed to inhibit its reproduction or survival.

Respiratory syncytial virus (RSV) A virus that forms masses, or syncytia, in tissue culture and that is responsible for severe respiratory diseases.

Retinal screening Regular eye examinations where the pupils are dilated to detect any early changes at the back of the eye, which could be signs of retinopathy.

Retinopathy A complication of diabetes that can lead to damage to the retina/blindness and results from damage to the blood vessels in the back of the eye due to many years of high blood glucose levels (a possible long-term complication of diabetes).

Retrovirus A type of RNA virus that reproduces by transcribing itself into DNA using reverse transcriptase. The resultant DNA inserts itself into a cell's DNA and is reproduced by the cell.

Reverse transcriptase (RT) An enzyme that catalyzes the formation of DNA using RNA as a template.

Reye syndrome Encephalopathy in children following an acute illness like influenza or chicken pox with symptoms of vomiting, agitation, and lethargy resulting in coma or death.

Rheumatic fever A childhood disease that may damage the heart valves or the outer lining of the heart.

Rheumatic heart disease Damage caused to the heart's valves by rheumatic fever, which is caused by streptococcal bacteria.

Rheumatoid factor An autoantibody found in the serum of most persons with rheumatoid arthritis.

Rigor Severe shaking chill.

Risk factor A condition, element, or activity that may adversely affect the heart.

Risk The likelihood that an individual will experience a certain event.

RNA (ribonucleic acid) A group of molecules similar in structure to a single strand of DNA. The function of RNA is to carry the information from the DNA in the cell's nucleus into the body of the cell to assemble proteins.

Roll back malaria Launched in 1998 by the WHO, the United Nations Children's Fund (UNICEF), the United Nations Development Programme (UNDP), and the World Bank, it aims to ensure that the Millennium Development Goal related to malaria – to halt and begin to reverse the incidence of malaria by 2015 – is achieved.

Rotavirus A retrovirus with a double-layer protein shell and a wheel-like appearance. Rotaviruses cause diarrhea, especially in infants.

Rubella (German measles) Viral infection that is milder than normal measles but as damaging to the fetus when it occurs early in pregnancy.

Sarcoidosis Rare autoimmune disease usually affecting the lungs.

Sarcoma A malignant tumor growing from connective tissues, such as cartilage, fat, muscle, or bone.

Saturated fat Fat that is found in foods from animal meats and skin, dairy products, and some vegetables.

Scabs Skin crusting over to form scans.

Scavenger cells Any of a diverse group of cells that have the capacity to engulf and destroy foreign material, dead tissues, or other cells.

Schistosomiasis Parasitic fluke infection in developing countries.

Schizogony Asexual reproductive stage of malaria parasites in red blood cells, for the development of a single trophozoite into numerous merozoites. A similar process happens in infected liver cells.

Schizont A developmental form of the malaria parasite in the liver-stage and blood-stage parasites that contains many merozoites.

SCID mouse A laboratory animal that, lacking an enzyme necessary to fashion an immune system of its own, can be turned into a model of the human immune system when injected with human cells or tissues.

Screening (for breast cancer) Looking for masses or suspicious areas in breast tissue on a periodic basis, usually with mammography.

Screening (for colon cancer) Looking for masses or suspicious areas in colon tissue on a periodic basis.

Seizure The sudden onset of a jerking and staring spell usually caused by fever. Also known as convulsions.

Sensor Test strip needed with your blood glucose monitor to test your blood glucose levels.

Sequelae Morbid conditions following as a consequence of a disease.

Sequencing It determines the order of nucleotides in a DNA or RNA molecule, or it also determines the order of amino acids in a protein.

Seroconversion Development of antibodies in the blood of an individual who previously did not have detectable antibodies.

Serology The scientific study of blood serum, other bodily fluids like cerebrospinal fluid, saliva, and semen, which may also contain antibodies after an infection.

Serosurvey Study measuring a person's risk of developing a particular disease.

Serous inflammation One type of inflammation producing a serous exudate.

Serum The clear liquid that separates from the blood when it is allowed to clot. This fluid retains any antibodies that were present in the whole blood.

Severe combined immune deficiency (SCID) Included in a group of rare, life-threatening disorders caused by at least 15 different single-gene defects that result in profound deficiencies in T- and B-lymphocyte function.

Shock It is a "circulatory collapse" and develops when blood pressure drops too low to maintain an adequate flow of blood to the body's tissues.

Sepsis It is an amplified, body-wide inflammatory response to severe bleeding, traumatic injury, or an infection caused by microbes with symptoms including fever, a drop in blood pressure, and mental confusion with lung and kidney failure.

Side effect Undesirable reaction resulting from immunization.

Sigmoidoscopy Inspection through a fiberoptic scope of the inside of the sigmoid colon, for diagnosing the cause of diarrhea, constipation, or abdominal pain and for identifying cancerous tissue.

Silent ischemia Ischemia not accompanied by chest pain.

Sinus node Cells specialized in the right atrium that produce the electrical impulses that cause the heart to contract.

Sinusitis Inflammation of the sinuses.

Skin conditions Any condition that affects the skin.

Skin lesion Lesions appearing on the skin.

Skin sores The occurrence of sores that are located on the skin.

Smallpox (variola) An acute, highly infectious, often fatal disease caused by a poxvirus and characterized by high fever and aches with subsequent widespread eruption of pimples that blister, produce pus, and form pockmarks.

Species Organisms in the same genus that have similar characteristics.

Sphygmomanometer The instrument used to measure blood pressure.

Spleen A lymphoid organ in the abdominal cavity that is an important center for immune system activities.

Splenectomy Removal of the spleen.

Splenomegaly An enlarged spleen, detected by physical examination. A common finding in malaria patients or other asymptomatic patients.

Sporozoite rate The proportion of female anopheline mosquitoes (expressed as a percentage) of a particular species that have sporozoites in their salivary glands or that are positive in immunologic tests to detect sporozoite antigens.

Sporozoite The delicate and spindle-shaped infective stage of the malaria parasite that is passed to the human host from the salivary glands of the mosquito to infect liver cells, disappearing from bloodstream within 30 min. Sporozoites are stages that are released into the hemocoel of the mosquito when the oocyst ruptures. Some eventually find their way to the salivary glands of the mosquito.

Stage I breast cancer The tumor is up to 2 cm in diameter and has not spread beyond the breast.

Stage IIA breast cancer The tumor is up to 2 cm and has spread to the axillary lymph nodes under the arm, or the tumor is between 2 and 5 cm and has not spread to the lymph nodes.

Stage IIB breast cancer The tumor is between 2 and 5 cm and has spread to the lymph nodes under the arm, or the tumor is larger than 5 cm in diameter and has not spread to the lymph nodes.

Stage IIIA breast cancer The tumor is larger than 5 cm in diameter and has spread to the lymph nodes under the arm, or the tumor is any size and has spread more extensively in the lymph nodes.

Stage IIIB breast cancer The tumor is any size and has extended to other tissues near the breast; the tumor may or may not have spread to the lymph nodes.

Stage IV breast cancer Cancer that has metastasized to other locations in the body, such as the lungs, liver, bones, or brain.

Staging A classification system for breast cancer based on the size of the tumor, whether the cancer has spread to the lymph nodes and whether the cancer has metastasis.

Statistical significance The probability that an event or difference occurred as the result of the intervention (vaccine) rather than by chance alone. This probability is determined by using statistical tests to evaluate collected data. Guidelines for defining significance are chosen before data collection begins.

Stem cell transplants A kind of passive immune therapy that transfers stem cells instead of antibodies to develop to all elements of the immune system, such as many types of lymphocytes and phagocytes.

Stenosis The narrowing or constriction of a blood vessel or valve in the heart.

Sterilizing immunity An immune response that completely prevents the establishment of an infection.

Steroids A large family of chemical substances, comprising many hormones, body constituents, and drugs; they are often immunosuppressive.

Stethoscope The instrument used to listen to the heart and other sounds in the body.

Strain A specific version of an organism. Many diseases, including HIV/AIDS and hepatitis, have multiple strains.

Stratification Separation of a study cohort into subgroups or strata according to specific characteristics.

Stroke The sudden disruption of blood flow to the brain.

Subacute inflammation A condition in between an acute and chronic inflammation.

Subclinical infection The presence of infection without symptoms. Also known as inapparent or asymptomatic infection.

Subcutaneous The fatty area under the skin. This is the area used when injecting insulin. Subcutaneous is the fatty area under the skin. This is the area used when injecting insulin.

Subunit vaccine A vaccine that uses merely one component of an infectious agent, rather than the whole, to stimulate an immune response.

Sudden death Death that occurs unexpectedly or immediately after onset of symptoms.

Sudden infant death syndrome (SIDS, "crib," or "cot" death) The sudden and unexpected death of a healthy infant under 1 year of age. A diagnosis of SIDS is made when an autopsy cannot determine another cause of death.

Sulfadoxine–pyrimethamine A drug used against malaria. Its value has been compromised by the emergence of drug-resistant malaria parasites.

Superantigens A class of antigens, including certain bacterial toxins, that unleash a massive and damaging immune response.

Suppressive treatment Treatment intended to prevent clinical symptoms and parasitemia through destruction of parasites in red blood cells.

Suppressor T cells A subset of T cells that turn off antibody production and other immune responses.

Suppurative inflammation One type of inflammation marked by pus formation.

Surrogate marker An indirect measure of disease progression.

Surveillance/follow-up An ongoing assessment by a patient's medical team, once treatment has been completed, to assess the cancer's remission and to look for any evidence of a cancer's return.

Susceptible Unprotected against disease.

Swollen glands Swelling of glands or lymph nodes.

Synapses Specialized junctions at which cells of the nervous system signal to one another and to nonneuronal cells, such as those of muscles and glands.

Synchronous cancer Multiple primary cancers occurring simultaneously.

Syndrome A set of signs and symptoms that tend to occur together and which reflect the presence of a particular disease or an increased chance of developing a particular disease.

Syphilis A sexually transmitted disease caused by *Treponema pallidum* bacteria. The condition is often asymptomatic in the early stages, but one or more sores may be present in the early stages. Untreated syphilis usually results in remission of visible symptoms, but further severe damage may occur to internal organs and other body tissues which can result in death.

T cell or T lymphocyte A small white blood cell that recognizes antigen fragments bound to cell surfaces by specialized antibody-like receptors. "T" stands for the thymus gland, where T cells develop and acquire their receptors.

T-cell receptor Complex protein molecule on the surfaces of T cells that recognizes bits of foreign antigen bound to self-MHC molecules.

T-lymphocyte proliferation assay A test used to measure the memory of T cells to antigens.

Tachycardia Rapid heartbeat.

Tachypnea Increased rate of breathing.

Tamoxifen A medication that interferes with the activity of the hormone estrogen to prevent it from fueling the growth of breast cancer. Tamoxifen is used to treat women with estrogen-receptor-positive breast cancer.

Target range The regulation of blood glucose levels with diet, exercise, and insulin. Before meals, the target range is 70–130 mg/dL, and 1–2 h after a meal, the target range is below 180 mg/dL.

TC cell Effector form of a cytotoxic T cell which induces apoptosis in cancerous or infected "self" cells.

TH cell Effector helper T cell which activates B cells and macrophages and releases cytokines to stimulate the immune system.

Temporal association Two or more events that occur around the same time but may be unrelated, chance occurrences.

Tent A device implanted in a vessel used to help keep it open.

Teratogenic Of relating to or causing developmental malformations.

Tetanus Toxin-producing bacterial disease marked by painful muscle spasms.

Tetracycline An antibiotic drug that can be used against malaria for treatment only, not prevention.

Thimerosal Thimerosal is a mercury-containing preservative used in some vaccines and other products since the 1930s. There is no convincing evidence of harm caused by the low concentrations of thimerosal in vaccines, except for minor reactions like redness and swelling at the injection site. However, in July 1999, the Public Health Service agencies, the American Academy of Pediatrics, and vaccine manufacturers agreed that thimerosal should be reduced or eliminated in vaccines as a precautionary measure. Today, all routinely recommended childhood vaccines manufactured for the US market contain either no thimerosal or only trace amounts with the exception of some flu vaccines. There are thimerosal-free influenza vaccines available.

Thrombocytopenia Low platelet count that can lead to impaired blood clotting and spontaneous bleeding.

Thrombolysis The breaking up of a blood clot.

Thrombolytic therapy The use of a medication that dissolves blood clots.

Thrombosis A blood clot formed in a blood vessel or in the heart.

Thymus A primary lymphoid organ, high in the chest, where T lymphocytes proliferate and mature.

TIL Tumor-infiltrating lymphocytes. These immune cells are extracted from the tumor tissue, treated in laboratory, and reinjected into the cancer patient.

Tinnitus Ringing sound in the ears, a common side effect of quinine treatment.

TIR domain A domain shared by the cytoplasmic portions of TLRs, IL-1R family members, and adaptor proteins that mediate signal transduction.

Tissue plasminogen activator (TPA) A medication used to dissolve blood clots.

Titer The quantity of a substance required to produce a reaction with a given volume of another substance or the amount of one substance required to correspond with a given amount of another substance.

TNM classification of malignant tumors (TNM) A cancer staging system that describes the extent of cancer in a patient's body literally describing tumor/nodes/metastasis. T describes the size of the tumor and whether it has invaded nearby tissue, N describes the number of regional lymph nodes that are involved, and M describes the presence of other metastases. This system is jointly maintained by the International Union Against Cancer (UICC) and the American Joint Committee on Cancer.

Tolerance The capacity of the body to become less responsive to a substance or a physiological insult. Tolerance to components of the self prevents or suppresses autoimmunity.

Toll-like receptor (TLRs) Pattern recognition receptors that detect pathogen-derived components and are composed of extracellular leucine-rich repeats and a cytoplasmic TIR domain.

Tonsils and adenoids Prominent oval masses of lymphoid tissues on either side of the throat.

Toxins Agents produced by plants and bacteria, normally very damaging to mammalian cells, that can be delivered directly to target cells by linking them to monoclonal antibodies or lymphokines.

Toxoid vaccine An inactivated and weakened version of the disease-causing toxin a microbe produces; it is still capable of inducing the formation of antibodies when injected.

Trans fat Vegetable oil that has been treated with hydrogen in order to make it more solid and give it a longer shelf life.

Transcriptomics It is the techniques used to identify mRNA from actively transcribed genes.

Transesophageal echocardiography (TEE) A diagnostic test that is used to measure the sound waves that bounce off of the heart.

Transgenic technology Technology used to deliberately alter the genome of an organism by the transfer of a gene or genes from another species or breed.

Transient ischemic attack (TIA) A stroke-like event that lasts for a short period of time and is caused by a blocked blood vessel.

Transmission-blocking vaccines Vaccines that prevent mosquitoes from becoming infected, even after feeding on an infected person.

Transverse myelitis Transverse myelitis is a demyelinating disorder of spinal cord with symptoms of general back pain followed by weakness in the feet and legs that moves upward. There is no cure and many patients are left with permanent disabilities or paralysis. May be associated with multiple sclerosis (MS).

Treatment monitoring An ongoing and frequent assessment by the medical team, during the time of treatment, to monitor how the patient is tolerating the treatment and how the cancer is responding.

Triglyceride A fatlike substance found in the blood.

Trophozoite A developmental form during the blood stage of malaria parasites. After merozoites have invaded the red blood cell, they develop into trophozoites (sometimes, early trophozoites are called "rings" or "ring-stage parasites"); trophozoites develop into schizonts.

Trypanosomiasis, East African A rare infectious disease caused by a parasite called *Trypanosoma brucei* rhodesiense and is transmitted through the bite of an infected tsetse fly. The infection causes an acute illness with symptoms occurring from days to weeks after infection. Death is relatively common, especially in untreated cases.

Tuberculosis Bacterial infection causing nodules forming, most commonly in the lung.

Tumor grade The characterization of a tumor based on how similar in appearance the cancer cells are to normal cells and on how many of those tumor cells are dividing. Tumor grade is one of many factors that, when used in combination, can indicate how aggressive a patient's cancer is.

Tumor necrosis factor (TNF) A type of cytokine produced primarily by monocytes and macrophages.

Trauma (physical trauma) Wound or injury caused by a physical force like the consequences of motor vehicle accidents, fires and burns, falls, drowning, gunshots, stabbings, and other conditions.

Tumor Tissue growth where the cells that make up the tissue have multiplied uncontrollably. A tumor can be benign (noncancerous) or malignant (cancerous).

Two-photon microscopy An imaging technique using high-powered laser microscopes to examine immune response in the nervous system.

Type 1 diabetes Insulin-dependent diabetes that requires lifelong insulin treatment; type 1 occurs when the pancreas does not make enough insulin, preventing your body from properly using blood glucose as energy.

Type 2 diabetes Non-insulin-dependent diabetes, a condition in which your body either does not make enough insulin or does not use it properly and cannot properly use blood glucose as energy; type 2 may be treated with oral medication but could eventually require insulin.

Typhoid fever Typhoid fever is a life-threatening illness caused by the bacterium *Salmonella* Typhi. Persons with typhoid fever carry the bacteria in their bloodstream and intestinal tract.

Ulcer A local defect in the mucosa.

Ulcerative colitis A disease where sores, or ulcers, form in the top layers of the lining of the large intestine. Inflammation usually occurs in the lower part of the colon and rectum.

Ulcerative inflammation One type of inflammation in which the surface near the necrosis leads to loss of tissue and creation of a local defect (ulcer).

Ultrasound A diagnostic tool used to measure high-frequency sound vibrations.

Urine test strips Method used before the availability of blood glucose testing to test for glucose. Urine testing only shows that your blood glucose level has been high, not what the level is or has been.

Urticaria The eruption of red marks on the skin that are usually accompanied by itching. This condition can be caused by an allergy (e.g., to food or drugs), stress, infection, or physical agents (e.g., heat or cold). Also known as hives.

Vaccination Injection of a killed or weakened infectious organism in order to prevent the disease.

Vaccine adverse event reporting system (VAERS) A database managed by the Centers for Disease Control and Prevention and the Food and Drug Administration. VAERS provides a mechanism for the collection and analysis of adverse events associated with vaccines currently licensed in the United States. Reports to VAERS can be made by the vaccine manufacturer, recipient, their parent/guardian, or healthcare provider. For more information on VAERS, call (800) 822-7967.

Vaccine safety datalink project (VSD) In order to increase knowledge about vaccine adverse events, the Centers for Disease Control and Prevention have formed partnerships with eight large Health Management Organizations (HMOs) to continually evaluate vaccine safety. The project contains data on more than six million people. Medical records are monitored for potential adverse events following immunization. The VSD project allows for planned vaccine safety studies as well as timely investigations of hypothesis.

Vaccine A substance that contains antigenic components from an infectious organism. By stimulating an immune response (but not disease), it protects against subsequent infection by that organism.

Vaccinia A virus related to the smallpox and cowpox viruses, which is used in smallpox vaccine.

Variable region That part of an antibody's structure that differs from one antibody to another.

Varicella (Chickenpox) An acute contagious disease characterized by papular and vesicular lesions.

VCAM-1 (vascular cell adhesion molecule-1) One of a number of molecules present on the surface of endothelial cells that controls cell adhesion and movement.

Vector competence The ability of a vector to transmit a disease.

Vector vaccines Vaccines made by inserting protective antigen genes into harmless bacteria or viruses (vectors). As the vectors multiply in the body, they expose the immune system to protective antigens, stimulating active immunity against the harmful organism.

Vector The organism, typically an insect, that transmits an infectious agent to its alternate host, typically a vertebrate; in human malaria, the vector of the parasite are mosquitoes, and the "carriers" or "hosts" are humans. In vaccine research, a bacterium or virus that does not cause disease in humans and is used in genetically engineered vaccines to transport genes coding for antigens into the body to induce an immune response.

Vein A blood vessel that carries blood to the heart from the body tissues.

Venous blood test Take a blood sample from a blood vessel (vein).

Ventricular fibrillation A condition in which the ventricles contract in rapid and unsynchronized rhythms and cannot pump blood into the body.

Ventricular tachycardia A condition in which the ventricles cause a very fast heartbeat.

Vertigo Dizziness.

Vesicular Characterized by small elevations of the skin containing fluid (blisters).

Viremia The presence of a virus in the blood.

Virulence The relative capacity of a pathogen to overcome body defenses.

Virulent Characterized by rapid course or severity.

Virus A tiny organism that multiples within cells and causes disease such as chickenpox, measles, mumps, rubella, pertussis, and hepatitis. Viruses are not affected by antibiotics, the drugs used to kill bacteria. A microorganism composed of a piece of genetic material – RNA or DNA – surrounded by a protein coat. To replicate, a virus must infect a cell and direct its cellular machinery to produce new viruses.

Visceral leishmaniasis A condition which is characterized by an infection of the viscera by leishmaniasis. A form of leishmaniasis involving the liver, spleen, and bone marrow.

Waning immunity The loss of protective antibodies over time.

Western blotting The identification of proteins or peptides that have been electrophoretically separated according to their size was then blotted and transferred to strips of nitrocellulose paper. The blots are then detected by radiolabeled antibody probes or by enzyme tag probes.

White blood cells Any of the blood cells that are colorless, lack hemoglobin, and contain a nucleus. They include the lymphocytes, dendritic cells, monocytes, neutrophils, eosinophils, and basophils, also called leukocytes.

Wire localization biopsy A type of biopsy performed when an abnormality can be seen on a mammogram but cannot be felt. A wire localization biopsy utilizes a mammogram to locate and identify the breast abnormality, after which a biopsy is performed.

X-ray A machine that uses radiation to produce pictures of the inside of the body.

Zoonosis A disease that naturally occurs in animals that can also occur in humans.

Zoonotic diseases Diseases spread from animals to people.

Zoophagy The process of feeding on animals (e.g., cattle).

Zoophilic Zoophilic mosquitoes are mosquitoes that prefer to take blood meals on animals.

Zoophilous Prefers to feed on animals.

References

Abbas A, Lichtman A, Pillai S (2012) Basic immunology functions and disorders of the immune system, 4th edn. Saunders/Elsevier, Philadelphia, p 40

Abdelmouttaleb I, Danchin N, Ilardo C, Aimone-Gastin I, Angioi M, Lozniewski A, Loubinoux J, Le Faou A, Gueant JL (1999) C-reactive protein and coronary artery disease: additional evidence of the implication of an inflammatory process in acute coronary syndromes. Am Heart J 137:346–351

Abernethy TJ, Avery OT (1941) The occurrence during acute infections of a protein not normally present in the blood. 1. Distribution of the reactive protein in patient's sera and the effect of calcium on the flocculation reaction with C polysaccharide of Pneumococcus. J Exp Med 73:173–182

Ablij H, Meinders A (2002) C-reactive protein: history and revival. Eur J Intern Med 13(7):412

Adhikary S, Virginia P, Kocieda VP, Yen JH et al (2012) Signaling through cannabinoid receptor 2 suppresses murine dendritic cell migration by inhibiting matrix metalloproteinase 9 expression. Blood 120:3741–3749

Afonso A, Hunt P, Cheesman S, Alves AC, Cunha CV, Rosário V, Cravo P (2006) Malaria parasites can develop stable resistance to artemisinin but lack mutations in candidate genes atp6 (encoding the sarcoplasmic and endoplasmic reticulum Ca2+ ATPase), tctp, mdr1, and cg10. Antimicrob Agents Chemother 50:480–489

Agrawal A, Shrive AK, Greenhough TJ, Volanakis JE (2001) Topology and structure of the C1q-binding site on C-reactive protein. J Immunol 166:3998–4004

Agrawal A, Gang TB, Rusiñol AE (2014) Recognition functions of pentameric C-reactive protein in cardiovascular disease. Mediators Inflamm 2014:319215. doi:10.1155/2014/319215. Epub 2014 May 19

Albert CM, Ma J, Rifai N, Stampfer MJ, Ridker PM (2002) Prospective study of C-reactive protein, homocysteine, and plasma lipid levels as predictors of sudden cardiac death. Circulation 105:2595–2599

Al-Hamodi Z et al (2014) Association of adipokines, leptin/adiponectin ratio and C-reactive protein with obesity and type 2 diabetes mellitus. Diabetol Metab Syndr 6:99

Ali S, Bassett JR (1995) Studies on the role of glycosylation in the origin of the electrophoretic variants for rat corticosteroid-binding globulin. Steroids 60(11):743–752

Alifrangis M, Lemnge MM, Moon R, Theisen M, Bygbjerg I, Ridley RG, Jakobsen PH (1999) IgG reactivities against recombinant Rhoptry-Associated Protein-1 (rRAP-1) are associated with mixed Plasmodium infections and protection against disease in Tanzanian children. Parasitology 119(Pt 4):337–342

Allin KH, Nordestgaard BG, Flyger H, Bojesen SE (2011) Elevated pre-treatment levels of plasma C-reactive protein are associated with poor prognosis after breast cancer: a cohort study. Breast Cancer Res 13:R55

Alsemgeest SPM (1994) Blood concentrations of acute-phase proteins in cattle as markers for disease. PhD thesis, Utrecht University, Utrecht, The Netherlands

Alsemgeest SPM, Taverne MAM (1993) Peripartum acute-phase protein serum amyloid-A concentration in plasma of cows and fetuses. Am J Vet Res 54:164–167

Alsemgeest SPM, Kalsbeek HC, Wensing T, Koeman JP, van Ederen AM, Gruys E (1994) Concentrations of serum amyloid-A (SAA) and haptoglobin (Hp) as parameters of inflammatory diseases in cattle. Vet Q 16:21–23

Alvar J, Cañavate C, Gutiérrez-Solar B, Jiménez M, Laguna F, López-Vélez R, Molina R, Moreno J (1997) Leishmania and human immunodeficiency virus coinfection: the first 10 years. Clin Microbiol Rev 10:298–319. [PMC free article] [PubMed]

Alves FP, Durlacher RR, Menezes MJ, Krieger H, Silva LH, Camargo EP (2002) High prevalence of asymptomatic Plasmodium vivax and Plasmodium falciparum infections in native Amazonian populations. Am J Trop Med Hyg 66:641–648

Al-Waili NS (2004) Natural honey lowers plasma glucose, C-reactive protein, homocysteine, and blood lipids in healthy, diabetic, and hyperlipidemic subjects: comparison with dextrose and sucrose. J Med Food 7:100e7

Amrock SM, Weitzman M (2014) Effect of increased leptin and C-reactive protein levels on mortality:

W. Ansar, S. Ghosh, *Biology of C Reactive Protein in Health and Disease*,
DOI 10.1007/978-81-322-2680-2

results from the National Health and Nutrition Examination Survey. Atherosclerosis 236(1):1–6. doi:10.1016/j.atherosclerosis.2014.06.009. Epub 2014 Jun 23

Andrade BB, Reis-Filho A, Souza-Neto SM, Clarêncio J, Camargo LMA, Barral A, Barral-Netto M (2010) Severe Plasmodium vivax malaria exhibits marked inflammatory imbalance. Malar J 9:13

Ansar W, Ghosh S (2013) C-reactive protein and the biology of disease. Immunol Res 56(1):131–142. doi:10.1007/s12026-013-8384-0

Ansar W, Bandyopadhyay SM, Chowdhury S, Habib SH, Mandal C (2006) Role of C-reactive protein in complement mediated hemolysis in Malaria. Glycoconj J 23(3–4):233–240

Ansar W, Habib SK, Roy S, Mandal C, Mandal C (2009a) Unraveling the C-reactive protein complement-cascade in destruction of red blood cells: potential pathological implications in Plasmodium falciparum malaria. Cell Physiol Biochem 23(1–3):175–190. doi:10.1159/000204106. Epub 2009 Feb 18

Ansar W, Mukhopadhyay S, Habib SK, Basu S, Saha B, Sen AK, Mandal CN, Mandal C (2009b) Disease-associated glycosylated molecular variants of human C-reactive protein activate complement-mediated hemolysis of erythrocytes in tuberculosis and Indian visceral leishmaniasis. Glycoconj J 26(9):1151–69. doi:10.1007/s10719-009-9236-y

Ansari NA, Sharma P, Salotra P (2007) Circulating nitric oxide and C-reactive protein levels in Indian kala azar patients: correlation with clinical outcome. Clin Immunol 122(3):343–348. Epub 2007 Jan 9

Ansari NA, Kumar R, Raj A, Salotra P (2008) Elevated levels of IgG3 and IgG4 subclass in paediatric cases of kala azar. Parasite Immunol 30(8):403–409. doi:10.1111/j.1365-3024.2008.01036.x

Arik Yilmaz E, Tanir G, Tuygun N, Taylan Ozkan A (2009) Visceral leishmaniasis in 13 pediatric patients in Turkey: treatment experience. Turkiye Parazitol Derg 33(4):259–262

Armah HB, Wilson NO, Sarfo BY, Powell MD, Bond VC, Anderson W, Adjei AA, Gyasi RK, Tettey Y, Wiredu EK, Tongren JE, Udhayakumar V, Stiles JK (2007) Cerebrospinal fluid and serum biomarkers of cerebral malaria mortality in Ghanaian children. Malar J 6:147

Armstrong PB, Swarnakar S, Srimal S, Misquith S, Hahn EA, Aimes RT, Quigley JP (1996) A cytolytic function for a sialic acid-binding lectin that is a member of the pentraxin family of proteins. J Biol Chem 271(25):14717–14721

Arnett HA, Viney JL (2010) Gatekeepers of intestinal inflammation. Inflamm Res 59(1):1–14

Arthur JM (2003) Proteomics. Curr Opin Nephrol Hypertens 12(4):423–430

Asai T, Mori M, Okada M, Uruno K, Yazawa S (1999) Elevated serum haptoglobin in pigs infected with porcine reproductive and respiratory syndrome virus. Vet Immunol Immunopathol 70:143–148

Asgari SA, Mokhtari G, Falahatkar S, Mansour-Ghanaei M, Roshani A, Zare A, Zamani M, Khosropanah I, Salehi M (2006) Diagnostic accuracy of C-reactive protein and erythrocyte sedimentation rate in patients with acute scrotum. Urol J 3(2):104–108

Autelitano DJ, Raineri L, Knight K, Bannister K, Rice GE (2012) Performance of a multianalyte test as an aid for the diagnosis of ovarian cancer in symptomatic women. J Transl Med 10:45

Aziz A, Rink M, Gakis G, Kluth LA, Dechet C, Miller F, Otto W, Gierth M, Denzinger S, Schwentner C, Stenzl A, Fisch M, Burger M, Fritsche HM (2014) Preoperative C-reactive protein in the serum: a prognostic biomarker for upper urinary tract urothelial carcinoma treated with radical nephroureterectomy. Urol Int 93(3):352–360. doi:10.1159/000362248, Epub 2014 Aug 19

Bachmann MF, Kopf M (2002) Balancing protective immunity and immunopathology. Curr Opin Immunol 14(4):413–419

Badolato R, Johnston JA, Wang JM, McVicar D, Xu LL, Oppenheim JJ, Kelvin DJ (1995) Serum amyloid A induces calcium mobilisation and chemotaxis in human monocytes by activating a pertussis toxin sensitive signaling pathway. J Immunol 155:4004–4010

Bae GS, Kim MS, Park KC, Bon et al (2012) Effect of biologically active fraction of Nardostachys jatamansi on cerulein-induced acute pancreatitis. World J Gastroenterol 18:3223–3234

Balci B, Tinaztepe K, Yilmaz E, Guçer S, Ozen S, Topaloğlu R et al (2002) MEFV gene mutations in familial Mediterranean fever phenotype II patients with renal amyloidosis in childhood: a retrospective clinicopathological and molecular study. Nephrol Dial Transplant 17:1921–1923

Ballou SP, Lozanski G (1992) Induction of inflammatory cytokine release from cultured human monocytes by C-reactive protein. Cytokine 4:361–368

Ballou SP, Lozanski FB, Hodder S, Rzewnicki DL, Mion LC, Sipe JD, Ford AB, Kushner I (1996) Quantitative and qualitative alterations of acute-phase proteins in healthy elderly persons. Age Ageing 25(3):224–230

Balmer P, Phillips HM, Maestre AE, McMonagle FA, Phillips RS (2000) The effect of nitric oxide on the growth of *Plasmodium falciparum*, *P. chabaudi* and *P. berghei* in vitro. Parasite Immunol 22:97–106

Bandyopadhyay SM, Mandal C (2008) Targeting glycoproteins or glycolipids and their metabolic pathways for antiparasite therapy. Adv Exp Med Biol 625:87–102

Bandyopadhyay S, Chatterjee M, Das T, Bandyopadhyay S, Sundar S, Mandal C (2004) Antibodies directed against O-acetylated sialoglycoconjugates accelerate complement activation in Leishmania donovani promastigotes. J Infect Dis 190(11):2010–2019. Epub 2004 Nov 3

Bang R, Marnell L, Mold C, Stein MP, Clos KT, Chivington-Buck C, Clos TW (2005) Analysis of binding sites in human C-reactive protein for FcεRI,

FcεRIIA, and C1q by site-directed mutagenesis. J Biol Chem 280:25095–25102

Barcus MJ, Basri H, Picarima H, Manyakori C, Sekartuti, Elyazar I, Bangs MJ, Maguire JD, Baird JK (2007) Demographic risk factors for severe and fatal vivax and falciparum malaria among hospital admissions in northeastern Indonesian Papua. Am J Trop Med Hyg 77:984–991

Bartel DP (2004) MicroRNAs: genomics, biogenesis, mechanism, and function. Cell 116:281–297

Barzilay JI, Abraham L, Heckbert SR, Cushman M, Kuller LH, Resnick HE, Tracy RP (2001) The relation of markers of inflammation to the development of glucose disorders in the elderly: the Cardiovascular Health Study. Diabetes 50:2384–2389

Bashir S, Shabbir I, Khan MU, Aasim M (2014) Role of C-reactive protein as a marker for microalbuminuria in type 2 diabetics. J Ayub Med Coll Abbottabad 26(1):32–34

Bastard JP, Maachi M, Lagathu C, Kim MJ, Caron M, Vidal H, Capeau J, Feve B (2006) Recent advances in the relationship between obesity, inflammation, and insulin resistance. Eur Cytokine Netw 17:4–12

Batanero E, Villalba M, Rodríguez R (1994) Glycosylation site of the major allergen from olive tree pollen. Allergenic implications of the carbohydrate moiety. Mol Immunol 31(1):31–37

Baumann H, Gauldie J (1994) The acute phase response. Immunol Today 15:74–80

Baumann M, Meri S (2004) Techniques for studying protein heterogeneity and post-translational modifications. Expert Rev Proteomics 1(2):207–217

Beckman L, Beckman G (1967) Individual and organ-specific variations of human acid phosphatase. Biochem Genet 1(2):145–153

Bee A, Culley FJ, Alkhalife IS, Bodman-Smith KB, Raynes JG, Bates PA (2001) Transformation of Leishmania mexicana metacyclic promastigotes to amastigote-like forms mediated by binding of human C-reactive protein. Parasitology 122:521–529

Beesley R, Filteau S, Tomkins A, Doherty T, Ayles H, Reid A, Ellman T, Parton S (2000) Impact of acute malaria on plasma concentrations of transferrin receptors. Trans R Soc Trop Med Hyg 94(3): 295–298

Bento AF, Leite DFP, Marcon R et al (2012) Evaluation of chemical mediators and cellular response during acute and chronic gut inflammatory response induced by dextran sodium sulfate in mice. Biochem Pharmacol 84:1459–1469

Bergman RN (1989) Lilly lecture 1989: toward physiological understanding of glucose tolerance: minimal-model approach. Diabetes 38:1512–1527

Bern C, Haque R, Chowdhury R, Ali M, Kurkjian KM, Vaz L, Amann J, Wahed MA, Wagatsuma Y, Breiman RF, Williamson J, Secor WE, Maguire JH (2007) The epidemiology of visceral leishmaniasis and asymptomatic leishmanial infection in a highly endemic Bangladeshi village. Am J Trop Med Hyg 76(5):909–914

Beverley SM, Turco SJ (1998) Lipophosphoglycan (LPG) and the identification of virulence genes in the protozoan parasite Leishmania. Trends Microbiol 6:35–40

Bharadwaj D, Stein MP, Volzer M, Mold C, Du Clos TW (1999) The major receptor for C-reactive protein on leukocytes is Fcγ receptor II. J Exp Med 190:585–590

Bharadwaj D, Mold C, Markham E, Du Clos TW (2001) Serum amyloid P component binds to Fc gamma receptors and opsonizes particles for phagocytosis. J Immunol 166:6735–6741

Bhatt AN, Mathur R, Farooque A, Verma A, Dwarakanath BS (2010) Cancer biomarkers – current perspectives. Indian J Med Res 132:129–149

Biasucci LM, Liuzzo G, Fantuzzi G et al (1999) Increasing levels of interleukin (IL)-1Ra and IL-6 during the first 2 days of hospitalisation in unstable angina are associated with increased risk of in hospital coronary events. Circulation 99:2079–2084

Biasucci LM, Liuzzo G, Della Bona R et al (2009) Different apparent prognostic value of hsCRP in type 2 diabetic and nondiabetic patients with acute coronary syndromes. Clin Chem 55:365–368

Bilgir O, Bilgir F, Topcuoglu T, Calan M, Calan O (2014) Comparison of high-sensitivity C-reactive protein and fetuin-A levels before and after treatment for subjects with subclinical hyperthyroidism. Endocrine 45(2):244–248

Black S, Kushner I, Samols D (2004) C-reactive protein. J Biol Chem 279(47):48487–48490. Epub 2004 Aug 26

Blackburn WD (1994) Validity of acute phase proteins as markers of disease activity. J Rheumatol 21(Suppl 42):9–13

Blake JG, Ridker PM (2001) Novel clinical markers of vascular wall inflammation. Circ Res 98:763–771

Blaschke F, Spanheimer R, Khan M, Law RE (2006) Vascular effects of TZDs: new implications. Vascul Pharmacol 45(1):3–18, Epub 2006 Jun 5

Bode JG, Albrecht U, Häussinger D, Heinrich PC, Schaper F (2012) Hepatic acute phase proteins – regulation by IL-6- and IL-1-type cytokines involving STAT3 and its crosstalk with NF-κB-dependent signaling. Eur J Cell Biol 91(6–7):496–505

Bodman-smith KB, Melendez AJ, Campbell I, Harrison PT, Allen JM, Raynes JG (2002a) C-reactive protein-mediated phagocytosis and phospholipase D signalling through the high-affinity receptor for immunoglobulin G (FcγRI). Immunology 107:252–260

Bodman-Smith KB, Mbuchi M, Culley FJ, Bates PA, Raynes JG (2002b) C-reactive protein-mediated phagocytosis of Leishmania donovani promastigotes does not alter parasite survival or macrophage responses. Parasite Immunol 24(9–10):447–454

Boer JP, Creasey AA, Chang A, Abbink JJ et al (1993) Alpha-2-macroglobulin functions as an inhibitor of fibrinolytic, clotting, and neutrophilic proteinases in sepsis: studies using a baboon model. Infect Immun 61(12):5035–5043

Bogdan C, Rollinghoff M, Diefenbach A (2000) Reactive oxygen and reactive nitrogen intermediates in innate and specific immunity. Curr Opin Immunol 12:64–76

Bone RC, Balk RA, Cerra FB, Dellinger RP, Fein AM, Knaus WA, Schein RM, Sibbald WJ (1992) Definitions for sepsis and organ failure and guidelines for the use of innovative therapies in sepsis. The ACCP/SCCM Consensus Conference Committee. American College of Chest Physicians/Society of Critical Care Medicine. Chest 101(6):1644–1655

Bonnefoy M, Ayzac L, Ingenbleek Y, Kostka T, Boisson RC, Bienvenu J (1998) Usefulness of the prognostic inflammatory and nutritional index (PINI) in hospitalized elderly patients. Int J Vitam Nutr Res 68:189–195

Bora D (1999) Epidemiology of visceral leishmaniasis in India. Natl Med J India 12(2):62–68

Boras E, Slevin M, Alexander MY, Aljohi A, Gilmore W, Ashworth J, Krupinski J, Potempa LA, Al Abdulkareem I, Elobeid A, Matou-Nasri S (2014) Monomeric C-reactive protein and Notch-3 co-operatively increase angiogenesis through PI3K signalling pathway. Cytokine 69(2):165–179. doi:10.1016/j.cyto.2014.05.027. Epub 2014 Jun 25

Bouree P, Botterel F, Lancon A (2000) Study of protein profile in the visceral leishmaniasis. J Egypt Soc Parasitol 30(3):885–893

Braga EM, Scopel KKG, Komatsu NT, da Silva-Nunes M, Ferreira MU (2005) Polymorphism in the Fcg receptor IIA and malaria morbidity. J Mol Genet Med 1:5–10

Bredt DS, Synder SH (1994) Nitric oxide: a physiologic messenger molecule. Ann Rev Biochem 63:175–195

Brito V, Alcaraz A, Augustovski F, Pichón-Riviere A, García-Martí S, Bardach A, Ciapponi A, Lopez A, Comandé D, Grupo de Evaluación de Tecnologías Sanitarias (2015) High sensitivity C protein as an independent risk factor in people with and without history of cardiovascular disease. Arch Cardiol Mex. pii: S1405-9940(14)00188-8. doi:10.1016/j.acmx.2014.11.005. [Epub ahead of print]

Bruce MC, Donnelly CA, Alpers MP, Galinski MR, Barnwell JW, Walliker D, Day KP (2000) Cross-species interactions between malaria parasites in humans. Science 287(5454):845–848

Bruijn M, Jansen EM, Klapwijk T, van der Lee JH, van Rijn RR, van Woensel JB, Bos AP (2013) Association between C-reactive protein levels and outcome in acute lung injury in children. Eur J Pediatr 172(8):1105–1110. doi:10.1007/s00431-013-2006-7. Epub 2013 May 3

Buckley DI, Fu R, Freeman M, Rogers K, Helfand M (2009) C-reactive protein as a risk factor for coronary heart disease: a systematic review and meta-analyses for the U.S. preventive services task force. Ann Intern Med 151:483e95

Burger W, Ewald C, Fennert EM (1998) Increase in C-reactive protein in the serum of piglets (pCRP) following ACTH or corticosteroid administration. J Vet Med B 45:1–6

Burgess RR, Thompson NE (2002) Advances in gentle immunoaffinity chromatography. Curr Opin Biotechnol 13(4):304–308

Burkot TR, Graves PM, Cattan JA, Wirtz RA, Gibson FD (1987) The efficiency of sporozoite transmission in the human malarias, Plasmodium falciparum and P. vivax. Bull World Health Organ 65:375–380

Bustamante A, Dominguez C, Rodriguez-Sureda V, Vilches A, Penalba A, Giralt D, García-Berrocoso T, Llombart V, Flores A, Rubiera M, Molina C, Alvarez-Sabín J, Montaner J (2013) Prognostic value of plasma chitotriosidase activity in acute stroke patients. Int J Stroke. doi:10.1111/ijs.12160

Carcereny E, Gallardo E, Navarro A, Gel B, Jansa S, Marrades R, Ramirez J, Casas F, Viñolas N, Monzó M (2007) MicroRNAs in human lung cancer and human lung embryogenesis: p 2–004. J Thorac Oncol 2(8):S483

Cardoso FS, Ricardo L, Gondar P, Deus JR, Horta D (2015) C-reactive protein may influence decisively the prescription of prophylactic antibiotics in acute pancreatitis: a population-based cohort study. Pancreas 44(3):404–408. doi:10.1097/MPA.0000000000000279

Castagna L, Zarzur J, Filipetti M, Landa C (1996) Isolation and partial characterization of N-acetyl-D-galactosamine-binding lectins from Epiphragmophora trenquelleonis snail. J Biochem 119(2):372–377

Castelli WP (1996) Lipids, risk factors and ischaemic heart disease. Atherosclerosis 124(Suppl):S1–S9

Cathcart ES, Shirahama T, Cohen AS (1967) Isolation and identification of a plasma component of amyloid. Biochim Biophys Acta 147:392–393

Cattani JA, Tulloch JL, Vrbova H, Jolley D, Gibson FD, Moir JS, Heywood PF, Alpers MP, Stevenson A, Clancy R (1986) The epidemiology of malaria in a population surrounding Madang, Papua New Guinea. Am J Trop Med Hyg 35:3–15

Cavaillon JM, Duff G (1999) Cytokines and the cellular mechanism of inflammation. The cytokine network and immune functions. Oxford University Press, New York, pp 251–261

Centers for Disease Control and Prevention [CDC], 2006. Diabetes data and trends [online]. URL: http://www.cdc.gov/diabetes/statistics

Chagnon A, Yao N, Carli P, Paris JF, Marlier S, Pierre C, Bussiere H (1992) C-reactive protein. Presse Med 21:217–218

Chakrabarty R, Mukherjee S, Lu HG, McGwire BS, Chang KP, Basu MK (1996) Kinetics of entry of virulent and avirulent strains of Leishmania donovani into macrophages: a possible role of virulence molecules (gp63 and LPG). J Parasitol 82(4):632–635

Chakraborty R, Chakraborty P, Basu MK (1998) Macrophage mannosyl fucosyl receptor: its role in invasion of virulent and avirulent L. donovani promastigotes. Biosci Rep 18(3):129–142

Chamanza R, Toussaint MJM, Hulskamp-Koch C, Fabri TH (2000) Serum amyloid A and transferrin in chicken. A preliminary investigation of using acute-phase variables to assess diseases in chickens. Vet Bull 70:284, abstr. 1601

Chang KP (1981) Leishmania donovani-macrophage binding mediated by surface glycoproteins/antigens: characterization in vitro by a radioisotopic assay. Mol Biochem Parasitol 4(1–2):67–76

Channon JY, Roberts MB, Blackwell JM (1984) A study of the differential respiratory burst activity elicited by promastigotes and amastigotes of Leishmania donovani in murine resident peritoneal macrophages. Immunology 53:345–355

Chappuis F, Sundar S, Hailu A, Ghalib H, Rijal S, Peeling RW, Alvar J, Boelaert M (2007) Visceral leishmaniasis: what are the needs for diagnosis, treatment and control? Nat Rev Microbiol 5(11):873–882

Chatterjee M, Chava AK, Kohla G, Pal S, Merling A, Hinderlich S, Unger U, Strasser P, Gerwig GJ, Kamerling JP, Vlasak R, Crocker PR, Schauer R, Schwartz-Albiez R, Mandal C (2003) Identification and characterization of adsorbed serum sialoglycans on Leishmania donovani promastigotes. Glycobiology 13:351–361

Chava AK, Chatterjee M, Gerwig GJ, Kamerling JP, Mandal C (2004) Identification of sialic acids on Leishmania donovani amastigotes. Biol Chem 385:59–66

Chava AK, Chatterjee M, Mandal C (2005) O-acetyl sialic acids in parasitic diseases. In: Yarema KJ (ed) Handbook of carbohydrate engineering. Chapter 3. Taylor and Francis Group, book division, Florida, pp 71–86 and references therein

Chen W, Wang JB, Abnet CC, Dawsey SM, Fan JH, Yin LY, Yin J, Taylor PR, Qiao YL, Freedman ND (2015) Association between C-reactive protein, incident liver cancer, and chronic liver disease mortality in the Linxian Nutrition Intervention Trials: a nested case–control study. Cancer Epidemiol Biomarkers Prev 24(2):386–392. doi:10.1158/1055-9965.EPI-14-1038, Epub 2015 Jan 22

Cheng Q, Xia W, Yang S, Ye P, Mei M, Song Y, Luo M, Li Q (2013) Association of serum pigment epithelium-derived factor with high-sensitivity C-reactive protein in women with polycystic ovary syndrome. J Endocrinol Investig 36(8):632–635

Cheung AT, Ree D, Kolls JK, Fuselier J, Coy DH, Bryer-Ash M (1998) An in vivo model for elucidation of the mechanism of tumor necrosis factor-alpha (TNF-alpha)-induced insulin resistance: evidence for differential regulation of insulin signaling by TNF-alpha. Endocrinology 139:4928–4935

Chockalingam PS, Glasson SS, Lohmander LS (2013) Tenascin-C levels in synovial fluid are elevated after injury to the human and canine joint and correlate with markers of inflammation and matrix degradation. Osteoarthr Cart 21(2):339–345

Choi MK, Kim SB, Kim KD, Ament JD (2014) Sequential changes of plasma C-reactive protein, erythrocyte sedimentation rate and white blood cell count in spine surgery: comparison between lumbar open discectomy and posterior lumbar interbody fusion. J Korean Neurosurg Soc 56(3):218–223. doi:10.3340/jkns.2014.56.3.218. Epub 2014 Sep 30

Christ-Crain M, Meier C, Guglielmetti M, Huber PR, Riesen W, Staub JJ, Müller B (2003) Elevated C-reactive protein and homocysteine values: cardiovascular risk factors in hypothyroidism? A cross-sectional and a double-blind, placebo-controlled trial. Atherosclerosis 166(2):379–386

Cid MC, Grant DS, Hoffman GS, Auerbach R, Fauci AS, Kleinman HK (1993) Identification of haptoglobin as an angiogenic factor in sera from patients with systemic vasculitis. J Clin Invest 91:977–985

Cioffi M, Rosa AD, Serao R, Picone I, Vietri MT (2015) Laboratory markers in ulcerative colitis: current insights and future advances. World J Gastrointest Pathophysiol 6(1):13–22

Cizmeci MN, Kara S, Kanburoglu MK, Simavli S, Duvan CI, Tatli MM (2014) Detection of cord blood hepcidin levels as a biomarker for early-onset neonatal sepsis. Med Hypotheses 82(3):310–312

Clack JW, Juhl M, Rice CA, Li J, Witzmann FA (2003) Proteomic analysis of transducin beta-subunit structural heterogeneity. Electrophoresis 24(19–20):3493–3499

Clarke NJ, Crow FW, Younkin S, Naylor S (2001) Analysis of in vivo-derived amyloid-beta polypeptides by on-line two-dimensional chromatography-mass spectrometry. Anal Biochem 298(1):32–39

Clyne B, Olshaker JS (1999) The C-reactive protein. J Emerg Med 17:1019–1025

Combariza JF, Lombana M, Pino LE, Arango M (2015) C-reactive protein and the MASCC risk index identify high-risk patients with febrile neutropenia and hematologic neoplasms. Support Care Cancer 23(4):1009–1013. doi:10.1007/s00520-014-2454-2, Epub 2014 Oct 2

Conroy AL, Liles WC, Molyneux ME, Rogerson SJ, Kain KC (2011) Performance characteristics of combinations of host biomarkers to identify women with occult placental malaria: a case-control study from Malawi. PLoS One 6:e28540

Cooper KA, Adelekan DA, Esimai AO, Northrop-Clewes CA, Thurnham DI (2002a) Elevated levels of nitric oxide and low levels of haptoglobin are associated with severe malarial anaemia in African children. Acta Trop 83(2):133–140

Cooper KA, Adelekan DA, Esimai AO, Northrop-Clewes CA, Thurnham DI (2002b) Lack of influence of red palm oil on severity of malaria infection in pre-school Nigerian children. Trans R Soc Trop Med Hyg 96(2):216–223

Corti A, Franzini M, Cianchetti S, Bergamini G, Lorenzini E, Melotti P, Paolicchi A, Paggiaro P, Pompella A (2012) Contribution by polymorphonucleate granulocytes to elevated gamma-glutamyltransferase in cystic fibrosis sputum. PLoS One 7:e34772

Cote N, Trout DR, Hayes MA (1998) Interaction of transforming growth factor-beta-1 with alpha-2-macroglobulin from normal and inflamed equine joints. Can J Vet Res 62:279–286

Counotte GHM, Toussaint MJM, van Ederen AM, Gruys E (2002) Food safety and acute phase response. Proceedings of the third European Coloquium on acute phase proteins. Department of Veterinary Pathology, Kaap Doorn, ISBN: 90-9015940-de Villiers WJ, Louw JP, Strachan A

Cox MJ, Kum DE, Tavul L, Narara A, Raiko A, Baisor M, Alpers MP, Medley GF, Day KP (1994) Dynamics of malaria parasitemia associated with febrile illness in children from a rural area of Madang, Papua New Guinea. Trans R Soc Trop Med Hyg 88:191–197

Cox-Singh J, Davis TM, Lee KS, Shamsul SS, Matusop A, Ratnam S, Rahman HA, Conway DJ, Singh B (2008) Plasmodium knowlesi malaria in humans is widely distributed and potentially life threatening. Clin Infect Dis 46:165–171

Cramer JP, Mockenhaupt FP, Ehrhardt S, Burkhardt J, Otchwemah RN, Dietz E, Gellert S, Bienzle U (2004) iNOS promoter variants and severe malaria in Ghanaian children. Trop Med Int Health 9:1074–1080

Creactive Protein Coronary Heart Disease Genetics Collaboration (CCGC), Wensley F, Gao P, Burgess S, Kaptoge S, Di Angelantonio E, Shah T, Engert JC, Clarke R, Davey-Smith G, Nordestgaard BG, Saleheen D, Samani NJ, Sandhu M, Anand S, Pepys MB, Smeeth L,Whittaker J, Casas JP, Thompson SG, Hingorani AD, Danesh J (2011) Association between C reactive protein and coronary heart disease: Mendelian randomisation analysis based on individual participant data. BMJ 342:d548. doi:10.1136/bmj.d548

Cuevas LE, Hart CA, Molyneux M, Mughogho GH (1988) C-reactive protein and bacterial meningitis. Ann Trop Paediatr 8(4):230–233

Culley FJ, Harris RA, Kaye PM, McAdam PWJ, Raynes JG (1996) C-reactive protein binds to a novel ligand on Leishmania donovani and increases uptake into human macrophages. J Immunol 156(12):4691–4696

Culley FJ, Thomson M, Raynes JG (1997) C-reactive protein increases C3 deposition on Leishmania donovani promastigotes in human serum. Biochem Soc Trans 25(2):286S

Culley FJ, Bodman-Smith KB, Ferguson MAJ, Nikolaev AV, Shantilal N, Raynes JG (2000) C-reactive protein binds to phosphorylated carbohydrates. Glycobiology 10(1):59–65, And references therein

Cybulsky MI, Gimbrone MA (1991) Endothelial expression of a mononuclear leukocyte adhesion molecule during atherogenesis. Science 251:788–791

Czarnywojtek A, Owecki M, Zgorzalewicz-Stachowiak M, Woliński K, Szczepanek-Parulska E, Budny B, Florek E, Waligórska-Stachura J, Miechowicz I, Bączyk M, Sawicka N, Dhir S, Ruchała M (2014) The role of serum C-reactive protein measured by high-sensitive method in thyroid disease. Arch Immunol Ther Exp (Warsz) 62(6):501–509

da Silva J Jr, (2006) National System in Health Surveillance: situation report: Rondônia. Brasflia: Ministério da Saúde, 24

Dai J, Tang K, Xiao W, Yu G, Zeng J, Li W, Zhang YQ, Xu H, Chen ZQ, Ye ZQ (2014) Prognostic significance of C-reactive protein in urological cancers: a systematic review and meta-analysis. Asian Pac J Cancer Prev 15(8):3369–3375

Dandona P (2008) Effects of antidiabetic and antihyperlipidemic agents on C-reactive protein. Mayo Clin Proc 83(3):333–342. doi:10.4065/83.3.333

Dandona P, Aljada A, O'donnell A, Dhindsa S, Garg R (2004) Insulin as an anti-inflammatory and antiatherosclerotic hormone. Metab Syndr Relat Disord 2(2):137–142. doi:10.1089/met.2004.2.137

Danesh J, Pepys MB (2009) C-reactive protein and coronary disease: is there a causal link? Circulation 120:2036–2039

Das UN (2001) Is obesity an inflammatory condition? Nutrition 17:953–966

Das T, Sen A, Kempf T, Pramanik SR, Mandal C, Mandal C (2003) Induction of glycosylation in human C-reactive protein under different pathological conditions. Biochem J 373(Pt 2):345–355

Das T, Mandal C, Mandal C (2004a) Variations in binding characteristics of glycosylated human C-reactive proteins in different pathological conditions. Glycoconj J 20(9):537–543

Das T, Mandal C, Mandal C (2004b) Protein A-a new ligand for human C-reactive protein. FEBS Lett 576:107–113

Dascombe MJ, Nahrevanian H (2003) Pharmacological assessment of the role of nitric oxide in mice infected with lethal and nonlethal species of malaria. Parasite Immunol 25:149–159

Dasu M, Jialal I (2010) Free fatty acids in the presence of high glucose amplify monocyte inflammation via Toll-like receptor. Am J Physiol Endocrinol Metab 300:E145–E154

Davies MJ (1996) Stability and instability: two faces of coronary atherosclerosis. Circulation 94:2013–2020

Davies H, Lomas L, Austen B (1999) Profiling of amyloid beta peptide variants using SELDI Protein Chip arrays. Biotechniques 27(6):1258–1261

Davis BG (2004) Biochemistry. Mimicking posttranslational modifications of proteins. Science 303(5657): 480–482

De Beer FC, Pepys MB (1982) Isolation of human C-reactive protein and serum amyloid P component. J Immunol Methods 50:17–31

De Beer FC, Dyck RF, Pepys MB (1982a) Solid-phase immunoradiometric assay for serum amyloid A protein using magnetisable cellulose particles. J Immunol Methods 54(2):213–221

De Beer FC, Shine B, Pepys MB (1982b) Radiometric ligand binding assay for C-reactive protein. Complexed C-reactive protein is not detectable in acute phase serum. Clin Exp Immunol 50(1):231–237

de Ferranti S, Rifai N (2002) C-reactive protein and cardiovascular disease: a review of risk prediction and interventions. Clin Chim Acta 317:1–15

de Kretser DM, Bensley JG, Pettilä V, Linko R, Hedger MP, Hayward S, Allan CA, McLachlan RI, Ludlow H, Phillips DJ (2013) Serum activin A and B levels predict outcome in patients with acute respiratory failure: a prospective cohort study. Crit Care 17(5):R263

del Prado GRL, García CH, Cea LM, Espinilla VF, Moreno MFM, Márquez AD, Polo MJP, García IA et al (2014) Malaria in developing countries. J Infect Dev Ctries 8(1):001–004. doi:10.3855/jidc.4610

Delange J, Langlois M, Ouyang J, Claeys G, De Buyzere M, Wuyts B (1998) Effect of haptoglobin phenotypes on growth of Streptococcus pyogenes. Clin Chem Lab Med 36:691–696

Deligeoroglou E, Vrachnis N, Athanasopoulos N, Iliodromiti Z, Sifakis S, Iliodromiti S, Siristatidis C, Creatsas G (2012) Mediators of chronic inflammation in polycystic ovarian syndrome. Gynecol Endocrinol 28(12):974–978

Delves P, Martin S, Burton D, Roitt IM (2006) Roitt's essential immunology, 11th edn. Wiley-Blackwell, Oxford

Deroost K, Tyberghein A, Lays N, Noppen S, Schwarzer E, Vanstreels E, Komuta M, Prato M, Lin JW, Pamplona A, Janse CJ, Arese P, Roskams T, Daelemans D, Opdenakker G, Van den Steen PE (2013) Hemozoin induces lung inflammation and correlates with malaria-associated acute respiratory distress syndrome. Am J Respir Cell Mol Biol 48(5):589–600. doi:10.1165/rcmb.2012-0450OC

Desjeux P (1999) Global control and Leishmania HIV co-infection. Clin Dermatol 17:317–325. [PubMed]

Desjeux P (2001) The increase in risk factors for leishmaniasis worldwide. Trans R Soc Trop Med Hyg 95:239–243. [PubMed]

DiCamelli R, Potempa LA, Siegel L, Suyehira L, Petras K, Gewurz H (1980) Binding reactivity of C-reactive protein for polycations. J Immunol 125:1933–1938

Dinarello CA (1983) Pathogenesis of fever during hemodialysis. Contrib Nephrol 36:90–99

Dinarello CA (1984) Interleukin-1 and the pathogenesis of the acute-phase response. N Engl J Med 311:1413–1418

Dinarello CA (1989) Interleukin-1 and its biologically related cytokines. Adv Immunol 44:153–205

Dinarello CA (1996) Biologic basis for interleukin-1 in disease. Blood 87:2095–2147

Dinarello CA (1997) Cytokines as endogenous pyrogens. In: Mackowiak PA (ed) Fever: basic mechanisms and management, 2nd edn. Lippincott-Raven, Philadelphia, pp 87–116

Dinarello CA, Marnoy SO, Rosenwasser LJ (1983) Role of arachidonate metabolism in the immunoregulatory function of human leukocytic pyrogen/lymphocyte-activating factor/interleukin 1. J Immunol 130(2):890–895

Dobrzycka B, Mackowiak-Matejczyk B, Terlikowska KM, Kulesza-Bronczyk B, Kinalski M, Terlikowski SJ (2013) Serum levels of IL-6, IL-8 and CRP as prognostic factors in epithelial ovarian cancer. Eur Cytokine Netw 24(3):106–113

Doehn C, Fornara P, Kausch I, Büttner H, Friedrich HJ, Jocham D (2001) Value of acute-phase proteins in the differential diagnosis of acute scrotum. Eur Urol 39(2):215–221

Doğanavşargil E, Gümüşdiş G (2003) Romatolojik hastalıkların tanısında labaratuar yöntemleri. Klinik Romatoloji El Kitabı:117–118

Dong Q, Wright JR (1996) Expression of C-reactive protein by alveolar macrophages. J Immunol 156:4815–4820

Dortay H, Schmöckel SM, Fettke J, Mueller-Roeber B (2011) Expression of human c-reactive protein in different systems and its purification from Leishmania tarentolae. Protein Expr Purif 78(1):55–60. doi:10.1016/j.pep.2011.03.010. Epub 2011 Apr 1

Du Clos TW (1989) C-reactive protein reacts with the U1 small nuclear ribonucleoprotein. J Immunol 143:2553–2559

Du Clos TW (2000) Function of C-reactive protein. Ann Med 32:274–278

Du Clos TW, Zlock LT, Rubin RL (1988) Analysis of the binding of C-reactive protein to histones and chromatin. J Immunol 141:4266–4270

Duffy LK, Bowyer RT, Testa JW, Faro JB (1996) Acute phase proteins and cytokines in Alaskan mammals as markers of chronic exposure to environmental pollutants. Am Fish Soc Symp 18:809–813

Duggan SN, Purcell C, Kilbane M, O'Keane M, McKenna M, Gaffney P, Ridgway PF, Boran G, Conlon KC (2015) An association between abnormal bone turnover, systemic inflammation, and osteoporosis in patients with chronic pancreatitis: a case-matched study. Am J Gastroenterol 110(2):336–345. doi:10.1038/ajg.2014.430. Epub 2015 Jan 27

Duncan BB, Schmidt MI, Offenbacher S, Wu KK, Savage PJ, Heiss G (1999) Factor VIII and other hemostasis variables are related to incident diabetes in adults. Diabetes Care 22:767–772

Dungan K, Binkley P, Osei K (2015) GlycA is a Novel Marker of Inflammation Among Non-Critically Ill Hospitalized Patients with Type 2 Diabetes. Inflammation [Epub ahead of print]

Dunmore SJ, Brown JE (2013) The role of adipokines in β-cell failure of type 2 diabetes. J Endocrinol 216:T37–T45 [PMID: 22991412 DOI: 10.1530/JOE-12-0278]

Eckersall PD, Young FJ, McComb C, Hogarth CJ, Safi S, Weber A, McDonald T, Nolan AM, Fitzpatrick JL (2001) Acute phase proteins in serum and milk from dairy cows with clinical mastitis. Vet Rec 148:35–41

El-Mesallamy HO, Abd El-Razek RS, El-Refaie TA (2013) Circulating high-sensitivity C-reactive protein and soluble CD40 ligand are inter-related in a cohort of women with polycystic ovary syndrome. Eur J Obstet Gynecol Reprod Biol 168(2):178–182

Emerging Risk Factors Collaboration, Kaptoge S, Di Angelantonio E, Pennells L, Wood AM, White IR, Gao P, Walker M, Thompson A, Sarwar N, Caslake M, Butterworth AS, Amouyel P, Assmann G, Bakker SJ, Barr EL, Barrett-Connor E, Benjamin EJ, Björkelund C, Brenner H, Brunner E, Clarke R, Cooper JA, Cremer P, Cushman M, Dagenais GR, D'Agostino RB Sr, Dankner R, Davey-Smith G, Deeg D, Dekker JM, Engström G, Folsom AR, Fowkes FG, Gallacher J, Gaziano JM, Giampaoli S, Gillum RF, Hofman A, Howard BV, Ingelsson E, Iso H, Jørgensen T, Kiechl S, Kitamura A, Kiyohara Y, Koenig W, Kromhout D, Kuller LH, Lawlor DA, Meade TW, Nissinen A, Nordestgaard BG, Onat A, Panagiotakos DB, Psaty BM, Rodriguez B, Rosengren A, Salomaa V, Kauhanen J, Salonen JT, Shaffer JA, Shea S, Ford I, Stehouwer CD, Strandberg TE, Tipping RW, Tosetto A,

Wassertheil-Smoller S, Wennberg P, Westendorp RG, Whincup PH, Wilhelmsen L, Woodward M, Lowe GD, Wareham NJ, Khaw KT, Sattar N, Packard CJ, Gudnason V, Ridker PM, Pepys MB, Thompson SG, Danesh J (2012) C-reactive protein, fibrinogen, and cardiovascular disease prediction. N Engl J Med 367(14):1310–1320. doi:10.1056/NEJMoa1107477

Emerging Risk Factors Collaboration, Di Angelantonio E, Gao P, Pennells L, Kaptoge S, Caslake M, Thompson A, Butterworth AS, Sarwar N, Wormser D, Saleheen D, Ballantyne CM, Psaty BM, Sundström J, Ridker PM, Nagel D, Gillum RF, Ford I, Ducimetiere P, Kiechl S, Koenig W, Dullaart RP, Assmann G, D'Agostino RB Sr, Dagenais GR, Cooper JA, Kromhout D, Onat A, Tipping RW, Gómez-de-la-Cámara A, Rosengren A, Sutherland SE, Gallacher J, Fowkes FG, Casiglia E, Hofman A, Salomaa V, Barrett-Connor E, Clarke R, Brunner E, Jukema JW, Simons LA, Sandhu M, Wareham NJ, Khaw KT, Kauhanen J, Salonen JT, Howard WJ, Nordestgaard BG, Wood AM, Thompson SG, Boekholdt SM, Sattar N, Packard C, Gudnason V, Danesh J (2012) Lipid-related markers and cardiovascular disease prediction. JAMA 307(23):2499–2506. doi:10.1001/jama.2012.6571

Erhart LM, Yingyuen K, Chuanak N, Buathong N, Laoboonchai A, Miller RS, Meshnick SR, Gasser RA Jr, Wongsrichanalai C (2004) Hematologic and clinical indices of malaria in a semi-immune population of western Thailand. Am J Trop Med Hyg 70:8–14

Eriksson B, Hellgren U, Rombo L (1989) Changes in erythrocyte sedimentation rate, C-reactive protein and hematological parameters in patients with acute malaria. Scand J Infect Dis 21(4):434–441

Esfahani A, Wong JM, Mirrahimi A, Srichaikul K, Jenkins DJ, Kendall CW (2009) The glycemic index: physiological significance. J Am Coll Nutr 28(Suppl):439S–445S

Esfahanian F, Zamani MM, Heshmat R, Moini nia F (2013) Effect of metformin compared with hypocaloric diet on serum C-reactive protein level and insulin resistance in obese and overweight women with polycystic ovary syndrome. J Obstet Gynaecol Res 39(4):806–813

Esposito G, Capoccia E, Sarnelli G, Scuderi C, Cirillo C, Cuomo R, Steardo L (2012) The antiprotozoal drug pentamidine ameliorates experimentally induced acute colitis in mice. J Neuroinflammation 9:277

Fang S, Wang Y, Sui D, Liu H, Ross MI, Gershenwald JE, Cormier JN, Royal RE, Lucci A, Schacherer CW, Gardner JM, Reveille JD, Bassett RL, Wang LE, Wei Q, Amos CI, Lee JE (2015) C-reactive protein as a marker of melanoma progression. J Clin Oncol 33(12):1389–1396. doi:10.1200/JCO.2014.58.0209, Epub 2015 Mar 16

Farzaneh-Far AF, Rudd J, Weissberg PL (2001) Myocardial ischemia: inflammatory mechanisms. Br Med Bull 59:55–68

Ferard G, Gaudias J, Bourguignat A, Ingenbleek Y (2002) C-reactive protein to transthyretin ratio for the early diagnosis and follow-up of postoperative infection. Clin Chem Lab Med 40:1334–1338

Ferreira A, Schofield L, Enea V, Schellekens H, van der Meide P, Collins WE, Nussenzweig R, Nussenzweig V (1986) Inhibition of development of exo-erythrocytic forms of malaria parasites by gamma interferon. Science 232:881–884

Festa A, D'Agostino R Jr, Howard G, Mykkanen L, Tracy RP, Haffner SM (2000) Chronic subclinical inflammation as part of the insulin resistance syndrome: the Insulin Resistance Atherosclerosis Study (IRAS). Circulation 102:42–47

Fichtlscherer S, Rosenberger G, Walter DH, Breuer S, Dimmeler S, Zeiher AM (2000) Elevated C-reactive protein levels and impaired endothelial vasoreactivity in patients with coronary artery disease. Circulation 102:1000–1006

Fidock DA, Nomura T, Talley AK, Cooper RA, Dzekunov SM, Ferdig MT, Ursos LM, Sidhu AB, Naude B, Deitsch KW, Su XZ, Wootton JC, Roepe PD, Wellems TE (2000) Mutations in the P. falciparum digestive vacuole transmembrane protein PfCRT and evidence for their role in chloroquine resistance. Mol Cell 6:861–871

Figeys D (2004) Combining different 'omics' technologies to map and validate protein-protein interactions in humans. Brief Funct Genomic Proteomic 2(4):357–365

Figeys D, McBroom LD, Moran MF (2001) Mass spectrometry for the study of protein-protein interactions. Methods 24(3):230–239

Ford ES (1999) Body mass index, diabetes, and C-reactive protein among U.S. Adults. Diabetes Care 22:1971–1977

Frohlich M, Imhof A, Berg G, Hutchinson WL, Pepys MB, Boeing H, Muche R, Brenner H, Koenig W (2000) Association between C-reactive protein and features of the metabolic syndrome: a population-based study. Diabetes Care 23:1835–1839

Fu T, Borensztajn J (2002) Macrophage uptake of low-density lipoprotein bound to aggregated C-reactive protein: possible mechanism of foam-cell formation in atherosclerotic lesions. Biochem J 366:195–201

Funke C, King DP, Brotheridge RM, Adelung D, Scott JL (1997) Harbor seal (phoca vitulina) C-reactive protein (C-RP): purification, characterization of specific monoclonal antibodies and development of an immuno- assay to measure serum C-RP concentrations. Vet Immunol Immunopathol 59:151–162

Gabay C, Kushner I (1999) Acute-phase proteins and other systemic responses to inflammation. N Engl J Med 340:448–454

Gaitonde S, Samols D, Kushner I (2008) C-reactive protein and systemic lupus erythematosus. Arthritis Rheum 59:1814–1820

Galve-de Rochemonteix B, Wiktorowicz K, Kushner I, Dayer JM (1993) C-reactive protein increases production of IL-1 alpha, IL-1 beta, and TNF-alpha, and expression of mRNA by human alveolar macrophages. J Leukoc Biol 53:439–445

Ganguly S, Das NK, Barbhuiya JN, Chatterjee M (2010) Post-kala-azar dermal leishmaniasis – an overview. Int J Dermatol 49(8):921–931. doi:10.1111/j.1365-4632.2010.04558.x

Ganie MA, Hassan S, Nisar S, Shamas N, Rashid A, Ahmed I, Douhat S, Mudassar S, Jan VM, Rashid F (2014) High-sensitivity C-reactive protein (hs-CRP) levels and its relationship with components of polycystic ovary syndrome in Indian adolescent women with polycystic ovary syndrome (PCOS). Gynecol Endocrinol 30(11):781–784. doi:10.3109/09513590.2014.924099. Epub 2014 Aug 19

Gardinassi LG, Dotz V, Hipgrave Ederveen A, de Almeida RP, Nery Costa CH, Costa DL, de Jesus AR, Mayboroda OA, Garcia GR, Wuhrer M, de Miranda Santos IK (2014) Clinical severity of visceral leishmaniasis is associated with changes in immunoglobulin g fc N-glycosylation. MBio 5(6):e01844. doi:10.1128/mBio.01844-14

Gasim S, Theander TG, ElHassan AM (2000) High levels of C-reactive protein in the peripheral blood during visceral leishmaniasis predict subsequent development of post kala-azar dermal leishmaniasis. Acta Trop 75(1):35–38

Genton B, D'Acremont V, Rare L, Baea K, Reeder JC, Alpers MP, Muller I (2008) Plasmodium vivax and mixed infections are associated with severe malaria in children: a prospective cohort study from Papua New Guinea. PLoS Med 5:e127

Gericke B, Raila J, Sehouli J, Haebel S, Könsgen D, Mustea A, Schweigert FJ (2005) Microheterogeneity of transthyretin in serum and ascitic fluid of ovarian cancer patients. BMC Cancer 5:133

Gerlach D, Wagner M, Schlott B, Zähringer U, Schmidt KH (2002) Chemical and physicochemical characterization of the sialic acid-specific lectin from Cepaea hortensis. FEMS Microbiol Lett 214(1):61–68

Gershov D, Kim S, Brot N, Elkon KB (2000) C-reactive protein binds to apoptotic cells, protects the cells from assembly of the terminal complement components, and sustains an anti-inflammatory innate immune response: implications for systemic autoimmunity. J Exp Med 192:1353–1363

Ghalib HW, Whittle JA, Kubin M et al (1995a) IL –12 enhances Th1-type responses in human Leishmania donovani infections. J Immunol 154:4623–4629

Ghalib HW, Piuvezam MR, Skeiky YA et al (1995b) Interleukin 10 production correlates with pathology in human Leishmania donovani infections. J Clin Invest 92:324–329

Giha HA, Nasr A, Ekström M, Israelsson E, Arambepola G, Arnot D, Theander TG, Troye-Blomberg M, Berzins K, Tornvall P, ElGhazali G (2009a) Hepcidin and Plasmodium falciparum malaria in anemic school children in Mali. Bull Soc Pathol Exot 102(4):219–20 [Article in French]

Giha HA, Nasr A, Iriemenam NC, Arnot D, Troye-Blomberg M, Theander TG, Berzins K, ElGhazali G, Pandey JP (2009b) Antigen-specific influence of GM/KM allotypes on IgG isotypes and association of GM allotypes with susceptibility to Plasmodium falciparum malaria. Malar J 22(8):306

Giha HA, ElGhazali G, Nasr A, Iriemenam NC, Berzins K, Troye-Blomberg M, Theander TG, Arnot D (2010a) Clustering of malaria treatment failure (TF) in Daraweesh: Hints for host genetic susceptibility to TF with emphasis on immune-modulating SNPs. Infect Genet Evol 10:481–486

Giha HA, Nasr A, Iriemenam NC, Balogun HA, Arnot D, Theander TG, Troye-Blomberg M, Berzins K, Elghazali G (2010b) Age-dependent association between IgG2 and IgG3 subclasses to Pf332-C231 antigen and protection from malaria, and induction of protective antibodies by sub-patent malaria infections, in Daraweesh. Vaccine 28:1732–1739

Giha HA, Nasr A, Ekström M, Israelsson E, Arambepola G, Arnot D, Theander TG, Troye-Blomberg M, Berzins K, Tornvall P, ElGhazali G (2010c) Association of a single nucleotide polymorphism in the C-reactive protein gene (−286) with susceptibility to Plasmodium falciparum malaria. Mol Med 16(1–2):27–33. doi:10.2119/molmed.2009.00136. Epub 2009 Nov 17

Giha HA, Nasr A, Iriemenam NC, Troye-Blomberg M, Berzins K, Pandey JP, ElGhazali G (2011) Associations of multi-locus polymorphisms in an immune network with susceptibility to uncomplicated Plasmodium falciparum malaria in Daraweesh village, Eastern Sudan. Infect Genet Evol 11:1674–1681

Giudetti AM, Cagnazzo R (2012) Beneficial effects of n-3 PUFA on chronic airway inflammatory diseases. Prostaglandins Other Lipid Mediat 99:57–67

Glass EJ, Craigmile SC, Springbett A, Preston PM, Kirvar E, Wilkie GM, Eckersall PD, Hall FR, Brown CG (2003) The protozoan parasite, Theileria annulata, induces a distinct acute phase protein response in cattle that is associated with pathology. Int J Parasitol 33:1409–1418

Global status report on noncommunicable diseases 2014 (2012) Fact sheet N°312. World Health Organization (WHO), Geneva

Godson DL, Bacaestrada ME, Vankessel AG, Hughes HPA, Morsy MA (1995) Regulation of bovine acute phase responses by recombinant interleukin-1. Can J Vet Res 59:249–255

Goitsuka R, Ohashi T (1990) IL-6 activity in feline infectious peritonitis. J Immunol 144:3195–3199

Goldberg RB (2000) Cardiovascular disease in diabetic patients. Med Clin North Am 84:81–93

Goldsby RA, Kindt TJ, Osborne BA (2007) Kuby's immunology, 6th edn. W.H. Freeman and Company, New York

Gomes MC (2002) C-reactive protein: a new golden marker of cardiovascular risk. Rev Port Cardiol 21(11):1329–1346

Gonzalez-Amaro R, Diaz-Gonzalez F, Sánchez- Madrid F (1998) Adhesion molecules in inflammatory diseases. Drugs 56:977–988

Goodarzi MT, Turner GA (1996) A lectin-binding assay for the rapid characterization of the glycosylation of purified glycoproteins. In: Walker JM (ed) The protein protocol handbooks. Humana Press, Totowa. doi:10.1007/978-1-60327-259-9_98

Gotoh K, Inoue M, Shiraishi K et al (2012) Spleen-derived interleukin-10 downregulates the severity of high-fat diet-induced non-alcoholic fatty pancreas disease. PLoS ONE 7:e53154

Gotschlich EC, Edelman GM (1967) Binding properties and specificity of C-reactive protein. Proc Natl Acad Sci U S A 57:706–712

Gould JM, Weiser JN (2001) Expression of C-reactive protein in the human respiratory tract. Infect Immun 69:1747–1754

Gouni-Berthold I, Berthold HK (2013) The role of niacin in lipid-lowering treatment: are we aiming too high? Curr Pharm Des 19(17):3094–3106

Goyal A, Terry MB, Jin Z, Siegel AB (2014) C-reactive protein and colorectal cancer mortality in U.S. adults. Cancer Epidemiol Biomarkers Prev 23(8):1609–1618. doi:10.1158/1055-9965.EPI-13-0577, Epub 2014 May 27

Graff JN, Beer TM, Liu B, Sonpavde G, Taioli E (2015) Pooled analysis of C-reactive protein levels and mortality in prostate cancer patients. Clin Genitourin Cancer 13(4):e217–e221. doi:10.1016/j.clgc.2015.01.011, Epub 2015 Jan 30

Graninger W, Thalhammer F, Hollenstein U, Zotter GM, Kremsner PG (1992) Serum protein concentrations in Plasmodium falciparum malaria. Acta Trop 52(2–3):121–128

Grant SG, Husi H (2001) Proteomics of multiprotein complexes: answering fundamental questions in neuroscience. Trends Biotechnol 19(10 Suppl):S49–S54

Grau AJ, Buggle F, Becher H, Werle E, Hacke W (1996) The association of leukocyte count, fibrinogen and C-reactive protein with vascular risk factors and ischemic vascular diseases. Thromb Res 82:245–255

Green PJ, Feizi T, Stoll MS, Thiel S, Prescott A, McConville MJ (1994) Recognition of the major cell surface glycoconjugates of Leishmania parasites by the human serum mannan-binding protein. Mol Biochem Parasitol 66:319–328

Greve B, Lehman LG, Lell B, Luckner D, Schmidt-Ott R, Kremsner PG (1999) High oxygen radical production is associated with fast parasite clearance in children with Plasmodium falciparum malaria. J Infect Dis 179(6):1584–1586

Griselli M, Herbert J, Hutchinson WL, Taylor KM, Sohail M, Krausz T et al (1999) C-reactive protein and complement are important mediators of tissue damage in acute myocardial infarction. J Exp Med 190:1733–1739

Grundy SM, Cleeman JI, Daniels SR, Donato KA, Eckel RH, Franklin BA et al (2005) Diagnosis and management of the metabolic syndrome: an American Heart Association/National Heart, Lung, and Blood Institute scientific statement. Circulation 112(17):2735–2752

Gruys E (2002) Acute phase proteins in bovine medicine. In: AVMA 2002 convention notes. Proceedings 2002 of the American Veterinary Medical Association, Nashville, pp 317–321

Gruys E, Snel FWJJ (1994) Animal models for reactive amyloidosis. Baillière's Clin Rheumatol 8:599–611

Gruys E, Obwolo MJ, Toussaint MJM (1994) Diagnostic significance of the major acute phase proteins in veterinary clinical chemistry: a review. Vet Bull 64:1009–1018

Gruys E, Toussaint MJM, Landman WJM, Tivapasi M, Chamanza R, van Veen L (1999) Infection, inflammation and stress inhibit growth. Mechanisms and non-specific assessment of the processes by acute phase proteins. In: Wensing T (ed) Production diseases in farm animals. 10th international conference, 1998. Wageningen Press, Wageningen, pp 72–87. ISBN: 90-74134-60-2

Gruys E, Toussaint MJM, Niewold TA, Koopmans SJ (2005) Acute phase reaction and acute phase proteins. J Zhejiang Univ Sci B 6(11):1045–1056

Grzymisławski M, Derc K, Sobieska M, Wiktorowicz K (2006) Microheterogeneity of acute phase proteins in patients with ulcerative colitis. World J Gastroenterol 12(32):5191–5195

Guan XJ, Song L, Han FF, Cui ZL, Chen X, Guo XJ, Xu WG (2013) Mesenchymal stem cells protect cigarette smoke-damaged lung and pulmonary function partly via VEGF-VEGF receptors. J Cell Biochem 114:323–335

Guy RA, Bolosevic M (1993) Comparson of receptors required for entry of Leishmania major amastigotes by macrophages. Infect Immunol 61:1553–1558

Guyton AC, Hall JE (2001) Resistance of the body to infection: chapter 1.Leukocytes, granulocytes, the monocyte-macrophage system and inflammation. Textbook of medical physiology. Tenth edn. Gopsons Papers Ltd., Noida, pp 397–399

Guzel S, Andican G, Seven A, Aslan M, Bolayirli M, Guzel EC et al (2012) Acute phase response and oxidative stress status in familial Mediterranean fever (FMF). Mod Rheumatol 22:431–437

Guzmán MA, Olguín MA (2014) Association between ferritin, high sensitivity c-reactive protein (hsCRP) and relative abundance of Hepcidin mRNA with the risk of type 2 diabetes in obese subjects. Nutr Hosp 30(3):577–584

Haag LM, Fischer A, Otto B, Plickert R, Ku¨ hl AA, Go¨ bel UB, Bereswill S, Heimesaat MM (2012) Campylobacter jejuni induces acute enterocolitis in gnotobiotic IL-10⁻/⁻ mice via Toll-like-receptor-2 and -4 signaling. PLoS One 7:1–11

Habeeb H, Ripper JR, Cohen A, Hinfey PB (2013) A case of imported severe plasmodium falciparum malaria in the emergency department and the current role of exchange transfusion treatment. J Emerg Med 44(2): e211–e215. doi:10.1016/j.jemermed.2012.02.051, Epub 2012 May 18

Hack EC, Wolbink GJ, Schalkwijk C, Speijer H, Hermens WT, van den Bosch H (1997) A role for secretory phospholipase A2 and C-reactive protein in the removal of injured cells. Immunol Today 18:111–115

Hage DS (1999) Affinity chromatography: a review of clinical applications. Clin Chem 45(5):593–615

Haig DM, McInnes CJ, Thomson J, Wood A, Bunyan K (1998) The orf virus OV 20.0 L gene product is involved in interferon resistance and inhibits an interferon-inducible, double-stranded RNA-dependent kinase. Immunology 93:335–340

Hallquist NA, Klasing KC (1994) Serotransferrin, ovotransferrin and metallothionein levels during an immune response in chickens. Comp Biochem Physiol Biochem Mol Biol 108:375–384

Halvani A, Tahghighi F, Nadooshan HH (2012) Evaluation of correlation between airway and serum inflammatory markers in asthmatic patients. Lung India 29(2):143–146. doi:10.4103/0970-2113.95317

Hanash S (2003) Disease proteomics. Nature 422(6928):226–232

Hancock WS (2002) Glycosylation, who cares? J Proteome Res 1(4):297

Hansen JE, Larsen VA, Bøg-Hansen TC (1984) The microheterogeneity of alpha 1-acid glycoprotein in inflammatory lung disease, cancer of the lung and normal health. Clin Chim Acta 138(1):41–47

Hansen JE, Iversen J, Lihme A, Bøg-Hansen TC (1987) Acute phase reaction, heterogeneity, and microheterogeneity of serum proteins as nonspecific tumor markers inlung cancer. Cancer 60(7):1630–1635

Hansen D, Dendale P, Beelen M, Jonkers RA, Mullens A, Corluy L et al (2010) Plasma adipokine and inflammatory marker concentrations are altered in obese, as opposed to non-obese, type 2 diabetes patients. Eur J Appl Physiol 109(3):397–404

Hardardottir I, Grunfeld C, Feingold KR (1994) Effects of endotoxin and cytokines on lipid metabolism. Curr Opin Lipidol 5:207–215

Harpaz R et al (1992) Serum cytokine profiles in experimental human malaria. Relationship to protection and disease course after challenge. J Clin Invest 90:515–523

Hautvast JL, Tolboom JJ, West CE, Kafwembe EM, Sauerwein RW, van Staveren WA (1998) Malaria is associated with reduced serum retinol levels in rural Zambian children. Int J Vitam Nutr Res 68(6):384–388

Hautvast JL, Tolboom JJ, Willems JL, Mwela CM, Monnens LA (2000) Consequences of infections for three-month length increment in young children in rural Zambia. Acta Paediatr 89(3):296–301

Havenaar EC, Drenth JP, van Ommen EC, van der Meer JW, van Dijk W (1995) Elevated serum level and altered glycosylation of alpha 1- acid glycoprotein in hyperimmunoglobulinemia D and periodic fever syndrome: evidence for persistent inflammation. Clin Immunol Immunopathol 76(3 Pt 1):279–284

Haverkate F, Thompson SG, Pyke SDM, Gallimore JR, Pepys MB (1997) Production of C-reactive protein and risk of coronary events in stable and unstable angina. Lancet 349:462–466

Hayashino Y, Jackson JL, Takumi H, Norio F, Fumiaki N, Shunichi F, Satoru T, Hitoshi I (2014a) Effects of exercise on C-reactive protein, inflammatory cytokine and adipokine in patients with type 2 diabetes: A meta-analysis of randomized controlled trials. Metobol Clinic Exp 63:431–440

Hayashino Y, Mashitani T, Tsujii S, Ishii H, for the Diabetes Distress and Care Registry at Tenri Study Group (2014b). Serum high-sensitivity C-reactive protein levels are associated with high risk of development, not progression, of diabetic nephropathy among Japanese Type 2 diabetic patients: a prospective cohort study (Diabetes Distress and Care Registry at Tenri [DDCRT7]). Diabetes Care 37

Haynes P, Miller I, Aebersold R, Gemeiner M, Eberini I, Lovati MR, Manzoni C, Vignati M, Gianazza E (1998) Proteins of rat serum: I. Establishing a reference two-dimensional electrophoresis map by immunodetection and microbore high performance liquid chromatography-electrospray mass spectrometry. Electrophoresis 19(8–9):1484–1492

Hayton K, Su XZ (2004) Genetic and biochemical aspects of drug resistance in malaria parasites. Curr Drug Targets Infect Disord 4:1–10

Heinrich PC, Castell JV, Andus T (1990) Interleukin-6 and the acute phase response. Biochem J 265:621–636

Heinrich PC, Behrmann I, Müller-Newen G, Schaper F, Graeve L (1998) Interleukin-6-type cytokine signaling through the gp130/Jak/STAT pathway. Biochem J 334:297–314

Herwaldt BL (2001) Leishmaniasis. In: Braunwald E, Fauci AS, Kasper DL, Hauser SL, Longo DL, Jameson JL (eds) Harrison's principles of internal medicine. McGraw Hill Companies, Inc., New York, pp 1213–1218

Hinderaker SG, Olsen BE, Lie RT, Bergsjø PB, Gasheka P, Bondevik GT, Ulvik R, Kvåle G (2002) Anemia in pregnancy in rural Tanzania: associations with micronutrients status and infections. Eur J Clin Nutr 56(3):192–199

Hirschfield GM, Pepys MB (2003) C-reactive protein and cardiovascular disease: new insights from an old molecule. QJM 96:793–807

Hirvonen J, Huszenicza G, Kulcsar M, Pyorala S (1999) Acute-phase response in dairy cows with acute postpartum metritis. Vet Bull 69:1198, abstr. 7665

Hochstrasser DF, Sanchez JC, Appel RD (2002) Proteomics and its trends facing nature's complexity. Proteomics 2(7):807–812

Höfner MC, Fosbery MW, Eckersall PD, Donaldson AI (1994) Haptoglobin response of cattle infected with foot-andmouth disease virus. Res Vet Sci 57:125–128

Hokama Y, Nakamura RM (1987) C-reactive protein – current status and future perspectives. J Clin Lab Anal 1:15–27

Hong SW, Lee HS, Jung KH et al (2012) Protective effect of fucoidan against acetaminophen-induced liver injury. Arch Pharm Res 35:1099–1105

Hopps E, Canino B, Caimi G (2011) Effects of exercise on inflammation markers in type 2 diabetic subjects. Acta Diabetol 48(3):183–189. doi:10.1007/s00592-011-0278-9, Epub 2011 Mar 24

Hrycaj P, Sobieska M, Mackiewicz S, Müller W (1993) Microheterogeneity of alpha 1 acid glycoprotein in rheumatoid arthritis: dependent on disease duration? Ann Rheum Dis 52(2):138–141

Hrycaj P, Wurm K, Mennet P, Müller W (1996) Microheterogeneity of acute-phase glycoproteins in patients with pulmonary sarcoidosis. Eur Respir J 9(2):313–318

Hsiao W, Herrel LA, Yu C, Kattan MW, Canter DJ, Carthon BC, Ogan K (2015) Master VA Nomograms incorporating serum C-reactive protein effectively predict mortality before and after surgical treatment of renal cell carcinoma. Int J Urol 22(3):264–270. doi:10.1111/iju.12672, Epub 2014 Nov 26

http://www.who.int/leishmaniasis/about_disease/en/

http://www.who.int/leishmaniasis/clinical_forms_leishmaniases/en/index2.html

Hu FB (2008) Globalization of food patterns and cardiovascular disease risk. Circulation 118(19):1913–1914

Huang Y, Feng JF, Liu JS, Chen QX (2015) Prognostic role of serum C-reactive protein in esophageal cancer: a systematic review and meta-analysis. Ther Clin Risk Manag 11:89–940. doi:10.2147/TCRM.S70954. eCollection 2015

Hulten C, Sandgren B, Skioldebrand E, Klingeborn B, Marhaug G, Forsberg M (1999) The acute phase protein serum amyloid A (SAA) as an inflammatory marker in equine influenza virus infection. Acta Vet Scand 40:323–333

Hurlimann J, Thorbecke GJ, Hochwald GM (1966) The liver as the site of C-reactive protein formation. J Exp Med 123:365–378

Hurt N, Smith T, Teuscher T, Tanner M (1994a) Do high levels of C-reactive protein in Tanzanian children indicate malaria morbidity. Clin Diagn Lab Immunol 1(4):437–444

Hurt N, Smith T, Tanner M, Mwankusye S, Bordmann G, Weiss NA, Teuscher T (1994b) Evaluation of C-reactive protein and haptoglobin as malaria episode markers in an area of high transmission in Africa. Trans R Soc Trop Med Hyg 88:182–186

Hurt DE, Widom J, Clardy J (2006) Structure of Plasmodium falciparum dihydroorotate dehydrogenase with a bound inhibitor. Acta Crystallogr D Biol Crystallogr 62(Pt 3):312–323, Epub 2006 Feb 22

Hutchinson WL, Noble GE, Hawkins PN, Pepys MB (1994) The pentraxin, C-reactive protein and serum amyloid P component, are cleared and catabolized by hepatocytes in vivo. J Clin Invest 94:1390–1396

Ibba F, Rossi G, Meazzi S, Giordano A, Paltrinieri S (2015) Serum concentration of high density lipoproteins (HDLs) in leishmaniotic dogs. Res Vet Sci 98:89–91. doi:10.1016/j.rvsc.2014.11.011, Epub 2014 Nov 18

Iijima S, Shiba K, Kimura M, Nagai K, Iwai T (2000) Changes of alpha1-acid glycoprotein microheterogeneity in acute inflammation stages analyzed by isoelectricfocusing using serum obtained postoperatively. Electrophoresis 21(4):753–759

Imrie H, Fowkes FJ, Michon P, Tavul L, Hume JC, Piper KP, Reeder JC, Day KP (2006) Haptoglobin levels are associated with haptoglobin genotype and {alpha}+-thalassemia in a malaria- endemic area. Am J Trop Med Hyg 74:965–971

Imrie H et al (2007) Low prevalence of an acute phase response in asymptomatic children from a malaria-endemic area of Papua New Guinea. Am J Trop Med Hyg 76:280–284

Ingenbleek Y, Bernstein LH (1999a) The nutritionally dependent adaptive dichotomy (NDAD) and stress hypermetabolism. J Clin Ligand Assay 22:259–267

Ingenbleek Y, Bernstein LH (1999b) The stressful condition as a nutritionally dependent adaptive dichotomy. Nutrition 15:305–320

Ingenbleek Y, Carpentier YA (1985) A prognostic inflammatory and nutritional index scoring critically ill patients. Int J Vitam Nutr Res 55:91–101

Ingenbleek Y, Young V (1994) Transthyretin (prealbumin) in health and disease: nutritional implications. Annu Rev Nutr 14:495–533

Inoue D, Ozaka M, Matsuyama M, Yamada I, Takano K, Saiura A, Ishii H (2015) Prognostic value of neutrophil-lymphocyte ratio and level of C-reactive protein in a large cohort of pancreaticcancer patients: a retrospective study in a single institute in Japan. Jpn J Clin Oncol 45(1):61–66. doi:10.1093/jjco/hyu159, Epub 2014 Oct 23

Israelsson E, Ekström M, Nasr A, Dolo A, Kearsley S, Arambepola G, Homann MV, Maiga B, Doumbo OK, Elghazali G, Giha HA, Troye-Blomberg M, Berzins K, Tornvall P (2009) Marked differences in CRP genotype frequencies between the Fulani and sympatric ethnic groups in Africa. Malar J 8:136. doi:10.1186/1475-2875-8-136

Iwaki D, Osaki T, Mizunoe Y, Wai SN, Iwanaga S, S K (1999) Functional and structural diversities of C-reactive proteins present in horseshoe crab hemolymph plasma. Eur J Biochem 264(2):314–326

Iwalewa EO, McGaw LJ, Naidoo V, Eloff JN (2007) Inflammation: the foundation of diseases and disorders. A review of phytomedicines of South African origin used to treat pain and inflammatory conditions. Afr J Biotechnol 6:2868–2885

Jabs WJ, Logering BA, Gerke P, Kreft B, Wolber EM, Klinger MH et al (2003) The kidney as a second site of human C-reactive protein formation in vivo. Eur J Immunol 33:152–161

Jakobsen PH, Bygbjerg IC, Theander TG, Hviid L, Fujioka H, Aikawa M, Bayoumi RA, Koch C (1997) Soluble haemoglobin is a marker of recent Plasmodium falciparum infections. Immunol Lett 59(1):35–42

Jakobsen PH, Mckay V, N'jie R, Olaleye BO, D'alessandro U, Zhang G-H, Eggelte TA, Koch C, Greenwood BM (1998) Decreased Antitoxic Activities among Children with Clinical Episodes of Malaria. Infect Immun 66(4):1654–1659

Jalkanen A, Kauko T, Koskinen JO, Waris ME, Airas L (2015) Elevated concentration of C-reactive protein is associated with pregnancy-related co-morbidities but not with relapse activity in multiple sclerosis. Neurol Sci 36(3):441–447. doi:10.1007/s10072-014-1980-5. Epub 2014 Oct 12

Jarczok MN, Koenig J, Mauss D, Fischer JE, Thayer JF (2014) Lower heart rate variability predicts increased level of C-reactive protein 4 years later in healthy, nonsmoking adults. J Intern Med 276(6):667–671. doi:10.1111/joim.12295. Epub 2014 Sep 8

Jensen MD (2008) Role of body fat distribution and the metabolic complications of obesity. J Clin Endocrinol Metab 93:S57–S63

Jin Y, Sun Y, Shi X, Zhao J, Shi L, Yu X (2014) Prognostic value of circulating C-reactive protein levels in patients with non-small cell lung cancer: a systematic review with meta-analysis. J Cancer Res Ther 10(Suppl):C160–C166. doi:10.4103/0973-1482.145854. Oncol Lett. 2015 Jan;9(1):425–429. Epub 2014 Nov 4

Johnson RW (1997) Inhibition of growth by pro-inflammatory cytokines: an integrated view. J Anim Sci 75:1244–1255

Johnson RW, Curtis SE, Dantzer R, Kelley KW (1993) Central and peripheral prostaglandins are involved in sickness behavior in birds. Physiol Behav 53:127–131

Jublanc C, Bruckert E, Giral P, Chapman MJ, Leenhardt L, Carreau V, Turpin G (2004) Relationship of circulating C-reactive protein levels to thyroid status and cardiovascular risk in hyperlipidemic euthyroid subjects: low free thyroxine is associated with elevated hsCRP. Atherosclerosis 172(1):7–11

Kahn SE (2008) The relative contributions of insulin resistance and beta-cell dysfunction in the pathophysiology of type 2 diabetes. Diabetologia 46:3–19

Kampoli AM, Tousoulis D, Briasoulis A, Latsios G, Papageorgiou N, Stefanadis C (2011) Potential pathogenic inflammatory mechanisms of endothelial dysfunction induced by type 2 diabetes mellitus. Curr Pharm Des 17:4147–4158

Kaplan MH, Volanakis JE (1974) Interactions of C-reactive protein complexes with the complement system, I: consumption of human complement associated with the reaction of C-reactive protein with pneumococcal C-polysaccharide and with the choline phosphatides, lecithin and sphingomyelin. J Immunol 112:2135–2147

Karbowska A, Boratyńska M, Klinger M (2009) Resistin: a pathogenic factor or a biomarker of metabolic disorders and inflammation. Postepy Hig Med Dosw (Online) 63:485–491

Karunaweera ND, Grau GE, Gamage P, Carter R, Mendis KN (1992) Dynamics of fever and serum levels of tumor necrosis factor are closely associated during clinical paroxysms in Plasmodium vivax malaria. Proc Natl Acad Sci U S A 89:3200–3203

Kathiresan S, Larson MG, Vasan RS, Guo CY, Gona P, Keaney JF Jr, Wilson PW, Newton-Cheh C, Musone SL, Camargo AL, Drake JA, Levy D, O'Donnell CJ, Hirschhorn JN, Benjamin EJ (2006) Contribution of clinical correlates and 13 C-reactive protein gene polymorphisms to interindividual variability in serum C-reactive protein level. Circulation 113:1415–1423

Katoh N, Nakagawa H (1999) Detection of haptoglobin in the high-density lipoprotein and the very high-density lipoprotein fractions from sera of calves with experimental pneumonia and cows with naturally occurring fatty liver. J Vet Med Sci 61:119–124

Kaul P, Malla N, Kaur S, Mahajan RC, Ganguly NK (2000) Evaluation of a 200-kDa amastigote-specific antigen of L. donovani by enzyme-linked immunosorbent assay (ELISA) for the diagnosis of visceral leishmaniasis. Trans R Soc Trop Med Hyg 94:173–175

Kaul K, Hodgkinson A, Tarr JM, Kohner EM, Chibber R (2010) Is inflammation a common retinal-renal-nerve pathogenic link in diabetes? Curr Diabetes Rev 6(5):294–303

Kausalya S, Kaur S, Malla N, Ganguly NK, Mahajan RC (1996) Microbicidal mechanisms of liver macrophages in experimental visceral leishmaniasis. APMIS 104:171–175

Kaya C, Akgül E, Pabuccu R (2010) C-reactive protein and homocysteine levels are associated with abnormal heart rate recovery in women with polycystic ovary syndrome. Fertil Steril 94(1):230–235

Keller CC, Kremsner PG, Hittner JB, Misukonis MA, Weinberg JB, Perkins DJ (2004) Elevated nitric oxide production in children with malarial anemia: hemozoin-induced nitric oxide synthase type 2 transcripts and nitric oxide in blood mononuclear cells. Infect Immun 72(8):4868–4873

Kelm S, Schauer R (1997) Sialic acids in molecular and cellular interactions. Int Rev Cytol 175:137–240

Kervinen H, Palosuo T, Manninen V, Tenkanen L, Vaarala O, Manttari M (2001) Joint effects of C-reactive protein and other risk factors on acute coronary events. Am Heart J 141:580–585

Khera A, McGuire DK, Murphy SA, Stanek HG, Das SR et al (2005) Race and gender differences in C-reactive protein levels. J Am Coll Cardiol 46:464–469. S0735-1097(05)01042-9 [pii];10.1016/j.jacc.2005.04.051 [doi]

Khorvash F, Naeini AE, Behjati M, Abdi F (2011) Visceral leishmaniasis in a patient with cutaneous lesions, negative Leishman-Donovan bodies and immunological test: A case report. J Res Med Sci 16(11):1507–1510

Kiernan UA, Tubbs KA, Nedelkov D, Niederkofler EE, McConnell E, Nelson RW (2003) Comparative urine protein phenotyping using mass spectrometric immunoassay. J Proteome Res 2(2):191–197

Kilicarslan A, Uysal A, Roach EC (2013) Acute phase reactants. Acta Medica 2:2–7

Kilpatrick L, McCawley L, Nachiappan V (1992) Alpha-1-antichymotrypsin inhibits the NADPH oxidase-enzyme complex in phorbol ester-stimulated neutrophil membranes. J Immunol 149:3059–3065

Kima PE, Constant SL, Hannum L, Colmenares M, Lee KS, Haberman AM, Shlomchik MJ, McMahon-Pratt D (2000) Internalisation of Leishmania mexicana complex amastigotes via the Fc receptor is required to sustain infection in murine cutaneous leishmaniasis. J Exp Med 191:1063–1067

Kimura M, Toth LA, Agostini H, Cady AB, Majde JA, Krueger JM (1995) Comparison of acute phase responses induced in rabbits by lipopolysaccharide and double-stranded RNA. Am J Physiol 36:1596–1605

Kindmark CO (1971) Stimulating effect of C-reactive protein on phagocytosis of various species of pathogenic bacteria. Clin Exp Immunol 8:941–948

King DE, Mainous AG 3rd, Buchanan TA, Pearson WS (2003) C-reactive protein and glycemic control in adults with diabetes. Diabetes Care 26:1535–1539

Kinoshita A, Onoda H, Imai N, Iwaku A, Oishi M, Tanaka K, Fushiya N, Koike K, Nishino H, Matsushima M, Tajiri H (2014) The addition of C-reactive protein to validated staging systems improves their prognostic ability in patients with hepatocellular carcinoma. Oncology 86(5–6):308–317. doi:10.1159/000360704, Epub 2014 Jun 7

Klopstein A, Nogueira ES, Quijorna IF et al (2012) Beneficial effects of αβ-crystallin in spinal cord contusion injury. J Neurosci 32:14478–14488

Kodama H, Arimitsu H (1999) Enhancement of phagocytic and chemokinetic activities of rainbow trout head kidney cells by C-reactive protein. Am J Vet Res 60:240–244

Komiya K, Ishii H, Teramoto S, Takahashi O, Eshima N, Yamaguchi O, Ebi N, Murakami J, Yamamoto H, Kadota J (2011) Diagnostic utility of C-reactive protein combined with brain natriuretic peptide in acutepulmonary edema: a cross sectional study. Respir Res 12:83. doi:10.1186/1465-9921-12-83

Koren-Morag N, Goldbourt U, Tanne D (2005) Relation between the metabolic syndrome and ischemic stroke or transient ischemic attack: a prospective cohort study in patients with atherosclerotic cardiovascular disease. Stroke 36(7):1366–1371

Kottgen E, Hell B, Kage A, Tauber R (1992) Lectin specificity and binding characteristics of human C-reactive protein. J Immunol 149:445–453

Kovacs A, Green F, Hansson LO, Lundman P, Samnegård A, Boquist S, Ericsson CG, Watkins H, Hamsten A, Tornvall P (2005) A novel common single nucleotide polymorphism in the promoter region of the C-reactive protein gene associated with the plasma concentration of C-reactive protein. Atherosclerosis 178:193–8

Kraft R, Ruchti C, Burkhardt AH, Cottier H (1992) Pathogenetic principles in the development of gut-derived infectious-toxic shock (GITS) and multiple organ failure. Cur Stud Haematol Blood Transfus 59:204–240

Kremsner PG, Winkler S, Wildling E, Prada J, Bienzle U, Graninger W, Nüssler AK (1996) High plasma levels of nitrogen oxides are associated with severe disease and correlate with rapid parasitological and clinical cure in Plasmodium falciparum malaria. Trans R Soc Trop Med Hyg 90(1):44–47

Krishnaswamy G, Smith JK, Mukkamala R et al (1998) Multifunctional cytokine expression by human coronary endothelium and glycocorticoids. Microvasc Res 55:189–200

Kumar M, Sarin SK (2009) Biomarkers of diseases in medicine. Current trends in science. Platinum Jubilee special. www.ias.ac.in/pubs/splpubs/pjubileebook/403.pdf

Kuribayashi K, Motoyama S, Sasaki T, Sato Y, Yoshino K, Wakita A, Imai K, Saito H, Minamiya Y (2014) C-reactive protein reduces the relative number of tumor-associated M2 macrophages and intratumoral angiogenesis in mice. Tohoku J Exp Med 233(4):249–255

Kurl S, Laukkanen JA, Niskanen L, Laaksonen D, Sivenius J (2006) Nyysso¨nen K, et al. Metabolic syndrome and the risk of stroke in middle-aged men. Stroke 37(3):806–811

Kuşcu F, Öztürk DB, Gül S, Babayiğit ML (2014) The epidemiology of malaria in Adana between 2002 and 2012. Turkiye Parazitol Derg 38(3):147–150. doi:10.5152/tpd.2014.3449

Kushner I (1982) The phenomenon of the acute phase response. Ann N Y Acad Sci 389:39–48

Kushner I (1988) The acute phase response: an overview. Methods Enzymol 163:373–383

Kushner I (1993) Regulation of the acute phase response by cytokines. Perspect Biol Med 36:611–622

Kushner I (2002) C-reactive protein and atherosclerosis. Science 297:520–521

Kushner I, Kaplan MH (1967) Antigenic analysis of extracts of human heart tissue: cardiac antigens with limited distribution in other organs. J Immunol 99:526–533

Kushner I, Mackiewicz A (1987) Acute phase proteins as disease markers. Dis Markers 5(1):1–11

Kushner I, Rakita L, Kaplan MH (1963) Studies of acute phase protein. II. Localization of Cx-reactive protein in heart in induced myocardial infarction in rabbits. J Clin Invest 42:286–292

Kushner I, Gewurz H, Benson MD (1981) C-reactive protein and the acute-phase response. J Lab Clin Med 97(6):739–749

Kusnierz-Cabala B, Gurda-Duda A, Dumnicka P, Panek J, Pawlica-Gosiewska D, Kulig J, Solnica B (2013) Plasma pentraxin 3 concentrations in patients with acute pancreatitis. Clin Lab 59(9–10):1003–1008

Kwiatkowski D, Cannon JG, Manogue KR, Cerami A, Dinarello CA, Greenwood BM (1989) Tumour necrosis factor production in falciparum malaria and its association with schizont rupture. Clin Exp Immunol 77:361–366

Kylänpää L, Rakonczay Z Jr, O'Reilly DA (2012) The clinical course of acute pancreatitis and the inflammatory mediators that drive it. Int J Inflamm 2012:1–10

Ladeia-Andrade S, Ferreira MU, de Carvalho ME, Curado I, Coura JR (2009) Age dependent acquisition of protective immunity to malaria in riverine populations of the Amazon Basin of Brazil. Am J Trop Med Hyg 80:452–459

Ladhani S, Lowe B, Cole AO, Kowuondo K, Newton CR (2002) Changes in white blood cells and platelets in children with falciparum malaria: relationship to disease outcome. Br J Haematol 119:839–847

Ladue TB, Weiner DL, Sipe J et al (1997) Analytical evaluation of particle-enhanced immunonephelometric assay for C-reactive protein, serum amyloid A, and mannose-binding protein in human serum (abs). Clin Chem 43:S240

Lal SP, Christopherson RI, dos Remedios CG (2002) Antibody arrays: an embryonic but rapidly growing technology. Drug Discov Today 7(18 Suppl):S143–S149

Lamers D, Famulla S, Wronkowitz N, Hartwig S, Lehr S, Ouwens DM, Eckardt K, Kaufman JM, Ryden M, Müller S, Hanisch FG, Ruige J, Arner P, Sell H, Eckel J (2011) Dipeptidyl peptidase 4 is a novel adipokine potentially linking obesity to the metabolic syndrome. Diabetes 60:1917–1925 [PMID: 21593202 DOI: 10.2337/db10-1707]

Langhans W (1996) Bacterial products and the control of ingestive behavior: clinical implications. Nutrition 12:303–315

Lasson A, Göransson J (1999) No microheterogenous changes of plasma C-reactive protein found in man during various diseases. Scand J Clin Lab Invest 59(4):293–304

Lazebnik LB, Boldyreva ON, Parfenov AI, Trubitsyna IE, Shcherbakov PL, Khomeriki SG, Kniazev OV, Sagynbaeva VÉ (2013) Cell adhesion molecules in evaluation of Crohn's disease therapy. Eksp Klin Gastroenterol 3:31–38

Ledue TB, Rifai N (2003) Preanalytic and analytic sources of variations in C-reactive protein measurement: Implications for cardiovascular disease risk assessment. Clin Chem 49:1258–1271

Lee RT, Takagahara I, Lee YC (2002) Mapping the binding areas of human C-reactive protein for phosphocholine and polycationic compounds. Relationship between the two types of binding sites. J Biol Chem 277:225–232

Lee JM, Kim SR, Yoo SJ, Hong OK, Son HS, Chang SA (2009) The relationship between adipokines, metabolic parameters and insulin resistance in patients with metabolic syndrome and type 2 diabetes. J Int Med Res 37:1803–1812

Lee Y, Kang D, Lee S-A (2014) Effect of dietary patterns on serum C-reactive protein level. Nutr Metab Cardiovasc Dis 24(9):1004–1011

Lei KJ, Liu T, Zon G, Soravia E, Liu T-Y, Goldman ND (1985) Genomic DNA sequence for human C-reactive protein. J Biol Chem 260:13377–13383

Lei W, Zhu CH, da Zeng X, Huang JA et al (2012) SOX40L: an important inflammatory mediator in adult bronchial asthma. Ann Acad Med Singapore 41:200–204

Leishman WB (2006) On the possibility of the occurrence of trypanosomiasis in India. 1903. Indian J Med Res 123(3):1252–1254; discussion 79

Li P, Mold C, Du Clos TW (1994) Sublytic complement attack exposes C-reactive protein binding sites on cell membranes. J Immunol 152(6):2995–3005

Li J, Chen J, Kirsner R (2007) Pathophysiology of acute wound healing. Clin Dermatol 25:9–18

Li X, Chen C, Gan F, Wang Y, Ding L, Hua W (2014) Plasma NT pro-BNP, hs-CRP and big-ET levels at admission as prognostic markers of survival in hospitalized patients with dilated cardiomyopathy: a single-center cohort study. BMC Cardiovasc Disord 14:67. doi:10.1186/1471-2261-14-67

Liao C, Yu Z, Guo W, Liu Q, Wu Y, Li Y, Bai L (2014) Prognostic value of circulating inflammatory factors in non-small cell lung cancer: a systematic review and meta-analysis. Cancer Biomark 14(6):469–481. doi:10.3233/CBM-140423

Libby P (1995) Molecular bases of the acute coronary syndromes. Circulation 91:2844–2850

Libby P (2007) Inflammatory mechanisms: the molecular basis of inflammation and disease. Nutr Rev 65:S140–S146

Lim A, Sengupta S, McComb ME, Théberge R, Wilson WG, Costello CE, Jacobsen DW (2003) In vitro and in vivo interactions of homocysteine with human plasma transthyretin. J Biol Chem 278(50):49707–49713. Epub 2003 Sep 24

Lima-Junior J d C, Rodrigues-da-Silva RN, Pereira VA, Storer FL, de Souza Perce-da-Silva D, Fabrino DL, Santos F, Banic DM, de Oliveira-Ferreira J (2012) Cells and mediators of inflammation (C-reactive protein, nitric oxide, platelets and neutrophils) in the acute and convalescent phases of uncomplicated Plasmodium vivax and Plasmodium falciparum infection. Mem Inst Oswaldo Cruz 107(8): 1035–1041

Lin Z, Crockett DK, Lim MS, Elenitoba-Johnson KS (2003) High-throughput analysis of protein/peptide complexes by immunoprecipitation and automated LC-MS/MS. J Biomol Tech 14(2):149–155

Lindenmann J, Fink-Neuboeck N, Koesslbacher M, Pichler M, Stojakovic T, Roller RE, Maier A, Anegg U, Smolle J, Smolle-Juettner FM (2014) The influence of elevated levels of C-reactive protein and hypoalbuminemia on survival in patients with advanced inoperable esophageal cancer undergoing palliative treatment. J Surg Oncol 110(6):645–650. doi:10.1002/jso.23711, Epub 2014 Jun 29

Linke RP, Bock V, Valet G, Rothe G (1991) Inhibition of the oxidative burst response of N-formyl peptide-stimulated neutrophils by serum amyloid-A protein. Biochem Biophys Res Commun 176:1100–1105

Lippincott's Illustrated Reviews: Immunology. Paperback: 384 pages. Williams & Wilkins, Lippincott, July 1, 2007. Language: English. ISBN 0-7817-9543-5. ISBN 978-0-7817-9543-2. Page 182

Liuzzo G, Biasucci LM, Gallimore JR et al (1994) The prognostic value of C-reactive protein and serum amyloid A protein in severe unstable angina. N Engl J Med 331:417–424

Livneh A (2006) Amyloidosis of familial Mediterraneanfever (FMF): insights to FMF phenotype II. Harefuah 145:743–745

Lloyd-Jones DM, Liu K, Lu T, Greenland P (2006) Narrative review: assessment of C-reactive protein in risk prediction for cardiovascular disease. Ann Intern Med 145:35–42

Lodge R, Diallo TO, Descoteaux A (2006) Leishmania donovani lipophosphoglycan blocks NADPH oxidase assembly at the phagosome membrane. Cell Microbiol 8:1922–1931

Loeffler DA, Under MC, Zamany M, Harel E, Paul MA, Baumann H, LeWitt PA (1999) Measurement of acute phase proteins in the Rat Brain: contribution of vascular contents. Neurochem Res 24(10):1313–1317

Lopez-Garcia E, Schulze MB, Meigs JB, Manson JE, Rifai N, Stampfer MJ et al (2005) Consumption of trans fatty acids is related to plasma biomarkers of inflammation and endothelial dysfunction. J Nutr 135:562e6

López Del Prado GR, Hernán García C, Moreno Cea L, Fernández Espinilla V, Muñoz Moreno MF, Delgado Márquez A, Polo Polo MJ, Andrés GI (2014) Malaria in developing countries. J Infect Dev Ctries 8(1):1–4. doi:10.3855/jidc.4610

Lumachi F, Basso SM, Santeufemia DA, Ermani M, Lo Re G, Chiara GB (2014) Preoperative serum C-reactive protein and its prognostic significance in patients with stage III-IV colorectal cancer. Anticancer Res 34(12):7263–7266

Lunney PC, Leong RW (2012) Review article: Ulcerative colitis, smoking and nicotine therapy. Aliment Pharmacol Ther 36:997–1008

Macintyre SS (1988) C-reactive protein. Meths in Enzymol 163:383–399

Macintyre SS, Kushner I, Samols D (1985) Secretion of C-reactive protein becomes more efficient during the course of the acute phase response. J Biol Chem 1985(260):4169–4173

Macintyre SS, Samols D, Dailey P (1994) Two carboxylesterases bind C-reactive protein within the endoplasmic reticulum and regulate its secretion during the acute phase response. J Biol Chem 269:24496–24503

Mackiewicz A, Pawłowski T, Mackiewicz-Pawłowska A, Wiktorowicz K, Mackiewicz S (1987a) Microheterogeneity forms of alpha 1-acid glycoprotein as indicators of rheumatoid arthritis activity. Clin Chim Acta 163(2):185–190

Mackiewicz A, Marcinkowska-Pieta R, Ballou S, Mackiewicz S, Kushner I (1987b) Microheterogeneity of alpha 1-acid glycoprotein in the detection of intercurrent infection in systemic lupus erythematosus. Arthritis Rheum 30(5):513–518

Mackiewicz A, Khan MA, Reynolds TL, van der Linden S, Kushner I (1989) Serum IgA, acute phase proteins, and glycosylation of alpha 1-acid glycoprotein in ankylosing spondylitis. Ann Rheum Dis 48(2):99–103

Macleod CM, Avery OT (1941) The occurrence during acute infections of a protein not normally present in the blood. II. Isolation and properties of the reactive protein. J Exp Med 73:183–190

Macy EM, Hayes TE, Tracy RP (1997) Variability in the measurement of C-reactive protein in healthy subjects: implications for reference intervals and epidemiological application. Clin Chem 43:52–58

Maes M, Bosmans E, Dejongh R, Kenis G, Van-Doolaeghe E, Naels H (1997) Increased serum IL-1 receptor antagonist concentrations in major depression and treatment resistant depression. Cytokines 9:853–858

Maina RN, Walsh D, Gaddy C, Hongo G, Waitumbi J, Otieno L, Jones D, Ogutu BR (2010) Impact of Plasmodium falciparum infection on haematological parameters in children living in Western Kenya. Malar J 9(Suppl 3):S4

Majno G, Joris I (1996) Cells, tissues and disease. Principles of general pathology. Blackwell Science, Cambridge, MA, pp 487–496

Malaria Fact sheet N°94. www.who.int/mediacentre/factsheets/fs094/en/

Malla N, Mahajan RC (2006) Pathophysiology of visceral leishmaniasis – some recent concepts. Indian J Med Res 123:267–274

Man SF, Van Eeden S, Sin DD (2012) Vascular risk in chronic obstructive pulmonary disease: role of inflammation and other mediators. Can J Cardiol 28:653–661

Mandal C, Sinha S, Mandal C (1999) Lectin like properties and differential sugar binding characteristics of C-reactive proteins purified from sera of normal and pollutant induced Labeo rohita. Glycoconj J 16(11):741–750

Mandal C, Srinivasan GV, Chowdhury S, Chandra S, Mandal C, Schauer R, Mandal C (2009) High level of sialate-O-acetyltransferase activity in lymphoblasts of childhood acute lymphoblastic leukaemia(ALL): enzyme characterization and correlation with disease status. Glycoconj J 26(1):57–73. doi:10.1007/s10719-008-9163-3. Epub 2008 Aug 3

Mann M, Jensen ON (2003) Proteomic analysis of post-translational modifications. Nat Biotechnol 21(3):255–261

Marcucci R, Valente S, Gori AM, Chiostri M, Paniccia R, Giusti B, Cau V, Lazzeri C, Gensini GF, Abbate R (2014) Global platelet hyperreactivity and elevated C-reactive protein levels predict long term mortality in STEMI patients. Thromb Res 134(4):884–888. doi:10.1016/j.thromres.2014.07.020. Epub 2014 Jul 24

Markowicz S, Delforge M, Necsoi C, De Wit S (2014) Cardiovascular risk evaluation of HIV-positive patients in a case-control study: comparison of the D:A:D and Framingham equations. J Int AIDS Soc 17(4 Suppl 3):19515. doi:10.7448/IAS.17.4.19515. eCollection 2014

Markt SC, Rider JR, Penney KL, Schumacher FR, Epstein MM, Fall K, Sesso HD, Stampfer MJ, Mucci LA (2014) Genetic variation across C-reactive protein and risk of prostate cancer. Prostate 74(10):1034–1042. doi:10.1002/pros.22820, Epub 2014 May 20

Marnell LL, Mold C, Volzer MA, Burlingame RW, Du Clos TW (1995) C-reactive protein binds to FcYRI in transfected COS cells. J Immunol 155:2185–2193

Marnell L, Mold C, Du Clos TW (2005) C-reactive protein: ligands, receptors and role in inflammation. Clin Immunol 117:104–111

Marsden PD, Jones TC (1985) Clinical manifestations, diagnosis and treatment of leishmaniasis. In: Chang KP, Bray RS (eds) Leishmaniasis. Elsevier Science Publishers, Amsterdam, pp 183–198

Marsh K (1998) Malaria disaster in Africa. Marsh 1998. Lancet 352(9132):924

Martelo OJ, Smoller M, Saladin TA (1969) Malaria in American soldiers. Arch Intern Med 123:383–387

Martinez-Subiela S, Strauss-Ayali D, Cerón JJ, Baneth G (2011) Acute phase protein response in experimental canine leishmaniasis. Vet Parasitol 180(3–4):197–202. doi:10.1016/j.vetpar.2011.03.032. Epub 2011 Mar 31

Martínez-Subiela S, García-Martínez JD, Tvarijonaviciute A, Tecles F, Caldin M, Bernal LJ, Cerón JJ (2013) Urinary C reactive protein levels in dogs with leishmaniasis at different stages of renal damage. Res Vet Sci 95(3):924–929. doi:10.1016/j.rvsc.2013.07.002. Epub 2013 Aug 6

Martocchia A, Cola S, Frugoni P, Indiano I, D'Urso R, Falaschi P (2010) The burden of comorbidity and the C-reactive protein levels in nonthyroidal illness syndrome with metabolic syndrome and atherosclerosis-related cardiovascular complications. Ann N Y Acad Sci 1193:164–166

Maudsley S, Rowe IF, de Beer FC, Munn EA, Herbert J, Feinstein A, Pepys MB (1987) Identification and isolation of two pentraxins from bovine serum. Clin Exp Immunol 67(3):662–673

Mayeux R (2004) Biomarkers: potential uses and limitations. NeuroRx 1:182–188

McCarty, M (2006) Historical perspective on C-reactive protein. Ann N Y Acad Sci 389. Wiley online library, 1982. doi:10.1111/j.1749-6632.1982.tb22121.x

McCarty MF (2004) Ampk activation may suppress hepatic production of C-reactive protein by nitric oxide synthetase. Med Hypotheses 63:328–333

McConville MJ, Turco SJ, Ferguson MA, Sacks DL (1992) Developmental modification of lipophosphoglycan during the differentiation of Leishmania major promastigotes to an infectious stage. EMBO J 11(10):3593–3600. Biomedica. 2002 Jun;22(2):167–177

McGuire W, D'Alessandro U, Olaleye BO, Thomson MC, Langerock P, Greenwood BM, Kwiatkowski D (1996) C-reactive protein and haptoglobin in the evaluation of a community-based malaria control programme. Trans R Soc Trop Med Hyg 90:10–14

McNair J, Kennedy DG, Bryson DG (1997) Evaluation of a competitive immunoassay for the detection of bovine haptoglobin. Res Vet Sci 63:145–149

McVey M, Tabuchi A, Kuebler WM (2012) Microparticles and acute lung injury. Am J Physiol Lung Cell Mol Physiol 303:L364–L381

Meigs JB, Hu FB, Rifai N, Manson JE (2004) Biomarkers of endothelial dysfunction and risk of type 2 diabetes mellitus. JAMA 291:1978–1986

Meigs JB, Larson MG, Fox CS, Keaney JF Jr, Vasan RS, Benjamin EJ (2007) Association of oxidative stress, insulin resistance, and diabetes risk phenotypes: the Framingham Offspring Study. Diabetes Care 30:2529–2535

Mekonnen G, Corban MT, Hung OY, Eshtehardi P, Eapen DJ, Al-Kassem H, Rasoul-Arzrumly E, Gogas BD, McDaniel MC, Pielak T, Thorball CW, Sperling L, Quyyumi AA, Samady H (2014) Plasma soluble urokinase-type plasminogen activator receptor level is independently associated with coronary microvascular function in patients with non-obstructive coronary artery disease. Atherosclerosis 239(1):55–60. doi:10.1016/j.atherosclerosis.2014.12.025. [Epub ahead of print]

Mellouk S, Maheswari RK, Rhodes Feuillette A, Beaudoin RL, Berbiguier N, Matile H, Landeau I, Pied S, Chigot JP, Friedman RM, Mazier D (1987) Inhibitory activity of interferon and interleukin 1 on the development of Plasmodium falciparum in human hepatocyte cultures. J Immunol 139:4192–4197

Menschikowski M, Hagelgans A, Fuessel S, Mareninova OA, Asatryan L, Wirth MP, Siegert G (2013) Serum amyloid A, phospholipase A_2-IIA and C-reactive protein as inflammatory biomarkers for prostate diseases. Inflamm Res 62(12):1063–1072. doi:10.1007/s00011-013-0665-5. Epub 2013 Sep 24

Mirrahimi A, Chiavaroli L, Srichaikul K, Augustin LS, Sievenpiper JL, Kendall CW, Jenkins DJ (2014) The role of glycemic index and glycemic load in cardiovascular disease and its risk factors: a review of the recent literature. Curr Atheroscler Rep 16(1):381. doi:10.1007/s11883-013-0381-1

Mirza S, Hossain M, Mathews C, Martinez P, Pino P, Gay JL, Rentfro A, McCormick JB, Fisher-Hoch SP (2012) Type 2-diabetes is associated with elevated levels of TNF-alpha, IL-6 and adiponectin and low levels of leptin in a population of Mexican Americans: a cross-sectional study. Cytokine 57(1):136–142

Mishra SK, Kang JH, Kim DK et al (2012) Orally administered aqueous extract of Inonotus obliquus ameliorates acute inflammation in dextran sulfate sodium (DSS)-induced colitis in mice. J Ethnopharmacol 143:524–532

Mitchell AJ, Hansen AM, Hee L, Ball HJ, Potter SM, Walker JC, Hunt NH (2005) Early cytokine production is associated with protection from murine cerebral malaria. Infect Immun 73(9):5645–5653

Mockenhaupt FP, Rong B, Günther M, Beck S, Till H, Kohne E, Thompson WN, Bienzle U (2000a) Anaemia in pregnant Ghanaian women: importance of malaria, iron deficiency, and haemoglobinopathies. Trans R Soc Trop Med Hyg 94(5):477–483

Mockenhaupt FP, Rong B, Till H, Eggelte TA, Beck S, Gyasi-Sarpong C, Thompson WN, Bienzle U (2000b) Submicroscopic Plasmodium falciparum infections in pregnancy in Ghana. Trop Med Int Health 5(3):167–173

Mockenhaupt FP, Rong B, Till H, Thompson WN, Bienzle U (2001) Short report: increased susceptibility to Plasmodium malariae in pregnant alpha(+)-thalassemic women. Am J Trop Med Hyg 64(1–2):6–8

Modvig S, Degn M, Horwitz H, Cramer SP, Larsson HB, Wanscher B, Sellebjerg F, Frederiksen JL (2013) Relationship between cerebrospinal fluid biomarkers for inflammation, demyelination and neurodegeneration in acute optic neuritis. PLoS One 8(10):e77163

Mohanty P, Hamouda W, Garg R, Aljada A, Ghanim H, Dandona P (2000) Glucose challenge stimulates reactive oxygen species (ROS) generation by leucocytes. J Clin Endocrinol Metab 85:2970–2973

Mohanty P, Ghanim H, Hamouda W, Aljada A, Garg R, Dandona P (2002) Both lipid and protein intakes stimulate increased generation of reactive oxygen species by polymorphonuclear leukocytes and mononuclear cells. Am J Clin Nutr 75:767–772

Mojiminiyi OA, Abdella NA, Al Arouj M, Ben Nakhi A (2007) Adiponectin, insulin resistance and clinical expression of the metabolic syndrome in patients with type 2 diabetes. Int J Obes 31:213–220

Mold C, Du Clos TW (2006) C-reactive protein increases cytokine responses to Streptococcus pneumoniae through interactions with Fc gamma receptors. J Immunol 176(12):7598–7604

Mold C, Rodgers CP, Richards RL, Alving CR, Gewurz H (1981) Interaction of C-reactive protein with liposomes. J Immunol 126:856–860

Moldawer LL, Copeland EM 3rd (1997) Proinflammatory cytokines, nutritional support, and the cachexia syndrome: interactions and therapeutic options. Cancer 79(9):1828–1839

Molloy MP, Witzmann FA (2002) Proteomics: technologies and applications. Brief Funct Genomic Proteomic 1(1):23–39

Mondal D, Khan MG (2011) Recent advances in post-kala-azar dermal leishmaniasis. Curr Opin Infect Dis 24(5):418–422. doi:10.1097/QCO.0b013e32834a8ba1

Mora-Jensen H, Jendholm J, Fossum A, Porse B, Borregaard N, Theilgaard-Monch K (2011) Technical advance: immunophenotypical characterization of human neutrophil differentiation. J Leukoc Biol 90:629–634

Morishita K, Aiboshi J, Kobayashi T et al (2012) Lipidomics analysis of mesenteric lymph after trauma and hemorrhagic shock. J Trauma Acute Care Surg 72:1541–1547

Morita A, Nakayama T, Doba N, Hinohara S, Mizutani T, Soma M (2007) Genotyping of triallelic SNPs using TaqMan PCR. Mol Cell Probes 21(3):171–176

Morley JJ, Kushner I (1982) Serum C-reactive protein levels in disease. Ann N Y Acad Sci 389:406–418

Morris HR, Chatterjee A, Panico M, Green BN, Almeida da Silva MA, Hartley BS (1989) Linked scans of peptides and protein digests: amino acid sequence determination of components of complex mixtures. Rapid Commun Mass Spectrom 3(4):110–116

Morrow DA, Rifai N, Antman L (2000) Serum amyloid-A predicts early mortality in acute coronary syndromes: a TIMI 11A substudy. J Am Coll Cardiol 35:358–363

Mortensen RF, Zhong W (2000) Regulation of phagocytic leukocyte activities by C reactive protein. J Leukoc Biol 67:495–500

Mortensen RF, Osmand AP, Lint TF, Gewurz H (1976) Interaction of C-reactive protein with lymphocytes and monocytes: complement-dependent adherence and phagocytosis. J Immunol 117(3):774–781

Moshage H (1997) Cytokines and the hepatic acute phase response. J Pathol 181(3):257–266

Mosser DM, Brittingham A (1997) Leishmania, macrophages and complement: a tale of subversion and expression. Parasitology 115:S9–S23

Mosser DM, Edelson PJ (1984) Activation of the alternative complement pathway by Leishmania promastigotes; parasite lysis and attachment to macrophages. J Immunol 132(3):1501–1505

Mosser DM, Rosenthal LA (1997) Leishmania-macrophage interactions: multiple receptors, multiple ligands and diverse cellular responses. Semin Cell Biol 4:315–322

Mourão AF, Santos MJ, Melo-Gomes J, Martins FM, Costa JA, Ramos F, Brito I, Duarte C, Figueira R, Figueiredo G, Furtado C, Lopes A, Oliveira M, Rodrigues A, Salgado M, Sousa M, Branco JC, Fonseca JE, Canhão H (2014) Using the Juvenile Arthritis Disease Activity Score based on erythrocyte sedimentation rate or C-reactive protein level: results from the Portuguese register. Arthritis Care Res (Hoboken) 66(4):585–591. doi:10.1002/acr.22215

Mroczko B, Groblewska M, Gryko M, Kedra B, Szmitkowski M (2010) Diagnostic usefulness of serum interleukin 6 (IL-6) and C-reactive protein (CRP) in the differentiation between pancreatic cancer and chronic pancreatitis. J Clin Lab Anal 24(4):256–261. doi:10.1002/jcla.20395

Mu AK, Bee PC, Lau YL, Chen Y (2014) Identification of protein markers in patients infected with Plasmodium knowlesi, Plasmodium falciparum and Plasmodium vivax. Int J Mol Sci 15(11):19952–19961. doi:10.3390/ijms151119952

Mugabo Y, Li L, Renier G (2010) The connection between C-reactive protein (CRP) and diabetic vasculopathy. Focus on preclinical findings. Curr Diabetes Rev 6(1):27–34

Mukhopadhyay S, Mandal C (2006) Glycobiology of Leishmania donovani. Indian J Med Res 123:203–220

Mukhopadhyay D, Dalton JE, Kaye PM, Chatterjee M (2014) Post kala-azar dermal leishmaniasis: an unresolved mystery. Trends Parasitol 30(2):65–74. doi:10.1016/j.pt.2013.12.004

Munford RS (2001) Statins and the acute-phase response. N Engl J Med 344:2016–2018

Murata H, Miyamoto T (1993) Bovine haptoglobin as a possible immunomodulator in the sera of transported calves. Br Vet J 149:277–283

Murray CJ, Rosenfeld LC, Lim SS, Andrews KG, Foreman KJ, Haring D, Fullman N, Naghavi M, Lozano R, Lopez AD (2012) Global malaria mortality between 1980 and 2010: a systematic analysis. Lancet 379:413–431

Muskus CE, Marín VM (2002) Metacyclogenesis: a basic process in the biology of Leishmania. Biomedica 22(2):167–177

Mutheneni SR, Upadhyayula SM, Kadiri MR, Nishing K (2014) Malaria prevalence in Arunachal Pradesh–a northeastern state of India. Am J Trop Med Hyg 91(6):1088–1093. doi:10.4269/ajtmh.14-0078, Epub 2014 Oct 20

Nagasaki T, Inaba M, Shirakawa K, Hiura Y, Tahara H, Kumeda Y, Ishikawa T, Ishimura E, Nishizawa Y (2007) Increased levels of C-reactive protein in hypothyroid patients and its correlation with arterial stiffness in the common carotid artery. Biomed Pharmacother 61(2–3):167–172, Epub 2007 Feb 21

Nagpurkar A, Mookerjea S (1981) A novel phosphorylcholine-binding protein from rat serum and its effect on heparin-lipoprotein complex formation in the presence of calcium. J Biol Chem 256:7440–7448

Nahrevanian H, Dascombe MJ (2001) Nitric oxide and reactive nitrogen intermediates during lethal and non-lethal strains of murine malaria. Parasite Immunol 23:491–501

Nahrevanian H, Dascombe MJ (2002) Expression of inducible nitric oxide synthase (iNOS) mRNA in target organs of lethal and non-lethal strains of murine malaria. Parasite Immunol 24:471–478

Nahrevanian H, Gholizadeh J, Farahmand M, Assmar M, Sharifi K, Ayatollahi Mousavi SA, Abolhassani M (2006) Nitric oxide induction as a novel immunoepidemiological target in malaria-infected patients from endemic areas of the Islamic Republic of Iran. Scand J Clin Lab Invest 66:201–210

Nahrevanian H, Gholizadeh J, Farahmand M, Assmar M (2008) Patterns of co-association of C-reactive protein and nitric oxide in malaria in endemic areas of Iran. Mem Inst Oswaldo Cruz 103(1):39–44

Naik P, Voller A (1984) Serum C-reactive protein levels and falciparum malaria. Trans R Soc Trop Med Hyg 78:812–813

Nair KS, Jaleel A, Asmann YW, Short KR, Raghavakaimal S (2004) Proteomic research: potential opportunities for clinical and physiological investigators. Am J Physiol Endocrinol Metab 286(6):E863–E874

Nakache E, Poulain N, Cand F et al (2000) Biopolymer and polymer nanoparticles and their biomedical applications. Handb Nanostruct Nanotechnol 5:577–635

Nakayama T, Sonoda S, Urano T, Yamada T, Okada M (1993) Monitoring both serum protein A and C-reactive protein as inflammatory markers in infectious diseases. Clin Chem 39:293–297

Narsaria N, Mohanty C, Das BK, Mishra SP, Prasad R (2012) Oxidative stress in children with severe malaria. J Trop Pediatr 58:147–150

Ndumele CE, Pradhan AD, Ridker PM (2006) Interrelationships between inflammation, C-reactive protein, and insulin resistance. J Cardiometab Syndr 1(3):190–196

Nebiker CA, Delko T, Zulewski H, Oertli D (2013) Acute thyroid swelling and elevated C reactive protein. BMJ Case Rep 2013. pii: bcr2013200427

Nedelkov D, Tubbs KA, Niederkofler EE, Kiernan UA, Nelson RW (2004) High-throughput comprehensive analysis of human plasma proteins: a step toward population proteomics. Anal Chem 76(6):1733–1737

Nelson RW, Nedelkov D, Tubbs KA (2000a) Biosensor chip mass spectrometry: a chip-based proteomics approach. Electrophoresis 21(6):1155–1163

Nelson RW, Nedelkov D, Tubbs KA (2000b) Biomolecular interaction analysis mass spectometry. BIA/MS can detect and characterize proteins in complex biological fluids at the low- to subfemtomole level. Anal Chem 72(11):404A–411A

Ni XF, Wu P, Wu CP, Ji M, Wu J, Gu XF, Jiang ZX (2015) Elevated serum C-reactive protein, carcinoembryonic antigen and N2 disease are poor prognostic indicators innon-small cell lung cancer. Asia Pac J Clin Oncol 11(4):e22–e30. doi:10.1111/ajco.12091, Epub 2014 May 30

Nicholas SB, Yuan J, Aminzadeh A et al (2012) Salutary effects of a novel oxidative stress modulator on adenine-induced chronic progressive tubulointerstitial nephropathy. Am J Transl Res 4:257–268

Niizuma T, Terada K, Kosaka Y, Daimon Y, Inoue M, Ogita S, Kataoka N, Tanaka K (2004) Elevated serum C-reactive protein in mumps orchitis. Pediatr Infect Dis J 23(10):971

Niikura M, Inoue S, Kobayashi F (2011) Role of interleukin-10 in malaria: focusing on coinfection with lethal and nonlethal murine malaria parasites. J Biomed Biotechnol 2011:383962. doi:10.1155/2011/383962, Epub 2011 Nov 13

Nillawar AN, Bardapurkar JS, Bardapurkar SJ (2012) High sensitive C-reactive protein as a systemic inflammatory marker and LDH-3 isoenzyme in chronic obstructive pulmonary disease. Lung India 29(1):24–29. doi:10.4103/0970-2113.92358

Nimptsch K, Aleksandrova K, Boeing H, Janke J, Lee YA, Jenab M, Bueno-de-Mesquita HB, Jansen EH, Tsilidis KK, Trichopoulou A, Weiderpass E, Wu C, Overvad K, Tjønneland A, Boutron-Ruault MC, Dossus L, Racine A, Kaaks R, Canzian F, Lagiou P, Trichopoulos D, Palli D, Agnoli C, Tumino R, Vineis P, Panico S, Johansson A, Van Guelpen B, Khaw KT, Wareham N, Peeters PH, Quirós JR, Venceslá García A, Molina-Montes E, Dorronsoro M, Chirlaque MD, Barricarte Gurrea A, Key TJ, Duarte-Salles T, Stepien M, Gunter MJ, Riboli E, Pischon T (2015) Association of CRP genetic variants with blood concentrations of C-reactive protein and colorectal cancer risk. Int J Cancer 136(5):1181–1192. doi:10.1002/ijc.29086. Epub 2014 Jul 28

Ninomiya JK, L'Italien G, Criqui MH, Whyte JL, Gamst A, Chen RS (2004) Association of the metabolic syndrome with history of myocardial infarction and stroke in the third national health and nutrition examination survey. Circulation 109(1):42–46

Nix WA, Zirwes R, Bangert V, Kaiser RP, Schilling M, Hostalek U, Obeid R (2015) Vitamin B status in patients with type 2 diabetes mellitus with and without incipient nephropathy. Diabetes Res Clin Pract 107:157–165

Nobili E, Salvado MD, Folkersen L et al (2012) Cysteinyl leukotriene signaling aggravates myocardial hypoxia in experimental atherosclerotic heart disease. PLoS One 7:e41786

Nowicka MA, Fritz-Rdzanek A, Grzybowski W, Walecka I, Niemiec KT, Jakimiuk AJ (2010) C-reactive protein as the indicator of severity in ovarian hyperstimulation syndrome. Gynecol Endocrinol 26(6):399–403

O'Donnell A, Allen SJ, Mgone CS, Martinson JJ, Clegg JB, Weatherall DJ (1998) Red cell morphology and malaria anaemia in children with Southeast-Asian ovalocytosis band 3 in Papua New Guinea. Br J Haematol 101:407–412

Obaldia N 3rd, Baro NK, Calzada JE, Santamaria AM, Daniels R, Wong W, Chang HH, Hamilton EJ, Arevalo-Herrera M, Herrera S, Wirth DF, Hartl DL, Marti M, Volkman SK (2015) Clonal outbreak of Plasmodium falciparum infection in eastern Panama. J Infect Dis 211(7):1087–1096. doi:10.1093/infdis/jiu575, Epub 2014 Oct 21

Obradovic MM, Trpkovic A, Bajic V, Soskic S, Jovanovic A, Stanimirovic J, Panic M, Isenovic ER (2015) Interrelatedness between C-reactive protein and oxidized low-density lipoprotein. Clin Chem Lab Med 53(1):29–34. doi:10.1515/cclm-2014-0590

Ogita M, Miyauchi K, Kasai T, Doi S, Wada H, Naito R, Konishi H, Tsuboi S, Dohi T, Tamura H, Okazaki S, Daida H (2015) Impact of preprocedural high-sensitive C-reactive protein levels on long-term clinical outcomes of patients with stable coronary artery disease and chronic kidney disease who were treated with drug-eluting stents. J Cardiol. pii: S0914-5087(14)00317-7. doi:10.1016/j.jjcc.2014.10.010

Okayama N, Hattori Y, Sadamitsu D, Nakano K (1999) Diagnosis of malaria by allele-specific PCR. Rinsho Ketsueki 40(4):299–304

Okusa MD, Li L (2012) Dendritic cells in acute kidney injury: cues from the microenvironment. Trans Am Clin Climatol Assoc 123:54–63

Oliveira EB, Gotschlich EC, Liu T-Y (1979) Primary structure of human C-reactive protein. J Biol Chem 254(2):489–502

Oliveira KG, von Zeidler SV, Lamas AZ, de Podestá JR, Sena A, Souza ED, Lenzi J, Lemos EM, Gouvea SA, Bissoli NS (2014) Relationship of inflammatory markers and pain in patients with head and neck cancer prior to anticancer therapy. Braz J Med Biol Res 47(7):600–604, Epub 2014 May 30

Oliveira-Ferreira J, Lacerda MV, Brasil P, Ladislau JL, Tauil PL, Daniel-Ribeiro CT (2010) Malaria in Brazil: an overview. Malar J 9:115. doi:10.1186/1475-2875-9-115

Olson RE (2000) Atherogenesis in children: implications for the prevention of atherosclerosis. Adv Pediatr 47:55–78

Omae K, Kondo T, Tanabe K (2015) High preoperative C-reactive protein values predict poor survival in patients on chronic hemodialysis undergoing nephrectomy for renal cancer. Urol Oncol 33(2):67.e9–67.e13. doi:10.1016/j.urolonc.2014.07.004. Epub 2014 Aug 12

Onat A, Ademoğlu E, Can G, Altay S, Karagöz A, Köroğlu B, Yüksel H (2015) Rheumatoid factor mediates excess serum lipoprotein(a) for independent association with type 2 diabetes in men. Anatol J Cardiol 15(10):782–788. doi:10.5152/akd.2014.5826, Epub 2014 Oct 31

Orre IJ, Murison R, Dahl AA, Ueland T, Aukrust P, Fosså SD (2009) Levels of circulating interleukin-1 receptor antagonist and C-reactive protein in long-term survivors of testicular cancer with chronic cancer-related fatigue. Brain Behav Immun 23(6):868–874

Ortega O, Cobo G, Rodríguez I, Camacho R, Gallar P, Mon C, Herrero JC, Ortiz M, Oliet A, Di Gioia C, Vigil A (2014) Lower plasma sodium is associated with a micro-inflammatory state among patients with advanced chronic kidney disease. Nephron Clin Pract 128(3–4):312–318. doi:10.1159/000368116. Epub 2014 Nov 28

Osmand AP, Friedenson B, Gewurz H, Painter RH, Hofmann T, Shelton E (1977) Characterization of C-reactive protein and the complement subcomponent C1t as homologous proteins displaying cyclic pentameric symmetry (pentraxins). Proc Nat Acad Sci USA 74:739–743

Oztoprak N, Aydemir H, Pişkin N, Seremet Keskin A, Araslı M, Gökmen A, Celebi G, Külekçi Uğur A, Taylan Özkan A (2010) An adult case of visceral leishmaniasis in a province of Black-Sea region, Turkey. Mikrobiyol Bul 44(4):671–677

Packer NH, Harrison MJ (1998) Glycobiology and proteomics: is mass spectrometry the Holy Grail? Electrophoresis 19(11):1872–1882

Paiva MS, Serrano CV Jr, Nicolau JC et al (2008) Differences in the inflammatory response between patients with and those without diabetes mellitus after coronary stenting. J Interv Cardiol 21:403–409

Pal S, Chatterjee M, Bhattacharya DK, Bandhyopadhyay S, Mandal C (2000) Identification and purification of cytolytic antibodies directed against O-acetylated sialic acid in childhood acute lymphoblastic leukemia. Glycobiology 10(6):539–549

Paltrinieri S, Ravicini S, Rossi G, Roura X (2010) Serum concentrations of the derivatives of reactive oxygen metabolites (d-ROMs) in dogs with leishmaniosis. Vet J 186(3):393–395. doi:10.1016/j.tvjl.2009.08.019. Epub 2009 Sep 15

Panter-Brick C, Lunn PG, Baker R, Todd A (2001) Elevated acute-phase protein in stunted Nepali children reporting low morbidity: different rural and urban profiles. Br J Nutr 85:125–131

Parashar S, Kella D, Reid KJ, Spertus JA, Tang F, Langberg J, Vaccarino V, Kontos MC, Lopes RD, Lloyd MS (2013) New-onset atrial fibrillation after acute myocardial infarction and its relation to admission biomarkers (from the TRIUMPH registry). Am J Cardiol 112(9):1390–1395

Parham P (2009) The immune system, 3rd edn. Garland Science. ISBN 10: 0815341466/ISBN 13: 9780815341468

Patel BV, Robbins MA, Topol EJ (2001) C-reactive protein: a "golden marker" for inflammation and coronary artery disease. Cleve Clin J Med 68:521–534

Paul I, Mandal C, Mandal C (1998) Effect of environmental pollutants on the C-reactive protein of a freshwater major carp, Catla catla. Dev Comp Immunol 22(5–6):519–532

Paul I, Mandal C, Allen AK, Mandal C (2001) Molecular variants of C-reactive proteins from the major carp Catla catla in fresh and polluted aquatic environments. Glycoconj J 18(7):547–556

Paul K, Boutain D, Agnew K, Thomas J, Hitti J (2008) The relationship between racial identity, income, stress and C-reactive protein among parous women: implications for preterm birth disparity research. J Natl Med Assoc 100(5):540–546

Pawłowski T, Mackiewicz A, Mackiewicz S (1986) Studies on microheterogeneity of acute-phase proteins in rheumatoid arthritis by using crossed affinoimmunoelectrophoresis with free concanavalin A. Behring Inst Mitt 80:11–15

Pepys MB (1981) C-reactive protein fifty years on. Lancet 1(8221):653–657

Pepys MB (1996) The acute phase response and C reactive protein. In: Weartherall DJ, Ledingham JGG, Warrel DA (eds) Oxford textbook of medicine, 3rd edn. Oxford University Press, Oxford, pp 1527–1533

Pepys MB, Baltz ML (1983) Acute phase proteins with special reference to C-reactive protein and related proteins (pentaxins) and serum amyloid A protein. Adv Immunol 34:141–212

Pepys MB, Hirschfield GM (2003) C-reactive protein: a critical update. J Clin Invest 111:1805–1812

Pepys MB, Dash AC, Fletcher TC, Richardson N, Munn EA (1978) Analogues in other mammals and fish of human plasma proteins C-reactive protein and amyloid P component. Nature 273:168–170

Pepys MB, Booth SE, Tennent GA, Butler PJG, Williams DG (1994) Binding of pentraxins to different nuclear-structures- C-reactive protein binds to small nuclear ribonucleoprotein-particles, serum amyloid-P component binds to chromatin and nucleoli. Clin Exp Immunol 97:152–157

Pepys MB, Gallimore JR, Lloyd J, Li Z, Graham D, Taylor GW et al (2012) Isolation and characterization of pharmaceutical grade human pentraxins, serum amyloid P component and C-reactive protein, for clinical use. J Immunol Methods 384:92–102

Percário S, Moreira DR, Gomes BA, Ferreira ME, Gonçalves AC, Laurindo PS, Vilhena TC, Dolabela MF, Green MD (2012) Oxidative stress in malaria. Int J Mol Sci 13(12):16346–16372. doi:10.3390/ijms131216346

Pereira IA, Borba EF (2008) The role of inflammation, humoral and cell mediated autoimmunity in the pathogenesis of atherosclerosis. Swiss Med Wkly 138:534–539

Peters W, Portus JH, Robinson BL (1975) The chemotherapy of rodent malaria, XXII The value of drug-resistant strains of P. berghei in screening for blood sch-izontocidal activity. Ann Trop Med Parasitol 69:155–171

Peters C, Aebischer A, Stiehorf YD, Fuchs M, Overath P (1995) The role of macrophage receptors in adhesion and uptake of Leishmania Mexicana amastigotes. J Cell Sci 108:3715–3724

Petersen HH, Nielsen JP, Heegaard PMH (2004) Application of acute phase protein measurements in veterinary clinical chemistry. Vet Res 35:163–187

Pfeffer A, Rogers KM (1993) Acute phase protein response, food intake, live weight change and lesions following intrathoracic injection of yeast in sheep. Res Vet Sci 55:360–366

Pfützner A, Forst T (2006) High-sensitivity C-reactive protein as cardiovascular risk marker in patients with diabetes mellitus. Diabetes Technol Ther 8(1):28–36

Pfützner A, Schöndorf T, Hanefeld M, Forst T (2010) High-sensitivity C-reactive protein predicts cardiovascular risk in diabetic and nondiabetic patients: effects of insulin-sensitizing treatment with pioglitazone. J Diabetes Sci Technol 4(3):706–716

Piciu A, Piciu D, Marlowe RJ, Irimie A (2013) Highly-sensitive C-reactive protein, a biomarker of cardiovascular disease risk, in radically-treated differentiated thyroid carcinoma patients after repeated thyroid hormone withholding. Exp Clin Endocrinol Diabetes 121(2):102–108

Pickup JC, Crook MA (1998) Is type II diabetes mellitus a disease of the innate immune system? Diabetologia 41:1241–1248

Pickup JC, Mattock MB, Chusney GD, Burt D (1997) NIDDM as a disease of the innate immune system: association of the acute-phase reactants and interleukin-6 with metabolic syndrome X. Diabetologia 40:1286–1292

Pied S, Nussler A, Pontent M, Miltgen F, Matile H, Lambert PH, Mazier D (1989) C-reactive protein protects against preerythrocytic stages of malaria. Infect Immun 57(1):278–282

Pietilla KO, Harmoinen AP, Jokiniitty J, Pastemack AI (1996) Serum C-reactive protein concentration in acute myocardial infarction and its relationship to mortality during 24 months of follow-up in patients under thrombolytic treatment. Eur Heart J 17:1345–1349

Pietropaolo M, Barinas-Mitchell E, Pieropaolo SL, Kuller LH, Trucco M (2000) Evidence of islet cell autoimmunity in elderly patients with type 2 diabetes. Diabetes 49:32–38

Plowe CV, Roper C, Barnwell JW, Happi CT, Joshi HH, Mbacham W, Meshnick SR, Mugittu K, Naidoo I, Price RN, Shafer RW, Sibley CH, Sutherland CJ, Zimmerman PA, Rosenthal PJ (2007) World Antimalarial Resistance Network (WARN) III: molecular markers for drug resistant malaria. Malar J 7:121

Poole EM, Lee IM, Ridker PM, Buring JE, Hankinson SE, Tworoger SS (2013) A prospective study of circulating C-reactive protein, interleukin-6, and tumor necrosis factor α receptor 2 levels and risk of ovarian cancer. Am J Epidemiol 178(8):1256–1264

Pos O, Oostendorp RA, van der Stelt ME, Scheper RJ, Van Dijk W (1990) Con A-nonreactive human alpha 1-acid glycoprotein (AGP) is more effective in modulation of lymphocyte proliferation than Con A-reactive AGP serum variants. Inflammation 14(2):133–141

Potempa LA, Maldonado BA, Laurent P, Zemel ES, Gewurz H (1983) Antigenic, electrophoretic and binding alterations of human C-reactive protein modified selectively in the absence of calcium. Mol Immunol 20(11):1165–1175

Pradhan AD, Manson JE, Rifai N, Buring JE, Ridker PM (2001) C-reactive protein, interleukin 6, and risk of developing type 2 diabetes mellitus. JAMA 286(3):327–334

Price DT, Loscalzo J (1999) Cellular adhesion molecules and atherogenesis. Am J Med 107:85–97

Price RN, Simpson JA, Nosten F, Luxemburger C, Hkirjaroen L, ter Kuile F, Chongsuphajaisiddhi T, White NJ (2001) Factors contributing to anemia after uncomplicated falciparum malaria. Am J Trop Med Hyg 65:614–622

Price RN, Tjitra E, Guerra CA, Yeung S, White NJ, Anstey NM (2007) Vivax malaria: neglected and not benign. Am J Trop Med Hyg 77:79–87

Pritchard DG, Volanakis JE (1985) Slutsky GM and Greenblatt CL C-reactive protein binds leishmanial excreted factors. Proc Soc Exp Biol Med 178(3):500–503

Proudfoot L, O'Donnell CA, Liew FY (1995) Glycoinositolphospholipids of Leishmania major inhibit nitric oxide synthesis and reduce leishmaniacidal activity in murine macrophages. Eur J Immunol 25:745–750

Pyorala S (2000) Hirvonen's thesis on acute phase response in dairy cattle. University of Helsinki. ISBN: 951-45-9106- http://ethesis.helsinki.fi/julkaisut/ela/kliin/hirvonen/

Qiu YD, Wang S, Yang Y et al (2012) Omega-3 polyunsaturated fatty acids promote liver regeneration after 90% hepatectomy in rats. World J Gastroenterol 18:3288–3295

Rang HP, Dale MM (1999) Local hormones, inflammation and allergy. In: Pharmacology. Second edn. Churchill Livingstone, ELBS, pp 243–279

Rasouli N, Kern PA (2008) Adipocytokines and the metabolic complications of obesity. J Clin Endocrinol Metab 93:S64–S73

Ratnam S, Mookerjea S (1998) The regulation of superoxide generation and nitric oxide synthesis by C-reactive protein. Immunology 94:560–568

Raynes J (1982) Variations in the relative proportions of microheterogeneous forms of plasma glycoproteins in pregnancy and disease. Biomed Pharmacother 36(2):77–86

Ready PD (2014) Epidemiology of visceral leishmaniasis. Clin Epidemiol 6:147–154. doi:10.2147/CLEP.S44267. eCollection 2014

Reang P, Gupta M, Kohli K (2006) Biological response modifiers in cancer. MedGenMed 8(4):33

Ree GH (1971) C-reactive protein in Gambian Africans with special reference to P. falciparum malaria. Trans R Soc Trop Med Hyg 65:574–580

Remaley AT, Kuhns DB, Basford RE, Glew RH, Kaplan SS (1984) Leishmanial phosphatase blocks neutrophil O-2 production. J Biol Chem 259(18):11173–11175

Reynolds JM, Martinez GJ, Nallaparaju KC et al (2012) Cutting edge: regulation of intestinal inflammation and barrier function by IL-17C. J Immunol 189:4226–4230

Rezaieyazdi Z, Sahebari M, Hatef MR, Abbasi B, Rafatpanah H, Afshari JT, Esmaily H (2011) Is there any correlation between high sensitive CRP and disease activity in systemic lupus erythematosus? Lupus 20(14):1494–1500

Richardson MD, Grey CA, Shankland GS (1991) Opsonic effect of C-reactive protein on phagocytosis and intracellular killing of virulent and attenuated strains of Candida albicans by human neutrophils. FEMS Microbiol Immunol 3:341–344

Ridker PM (1999) Evaluating novel cardiovascular risk factors: can we better predict heart attacks? Ann Intern Med 130:933–939

Ridker PM (2001) High-sensitivity C-reactive protein: potential adjunct for global risk assessment in the primary prevention of cardiovascular disease. Circulation 103:1813–1818

Ridker PM (2003) Mini-review: expert opinions, clinical application of C-reactive protein for cardiovascular disease detection and prevention. Circulation 107:363–369

Ridker PM (2009) C-reactive protein: eighty years from discovery to emergence as a major risk marker for cardiovascular disease. Clin Chem 55(2):209–215. doi:10.1373/clinchem.2008.119214, Epub 2008 Dec 18

Ridker PM, Cushman M, Stampfer MJ, Tracy RP, Hennekens CH (1997) Inflammation, aspirin, and the risk of cardiovascular disease in apparently healthy men. N Engl J Med 336(14):973–979

Ridker PM, Glynn RJ, Hennekens CH (1998a) C-reactive protein adds to the predictive value of total and HDL cholesterol in determining risk of first myocardial infarction. Circulation 97:2007–2011

Ridker PM, Buring JE, Shih J, Matias M, Hennekens CH (1998b) Prospective study of Creactive protein and the risk of future cardiovascular events among apparently healthy women. Circulation 98(8):731–733

Ridker PM, Rifai N, Pfeffer MA, Sacks F, Braunwald E (1999) Long-term effects of pravastatin on plasma concentration of C-reactive protein. The Cholesterol and Recurrent Events (CARE) Investigators. Circulation 100:230–235

Ridker PM, Rifai N, Pfeffer M et al (2000) Elevation of tumor necrosis factor-a and increased risk of recurrent coronary events after myocardial infarction. Circulation 101:2149–2154

Ridker PM, Stampfer MJ, Rifai N (2001a) Novel risk factors for systemic atherosclerosis: a comparison of C-reactive protein, fibrinogen, homocysteine, lipoprotein(a), and standard cholesterol screening as predictors of peripheral arterial disease. JAMA 285:2481–2485

Ridker PM, Rifai N, Lowenthal SP (2001b) Rapid reduction of C-reactive protein with cerivastatin among 785 patients with primary hypercholesterolemia. Circulation 103:1191–1193

Ridker PM, Rifai N, Clearfield M, Downs JR, Weis SE, Miles JS et al (2001c) Measurement of C-reactive protein for the targeting of stain therapy in the primary prevention of acute coronary events. N Engl J Med 344:1959–1965

Ridker PM, Rifai N, Clearfield M, Downs JR, Weis SE, Miles JS, Gotto AM Jr, Air Force/Texas Coronary Atherosclerosis Prevention Study Investigators (2001d) Measurement of C-reactive protein for the targeting of statin therapy in the primary prevention of acute coronary events. N Engl J Med 344(26):1959–1965

Ridker PM, Danielson E, Fonseca FAH, Genest J, Gotto AM, Kastelein JJP, for the Jupiter Study Group et al (2008) Rosuvastatin to prevent vascular events in men and women with elevated C – reactive protein. N Engl J Med 359:2195–2207

Riley EM (1999) Is T-cell priming required for initiation of pathology in malaria infections? Immunol Today 20(5):228–233

Ristich K (2002) The race to develop tools to pinpoint arterial inflammation. June 19. Heartwire from Medscape

Ritchie RF, Palomaki GE, Neveux LM, Navolotskaia O, Ledue TB, Craig WY (1999) Reference distributions for the negative acute-phase serum proteins, albumin, transferrin, and transthyretin: a practical, simple and clinically relevant approach in a large cohort. J Clin Lab Anal 13(6):273–279. doi:10.1002/(SICI)1098-2825(1999)13:6<273::AID-JCLA4>3.0.CO;2-X

Rittig MG, Bogdan C (2000) Leishmania–host-cell interaction: complexities and alternative views. Parasitol Today 16:292–297

Rivzi FS, Ouaissi MA, Marty B, Santoro F, Capron A (1988) The major surface protein of Leishmania promastigotes is a fibronectin -like molecule. Eur J Immunol 18:473–476

Robey FA, Liu T-Y (1981) Limulin – a C-reactive protein from Limulus polyphemus. J Biol Chem 256:969–975

Robey FA, Jones KD, Tanaka T, Liu TY (1984) Binding of C-reactive protein to chromatin and nucleosome core particles: a possible physiological role of C-reactive protein. J Biol Chem 259:7311–7316

Rocha P, Morgan CJ, Templeton AJ, Pond GR, Naik G, Sonpavde G (2014) Prognostic impact of C-reactive protein in metastatic prostate cancer: a systematic review and meta-analysis. Oncol Res Treat 37(12):772–776. doi:10.1159/000369545, Epub 2014 Nov 20

Rodriguez-Gil JL, Takita C, Wright J, Reis IM, Zhao W, Lally BE, Hu JJ (2014) Inflammatory biomarker C-reactive protein and radiotherapy-induced early adverse skin reactions in patients with breast cancer. Cancer Epidemiol Biomarkers Prev 23(9):1873–1883. doi:10.1158/1055-9965.EPI-14-0263, Epub 2014 Jun 10

Rodriguez-Moran M, Guerrero-Romero F (1999) Increased levels of C-reactive protein in noncontrolled type II diabetic subjects. J Diabetes Complications 13:211–215

Rosales FJ, Topping JD, Smith JE, Shankar AH, Ross AC (2000) Relation of serum retinol to acute phase proteins and malarial morbidity in Papua New Guinea children. Am J Clin Nutr 71(6):1582–1588

Ross R (1976) The pathogenesis of atherosclerosis. N Engl J Med 295(369–77):420–425

Ross R (1999) Atherosclerosis – an inflammatory disease. N Engl J Med 340:123–135

Rossi G, Ibba F, Meazzi S, Giordano A, Paltrinieri S (2014) Paraoxonase activity as a tool for clinical monitoring of dogs treated for canine leishmaniasis. Vet J 199(1):143–149. doi:10.1016/j.tvjl.2013.10.007. Epub 2013 Oct 11

Rus HG, Vlaicu R, Niculescu F (1996) Interleukin-6 and interleukin-8 protein and gene expression in human arterial atherosclerotic wall. Atherosclerosis 127:263–271

Russell RB, Alber F, Aloy P, Davis FP, Korkin D, Pichaud M, Topf M, Sali A (2004) A structural perspective on protein-protein interactions. Curr Opin Struct Biol 14(3):313–324

Ryo M, Nakamura T, Kihara S, Kumada M, Shibazaki S, Takahashi M, Nagai M, Matsuzawa Y, Funahashi T (2004) Adiponectin as a biomarker of the metabolic syndrome. Circ J 68:975–981

Salonen EM, Vatrio T, Hedman K, Vaheri A (1984) Binding of fibronectin by the acute phase reactant C-reactive protein. J Biol Chem 259:1496–1501

Sasidevi A, Vellanki P, Kunselman AR, Raja-Khan N, Dunaif A, Legro RS (2013) Familial aggregation of circulating C-reactive protein in polycystic ovary syndrome. Hum Reprod 28(3):770–776. doi:10.1093/humrep/des416. Epub 2012 Dec

Savage CD, Castejon GL, Denes A et al (2012) NLRP3-inflammasome activating DAMPs stimulate an inflammatory response in glia in the absence of priming which contributes to brain inflammation after injury. Front Immunol 3:288–298

Schmidt MI, Duncan BB, Sharrett AR, Lindberg G, Savage PJ, Offenbacher S, Azambuja MI, Tracy RP, Heiss G (1999) Markers of inflammation and prediction of diabetes mellitus in adults (atherosclerosis risk in communities study): a cohort study. Lancet 353:1649–1652

Schmidt NW, Podyminogin RL, Butler NS, Badovinac VP, Tucker BJ, Bahjat KS, Lauer P, Reyes-Sandoval A, Hutchings CL, Moore AC, Gilbert SC, Hill AV, Bartholomay LC, Harty JT (2008) Memory CD8 T cell responses exceeding a large but definable threshold provide long-term immunity to malaria. Proc Natl Acad Sci U S A 105(37):14017–14022. doi:10.1073/pnas.0805452105, Epub 2008 Sep 9

Schmieg J, Yang G, Franck RW, Tsuji M (2003) Superior protection against malaria and melanoma metastases by a C-glycoside analogue of the natural killer T cell ligand alpha-Galactosylceramide. J Exp Med 198(11):1631–1641

Schnoeller TJ, Steinestel J, Steinestel K, Jentzmik F, Schrader AJ (2015) Do preoperative serum C-reactive protein levels predict the definitive pathological stage in patients with clinically localized prostate cancer? Int Urol Nephrol 47(5):765–770. doi:10.1007/s11255-015-0952-x, Epub 2015 Mar 19

Schultz DR, Arnold PL (1990) Properties of four acute phase proteins: C-reactive protein, serum amyloid A protein, a1-acid glycoprotein and fibrinogen. Semin Arthritis Rheum 20:129–147

Schuster M, Nechansky A, Kircheis R (2006) Cancer immunotherapy. Biotechnol J 1(2):138–147

Schweigert FJ (2005) Characterisation of protein microheterogeneity and protein complexes using on-chip immunoaffinity purification-mass spectrometry. Brief Funct Genomic Proteomic 4(1):7–15

Shantsila E, Kamphuisen PW, Lip GY (2010) Circulating microparticles in cardiovascular disease: implications for atherogenesis and atherothrombosis. J Thromb Haemost 8:2358–2368

Sharma U, Singh S (2009) Immunobiology of leishmaniasis. Indian J Exp Biol 47:412–423

Sharma V, Chatterjee M, Mandal C, Sen S, Basu D (1998) Rapid diagnosis of Indian visceral leishmaniasis using achatinin H, a 9-O-acetylated sialic acid binding lectin. Am J Trop Med Hyg 58:551–554

Sharma S, Ghalaut VS, Dixit R, Kumar S, George PJ (2013) Microalbuminuria and C-reactive protein as a predictor of coronary artery disease in patients of acute chest pain. J Cardiovasc Dis Res 4(1):37–39

Shibutani M, Maeda K, Nagahara H, Ohtani H, Sugano K, Ikeya T, Kimura K, Amano R, Kubo N, Tanaka H, Muguruma K, Ohira M, Hirakawa K (2014) Elevated preoperative serum C-reactive protein levels are associated with poor survival in patients with colorectal cancer. Hepatogastroenterology 61(136): 2236–2240

Shibutani M, Maeda K, Nagahara H, Noda E, Ohtani H, Nishiguchi Y, Hirakawa K (2015) Prognostic significance of the preoperative serum C-reactive protein level in patients with stage IV colorectal cancer. Surg Today 45(3):315–321. doi:10.1007/s00595-014-0909-1, Epub 2014 May 10

Shimizu K, Konno S, Ozaki M et al (2012) Dehydroxymethylepoxyquinomicin (DHMEQ), a novel NF-kappaB inhibitor, inhibits allergic inflammation and airway remodelling in murine models of asthma. Clin Exp Allergy 42:1273–1281

Shin JH, Kim CJ, Jeon EJ, Sung CO, Shin HJ, Choi J, Yu E (2015) Overexpression of C-reactive protein as a poor prognostic marker of resectable hepatocellular carcinomas. J Pathol Trans Med 49(2):105–111. doi:10.4132/jptm.2015.01.19, Epub 2015 Mar 12

Shishido Y, Fujitani K, Yamamoto K, Hirao M, Tsujinaka T, Sekimoto M (2016) C-reactive protein on postoperative day 3 as a predictor of infectious complications following gastric cancer resection. Gastric Cancer 19(1):293–301. doi:10.1007/s10120-014-0455-y, Epub 2015 Jan 6

Shukla AK, Schauer R (1982) Fluorimetric determination of unsubstituted and 9(8)-O-acetylated sialic acids in erythrocyte membranes. Hoppe Seylers Z Physiol Chem 363(3):255–262

Sicking I, Edlund K, Wesbuer E, Weyer V, Battista MJ, Lebrecht A, Solbach C, Grinberg M, Lotz J, Hoffmann G, Rahnenführer J, Hengstler JG, Schmidt M (2014) Prognostic influence of pre-operative C-reactive protein in node-negative breast cancer patients. PLoS One 9(10):e111306. doi:10.1371/journal.pone.0111306. eCollection 2014

Silverstein SC (1977) Endocytic uptake of particles by mononuclear phagocytes and the penetration of obligate intracellular parasites. Am J Trop Med Hyg 26:161–169

Singh S (2014) Changing trends in the epidemiology, clinical presentation, and diagnosis of Leishmania-HIV co-infection in India. Int J Infect Dis 29C:103–112. doi:10.1016/j.ijid.2014.07.011

Singh S, Sharma U, Mishra J (2011) Post-kala-azar dermal leishmaniasis: recent developments. Int J Dermatol 50(9):1099–1108. doi:10.1111/j.1365-4632.2011.04925.x

Sinha S, Mandal C (1996) Microheterogeneity of C-reactive protein in the sera of fish Labeo rohita induced by metal pollutants. Biochem Biophys Res Commun 226(3):681–687

Sinha D, Mandal C, Bhattacharya DK (1999a) Identification of 9-O acetyl sialoglycoconjugates (9-OAcSGs) as biomarkers in childhood acute lymphoblasticleukemia using a lectin, AchatininH, as a probe. Leukemia 13(1):119–125

Sinha D, Bhattacharya DK, Mandal C (1999b) A colorimetric assay to evaluate the chemotherapeutic response of children with acute lymphoblastic leukemia (ALL) employing achatininH: a 9-O-acetyl sialic acid binding lectin. Leuk Res 23(9):803–809

Sinha D, Mandal C, Bhattacharya DK (1999c) Development of a simple, blood based lymphoproliferation assay to assess the clinical status of patients with acute lymphoblastic leukemia. Leuk Res 23(5):433–439

Sinha D, Mandal C, Bhattacharya DK (1999d) A novel method for prognostic evaluation of childhood acute lymphoblastic leukemia. Leukemia 13(2):309–312

Sinha S, Mandal C, Allen AK, Mandal C (2001) Acute phase response of C-reactive protein of Labeo rohita to aquatic pollutants is accompanied by the appearance of distinct molecular forms. Arch Biochem Biophys 396(2):139–150

Sipe JD (1995) Acute-phase proteins in osteoarthritis. Semin Arthritis Rheum 25:75–86

Skinner JG, Roberts L (1994) Haptoglobin as an indicator of infection in sheep. Vet Rec 134:33–36

Sloneker JH (1972) Gas–liquid chromatography of alditol acetates. In: Whistler RL, BeMiller JN (eds) Methods in carbohydrate chemistry, vol VI. Academic Press, New York, pp 20–24

Smith JK, Whitby LG (1968) The heterogeneity of prostatic acid phosphatase. Biochim Biophys Acta 151(3):607–618

Smith DJ, Roberts D (1994) Effects of high volume and/or intense exercise on selected blood chemistry parameters. Clin Biochem 27:435–440

Spiro RC, Parsons WG, Perry SK, Caulfield JP, Hein A, Reisfeld RA, Harper JR, Austen KF, Stevens RL (1986) Inhibition of post-translational modification and surface expression of a melanoma-associated chondroitin sulfate proteoglycan by diethylcarbamazine or ammonium chloride. J Biol Chem 261(11):5121–5129

Spranger J, Kroke A, Möhlig M et al (2003) Inflammatory cytokines and the risk to develop type 2 diabetes results of the prospective population based European Prospective Investigation into Cancer and Nutrition (EPIC)-Potsdam Study. Diabetes 52:812–817

Srinivasan N, White HE, Emsley J, Wood SP, Pepys MB, Blundell TL (1994) Comparative analyses of pentraxins: Implications for protomer assembly and ligand binding. Structure 2(11):1017–1027

Stauga S, Hahn A, Brattig NW, Fischer-Herr J, Baldus S, Burchard GD, Cramer JP (2013) Clinical relevance of different biomarkers in imported plasmodium falciparum malaria in adults: a case control study. Malar J 12:246

Stein MP, Edberg JC, Kimberly RP, Mangan EK, Bharadwaj D, Mold C, Du Clos TW (2000) C-reactive protein binding to FcgammaRIIa on human monocytes and neutrophils is allele-specific. J Clin Invest 105:369–376

Stevens L, Pathak S, Nunes QM, Pandanaboyana S, Macutkiewicz C, Smart N, Smith AM (2015) Prognostic significance of pre-operative C-reactive protein and the neutrophil-lymphocyte ratio in resectable pancreatic cancer: a systematic review. HPB Oxford 17(4):285–291. doi:10.1111/hpb.12355, Epub 2014 Nov 27

Strati I, Copelovici Y, Cajal N, Marinescu G, Vulcan V (1987) Presence of C-reactive protein in the development of acute mumps infection. Preliminary report. Virologie 38(2):121–125

Strati I, Copelovici Y, Selim G (1993) [C-reactive protein (Crp) in different clinical forms of mumps infection]. Rev Roum Virol 44(3–4):305–306

Stumvoll M, Goldstein B, van Haeften T (2005) Type 2 diabetes: principles of pathogenesis and therapy. Lancet 365:1333–1346

Su SC, Pei D, Hsieh CH, Hsiao FC, Wu CZ, Hung YJ (2011) Circulating pro-inflammatory cytokines and adiponectin in young men with type 2 diabetes. Acta Diabetol 48(2):113–119. doi:10.1007/s00592-009-0171-y, Epub 2010 Jan 8

Su HX, Zhou HH, Wang MY, Cheng J, Zhang SC, Hui F, Chen XZ, Liu SH, Liu QJ, Zhu ZJ, Hu QR, Wu Y, Ji SR (2014) Mutations of C-reactive protein (CRP) -286 SNP, APC and p53 in colorectal cancer: implication for a CRP-Wnt crosstalk. PLoS One 9(7):e102418. doi:10.1371/journal.pone.0102418. eCollection 2014

Sugita S, Hozumi T, Yamakawa K, Goto T, Kondo T 2015 White blood cell count and C-reactive protein variations following posterior surgery with intraoperative radiotherapy for spinal metastasis. J Spinal Disord Tech. [Epub ahead of print]

Sundar S, Rai M (2002) Advances in the treatment of leishmaniasis. Curr Opin Infect Dis 15(6):593–598

Sundar S, Singh A, Chakravarty J, Rai M (2015) Efficacy and safety of miltefosine in treatment of post-kala-azar dermal leishmaniasis. ScientificWorldJournal 2015:414378. doi:10.1155/2015/414378. Epub 2015 Jan 1

Sunder S, Pai K, Sahu M, Kumar V, Murray HW (2002) Immunochromatographic strip-test detection of anti-K39 antibody in Indian visceral leishmaniasis. Ann Trop Med Parasitol 96:19–23

Sung KC, Rhee EJ, Kim H, Park JB, Kim YK, Rosenson RS (2012) Prevalence of low LDL-cholesterol levels and elevated high-sensitivity C-reactive protein levels in apparently healthy Korean adults. Nutr Metab Cardiovasc Dis 22:1061e6

Suresh MV, Singh SK, Ferguson DA Jr, Agrawal A (2007) Human C-reactive protein protects mice from Streptococcus pneumoniae infection without binding to pneumococcal C-polysaccharide. J Immunol 178(2):1158–1163

Swai H, Semete B, Kalombo L, Chelule P (2008) Potential of treating tuberculosis with a polymeric nano-drug delivery system. J Control Release 132:e48

Sydor JR, Scalf M, Sideris S, Mao GD, Pandey Y, Tan M, Mariano M, Moran MF, Nock S, Wagner P (2003) Chip-based analysis of protein-protein interactions by fluorescence detection and on-chip immunoprecipitation combined with microLC-MS/MS analysis. Anal Chem 75(22):6163–6170

Szalai AJ, McCrory MA (2002) Varied biologic functions of C-reactive protein: lessons learned from transgenic mice. Immunol Res 26(1–3):279–287

Szalai AJ, Norcum MT, Bly JE, Clem LW (1992) Isolation of an acute-phase phosphorylcholine-reactive pentraxin from channel catfish (Ictalurus punctatus). Comp Biochem Physiol B 102(3):535–543

Szalai AJ, Briles DE, Volanakis JE (1995) Human C-reactive protein is protective against fatal Streptococcus pneumoniae infection in transgenic mice. J Immunol 155(5):2557–2563

Szalai AJ, Alarcón GS, Calvo-Alén J, Toloza SM, McCrory MA, Edberg JC, McGwin G Jr, Bastian HM, Fessler BJ, Vilá LM, Kimberly RP, Reveille JD (2005a) Systemic lupus erythematosus in a multiethnic US Cohort (LUMINA). XXX: association between C-reactive protein (CRP) gene polymorphisms and vascular events. Rheumatology (Oxford) 44(7):864–868. Epub 2005 Mar 29

Szalai AJ, Wu J, Lange EM, McCrory MA, Langefeld CD, Williams A, Zakharkin SO, George V, Allison DB, Cooper GS, Xie F, Fan Z, Edberg JC, Kimberly RP (2005b) Single-nucleotide polymorphisms in the

C-reactive protein (CRP) gene promoter that affect transcription factor binding, alter transcriptional activity, and associate with differences in baseline serum CRP level. J Mol Med (Berl) 83(6):440–447. Epub 2005 Mar 19

Szturmowicz M, Rudziński P, Kacprzak A, Langfort R, Bestry I, Broniarek-Samson B, Orłowski T (2014) Prognostic value of serum C-reactive protein (CRP) and cytokeratin 19 fragments (Cyfra 21–1) but not carcinoembryonic antigen (CEA) in surgically treated patients with non-small cell lung cancer. Pneumonol Alergol Pol 82(5):422–429. doi:10.5603/PiAP.2014.0055

Takahashi H, Iinuma S, Honma M, Iizuka H (2014) Increased serum C-reactive protein level in Japanese patients of psoriasis with cardio- and cerebrovascular disease. J Dermatol 41(11):981–985. doi:10.1111/1346-8138.12632. Epub 2014 Oct 8

Takeda K, Kaisho T, Akira S (2003) Toll-like receptors. Annu Rev Immunol 21:335–376

Talisuna AO, Bloland P, D'Alessandro U (2004) History, dynamics and public health importance of malaria parasite resistance. Clin Microbiol Rev 17:235–254

Tanigaki K, Vongpatanasin W, Barrera JA, Atochin DN, Huang PL, Bonvini E, Shaul PW, Mineo C (2013) C-reactive protein causes insulin resistance in mice through Fcγ receptor IIB-mediated inhibition of skeletal muscle glucose delivery. Diabetes 62(3):721–731. doi:10.2337/db12-0133, Epub 2012 Oct 15

Tanir G, Taylan Ozkan A, Dağlar E (2006) Pediatric visceral Leishmaniasis in Turkey. Pediatr Int 48(1):66–69

Tant L, Reygaerts T, Badot V, Soyfoo MS, Margaux J (2014) The new classification criteria for spondylarthritis: implications in clinical practice. Rev Med Brux 35(4):223–227

Taylor WR, Widjaja H, Basri H, Ohrt C, Taufik T, Tjitra E, Baso S, Fryauff D, Hoffman SL, Richie TL (2008) Changes in the total leukocyte and platelet counts in Papuan and non Papuan adults from northeast Papua infected with acute Plasmodium vivax or uncomplicated Plasmodium falciparum malaria. Malar J 7:259

Tchernof A, Nolan A, Sites CK, Ades PA, Poehlman ET (2002) Weight loss reduces C-reactive protein levels in obese postmenopausal women. Circulation 105(5):564–569

Teishima J, Kobatake K, Hayashi T, Seno Y, Ikeda K, Nagamatsu H, Hieda K, Shoji K, Miyamoto K, Inoue S, Kobayashi K, Ohara S, Kajiwara M, Matsubara A (2014) Prognostic significance of C-reactive protein in patients with intermediate-risk metastatic renal cell carcinoma treated with molecular targeted therapy. Oncol Lett 8(2):881–885, Epub 2014 Jun 2

Teixeira PC, Cutler P, Vuilleumier N (2012) Autoantibodies to apolipoprotein a-1 in cardiovascular diseases: current perspectives. Clin Dev Immunol 2012:868251

Ter Borg EJ, Horst G, Limburg PC, van Rijswijk MH, Kallenberg CG (1990) C-reactive protein levels during disease exacerbations and infections in systemic lupus erythematosus: a prospective longitudinal study. J Rheumatol 17:1642–1648

Testelmans D, Tamisier R, Barone-Rochette G, Baguet JP, Roux-Lombard P, Pépin JL, Lévy P (2013) Profile of circulating cytokines: impact of OSA, obesity and acute cardiovascular events. Cytokine 62(2):210–216. doi:10.1016/j.cyto.2013.02.021

Thakur CP (2000) Socioeconomics of visceral leishmaniasis in Bihar (India). Trans R Soc Trop Med Hyg 94:156–157. [PubMed]

Thomas-Rudolph D, Du Clos TW, Snapper CM, Mold C (2007) C-reactive protein enhances immunity to Streptococcus pneumoniae by targeting uptake to Fc gamma R on dendritic cells. J Immunol 178(11):7283–7291

Thompson SG, Kienast J, Pyke SDM, Haverkate F, van de Loo JCW (1995) Hemostatic factors and the risk of myocardial infarction or sudden death in patients with angina pectoris. N Engl J Med 332:635–641

Thompson D, Pepys MB, Wood SP (1999) The physiological structure of human C-reactive protein and its complex with phosphocholine. Structure 7:169–177

Thurner EM, Krenn-Pilko S, Langsenlehner U, Stojakovic T, Pichler M, Gerger A, Kapp KS, Langsenlehner T (2015) The elevated C-reactive protein level is associated with poor prognosis in prostate cancer patients treated with radiotherapy. Eur J Cancer 51(5):610–619. doi:10.1016/j.ejca.2015.01.002. Epub 2015 Jan 21

Tillett WS, Francis T (1930) Serological reactions in pneumonia with a non-protein somatic fraction of pneumococcus. J Exp Med 52(4):561–571

Tirziu E (2009) Acute-phase proteins in immune response. Lucrări stiinłifice medicină veterinară XLII(1), timisoara

Tohjo H, Miyoshi F, Uchida E, Niiyama M, Syuto B, Moritsu Y, Ichikawa S, Takeauchi M (1995) Polyacrylamide gel electrophoretic patterns of chicken serum in acute inflammation induced by intramuscular injection of turpentine. Poult Sci 74:648–655

Tohjo H, Yadatsu M, Uchida E, Niiyama M, Syuto B, Moritsu Y, Ichikawa S, Takeuchi M (1996) Polyacrylamide gel electrophoretic serum protein patterns of acute inflammation induced by intramuscular injection of turpentine in young broiler chickens. J Vet Med Sci 58:267–268

Tolson J, Bogumil R, Brunst E, Beck H, Elsner R, Humeny A, Kratzin H, Deeg M, Kuczyk M, Mueller GA, Mueller CA, Flad T (2004) Serum protein profiling by SELDI mass spectrometry: detection of multiple variants of serum amyloid alpha in renal cancer patients. Lab Invest 84(7):845–856

Toriola AT, Grankvist K, Agborsangaya CB, Lukanova A, Lehtinen M, Surcel HM (2011) Changes in prediagnostic serum C-reactive protein concentrations and ovarian cancer risk: a longitudinal study. Ann Oncol 22(8):1916–1921

Tousoulis D, Papageorgiou N, Androulakis E, Siasos G, Latsios G, Tentolouris K, Stefanadis C (2013) Diabetes Mellitus-Associated Vascular Impairment: Novel Circulating Biomarkers and Therapeutic Approaches. J Am Coll Cardiol 62:667–676

Toussaint MJM (2000) Acute phase protein in different species measured as a tool to assess animal health. Eur Colloquium Rep 1:1–3

Toussaint MJM, van Ederen AM, Gruys E (1995) Implication of clinical pathology in assessment of animal health and in animal production and meat inspection. Comp Haematol Int 5:149–157

Toussaint MJM, van Ederen AM, Hulskamp-Koch CAM, Gruys E (1997) Measurement of acute phase proteins in porcine blood as a tool for clinical pathology in pigs. Comp Haematol Int 7:182

Toussaint MJM, Eckersall PD, Alava M, Madec F, Meloen RH, Gruys E (2000a) Acute phase protein assays as tool in assessment of health in pigs. Proceedings of the ISACB congress Toulouse. Vet Med Rev 151:780

Toussaint MJM, Lipperheide C, Eckersall PD, Alava M, Jobert JL, Heegaard PMH, Meloen RH, Madec F (2000b) Assessment of health in pigs by acute phase protein assays. In: Tielen MJM, Voets MT (eds) Proceedings of the Xth international congress on animal hygiene, vol 1. Animal Health Service Centre, Boxtel, pp 139–143. ISBN: 90-71649-04-0

Trabert B, Pinto L, Hartge P, Kemp T, Black A, Sherman ME, Brinton LA, Pfeiffer RM, Shiels MS, Chaturvedi AK, Hildesheim A, Wentzensen N (2014) Prediagnostic serum levels of inflammation markers and risk of ovarian cancer in the prostate, lung, colorectal and ovarian cancer (PLCO) screening trial. Gynecol Oncol 135(2):297–304

Troppan KT, Schlick K, Deutsch A, Melchardt T, Egle A, Stojakovic T, Beham-Schmid C, Weiss L, Neureiter D, Wenzl K, Greil R, Neumeister P, Pichler M (2014) C-reactive protein level is a prognostic indicator for survival and improves the predictive ability of the R-IPI score in diffuse large B-cell lymphoma patients. Br J Cancer 111(1):55–60. doi:10.1038/bjc.2014.277, Epub 2014 May 29

Tseng J, Mortensen RF (1988) Binding of human C-reactive protein (CRP) to plasma fibronectin occurs via the phosphorylcholine-binding site. Mol Immunol 25:679–686

Tsui H, Paltser G, Chan Y, Dorfman R, Dosch HM (2011) 'Sensing' the link between type 1 and type 2 diabetes. Diabetes Metab Res Rev 27:913–918, PMID: 22069284

Tubbs KA, Nedelkov D, Nelson RW (2001) Detection and quantification of beta-2-microglobulin using mass spectrometric immunoassay. Anal Biochem 289(1):26–35

Tuomi T, Groop LC, Zimmet PZ, Rowley MJ, Knowles W, Mackay IR (1993) Antibodies to glutamic acid decarboxylase reveal latent autoimmune diabetes mellitus in adults with a non-insulin-dependent onset of disease. Diabetes 42:359–362

Tuomilehto J, Zimmet P, Mackay IR, Koskela P, Vidgren G, Toivanen L, Tuomilehto-Wolf E, Kohtamaki K, Stengard J, Rowley MJ (1994) Antibodies to glutamic acid decarboxylase as predictors of insulin-dependent diabetes mellitus before clinical onset of disease. Lancet 343:1383–1385

Turner RC, Millns H, Neil HA, Stratton IM, Manley SE, Matthews DR, Holman RR (1998) Risk factors for coronary artery disease in non-insulin dependent diabetes mellitus: United Kingdom prospective diabetes study (UKPDS 23). BMJ 316:823–828

Tuzcu A, Bahceci M, Gokalp D, Tuzun Y, Gunes K (2005) Subclinical hypothyroidism may be associated with elevated high-sensitive c-reactive protein (low grade inflammation) and fasting hyperinsulinemia. Endocr J 52(1):89–94

Tyers M, Mann M (2003) From genomics to proteomics. Nature 422(6928):193–197

Ulenbruck G, Soelter J, Janssen E, Haupt H (1981) 2 new, additional combining sites of C-reactive protein-lectin specificity of the anti-galactan type and anti-hemocyanin reactivity. Hoppe-Seyler's Z Physiol Chem 362:1167–1169

Uysal KT, Wiesbrock SM, Marino MW, Hotamisligil GS (1997) Protection from obesity-induced insulin resistance in mice lacking TNF-alpha function. Nature 389:610–614

Valente AJ, Rozek MM, Sprague EA, Schwartz CJ (1992) Mechanisms in intimal monocytes-macrophage recruitment. A special role for monocyte chemotactic protein-1. Circulation 86:III-20–III-25

van Duivenvoorde LM, Voorberg-van der Wel A, van der Werff NM, Braskamp G, Remarque EJ, Kondova I, Kocken CHM, Thomas AW (2010) Suppression of plasmodium cynomolgi in Rhesus Macaques by coinfection with Babesia microti. Infect Immun 78(3):1032–1039

van Miert AS (1995) Pro-inflammatory cytokines in a ruminant model: pathophysiological, pharmacological, and therapeutic aspects. Vet Q 17(2):41–50

van Reeth K, Nauwynck H, Pensaert M (1998) Bronchoalveolar interferon-alpha, tumor necrosis factor-alpha, interleukin-1, and inflammation during acute influenza in pigs: a possible model for humans? J Infect Dis 177:1076–1079

Varki A (2009) Multiple changes in sialic acid biology during human evolution. Glycoconj J 26(3):231–245. doi:10.1007/s10719-008-9183-z, Epub 2008 Sep 7

Ventre J, Doebber T, Wu M, MacNaul K, Stevens K, Pasparakis M, Kollias G, Moller DE (1997) Targeted disruption of the tumor necrosis factor-alpha gene. metabolic consequences in obese and nonobese mice. Diabetes 46:1526–1531

Verhoef H, West CE, Ndeto P, Burema J, Beguin Y, Kok FJ (2001) Serum transferrin receptor concentration indicates increased erythropoiesis in Kenyan children with asymptomatic malaria. Am J Clin Nutr 74:767–775

Vogt B, Führnrohr B, Müller R, Sheriff A (2007) CRP and the disposal of dying cells: consequences for systemic lupus erythematosus and rheumatoid arthritis. Autoimmunity 40:295–298

Voils SA, Cooper-DeHoff RM (2014) Association between high sensitivity C-reactive protein and metabolic syndrome in subjects completing the National Health and Nutrition Examination Survey (NHANES) 2009–10. Diabetol Metab Syndr Clini Res Rev 8:88–90

Volanakis JE (2001) Human C-reactive protein: expression, structure, and function. Mol Immunol 38:189–197

Volanakis JE, Kaplan MH (1971) Specificity of C-reactive protein for choline phosphate residues of pneumococcal C-polysaccharide. Proc Soc Exp Biol Med 136:612–614

Volanakis JE, Narkates AJ (1981) Interaction of C-reactive protein with artificial phosphatydilcholine bilayers and complement. J Immunol 126:1820–1825

Volanakis JE, Wirtz KWA (1979) Interaction of C-reactive protein with artificial phosphatydilcholine bilayers. Nature 281:155–157

Volanakis JE, Clements WL, Schrohenlohr RE (1978) C-reactive protein: purification by affinity chromatography and physiochemical characterization. J I Methods 23:285–295

Wang JM, Sica A, Peri G, Walter S, Padura IM, Libby P, Ceska M, Lindley I, Colotta F, Mantovani A (1991) Expression of monocyte chemotactic protein-1 and interleukin-8 by cytokine-activated human vascular smooth muscle cells. Arterioscler Thromb 11(5):1166–1174

Wang HW, Wu Y, Chen Y, Sui SF (2002) Polymorphism of structural forms of C-reactive protein. Int J Mol Med 9(6):665–671

Wang X, Bao W, Liu J, Ouyang YY, Wang D, Rong S, Xiao X, Shan ZL, Zhang Y, Yao P, Liu LG (2013) Inflammatory markers and risk of type 2 diabetes: a systematic review and meta-analysis. Diabetes Care 36(1):166–175. doi:10.2337/dc12-0702

Wang G, Li N, Cao L, Chen S, Guo L, Su K, Li F, Cui H, He J, Dai M, Zou Y, Sun D, Wu S (2014) [Association between the level of high sensitivity C-reactive protein and risk of breast cancer among non-diabetic females: a prospective study in Kailuan group]. Zhonghua Zhong Liu Za Zhi 36(12):944–948

Ware LB, Koyama T, Zhao Z, Janz DR, Wickersham N, Bernard GR, May AK, Calfee CS, Matthay MA (2013) Biomarkers of lung epithelial injury and inflammation distinguish severe sepsis patients with acute respiratory distress syndrome. Crit Care 17(5):R253

Wasunna KM, Raynes JG, Were JB, Muigai R, Sherwood J, Gachihi G, Carpenter L, McAdam KP (1995) Acute phase protein concentrations predict parasite clearance rate during therapy for visceral leishmaniasis. Trans R Soc Trop Med Hyg 89(6):678–681

Wei M, Gaskill SP, Haffner SM, Stern MP (1998) Effects of diabetes and level of glycemia on all-cause mortality: the San Antonio Heart Study. Diabetes Care 21:1167–1172

Werling D, Sutter F, Arnold M, Kun G, Tooten PCJ, Gruys E, Kreuzer M, Langhans W (1996) Characterisation of the acute phase response of heifers to a prolonged low dose infusion of lipopolysaccharide. Res Vet Sci 61:252–257

Westhorpe CL, Schneider HG, Dunne M, Middleton T, Sundararajan V, Spelman T, Carter V, Crowe SM, Dart A, Mijch A, Kotsanas D, Woolley I (2014) C-reactive protein as a predictor of cardiovascular risk in HIV-infected individuals. Sex Health 11(6):580–582. doi:10.1071/SH14130

Wethal T, Haugnes HS, Kjekshus J, Småstuen MC, Ueland T, Aukrust P, Fosså SD (2010) C-reactive protein; a potential marker of second cancer and cardiovascular disease in testicular cancer survivors? Eur J Cancer 46(18):3425–3433

WHO (2011a) World Malaria Report. World Health Organization, Geneva

WHO (2011b) Malaria. In WHO International Travel and Health 2011. World Health Organization, Geneva, pp 144–167

WHO: Leishmaniasis Fact sheet N°375. www.who.int/mediacentre/factsheets/fs375/en/

Wigmore SJ, Fearon KC, Maingay JP, Lai PB, Ross JA (1997) Interleukin-8 can mediate acute-phase protein production by isolated human hepatocytes. Am J Physiol 273:720–726

Wilmanski J, Siddiqi M, Deitch EA, Spolarics Z (2005) Augmented IL-10 production and redox-dependent signaling pathways in glucose-6-phosphate dehydrogenase-deficient mouse peritoneal macrophages. J Leukoc Biol 78(1):85–94. Epub 2005 Apr 7

Wilson ME, Pearson RD (1986) Evidence that Leishmania donovani utilises a mannose receptor on human mononuclear phagocytes to establish intracellular infection. J Immunol 136:4681–4688

Wilson ME, Pearson RD (1988) Roles of CR3 and mannose receptors in the attachment and ingestion of Leishmania donovani by human mononuclear phagocytes. Infect Immun 56:363–369

Wilson ME, Sandor M, Blum AM et al (1996) Local suppression of IFN-gamma in hepatic granulomas correlates with tissue-specific replication of Leishmania chagasi. J Immunol 156:2231–2239

Wilson RG, Stevens BW, Guo AY, Russell CN, Thornton A, Cohen MA, Sturgeon HC, Giallourakis C, Khalili H, Nguyen DD, Yajnik V, Sauk J, Ananthakrishnan AN (2015) High C-reactive protein is associated with poor sleep quality independent of nocturnal symptoms in patients with inflammatory bowel disease. Dig Dis Sci 60(7):2136–2143. doi:10.1007/s10620-015-3580-5, Epub 2015 Feb 21

Winters RA, Murray HW (1992) Malaria - the mime revisited: fifteen more years of experience at a New York City teaching hospital. Am J Med 93:243–246

Wongsrichanalai C, Pickard AL, Wernsdorfer WH, Meshnick SR (2002) Epidemiology of drug-resistant malaria. Lancet Infect Dis 2:209–218

Woo P, Korenberg JR, Whitehead AS (1985) Characterization of genomic and complementary DNA sequence of human C-reactive protein, and comparison with the complementary DNA sequence of serum amyloid P component. J Biol Chem 260:13384–13388

World Health Organization (1995) Report on the consultative meeting on Leishmania/HIV Co-infection,

Rome, 6–7 September 1994. Document WHO/LEISH/95.35. WHO, Geneva

World Health Organization (WHO) (1996) Assessment of therapeutic efficacy of antimalarial drugs for uncomplicated Malaria in areas with intense transmission. WHO, Geneva

World Health Organization Control of the leishmaniasis: report of a meeting of the WHO Expert Committee on the Control of Leishmaniases, Geneva, March 22–26, 2010. World Health Organ Tech Rep Ser 949:1–186

Wu T, Dorn JP, Donahue RP, Sempos CT, Trevisan M (2002) Associations of serum C-reactive protein with fasting insulin, glucose, and glycosylated hemoglobin: the Third National Health and Nutrition Examination Survey, 1988–1994. Am J Epidemiol 155:65–71

www.biolegend.com

Xu Y, Whitmer K (2006) C-reactive protein and cardiovascular disease in people with diabetes: high-sensitivity CRP testing can help assess risk for future cardiovascular disease events in this population. Am J Nurs 106(8):66–72

Xu Y, Barregard L, Nielsen J, Gudmundsson A, Wierzbicka A, Axmon A, Jönsson BA, Kåredal M, Albin M (2013) Effects of diesel exposure on lung function and inflammation biomarkers from airway and peripheral blood of healthy volunteers in a chamber study. Part Fibre Toxicol 10:60. doi:10.1186/1743-8977-10-60

Xu L, Zhao Q, Huang S, Li S, Wang J, Li Q (2015) Serum C-reactive protein acted as a prognostic biomarker for overall survival in metastatic prostate cancer patients. Tumour Biol 36(2):669–673. doi:10.1007/s13277-014-2670-x, Epub 2014 Oct 7

Yasojima K, Schwab C, McGeer EG, McGeer PL (2001) Generation of C-reactive protein and complement components in atherosclerotic plaques. Am J Pathol 158:1039–1051

Yeh ET, Willerson JT (2003) Coming of age of C-reactive protein: using inflammation markers in cardiology. Circulation 107:370–372

Yeh ET, Anderson HV, Pasceri V, Willerson JT (2001) C-reactive protein: linking inflammation to cardiovascular complications. Circulation 104:974–975

Yeo ES, Hwang JY, Park JE, Choi YJ, Huh KB, Kim WY (2010) Tumor necrosis factor (TNF-alpha) and C-reactive protein (CRP) are positively associated with the risk of chronic kidney disease in patients with type 2 diabetes. Yonsei Med J 51(4):519–525

Yi JH, Wang D, Li ZY, Hu J, Niu XF, Liu XL (2014) C-reactive protein as a prognostic factor for human osteosarcoma: a meta-analysis and literature review. PLoS One 9(5):e94632. doi:10.1371/journal.pone.0094632. eCollection 2014

Ying SC, Gewurz H, Kinoshita CM, Potempa LA, Siegel JN (1989) Identification and partial characterization of multiple native and neoantigenic epitopes of human C-reactive protein by using monoclonal antibodies. J Immunol 143:221–228

You L, Sheng Jun F, Dong Li S, Hu X, Li Y (2015) Preoperative C-reactive protein as a prognostic predictor for upper tract urothelial carcinoma: A systematic review and meta-analysis. Mol Clin Oncol 3(4):924–928. doi:10.3892/mco.2015.553

Yu YT, Ho CT, Hsu HS, Li CI, Davidson LE, Liu CS, Li TC, Shih CM, Lin CC, Lin WY (2013) Subclinical hypothyroidism is associated with elevated high-sensitive C-reactive protein among adult Taiwanese. Endocrine 44(3):716–722

Yudkin JS, Stehouwer CD, Emeis JJ, Coppack SW (1999) C-reactive protein in healthy subjects: associations with obesity, insulin resistance, and endothelial dysfunction: a potential role for cytokines originating from adipose tissue? Arterioscler Thromb Vasc Biol 19:972–978

Yue CC, Muller-Greven J, Dailey P, Lozanski G, Anderson V, Macintyre S (1996) Identification of a human C-reactive protein binding site in two hepatic carboxylesterases capable of retaining C-reactive protein within the endoplasmic reticulum. J Biol Chem 271:22245–22250

Yung S, Chan T (2012) Autoantibodies and resident renal cells in the pathogenesis of lupus nephritis: getting to know the unknown. Clin Dev Immunol 2012:139365

Zeng YC, Wu R, Xiao YP, Chi F, Xue M, Zhang ZY, Xing R, Zhong WZ, Wang SL, Tian X, Chen W, Chen JJ, Wu LN (2015) Serum C-reactive protein predicts poor prognosis in patients with locoregionally advanced nasopharyngeal carcinoma treated with chemoradiotherapy. Curr Oncol 22(1):20–24. doi:10.3747/co.22.2178

Zheng H, Fletcher D, Kozak W, Jiang M, Hofmann KJ, Conn CA, Soszynski D, Grabiec C, Trumbauer ME, Shaw A et al (1995) Resistance to fever induction and impaired acute-phase response in interleukin-1 *b*-deficient mice. Immunity 3(1):9–19

Zheng TL, Cao K, Liang C, Zhang K, Guo HZ, Li DP, Zhao S (2014) Prognostic value of C-reactive protein in esophageal cancer: a meta-analysis. Asian Pac J Cancer Prev 15(19):8075–8081

Zhou B, Shu B, Yang J, Liu J, Xi T, Xing Y (2014) C-reactive protein, interleukin-6 and the risk of colorectal cancer: a meta-analysis. Cancer Causes Control 25(10):1397–1405. doi:10.1007/s10552-014-0445-8, Epub 2014 Jul 23

Zhu L, Chen G, Xia Q, Tang W, Huang Z, Wang M, Guo J, Yang X, Zhou Y (2010) Use of band cell percentage as an early predictor of death and ICU admission in severe acute pancreatitis. Hepatogastroenterology 57:1543–1548

Zhu W, London NR, Gibson CC et al (2012) Interleukin receptor activates a MYD88–ARNO–ARF6 cascade to disrupt vascular stability. Nature 492: 252–255

Zijlstra EE, Musa AM, Khalil EAG, El Hassan IM, El-Hassan AM (2003) Post-kala-azar dermal leishmaniasis. Lancet Infect Dis 3(2):87–98. doi:10.1016/s1473-3099(03)00517-6

Zimmermann O, Li K, Zaczkiewicz M et al (2014) C-reactive protein in human atherogenesis: facts and fiction. Mediat Inflamm 2014:561428. Epub 2014 Apr 1